T0213736

Lecture Notes in Computer Science 10114

Commenced Publication in 1973
Founding and Former Series Editors:
Gerhard Goos, Juris Hartmanis, and Jan van Leeuwen

More information about this series at http://www.springer.com/series/7412

Shang-Hong Lai · Vincent Lepetit
Ko Nishino · Yoichi Sato (Eds.)

Computer Vision – ACCV 2016

13th Asian Conference on Computer Vision
Taipei, Taiwan, November 20–24, 2016
Revised Selected Papers, Part IV

 Springer

Editors

Shang-Hong Lai
National Tsing Hua University
Hsinchu
Taiwan

Ko Nishino
Drexel University
Philadelphia, PA
USA

Vincent Lepetit
Graz University of Technology
Graz
Austria

Yoichi Sato
The University of Tokyo
Tokyo
Japan

ISSN 0302-9743   ISSN 1611-3349   (electronic)
Lecture Notes in Computer Science
ISBN 978-3-319-54189-1   ISBN 978-3-319-54190-7   (eBook)
DOI 10.1007/978-3-319-54190-7

Library of Congress Control Number: 2017932642

LNCS Sublibrary: SL6 – Image Processing, Computer Vision, Pattern Recognition, and Graphics

Printed on acid-free paper

This Springer imprint is published by Springer Nature
The registered company is Springer International Publishing AG
The registered company address is: Gewerbestrasse 11, 6330 Cham, Switzerland

Preface

Welcome to the 2016 edition of the Asian Conference on Computer Vision in Taipei. ACCV 2016 received a total number of 590 submissions, of which 479 papers went through a review process after excluding papers rejected without review because of violation of the ACCV submission guidelines or being withdrawn before review. The papers were submitted from diverse regions with 69% from Asia, 19% from Europe, and 12% from North America.

The program chairs assembled a geographically diverse team of 39 area chairs who handled nine to 15 papers each. Area chairs were selected to provide a broad range of expertise, to balance junior and senior members, and to represent a variety of geographical locations. Area chairs recommended reviewers for papers, and each paper received at least three reviews from the 631 reviewers who participated in the process. Paper decisions were finalized at an area chair meeting held in Taipei during August 13–14, 2016. At this meeting, the area chairs worked in threes to reach collective decisions about acceptance, and in panels of nine or 12 to decide on the oral/poster distinction. The total number of papers accepted was 143 (an overall acceptance rate of 24%). Of these, 33 were selected for oral presentations and 110 were selected for poster presentations.

We wish to thank all members of the local arrangements team for helping us run the area chair meeting smoothly. We also wish to extend our immense gratitude to the area chairs and reviewers for their generous participation in the process. The conference would not have been possible without this huge voluntary investment of time and effort. We acknowledge particularly the contribution of 29 reviewers designated as "Outstanding Reviewers" who were nominated by the area chairs and program chairs for having provided a large number of helpful, high-quality reviews. Last but not the least, we would like to show our deepest gratitude to all of the emergency reviewers who kindly responded to our last-minute request and provided thorough reviews for papers with missing reviews. Finally, we wish all the attendees a highly simulating, informative, and enjoyable conference.

January 2017

Shang-Hong Lai
Vincent Lepetit
Ko Nishino
Yoichi Sato

Organization

ACCV 2016 Organizers

Steering Committee

Michael Brown	National University of Singapore, Singapore
Katsu Ikeuchi	University of Tokyo, Japan
In-So Kweon	KAIST, Korea
Tienlu Tan	Chinese Academy of Sciences, China
Yasushi Yagi	Osaka University, Japan

Honorary Chairs

Thomas Huang	University of Illinois at Urbana-Champaign, USA
Wen-Hsiang Tsai	National Chiao Tung University, Taiwan, ROC

General Chairs

Yi-Ping Hung	National Taiwan University, Taiwan, ROC
Ming-Hsuan Yang	University of California at Merced, USA
Hongbin Zha	Peking University, China

Program Chairs

Shang-Hong Lai	National Tsing Hua University, Taiwan, ROC
Vincent Lepetit	TU Graz, Austria
Ko Nishino	Drexel University, USA
Yoichi Sato	University of Tokyo, Japan

Publicity Chairs

Ming-Ming Cheng	Nankai University, China
Jen-Hui Chuang	National Chiao Tung University, Taiwan, ROC
Seon Joo Kim	Yonsei University, Korea

Local Arrangements Chairs

Yung-Yu Chuang	National Taiwan University, Taiwan, ROC
Yen-Yu Lin	Academia Sinica, Taiwan, ROC
Sheng-Wen Shih	National Chi Nan University, Taiwan, ROC
Yu-Chiang Frank Wang	Academia Sinica, Taiwan, ROC

Workshops Chairs

Chu-Song Chen	Academia Sinica, Taiwan, ROC
Jiwen Lu	Tsinghua University, China
Kai-Kuang Ma	Nanyang Technological University, Singapore

Tutorial Chairs

Bernard Ghanem	King Abdullah University of Science and Technology, Saudi Arabia
Fay Huang	National Ilan University, Taiwan, ROC
Yukiko Kenmochi	Université Paris-Est, France

Exhibition and Demo Chairs

Gee-Sern Hsu	National Taiwan University of Science and Technology, Taiwan, ROC
Xue Mei	Toyota Research Institute, USA

Publication Chairs

Chih-Yi Chiu	National Chiayi University, Taiwan, ROC
Jenn-Jier (James) Lien	National Cheng Kung University, Taiwan, ROC
Huei-Yung Lin	National Chung Cheng University, Taiwan, ROC

Industry Chairs

Winston Hsu	National Taiwan University, Taiwan, ROC
Fatih Porikli	Australian National University, Australia
Li Xu	SenseTime Group Limited, Hong Kong, SAR China

Finance Chairs

Yong-Sheng Chen	National Chiao Tung University, Taiwan, ROC
Ming-Sui Lee	National Taiwan University, Taiwan, ROC

Registration Chairs

Kuan-Wen Chen	National Chiao Tung University, Taiwan, ROC
Wen-Huang Cheng	Academia Sinica, Taiwan, ROC
Min Sun	National Tsing Hua University, Taiwan, ROC

Web Chairs

Hwann-Tzong Chen	National Tsing Hua University, Taiwan, ROC
Ju-Chun Ko	National Taipei University of Technology, Taiwan, ROC
Neng-Hao Yu	National Chengchi University, Taiwan, ROC

Area Chairs

Narendra Ahuja	UIUC
Michael Brown	National University of Singapore
Yung-Yu Chuang	National Taiwan University, Taiwan, ROC
Pau-Choo Chung	National Cheng Kung University, Taiwan, ROC
Larry Davis	University of Maryland, USA

Contents – Part IV

Image Alignment

3D Computer Vision

Computational Photography
and Image Processing

Blind Image Quality Assessment Based on Natural Redundancy Statistics

Jia Yan$^{(\boxtimes)}$, Weixia Zhang, and Tianpeng Feng

School of Electronic and Information, Wuhan University, Wuhan, China
yanjia2011@gmail.com

Abstract. Blind image quality assessment (BIQA) aims to evaluate the perceptual quality of a distorted image without information regarding its reference image and the distortion type. Existing BIQA methods usually predict the image quality by employing natural scene statistic (NSS), which is derived from the statistical distributions of image coefficients by reducing the redundancies in a transformed domain. Contrary to these methods, we directly measure the redundancy existing in a natural image and compute the natural redundancy statistics (NRS) to capture the distortion degree. Specially, we utilize the singular value decomposition (SVD) and asymmetric generalized Gaussian distribution (AGGD) modeling to obtain NRS from opponent color spaces, and learn a regression model to map the NRS features to the subjective quality score. Extensive experiments demonstrate very competitive quality prediction performance and generalization ability of the proposed method.

1 Introduction

An increasingly large number of digital images are being produced by professional and casual users as digital cameras have become ubiquitous in smart phones, tablets, and stand-alone units. Since in nearly every instance it is desirable to produce clear, crisp images free of excessive noise, blur or other artifacts, the development of methods for automatically assessing the perceptual quality of images has become an important goal of model and algorithm developers. Such tools would greatly facilitate the sorting and culling of the large volumes of images. Current objective image quality assessment (IQA) methods fall into three general categories: Full Reference (FR), Reduced Reference (RR) and No Reference (NR) [9]. FR and RR models require that all or part of the information about the reference image be available. However, in most application scenarios information regarding a reference image is inaccessible. Hence effective NR models may ultimately prove to be more viable.

Early NR-IQA approaches extract distortion-specific features for quality prediction, based on a model of the presumed distortion type(s), such as ringing [10], blur [2,26] or compression [8]. Hence, the application scope of these methods is rather limited. Recent studies on NR-IQA have focused on the so-called blind image quality assessment (BIQA) problem, where prior knowledge of the distortion types is unavailable. A survey about recent BIQA methods can be

© Springer International Publishing AG 2017
S.-H. Lai et al. (Eds.): ACCV 2016, Part IV, LNCS 10114, pp. 3–18, 2017.
DOI: 10.1007/978-3-319-54190-7_1

found in [11]. The BIQA can be mainly categorized into two classes. The first is to extract effective features from distorted images followed by training a regression module using those features, called "opinion-aware" methods. Most of these methods employ natural scene statistics (NSS), which is based on the hypothesis that natural scenes possess certain statistical properties that are altered in the presence of distortion, rendering them unnatural. To capture the quality aware representation, the NSS-based BIQA algorithms have been investigated in both the spatial domain [13,27] and the transform domain such as wavelet (BIQI [15], DIIVINE [16]) and DCT domain (BLIINDS [20], BLIINDS-II [21]). Since natural images are high dimensional signals that contain a rich amount of redundancies, the extraction of statistical features in the transform domain may be viewed as a process of removing the redundancies to reveal the low dimensional manifold space of image perceptual quality.

The second class of BIQA metrics operates without human scores, called "opinion-aware" methods. The natural image quality evaluator (NIQE) model proposed by Mittal et al. [14] extracts a set of local features from an image, then fits the feature vectors to a multivariate Gaussian (MVG) model. The quality of a test image is then predicted by the distance between its MVG model and the MVG model learned from a corpus of pristine naturalistic images. Quality-aware clustering (QAC) [28] works by learning a set of quality-aware centroids to act as a codebook to compute the quality levels of image patches and infer the quality score of the overall image. In [29], Zhang et al. learn a multivariate Gaussian model of image patches from a collection of pristine natural images. Using the learned multivariate Gaussian model, a Bhattacharyya-like distance is used to measure the quality of the distorted image. One distinct property of opinion-unaware BIQA methods is that they have the potential to deliver higher generalization capability than their opinion-aware counterparts due to the fact that they do not depend on training samples of distorted images and associated subjective quality scores.

Contrary to the above opinion-aware BIQA methods which work in redundancy reduced domains, we calculate the image statistics directly on the image redundancy as the quality aware features. We find that measuring image redundancy by using the simple singular value decomposition (SVD) can effectively avoid the influence of different image contents and capture the distortion degree. SVD has been successfully used to solve IQA problems. The method in [4,22,23] evaluates quality of each image block based on the error in singular values of the block. The overall image quality score is computed as the average absolute difference between each blocks error and the median error over all blocks. Sang et al. [22] Proposed two blind image quality assessment metrics based on the area and exponent of reciprocal singular value curves on the finding that the bending degree of the reciprocal singular value curve varies with distortion type and severity. Based on the observation that a decreasing rate of singular values is highly correlated to a degree of distortions regardless of their type, the approach proposed in [4] utilizes the decreasing rate of singular values to model a simple and reliable NR-IQA method. Instead of singular values used in [23], the method proposed in [12] projects the distorted image on the singular vectors of the original image and uses a referee matrix of the distorted image to assess quality. Narwaria et al. [17] also proposed to address the problem with the use of singular vectors out of SVD as features for quantifying major structural information in images and then support vector regression (SVR) for automatic prediction of image quality. In [18], both singular vectors and values are detected as a more reasonable and comprehensive approach.

It is worthwhile to note that neither singular values nor singular vectors are used as features in this paper, SVD is only utilized to obtain the image redundancy by repeated reconstruction. On the observation that the statistics of a distorted image is measurably different from those of pristine images in redundancy domain, we generate asymmetric generalized Gaussian distribution model to extract quality aware features derived from multiple color channels. The features are fed to a support vector machine (SVM) regression model to predict quality score of input image. The novelty of our work lies in that we propose to directly use natural redundancy statistics features for BIQA model learning and it achieves good performance in terms of quality prediction accuracy and database generalization.

The rest of the paper is organized as follows. Section 2 analyzes the natural redundancy statistics and presents the framework of the new image quality assessment model. Experiments conducted on public IQA databases are presented and analyzed in Sect. 3. Section 4 concludes the paper.

2   Natural Redundancy Statistics for BIQA

2.1   Relationship Between Redundancy Images and Image Quality

For an input gray image I, we describe the redundancy by using SVD and image reconstruction. SVD decomposes an image into several basis images and corresponding singular values. In other words, an image is decomposed into left singular vector matrix U and right singular vector matrix V and the corresponding diagonal singular value matrix $\sum$, i.e.,

$$
\begin{aligned}
I &= \mathrm{U} \sum \mathrm{V}^{\mathrm{T}} \\
\mathrm{U} &= \begin{bmatrix} u_1 \ u_2 \ ... \ u_r \end{bmatrix} \\
\sum &= \mathrm{diag}\left(\sigma_1, \sigma_2, ..., \sigma_t\right) \\
\mathrm{V} &= \begin{bmatrix} v_1 \ v_2 \ ... \ v_c \end{bmatrix}
\end{aligned}
\tag{1}
$$

where I is an image whose size is $r \times c$, t is $min\,(r, c)$, u_i and v_j are column vectors, and σ_k is a singular value ($i = 1, 2, ..., r, j = 1, 2, ..., c, k = 1, 2, ..., t$). The singular values are sorted in a descending order, i.e., $\sigma_1 \geq \sigma_2 \geq ... \geq \sigma_t$. In practice, the last singular values of I can be small, but not necessarily zero. And, by using singular vectors and singular values, images can be represented as

$$
I = \sum_k \sigma_k u_k v_k{}^T
\tag{2}
$$

Here, singular vectors u_k and v_k constitute basis images whereas singular value σ_k corresponds to their significance. Given a threshold ε $(0 < \varepsilon < 1)$, let $\sum' = \mathrm{diag}(\sigma_1', \sigma_2', ..., \sigma_t')$, where $\sigma_k' = \sigma_k$, if $k < \varepsilon t$ and $\sigma_k' = 0$ if $k \geq \varepsilon t$. So, by setting to zero some singular values, we can obtain a new image I':

$$I' = \sum_k \sigma_k' u_k v_k^T \tag{3}$$

This useful tool of linear algebra has been successfully applied to image denoising problem by estimating the best value of the threshold [5]. The characteristics of either singular values σ_k or singular vector u_k and v_k have been thoroughly analyzed to measure the distortion degree and predict images quality [4,12,17,18,22,23]. But to the best of our knowledge, the redundancy image $RI = I - I'$ has been little investigated in solving BIQA problem. By setting different threshold ε, we could calculate several RIs from I. In Fig. 1, we show the RIs of a reference image selected from the LIVE database [24]. Figure 1(a) is a natural undistorted image, and the Fig. 1(b)–(i) are the redundancy images (displayed in pseudo-color) by setting the value of ε from 0.1 to 0.8 stepped by 0.1.

From the RIs, we can easily notice that the higher ε corresponds to the less contents retained in RI. By setting to zero the "non significant" singular values, in effect we perform a lossy compression on Fig. 1(a). The amount of "loss" is controlled by the threshold ε. We plot the histograms of $RI1 - 8$ together in Fig. 1(j), the ordinate denotes the percentage (%). The histograms reveal a number of interesting findings. First, the histogram of each RI closely follows Gaussian-like distribution. Secondly, the shapes of these distributions are different, such as the sharpness and the width. We have observed this statistical phenomenon to broadly hold on natural undistorted images. Figure 2 shows other reference images from LIVE database and their histograms of RIs. The variation of these histogram distributions share almost the same law, this characteristic is independent of the image content. The shapes of the distributions are different between different images, so we cannot use the statistics of a single redundancy image to describe the quality of the image. Instead, we use all the RIs.

We have found that by introducing distortions to an image, not only the shapes of its RIs histograms but also the similarity of these histograms is changed. We use an example to demonstrate this fact. Figure 3 shows five images selected from LIVE database, which are distorted versions of Fig. 1(a), and the corresponding histograms are plotted. When distortions are introduced, the distributions are affected. For example, white noise, JPEG and JPEG2K make some distributions highly picked, and create a more Laplacian appearance. When it comes to the similarity, some distributions are almost the same either in the Gaussian blur or fast fading distortion, and there are not eight individual distributions in these distortions. The loss of perceptual quality affects the generated redundancy images in different degrees, by generating features based on these redundancy images, presence and severity of the distortions are objectively sensible.

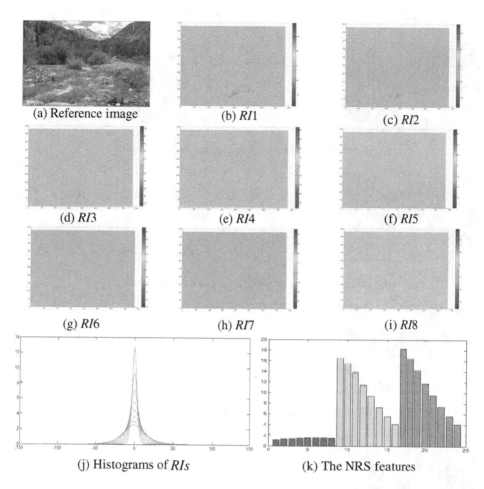

(a) Reference image   (b) *RI*1   (c) *RI*2

(d) *RI*3   (e) *RI*4   (f) *RI*5

(g) *RI*6   (h) *RI*7   (i) *RI*8

(j) Histograms of *RIs*   (k) The NRS features

Fig. 1. The *RIs* of a reference image with the histograms and the extracted NRS features. (Color figure online)

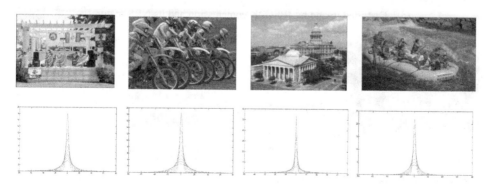

Fig. 2. Four random reference images from LIVE database and corresponding histograms of their *RIs*.

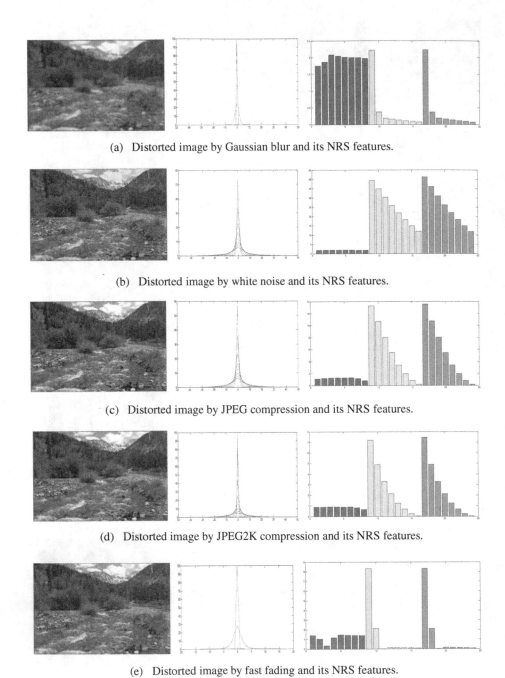

(a) Distorted image by Gaussian blur and its NRS features.

(b) Distorted image by white noise and its NRS features.

(c) Distorted image by JPEG compression and its NRS features.

(d) Distorted image by JPEG2K compression and its NRS features.

(e) Distorted image by fast fading and its NRS features.

Fig. 3. Histograms of RIs for various distorted versions of Fig. 1(a) and their NRS features. Distortions are from the LIVE database. (Color figure online)

2.2 Natural Redundancy Statistics

In order to measure the distributions of RIs, we use the asymmetric generalized Gaussian distribution (AGGD) to effectively capture a broader spectrum of distorted image statistics, where the AGGD with zero mean is given by:

$$f(x; v, \sigma_l^2, \sigma_r^2) = \begin{cases} \dfrac{v}{(\beta_l + \beta_r)\,\Gamma(1/v)} \exp\left(-\left(\dfrac{-x}{\beta_l}\right)^v\right) & x < 0 \\[2ex] \dfrac{v}{(\beta_l + \beta_r)\,\Gamma(1/v)} \exp\left(-\left(\dfrac{x}{\beta_r}\right)^v\right) & x \geq 0 \end{cases} \tag{4}$$

where

$$\beta_l = \sigma_l \sqrt{\frac{\Gamma(1/v)}{\Gamma(3/v)}} \tag{5}$$

$$\beta_r = \sigma_r \sqrt{\frac{\Gamma(1/v)}{\Gamma(3/v)}} \tag{6}$$

And $\Gamma(\bullet)$ is the gamma function. The shape parameter controls the 'shape' of the distribution while σ_l^2 and σ_r^2 are scale parameters that control the spread on each side of the mode, respectively. The parameters of $(v, \sigma_l^2, \sigma_r^2)$ are estimated using the moment-matching based approach proposed in [7]. These three parameters can completely characterize the statistical information of the distribution of one redundancy image. Note that the AGGD model has also been successfully used in the NSS based works [13, 29].

The parameters are extracted over all the eight redundancy images and this yields 24 features that extracted from each image. Because the features are computed based on the statistics of redundancy images, we refer to these features by natural redundancy statistics (NRS) features. In Fig. 1(k), we give the features of the undistorted image, denoted by f, and the features of distorted versions are also shown in Fig. 3. Note that we put the parameters of eight distributions together ($f1 - f8$, denoted by the red bars), and parameters σ_l^2 are followed ($f9 - f16$, denoted by the yellow bars), the last are σ_r^2 ($f17 - f24$), denoted by the green bars). From the NRS features, we can more easily tell the difference between the reference image and the distorted ones than from the histograms. For example, the scale parameters ($f9 - f24$) are much smaller in Gaussian blur distortion than in the reference image, and in the fast fading distortion, the shape parameters ($f1 - f8$) change greatly and are different.

Most assessment methods are designed for the luminance component, and color space information is not properly used. To exploit the statistics of distortions in color space, we transform RGB images into a perceptually relevant opponent color space [3, 29] to computing the NRS features:

$$\begin{bmatrix} O_1 \\ O_2 \\ O_3 \end{bmatrix} = \begin{bmatrix} 0.06 & 0.63 & 0.27 \\ 0.3 & 0.04 & -0.35 \\ 0.34 & -0.6 & 0.17 \end{bmatrix} \begin{bmatrix} R \\ G \\ B \end{bmatrix} \tag{7}$$

2.3   Quality Evaluation

With the NRS feature vector f of an image, a regression function could be learned to map f to the image subjective quality score. Therefore, we need a set of training images, whose subjective quality scores are available. Such a training dataset can be extracted from several existing IQA database. Then, we can construct a training set. The framework is generic enough to allow for the use of any regressor. In our implementation, a SVM regressor (SVR) is used. SVR has previously been applied to image quality assessment problems [13,17,18]. In this work, we utilize the LIBSVM package [1] to implement the SVR with a linear kernel [25]. Once the regression model is learned, we can use it to estimate the perceptual quality of any input image. The flowchart of the proposed NRS based BIQA method is illustrated in Fig. 4. Note that the feature extraction should be repeated several times for each subblock in each color space.

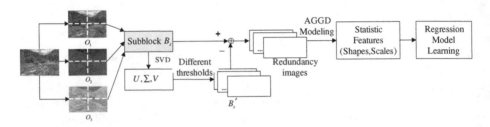

Fig. 4. Flowchart of the proposed BIQA method by statistics of natural redundancy. (Color figure online)

3   Experimental Results

3.1   Implementation Details

For comparison to state-of-the-arts, we utilize public database such as LIVE [24], CSIQ [6], and TID2013 [19]. These database contain distorted images and corresponding mean opinion score (MOS) or difference mean opinion score (DMOS). The LIVE database consists of 779 distorted images degenerated from 29 reference images with 5 distortions at various levels. The distortions are JPEG2000 compression (JP2K), JPEG compression, white noise (WN), Gaussian blur (GB), and fast fading (FF). The CSIQ database includes 900 distorted images degraded from 30 original images by 6 types of distortions at 6 levels. The distortions are JP2K, JPEG, WN, GB, global contrast decrements (GCD) and pink noise (PN). As to TID2013, a total of 3000 distorted images are generated by applying 24 distortion operations at 5 levels to 25 pristine images. It is worth noting that the TID2013 database include images with multiple distortions.

We compare the proposed BIQA method with four state-of-the-art opinion-aware BIQA methods, including BIQI [15], BRISQUE [13], BLIINDS2 [21] and DIIVINE [16] and three state-of-the-art opinion-unaware methods: NIQE [14],

QAC [28] and IL-NIQE [29] are also included because of their excellent generalization ability. We refer to our method by the NRS evaluator (NRSE) for short. There are some parameters in the NRSE: the number of the redundancy images in each color channel, the thresholds and the number of subblocks. In experiments, we use eight redundancy images generated by setting the thresholds from 0.1 to 0.8 stepped by 0.1. Each color channel is divided into 2×2 subblocks. Thus, a total of $24 \times 2 \times 2 \times 3 = 288$ features are used to perform quality assessment. For example, $f1 - f96$ are computed from O_1. Performance comparison of our method with respect to variation in the number of redundancy images is also given in Sect. 3.4.

Performance is evaluated by two common measures: the Spearman rank order correlation coefficients (SROCC) which measures the prediction monotonicity and the Pearson correlation coefficients (PCC) which is related to the prediction linearity. If both measurements are close to 1, it means that a prediction method shows high correlation to human scores. Since opinion-aware methods require distorted images to learn the model, we partitioned each database into a training subset and a testing subset: 80% of the images were used for training and the remaining 20% were used for test. During our experiments, the training and test sets are split according to the reference image to guarantee the independency of the image content in training set and test set. This splitting is repeated for 1,000 times and the median results are used to evaluate the final performance.

3.2 Performance on Individual Database

We first evaluate the overall performance of the competing BIQA models on each of the three database. The results are listed in Table 1. Note that we consider all distortion types involved in each database, not only the common four distortion types to the three database. The top two results are highlighted with boldface for each column. From Table 1, we can observe that NRSE shows good performance on the three database. When the entire LIVE database is considered, the top two methods are NRSE and BRISQUE. The advantages of the proposed NRSE over the other BIQA models are significant on the CSIQ and TID2013 database, NRSE performs the best in terms of SRC and PCC scores, followed by IL-NIQE.

To more comprehensively evaluate BIQA's ability to predict image quality degradations caused by specific types of distortions, in this experiment, we examine the performance of the competing methods on each type of distortions. We tabulate the performance of the methods for each individual distortion on LIVE and CSIQ databases in Tables 2 and 3, respectively. To save space, only the results of SRC index are shown in these tables. In LIVE database, NRSE shows top performance on JP2K, GBLUR and FF. As to CSIQ database, NRSE performs best on all distortions except JPEG. More specifically, methods based on wavelet features (BIQI and DIIVINE)achieve low SRC scores on JP2K distortion, while the DCT based BLIINDS-II, IL-NIQE and the proposed NRSE show excellent performance. For most commonly encountered distortion types, such as JPEG, JP2K, GBLUR and WN, most of the methods can achieve a SRC more than 0.9. As to FF distortion, it is hard to capture the intrinsic characteristic

Table 1. The performance comparison on the three database.

Methods	LIVE		CSIQ		TID2013	
	SRC	PCC	SRC	PCC	SRC	PCC
BIQI	0.816	0.836	0.078	0.223	0.302	0.351
DIIVINE	0.893	0.891	0.748	0.791	0.550	0.645
BLIINDS-II	0.920	0.923	0.791	0.841	0.548	0.635
BRISQUE	**0.931**	**0.932**	0.756	0.809	**0.561**	0.637
NIQE	0.914	0.919	0.609	0.710	0.317	0.426
QAC	0.885	0.895	0.498	0.671	0.390	0.495
IL-NIQE	0.896	0.897	**0.813**	**0.849**	0.521	**0.648**
NRSE	**0.936**	**0.933**	**0.901**	**0.902**	**0.712**	**0.752**

Table 2. Median SRC of the BIQA methods on LIVE database.

Methods	JPEG	JP2K	GBLUR	WN	FF	ALL
BIQI	0.890	0.794	0.845	0.925	0.703	0.816
DIIVINE	0.892	0.853	0.880	0.962	0.833	0.893
BLIINDS-II	**0.942**	**0.927**	0.913	**0.978**	**0.886**	0.920
BRISQUE	**0.961**	0.920	**0.959**	0.967	0.881	**0.931**
NIQE	0.924	0.910	0.930	0.966	0.858	0.914
QAC	0.922	0.888	0.932	0.930	0.823	0.885
IL-NIQE	0.931	0.883	0.918	0.972	0.827	0.896
NRSE	0.939	**0.950**	**0.972**	**0.977**	**0.908**	**0.936**

Table 3. Median SRC of the BIQA methods on CSIQ database.

Methods	JPEG	JP2K	GBLUR	WN	PN	GCD	All
BIQI	0.870	0.692	0.741	0.873	0.381	**0.546**	0.078
DIIVINE	0.805	0.809	0.880	0.859	0.155	0.388	0.748
BLIINDS-II	0.912	0.909	**0.912**	0.829	0.318	0.020	0.791
BRISQUE	0.908	0.856	0.911	**0.923**	0.223	0.031	0.756
NIQE	0.879	0.897	0.905	0.805	0.277	0.213	0.609
QAC	**0.917**	0.871	0.823	0.811	0.001	0.219	0.498
IL-NIQE	0.890	**0.911**	0.837	0.845	**0.867**	0.512	**0.813**
NRSE	**0.915**	**0.928**	0.912	**0.942**	**0.935**	0.825	0.901

for quality prediction because FF simultaneously introduces structure shifting, ringing and color contamination [24]. All methods fail to give a high SRC score, and our NRSE behaves the best (0.908). Our NRSE performs the best in CSIQ database, especially on images of PN and GCD.

We further test the performance of NRSE, BRISQUE and DIIVINE, which are the best three methods on the entire TID2013 database, on more challenging distortion types existed in TID2013 database. The SRC results are shown in Table 4. Note that the results of NRSE are given under three partition proportions: distorted images associated with 40%, 60%, and 80% of the reference images are used for training and the rest are used for testing. The superiority of NRSE to other methods is clearly observable, especially on some distortions

Table 4. Median SRC with more distortion types on TID2013 database.

| | DIIVINE | BRISQUE | NRSE | | |
	80%	80%	40%	60%	80%
Additive Gaussian noise	0.711	0.766	0.916	0.932	0.945
Additive noise more in color	0.442	0.581	0.831	0.834	0.840
Spatially correlated noise	0.826	0.823	0.920	0.941	0.962
Masked noise	0.133	0.213	0.778	0.841	0.873
High frequency noise	0.816	0.876	0.917	0.926	0.943
Impulse noise	0.787	0.809	0.911	0.912	0.918
Quantization noise	0.553	0.712	0.726	0.846	0.884
Gaussian blur	0.923	0.870	0.874	0.909	0.914
Image denoising	0.720	0.509	0.862	0.879	0.909
JPEG compression	0.723	0.781	0.887	0.916	0.939
JPEG2000 compression	0.871	0.798	0.831	0.867	0.883
JPEG transmission errors	0.314	0.217	0.599	0.689	0.717
JPEG2000 transmission errors	0.699	0.712	0.634	0.689	0.701
Non eccentricity pattern noise	0.144	0.109	0.035	0.082	0.248
Local block-wise distortions	0.312	0.245	0.274	0.356	0.497
Mean shift	0.192	0.115	0.273	0.439	0.538
Contrast change	0.340	0.063	0.626	0.743	0.802
Change of color saturation	0.223	0.096	0.766	0.899	0.921
Multiplicative Gaussian noise	0.691	0.632	0.901	0.925	0.935
Comfort noise	0.268	0.172	0.615	0.740	0.804
Lossy compression of noisy images	0.687	0.523	0.889	0.921	0.939
Image color quantization with dither	0.813	0.865	0.775	0.865	0.877
Chromatic aberrations	0.773	0.723	0.755	0.794	0.849
Sparse sampling and reconstruction	0.851	0.811	0.891	0.907	0.916
All	0.550	0.561	0.679	0.709	0.712

Table 5. Median SRC of cross-database performance evaluation.

	Trained on live		Trained on TID2013	
	CSIQ	TID2013	LIVE	CSIQ
BIQI	0.619	**0.394**	0.047	0.010
DIIVINE	0.596	0.355	0.042	0.146
BLIINDS-II	0.577	0.393	0.076	0.456
BRISQUE	0.557	0.367	0.088	**0.639**
NRSE	**0.682**	0.373	**0.755**	0.494
NIQE	0.627	0.311	**0.906**	0.627
QAC	0.490	0.372	0.868	0.490
IL-NIQE	**0.815**	**0.494**	0.898	**0.815**

(the rows shaded in Table 4) that other BIQA methods fail to give acceptable performance. Examples of these distortions include masked noise, JPEG transmission errors, contrast change, change of color saturation and comfort noise. The satisfying results of NRSE are due to the fact that our method takes the color space information into account. In Fig. 5, we show the NRS features of a reference image and its two distorted images (contrast change and change of color saturation). From the NRS features we can easily tell the difference between the reference image and the distorted images.

3.3 Cross-Database Performance Evaluation

In the above experiments, the training and test samples were drawn from the same database. Now, we test the generalization capability of opinion-aware models by training them on one database, then testing them on other database. Of course, for opinion-unaware methods (NIQE, QAC and IL-NIQE), the training step is not needed. The SRC score is used for evaluation, and the results are presented in Table 5. The top result is highlighted in bold individually for opinion-aware and opinion-unaware methods. Again, the proposed NRSE performs very well. It achieves the best SRC scores in 2 out of the 4 tests among the 5 opinion-aware methods, and sometimes its results are better than the opinion-unaware methods. Note that when trained on TID2013 and then applied to LIVE database, the opinion-aware methods deliver poor performance, only our NRSE achieves a satisfying result. From Table 5, we can also draw a conclusion: opinion-unaware methods exhibit clear generalization performance advantages over their opinion-aware counterparts.

3.4 Sensitivity to Parameter Variations

In this sub-section, we further show that NRSEs performance is not sensitive to the variations of parameters. We conducted experiments to test NRSE on LIVE

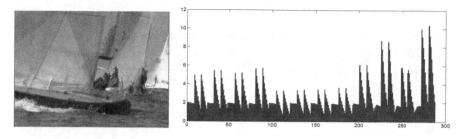

(a) A reference image from TID2013 and its NRS features.

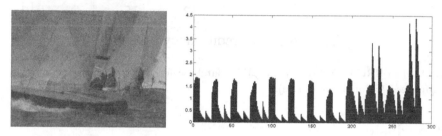

(b) Distorted image by contrast change and its NRS features.

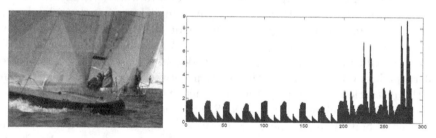

(c) Distorted image by change of color saturation and its NRS features.

Fig. 5. The NRS features of a reference image and its two distorted images. Distortions are from the TID2013 database. (Color figure online)

database by varying the key parameters: the number of the redundancy images RIs. Different number of RIs are obtained by fixing the range of the threshold and varying the step size. For example, ten RIs are generated by setting the thresholds from 0.1 to 0.8 stepped by 0.08. The results are presented in Fig. 6. SRC is used as the metric. As observed, the performance of NRSE remains relatively stable with respect to variation in the number of redundancy images used to compute the features.

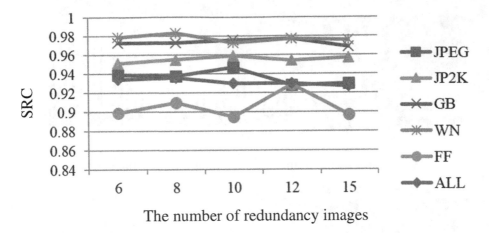

The number of redundancy images

Fig. 6. Median SRC of NRSE on the LIVE database for different numbers of redundancy images. (Color figure online)

4   Conclusion

Existing BIQA models typically extract the quality-aware statistical features in a redundancy reduced domain. However, few BIQA models exploit the redundancy information existing in the images. Here we made the first attempt to directly use the redundancy images to obtain the natural redundancy statistics (NRS) features. We utilize the simple singular value decomposition (SVD) to generate several redundancy images. Since the relation between redundancy images can reflect the distortion, we employ the asymmetric generalized Gaussian distribution (AGGD) modeling to obtain NRS, and the opponent color spaces are also utilized. Then, we map the NRS features to the subjective quality score by a regression model. Experimental results demonstrate that our method leads to highly competitive performance with many state-of-the-art BIQA methods in terms of quality prediction accuracy and generalization ability (i.e., across-database prediction capability).

References

1. Chang, C.C., Lin, C.J.: LIBSVM: a library for support vector machines. ACM Trans. Intell. Syst. Technol. (TIST) **2**(3), 27 (2011)
2. Ferzli, R., Karam, L.J.: A no-reference objective image sharpness metric based on the notion of just noticeable blur (JNB). IEEE Trans. Image Process. **18**(4), 717–728 (2009)
3. Geusebroek, J.M., Van den Boomgaard, R., Smeulders, A.W.M.: Color invariance. IEEE Trans. Pattern Anal. Mach. Intell. **23**(12), 1338–1350 (2001)
4. Kim, J., Eun, H., Kim, C.: No-reference image quality assessment based on singular value decomposition without learning. In: Ho, Y.-S., Sang, J., Ro, Y.M., Kim, J., Wu, F. (eds.) PCM 2015. LNCS, vol. 9315, pp. 506–515. Springer, Cham (2015). doi:10.1007/978-3-319-24078-7_51

5. Konstantinides, K., Natarajan, B., Yovanof, G.S.: Noise estimation and filtering using block-based singular value decomposition. IEEE Trans. Image Process. Publ. IEEE Signal Process. Soc. **6**(3), 479–483 (1996)
6. Larson, E.C., Chandler, D.M.: Most apparent distortion: full-reference image quality assessment and the role of strategy. J. Electron. Imaging **19**(1), 011006-1–011006-21 (2010)
7. Lasmar, N.E., Stitou, Y., Berthoumieu, Y.: Multiscale skewed heavy tailed model for texture analysis. In: 2009 16th IEEE International Conference on Image Processing (ICIP), pp. 2281–2284. IEEE (2009)
8. Liang, L., Wang, S., Chen, J.: No-reference perceptual image quality metric using gradient profiles for JPEG2000. Signal Process. Image Commun. **25**(7), 502–516 (2010)
9. Lin, W., Kuo, C.C.J.: Perceptual visual quality metrics: a survey. J. Vis. Commun. Image Represent. **22**(4), 297–312 (2011)
10. Liu, H., Klomp, N., Heynderickx, I.: A no-reference metric for perceived ringing artifacts in images. IEEE Trans. Circ. Syst. Video Technol. **20**(4), 529–539 (2010)
11. Manap, R.A., Shao, L.: Non-distortion-specific no-reference image quality assessment: a survey. Inf. Sci. **301**, 141–160 (2015)
12. Mansouri, A., Aznaveh, A.M., Torkamani-Azar, F.: Image quality assessment using the singular value decomposition theorem. Opt. Rev. **16**(2), 49–53 (2009)
13. Mittal, A., Moorthy, A.K., Bovik, A.C.: No-reference image quality assessment in the spatial domain. IEEE Trans. Image Process. **21**(12), 4695–4708 (2012)
14. Mittal, A., Soundararajan, R., Bovik, A.C.: Making a completely blind image quality analyzer. IEEE Signal Process. Lett. **20**(3), 209–212 (2013)
15. Moorthy, A.K., Bovik, A.C.: A two-step framework for constructing blind image quality indices. IEEE Signal Process. Lett. **17**(5), 513–516 (2010)
16. Moorthy, A.K., Bovik, A.C.: Blind image quality assessment: from natural scene statistics to perceptual quality. IEEE Trans. Image Process. **20**(12), 3350–3364 (2011)
17. Narwaria, M., Lin, W.: Objective image quality assessment based on support vector regression. IEEE Trans. Neural Netw. **21**(3), 515–519 (2010)
18. Narwaria, M., Lin, W.: SVD-based quality metric for image and video using machine learning. IEEE Trans. Syst. Man Cybern. Part B (Cybern.) **42**(2), 347–364 (2012)
19. Ponomarenko, N., Ieremeiev, O., Lukin, V.: Color image database TID2013: peculiarities and preliminary results. In: 2013 4th European Workshop on Visual Information Processing (EUVIP), pp. 106–111. IEEE (2013)
20. Saad, M.A., Bovik, A.C., Charrier, C.: A DCT statistics-based blind image quality index. IEEE Signal Process. Lett. **17**(6), 583–586 (2010)
21. Saad, M.A., Bovik, A.C., Charrier, C.: Blind image quality assessment: a natural scene statistics approach in the DCT domain. IEEE Trans. Image Process. **21**(8), 3339–3352 (2012)
22. Sang, Q., Wu, X., Li, C.: Blind image quality assessment using a reciprocal singular value curve. Signal Process. Image Commun. **29**(10), 1149–1157 (2014)
23. Shnayderman, A., Gusev, A., Eskicioglu, A.M.: An SVD-based grayscale image quality measure for local and global assessment. IEEE Trans. Image process. **15**(2), 422–429 (2006)
24. Sheikh, H.R., Sabir, M.F., Bovik, A.C.: A statistical evaluation of recent full reference image quality assessment algorithms. IEEE Trans. Image Process. **15**(11), 3440–3451 (2006)

25. Smola, A.J., Schölkopf, B.: A tutorial on support vector regression. Stat. Comput. **14**(3), 199–222 (2004)
26. Varadarajan, S., Karam, L.J.: An improved perception-based no-reference objective image sharpness metric using iterative edge refinement. In: 2008 15th IEEE International Conference on Image Processing, pp. 401–404 (2008)
27. Xue, W., Mou, X., Zhang, L.: Blind image quality assessment using joint statistics of gradient magnitude and Laplacian features. IEEE Trans. Image Process. **23**(11), 4850–4862 (2014)
28. Xue, W., Zhang, L., Mou, X.: Learning without human scores for blind image quality assessment. In: Proceedings of the IEEE Conference on Computer Vision and Pattern Recognition, pp. 995–1002(2013)
29. Zhang, L., Zhang, L., Bovik, A.C.: A feature-enriched completely blind image quality evaluator. IEEE Trans. Image Process. **24**(8), 2579–2591 (2015)

Sparse Coding on Cascaded Residuals

Tong Zhang[1]([✉]) and Fatih Porikli[1,2]

[1] Australian National University, Canberra, Australia
{tong.zhang,fatih.porikli}@anu.edu.au
[2] Data61/CSIRO, Eveleigh, Australia

Abstract. This paper seeks to combine dictionary learning and hierarchical image representation in a principled way. To make dictionary atoms capturing additional information from extended receptive fields and attain improved descriptive capacity, we present a two-pass multiresolution cascade framework for dictionary learning and sparse coding. The cascade allows collaborative reconstructions at different resolutions using the same dimensional dictionary atoms. Our jointly learned dictionary comprises atoms that adapt to the information available at the coarsest layer where the support of atoms reaches their maximum range and the residual images where the supplementary details progressively refine the reconstruction objective. The residual at a layer is computed by the difference between the aggregated reconstructions of the previous layers and the downsampled original image at that layer. Our method generates more flexible and accurate representations using much less number of coefficients. Its computational efficiency stems from encoding at the coarsest resolution, which is minuscule, and encoding the residuals, which are relatively much sparse. Our extensive experiments on multiple datasets demonstrate that this new method is powerful in image coding, denoising, inpainting and artifact removal tasks outperforming the state-of-the-art techniques.

1 Introduction

Sparse representation promises noise resilience by assigning representation coefficients from dictionary atoms characterizing the clean data distribution, improved classification performance by attaining discriminative features, robustness by preventing the model from overfitting data, and semantic interpretation by allowing atoms to associate with meaningful attributes. Computer vision applications include image compression, regularization in reverse problems, feature extraction, recognition, interpolation for incomplete data, and more [1–6].

An overcomplete dictionary that leads to sparse representations can either be chosen from a predetermined set of functions or designed by adapting its content to fit a given set of samples. The performance of the predetermined dictionaries, such as overcomplete Discrete Cosine Transform (DCT) [7], wavelets [8], curvelets [9], contourlets [10], shearlets [11] and other analytic forms, depends on how suitable they are to sparsely describe the samples in question. On the other hand, the learned dictionaries are data driven and tailored for distinct

© Springer International Publishing AG 2017
S.-H. Lai et al. (Eds.): ACCV 2016, Part IV, LNCS 10114, pp. 19–34, 2017.
DOI: 10.1007/978-3-319-54190-7_2

applications. Noteworthy algorithms of this type include the Method of Optimal Directions (MOD) [12], generalized PCA [13], KSVD [2], Online Dictionary Learning (ODL) [4,14]. The learned dictionaries adapt better compared to analytic ones and provide improved performance.

In general, image based dictionary learning and sparse encoding tasks are formulated as an optimization problem

$$\arg\min_{\mathbf{D},\mathbf{X}} \|\mathbf{Y} - \mathbf{D}\mathbf{X}\|_F^2 \qquad \text{s.t. } \|\mathbf{x}_i\|_0 \leq T , \tag{1}$$

or its equivalent form

$$\arg\min_{\mathbf{X},\mathbf{D}} \sum_i \|\mathbf{x}_i\|_0 \qquad \text{s.t. } \|\mathbf{Y} - \mathbf{D}\mathbf{X}\|_F^2 \leq \epsilon \tag{2}$$

where $\mathbf{Y} \in \mathbb{R}^{n \times k}$ is k image patches with dimension n, $\mathbf{X} \in \mathbb{R}^{m \times k}$ denotes the coefficients of corresponding images, $\mathbf{D} \in \mathbb{R}^{n \times m}$ is an overcomplete matrix, T is the number of coefficient used to describe the images, and ϵ is the error tolerance such that once the reconstruction error is smaller than the tolerance the pursuit will be terminated. The sparsity is achieved because $n \ll m$ and $T \ll m$. For an extended discussion on the solutions of above objectives, see Sect. 2.

For mathematical convenience, dictionary learning methods often employ in uniform spaces, e.g. in the vector space of 8×8 image patches. In other words, same scale blocks are pulled from overlapping or non-overlapping image patches on a dense grid and a single-scale dictionary is learned. However, dictionary atoms learned in this fashion tend to be myopic and blind to global context since such fixed-scale patches only contain local information within their small support. Simply increasing the patch size results in adverse outcomes, i.e. decreased the flexibility of the dictionary to fit data and increased computational complexity. Moreover, optimal patch size varies depending on the underlying texture information. For example, finer partitioning by smaller blocks is preferable for textured regions, yet larger blocks would suit better for smooth areas. Suppose the image to be encoded is a 256×256 flat (e.g. all pixels have the same value) image. Using the conventional 8×8 overlapping blocks would require more than 60 K coefficients, yet the same image can be represented using a small number of coefficients of larger patches, even only a single coefficient in the ideal case of the patch has the size of the image.

As an alternative, multi-scale methods aim to learn dictionaries at different image resolutions for the same patch size using shearlets, wavelets, and Laplacian pyramid [4,5,15–17]. A major drawback of these methods is that each layer in the pyramid is either processed independently or in small frequency bands; thus reconstruction errors of coarser layers are projected directly on the finest layer. Such errors cannot be compensated by other layers. This implies, to attain a satisfactory quality, all layers need to be constructed accurately. Instead of learning in different image resolutions, [18] first builds separate dictionaries for quadtree partitioned patches and then zero-pad smaller patches to the largest scale. However, the size of the dictionary learned in this fashion is proportional to the

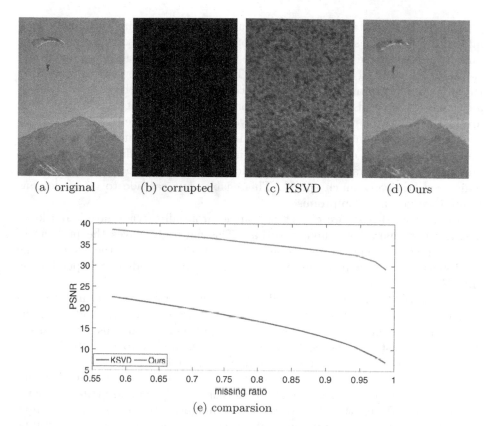

(a) original   (b) corrupted   (c) KSVD   (d) Ours

(e) comparsion

Fig. 1. (a) Original image. (b) Corrupt image where 93% the original pixels are removed. (c) Reconstruction result of KSVD, PSNR is 11.80 dB. (d) Reconstruction of our method, PSNR is 33.34 dB. (e) Reconstructed image quality vs. the rate of missing coefficients. Red: our method, blue: KSVD. As visible, our method is significantly superior to KSVD. (Color figure online)

maximum patch size, which prohibits its applicability due to heavy computational load and inflated memory requirements.

Existing multi-scale dictionary learning methods overlook the redundancy between the layers. As a consequence, larger dictionaries are required, and a high number of coefficients are spent unnecessarily on smooth areas. To the best of our knowledge, no method offers a systematic solution where encodings of the coarser scales progressively enhance the reconstructions of the finer layers.

Our Contributions
Aiming to address the above shortcomings and allow dictionary atoms to access larger context for improved descriptive capacity, here we present a computationally efficient cascade framework that employs multi-resolution residual maps for dictionary learning and sparse coding.

To this end, we start with building an image pyramid using bicubic interpolation. In the first-pass, we learn a dictionary from the coarsest resolution layer and obtain the sparse representation. We upsample the reconstructed image and compute the residual in the next layer. The residual at a level is computed by the difference between the aggregated reconstructions from the coarser layers in a cascade fashion and the downsampled original image at that layer. Dictionaries are learned from the residual in every layer. We use the same patch size yet different resolution input images, which is instrumental in reducing computations and capturing larger context through. The computational efficiency stems from encoding at the coarsest resolution, which is tiny, and encoding the residuals, which are relatively much sparse. This enables our cascade to go as deep as needed without any compromise.

In the second-pass, we collect all patches from all cascade layers and learn a single dictionary for a final encoding. This naturally solves the problem of determining how many atoms to be assigned at a layer. Thus, the atoms in the dictionary have the same dimension still their receptive fields vary depending on the layer.

Compared to existing multi-scale approaches operating indiscriminately on image pyramids or wavelets, our dictionary comprises atoms that adapt to the information available at each layer. The supplementary details progressively refine our reconstruction objective. This allows our method to generate a flexible image representation using much less number of coefficients.

Our extensive experiments on large datasets demonstrate that this new method is powerful in image coding, denoising, inpainting and artifact removal tasks outperforming the state-of-the-art techniques. Figure 1 shows a sample inpainting result from our method where 93% of pixels are missing. As visible, by taking the take advantage of the multi-resolution cascade, we can recover even the very large missing areas.

2   Related Work

The nature of dictionary learning objective makes it an NP-hard problem since neither the dictionary nor the coefficients are known. To handle this, most dictionary learning algorithms alternate between sparse coding and dictionary updating steps iteratively by fixing one while optimizing the other. For example, MOD updates the dictionary by solving an analytic solution of the quadratic problem using as Moore-Penrose pseudo-inverse; KSVD incorporates k-means clustering and singular value decomposition by refining coefficients and dictionary atoms recursively; ODL updates the dictionary by using the first-order stochastic gradient descent in small batches. Adding to the complexity, sparse coding itself is an NP-hard problem due to the ℓ_0 norm. It is often approximated by greedy schemes such as Matching pursuit (MP) [19] and Orthogonal Matching Pursuit (OMP) [20]. Another popular solution is to replace the ℓ^0-norm with an ℓ^p-norm with $p <= 1$. When $p = 1$, the solution can be approximated by Basis Pursuit [21], FoCUSS [22], and Least Angle Regression (LARS) [23] to count a few.

Multi-scale methods for coding have been widely studied in the past. Wavelets have become the premier multi-scale analysis tool in signal processing and many wavelets alike methods such as bandlets [24], contourlets [10], curvelets [9] as well as decomposition methods including wavelet pyramid [25], steerable pyramid [26], and Laplacian pyramid [27] have emerged. These methods aim to improve upon the pure spatial frequency analysis of Fourier transform by providing resolution in both spatial frequency and spatial location.

Nevertheless, there have been few attempts to learn multi-scale dictionaries. In [18], use of a quadtree structure was proposed. Dictionaries with different atom dimensions are learned for different levels of the quadtree, and then concatenated together by zero-padding smaller atoms in a dyadic fashion. Unfortunately, the number of scales and the maximum dimension of dictionary atoms are restricted due to the heavy computational and memory requirements. Besides, this approach does not take the advantage of the coarse-scale information that may be more suitable to represent patches using atoms of the same size.

In order to overcome computational issues, Ophir *et al.* [5] learned sub-dictionaries in the wavelet transform domain by exploiting the sparsity between the wavelets coefficients. This work leverages frequency selectivity of the individual levels of a wavelet pyramid to remove redundancy in the learned representations. However, separate dictionaries are learned for the directional sub-bands, which tends to generate inferior performance when compared to single-scale KSVD in denosing task. Their following work [6] exploited multi-scale analysis and single-scale dictionary learning, and merged both outputs by weighted joint sparse coding. Since the fused dictionary is several times larger than the single-scale version, the computational complexity is high. Besides, the denoising performance is sensitive to the size and category of images. A similar work [4] built multi-resolution dictionaries on wavelet pyramid by employing k-means clustering before the ODL step. For each resolution, it clusters the patches of three sub-bands, and concatenates all dictionary atoms. Even though denoising performance improves due to non-local clustering on sub-bands, each layer requires a large dictionary, which reflects on the computationally load.

Multi-resolution sparse representations are also employed for image fusion and super-resolution. Liu *et al.* [16] fused two images by obtaining sparse coefficients for high-pass and low-pass frequency bands by OMP given the pre-trained dictionary. The fused coefficient columns in each band are chosen by maximal ℓ_1 norm of corresponding coefficients. Towards the same goal, Yin *et al.* [17] merged two coefficient vectors, however, the fused coefficient columns are selected by ℓ_2 norm. Instead of training sub-dictionaries independently, they learn $3S + 1$ sub-dictionaries jointly, which means the dimension of the matrix is $(3S+1)n \times k$, thus the learning stage is computationally expensive. In [15] proposed a multi-scale approach to super-resolve diffusion weighted images where the low-resolution dictionary is based on the shearlet transform and the high-resolution one is based on intensity. In [28] sparse representation was used to build a model for image interpolation. This model describes each patch as a linear combination of similar non-local patch neighbors, and every patch is sparse represented with a specific dictionary. In order to decrease the coherence of the representation basis, it clusters patches into multiple groups and learns multiple local PCA dictionaries.

3   Sparse Coding on Cascade Layers

A flow diagram of our framework is shown in Fig. 2 for a sample 4-layer cascade. Given an image $\mathbf{Y}$, we first construct an image pyramid $\mathbf{Y} = \{\mathbf{Y}_0, \mathbf{Y}_1, ... \mathbf{Y}_N\}$ by bicubic downsampling. Here, $\mathbf{Y}_0$ is the finest (original) resolution and $\mathbf{Y}_N$ is the coarsest resolution. Other options for the image pyramid are Gaussian pyramid, Laplacian pyramid, bilinear interpolation, and subsampling. Images resampled with bicubic interpolation are smoother and have fewer interpolation artifacts. In contrast, bicubic interpolation considers larger support.

We employ a two-pass scheme wherein the first-pass we obtain residuals from layer-wise dictionaries, and in the second-pass, we learn a single, global dictionary that extracts and refines the atoms from the dictionaries generated in the first-pass.

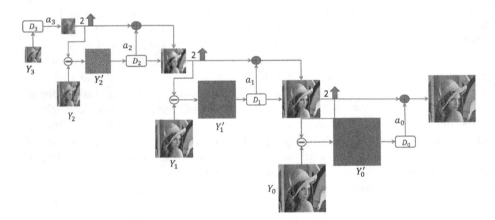

Fig. 2. First-pass of our method for a 4-layer cascade. $\mathbf{Y}_0$ is the original image, $\{\mathbf{Y}_3, ..., \mathbf{Y}_0\}$ denote each layer of the image $\mathbf{Y}_3$ pyramid, and $\{\mathbf{D}_3, ..., \mathbf{D}_0\}$ are the dictionaries. $\mathbf{D}_3$ is learned from the downsampled image, remainig dictionaries are learned from the residuals $\{\mathbf{Y}_2', \mathbf{Y}_1', \mathbf{Y}_0'\}$. α_n are the coefficients used to reconstruct each layer.

First-pass

We start at the coarsest layer N in the cascade. After learning the layer dictionary and finding the sparse coefficients, we propagate consecutively the reconstructed images to the finer layers. In the coarsest layer we process the downsampled image, in the consecutive layers we encode and decode the residuals. In each layer, we use same size $b \times b$ patches. A patch in the layer n corresponds

---

Algorithm 1. Cascade sparse coding

Input:
1: N (the highest pyramid layer), $\mathbf{Y}$(image),
2: T_n (number of coefficient used in layer n)
Output: $\mathbf{Y}', \hat{\mathbf{Y}}, \hat{\mathbf{D}}_{global}$
3: $\mathbf{Y}_n \leftarrow$ subsampling$(\mathbf{Y}, 2^n)$
4: **for** $n = \{N, N-1, \cdots, 0\}$ **do**
5:    **if** n = N **then**
6:       $\mathbf{Y}'_n \leftarrow \mathbf{Y}_n$
7:    **else**
8:       $\mathbf{Y}'_n \leftarrow \mathbf{Y}_n -$ upsample$(\hat{\mathbf{Y}}_{n+1}, 2)$
9:    Perform KSVD to learn dictionary $\hat{\mathbf{D}}_n$ and encode $\mathbf{Y}'_n$
10:    $\forall ij \ \{\hat{\mathbf{x}}_n^{ij}, \hat{\mathbf{D}}_n\} \leftarrow \underset{\mathbf{x}_n^{ij}, \mathbf{D}_n}{\arg\min} \sum_{ij} \|\mathbf{R}_{ij}\mathbf{Y}'_n \quad \mathbf{D}_n\mathbf{x}_n^{ij}\|_2^2$ s.t $\|\mathbf{x}_n^{ij}\|_0 \leq T_n$
11:    **if** n = N **then**
12:       $\hat{\mathbf{Y}}_n \leftarrow (\sum_{ij} \mathbf{R}_{ij}^T \mathbf{R}_{ij})^{-1}(\sum_{ij} \mathbf{R}_{ij}^T \hat{\mathbf{D}}_n \hat{\mathbf{x}}_n^{ij})$
13:    **else**
14:       $\hat{\mathbf{Y}}_n \leftarrow (\sum_{ij} \mathbf{R}_{ij}^T \mathbf{R}_{ij})^{-1}(\sum_{ij} \mathbf{R}_{ij}^T \hat{\mathbf{D}}_n \hat{\mathbf{x}}_n^{ij}) +$ upsample$(\hat{\mathbf{Y}}_{n+1}, 2)$
15: $\mathbf{Y}' \leftarrow \{\mathbf{Y}'_N, \mathbf{Y}'_{N-1} \cdots, \mathbf{Y}'_0\}$
16: $\forall ij \ \hat{\mathbf{D}}_{global} \leftarrow \underset{\mathbf{D}}{\arg\min} \sum_{ij} \|\mathbf{R}_{ij}\mathbf{Y}' - \mathbf{D}\mathbf{x}^{ij}\|_2^2$ s.t $\|\mathbf{x}^{ij}\|_0 \leq T$
17: **Reconstruction:**
18: $\hat{\mathbf{Y}} \leftarrow 0$
19: **for** $n = \{N, N-1, \cdots, 0\}$ **do**
20:    $\mathbf{Y}'_n = \mathbf{Y}_n -$ upsample$(\hat{\mathbf{Y}}, 2)$
21:    $\forall ij \ \{\hat{\mathbf{x}}_n^{ij}\} \leftarrow \underset{\mathbf{x}_n^{ij}}{\arg\min} \sum_{ij} \|\mathbf{R}_{ij}\mathbf{Y}'_n - \hat{\mathbf{D}}_{global}\mathbf{x}_n^{ij}\|_2^2$ s.t $\|\mathbf{x}_n^{ij}\|_0 \leq T_n$
22:    $\hat{\mathbf{Y}} \leftarrow (\sum_{ij} \mathbf{R}_{ij}^T \mathbf{R}_{ij})^{-1}(\sum_{ij} \mathbf{R}_{ij}^T \hat{\mathbf{D}}_{global}\hat{\mathbf{x}}_n^{ij}) +$ upsample$(\hat{\mathbf{Y}}, 2)$
23: **return**

---

to a $(b \times 2^n) \times (b \times 2^n)$ area in the original image. Algorithm 1 summarizes the first-pass.

Dictionary learning: We learn a dictionary at the coarsest layer and use it to reconstruct the downsampled image. This layer's dictionary $\hat{\mathbf{D}}_N$ is produced by minimizing the objective function using the coarsest resolution image patches

$$\underset{\mathbf{D}_N, \mathbf{x}_N^{ij}}{\arg\min} \sum_{ij} \|\mathbf{R}_{ij}\mathbf{Y}_N - \mathbf{D}_N\mathbf{x}_N^{ij}\|_2^2 + \lambda\|\mathbf{x}_N^{ij}\|_0 \tag{3}$$

where the operator $\mathbf{R}_{ij}$ is a binary matrix that extracts a square patch of size $b \times b$ at location (i, j) in the image then arranges the patch in a column vector form. The parameter λ balances the data fidelity term and the regularization term, and $\mathbf{x}_N^{ij}$ denotes the coefficients for the patch (i, j).

We initialize the dictionary $\mathbf{D}_N$ with a DCT basis by extracting several atoms from the DCT bases and applying Kronecker product. It is possible to choose any dictionary update methods such as KSVD [2], approximate KSVD [29],

MOD [12], and ODL [14]. Both ODL and approximate KSVD can achieve the same PSNR with less coefficients. In order to reveal the strength of our method, we choose the original KSVD to update the dictionaries. Therefore, we do a sequence of rank-one approximations that update both the dictionary atoms and the coefficients.

Iteratively, we first fix all coefficients $\hat{\mathbf{x}}_N^{ij}$ and select each dictionary atom one by one $\mathbf{d}_N^l$, $l = \{1, 2, \cdots, k\}$. For any atom $\mathbf{d}_N^l$, we extract the patches, which are composed by the atom $(i, j) \in \mathbf{d}_N^l$, to compute its residual. The corresponding coefficients are denoted as $\mathbf{x}_N^{ij}(l)$, which are the non-zero entries of the l-th row of coefficient matrix

$$\mathbf{e}_N^{ij}(l) = \mathbf{R}_{ij}\mathbf{Y}_N - \hat{\mathbf{D}}_N\mathbf{x}_N^{ij} + \mathbf{d}_N^l\mathbf{x}_N^{ij}(l). \tag{4}$$

We arrange all $\mathbf{e}_N^{ij}(l)$ as the columns of the overall representation error matrix $\mathbf{E}_N^l$. Then, we update the atom $\hat{\mathbf{d}}_N^l$ and the l-th row $\hat{\mathbf{x}}_N(l)$ by

$$\{\hat{\mathbf{d}}_N^l, \hat{\mathbf{x}}_N(l)\} = \arg\min_{\mathbf{d}, \mathbf{x}} \|\mathbf{E}_N^l - \mathbf{d}\mathbf{x}\|_F^2. \tag{5}$$

Finally, we perform SVD decomposition on the error matrix, and update the l-th dictionary atom $\hat{\mathbf{d}}_N^l$ by the first column of $\mathbf{U}$, where $\mathbf{E}_N^l = \mathbf{U}\Sigma\mathbf{V}^T$. The coefficient vector $\hat{\mathbf{x}}_N(l)$ is the first column of matrix $\Sigma(1, 1)\mathbf{V}$. In every iteration all dictionary atoms and coefficients are updated simultaneously.

Sparse coding: After getting the updated dictionary, sparse coding is done with the Orthogonal Matching Pursuit (OMP), a greedy algorithm that is computationally efficient [30]. The sparse coding stops when the number of coefficient reaches the upper limit T_N or the reconstruction error becomes less than threshold

$$\hat{\mathbf{x}}_N^{ij} = \arg\min_{\mathbf{x}_N^{ij}} \sum_{ij} \|\mathbf{R}_{ij}\mathbf{Y}_N' - \hat{\mathbf{D}}_n\mathbf{x}_n^{ij}\|_2^2 \quad \text{s.t.} \quad \|\mathbf{x}_n^{ij}\|_0 \leq T_N. \tag{6}$$

Putting the updated coefficient matrix $\hat{\mathbf{x}}_N^{ij}$ back into Dictionary learning to update the dictionary and coefficient until reaching the iteration times.

Residuals: In each layer, we use at most T_n active coefficients for each patch to reconstruct the image and then compute the residual. The number of coefficients governs how strong the residual to emerge. Larger values of T_n generates a more accurate reconstructed image. Thus, the total energy of residuals will diminish. Smaller values of T_n cause the residual to increase, not only due to sparse coding but also resampling across layers. Since the dictionary is designed to represent a wide spectrum of patterns to keep the encodings as sparse as possible, T_n should be small. The reconstructed image is a weighted average of the patches that contain the same pixel

$$\hat{\mathbf{Y}}_N = (\sum_{ij} \mathbf{R}_{ij}^T\mathbf{R}_{ij})^{-1}(\sum_{ij} \mathbf{R}_{ij}^T\hat{\mathbf{D}}_N\hat{\mathbf{x}}_N^{ij}). \tag{7}$$

After decoding based on the dictionary $\hat{\mathbf{D}}_N$, we obtain the residual image $\mathbf{Y}'_{N-1}$ by subtracting the upsampled reconstruction $\mathbf{U}(\hat{\mathbf{Y}}_N)$ from the next layer image $\mathbf{Y}_{N-1}$, e.g. $\mathbf{Y}'_{N-1} = \mathbf{Y}_{N-1} - \mathbf{U}(\hat{\mathbf{Y}}_N)$. Here, $\mathbf{U}(\cdot)$ denotes the bicubic upsampling operator. As the procedure of dictionary learning ans sparse coding in the N-th layer, we reconstruct residual $\hat{\mathbf{Y}}'_{N-1}$ by training a residual dictionary $\mathbf{D}_{N-1}$ from the residual image itself. We keep encoding and decoding on residuals up to the finest layer. The cascade residual dictionary learning and reconstruction can be expressed as follows:

$$\{\hat{\mathbf{x}}_n^{ij}, \hat{\mathbf{D}}_n\} = \arg\min_{\mathbf{x}_n^{ij}, \mathbf{D}_n} \sum_{ij} \|\mathbf{R}_{ij}\mathbf{Y}'_n - \mathbf{D}_n\mathbf{x}_n^{ij}\|_2^2 \quad \text{s.t.} \quad \|\mathbf{x}_n^{ij}\|_0 \leq T_n, \tag{8}$$

where residual image is

$$\mathbf{Y}'_n = \begin{cases} \mathbf{Y}_n - \mathbf{U}(\hat{\mathbf{Y}}_{n+1}), & 0 \leq n < N \\ \mathbf{Y}_N, & n = N, \end{cases} \tag{9}$$

and the reconstructed residual is

$$\hat{\mathbf{Y}}_n = \begin{cases} (\sum_{ij} \mathbf{R}_{ij}^T\mathbf{R}_{ij})^{-1}(\sum_{ij} \mathbf{R}_{ij}^T\hat{\mathbf{D}}_n\hat{\mathbf{x}}_n^{ij}) + \mathbf{U}(\hat{\mathbf{Y}}_{n+1}), & 0 \leq n < N \\ (\sum_{ij} \mathbf{R}_{ij}^T\mathbf{R}_{ij})^{-1}(\sum_{ij} \mathbf{R}_{ij}^T\hat{\mathbf{D}}_n\hat{\mathbf{x}}_n^{ij}) & n = N. \end{cases} \tag{10}$$

The more coefficients used, which reduce the error caused by sparse representation. Since we are pursuing sparser representation, less number of coefficient would be better.

Second-pass

In each layer the more atoms we use, the better quality can be achieved. However, this would not be the best use of the limited number of atoms. For instance, image patches from the coarsest layer are limited both in quantity and variety. The residual images are relatively sparse which imply they do not require many

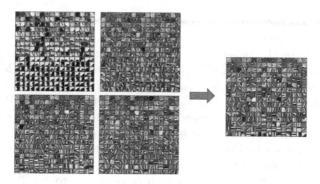

Fig. 3. Left: Four dictionaries of the different levels learned in the first pass (clockwise from the upper-left: the coarsest level, the second, the third, and the finest level). Right: The unifying dictionary learned in the second pass.

dictionary atoms. However, it is not straightforward to determine the optimal number of atoms for each dictionary since finer level residuals depend on coarser ones. Rather than keeping all dictionaries, we train a global dictionary $\mathbf{D}$ using patches from $\mathbf{Y}' = \{\mathbf{Y}_N, \mathbf{Y}'_{N-1}, \cdots, \mathbf{Y}'_0\}$. As illustrated in Fig. 3, the dictionaries learned from $\mathbf{Y}'$ in the first pass are redundant. The overall dictionary is less repetitive and more general to reconstruct all four layers. This allows us to select most useful atoms automatically without making sub-optimal layer-wise decisions. Notice that, in this procedure the number of coefficient can be arbitrarily chosen depending on the target quality of each layer.

4   Experimental Analysis

To demonstrate the flexibility of our method, we evaluate its performance on three different and popular image processing tasks: image coding, image denoising, and image inpainting. Our method is shown to generate the best image inpainting results and provide the most compact set of coding coefficients.

4.1   Image Coding

We compare our method with five state-of-the-art dictionary learning algorithms including both single and multi-scale methods: approximate KSVD (a-KSVD) [29], ODL [14], KSVD [2] Multi-scale KSVD [18], Multi-scale KSVD using wavelets (Multi-wavelets) [5].

For objectiveness, we use the same number of dictionary atoms for our and all other methods. Notice that, a larger dictionary would generate a sparser representations. We employ 4-times over-complete dictionaries, i.e. $\mathbf{D} \in \mathbb{R}^{64 \times 256}$ except for the Multi-wavelets where the dictionary in each sub-band has as many atoms as our dictionary (in favor of Multi-wavelets).

For a comprehensive evaluation, we build five different image datasets, each contains 50 samples of a specific class of images: animals, landscape, texture, face, and fingerprint.

Figure 4 depicts the number of coefficients per pixel vs. PSNR as the function of number of coefficient per each pixel. Each point is the average score for the corresponding method. As seen, our method is the best performing algorithm among the state-of-the-art. In all five image datasets, it achieves higher PSNR scores with significantly much less number of coefficients. In these experiments, the patches are extracted by 1-pixel overlapping in all images. We use 8×8 blocks on each layer, and the cascade comprises 4 layers. Since the blocks in every layer have the same size, the lower resolution blocks efficiently represent larger receptive fields when they are upsampling onto a higher resolution.

From another perspective, when decoding on the coarsest resolution, our method first employs 8×8 blocks, which corresponds to $8 * 2^{n-1} \times 8 * 2^{n-1}$ region on the finest resolution using the same dictionary atoms. Since there is a single global dictionary after the second pass, all layers share the same atoms. This resembles the quadtree structure, however, our method is not limited by the

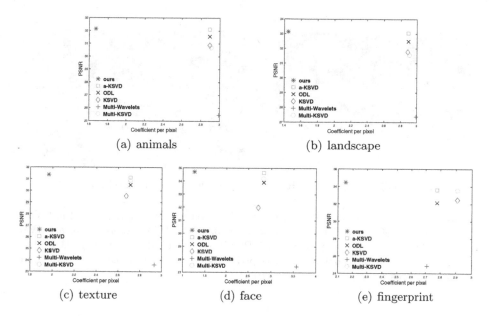

Fig. 4. Reconstruction results on different 5 different image datasets. The horizontal axis represents the number of coefficient per pixel and the vertical axis is the quality in terms of PSNR (dB).

size of the dictionary (dimension of patches - atoms - and number of atoms) and it is as fast as single-scale dictionary learning and sparse coding. For Multiscale KSVD, the maximum dimension of dictionary atom can be 8 and only 2 scales can be performed. Thus, we extracted 128 atoms at each scale.

Compared with other algorithms, our method can save an outstanding 55.6%, 42.23% and 49.95% coefficients for the face, animals, and landscape datasets, respectively. For the image classes where spatial texture is dominant, our method is also superior by decreasing the number of coefficient by 27.74% and 22.38% for the texture and fingerprint datasets. Sample image coding results for qualitative assessment are given in Fig. 5. As shown, a-KSVD image coding is inferior to our even though a-KSVD uses more coefficients.

4.2   Image Denoising

We also analyze the image denoising performance of our method. We make comparison with five dictionary learning algorithms. We note that the state-of-the-art is collaborative and non-local techniques, such as BM3D [31], LSSC [1], yet we do not engineer a collaborative scheme. Our goal here is to understand how our method compares to other dictionary learning methods.

We minimize the cost function in Eq. (11) for denoising. We use the difference between the downsampled input image and aggregated reconstructions at each layer to terminate the OMP.

(a) a-KSVD:28.68 db PSNR   (b) Our method: 32.62 db PSNR

Fig. 5. Image coding results the comparison between a-KSVD and our method. Our method uses 1309035 coefficients and achieves 32.62 dB PSNR score while a-KSVD uses 1332286 coefficients to get 28.65 dB PSNR. our method is almost **4 dB** better. Enlarged red regions are shown on the top-right corner of each image. As visible, our method produces more accurate reconstructions.

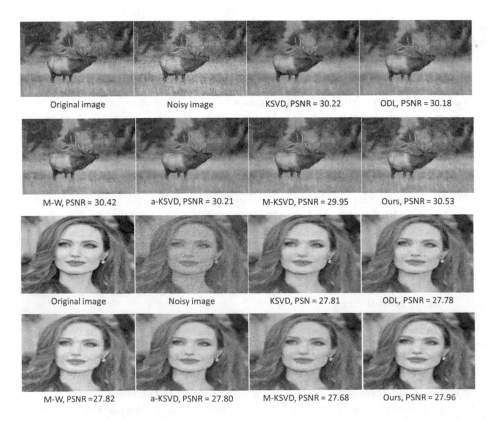

Fig. 6. Denoised face images. Additive zero-mean Gaussian noise with $\sigma = 30$.

$$\arg\min_{\mathbf{x}_i} \sum_{ij} \|\mathbf{x}_n^{ij}\|_0$$

$$\text{s.t.} \|\mathbf{R}_{ij}\mathbf{Y}_n - \mathbf{D}_n\mathbf{x}_n^{ij} + \mathbf{R}_{ij}\mathbf{U}(\hat{\mathbf{Y}}_{n+1})\|_2^2 \leq C\sigma \tag{11}$$

Above, the reconstructed residual $\hat{\mathbf{Y}}_{n+1}$ is defined as in Eq. (10), and σ is chosen according to the variance of the noise. As before, we choose the 4-layer cascade and 8×8 patch size. The parameters of KSVD and multi-scale wavelets are set as recommenced by original authors. We fixed all parameters for all test images. As shown in Fig. 6, our method achieves higher PSNR scores than the state-of-the-art. In addition, it can render finer details more accurately.

4.3 Image Inpainting

Image inpainting is often used for restoration of damaged photographs and removal of specific artifacts such as missing pixels. Previous dictionary learning based algorithms work when the missing area is small and smaller than the dimension of dictionary atoms.

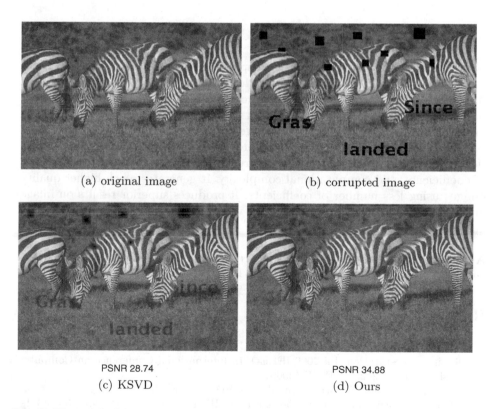

| (a) original image | (b) corrupted image |

PSNR 28.74   PSNR 34.88

(c) KSVD   (d) Ours

Fig. 7. The original image is corrupted with large artifacts. The sizes of the artifacts range from 8 to 32 pixels. Our method efficiently removes the artifacts.

As demonstrated in Fig. 1 our method can restore the missing image regions that are remarkably much larger than the dimension of dictionary atoms, outperforming the state-of-the-art methods. By reconstructing the image starting at the coarsest layer, we can fix completely missing regions. The larger the missing area, the smoother the restored image becomes. In comparison, single-scale based methods fail completely.

Given the mask $\mathbf{M}$ of missing pixels, our formulation in each layer is

$$\hat{\mathbf{x}}_n^{ij} = \arg\min_{\mathbf{x}_n} \sum_{ij} \|\mathbf{R}_{ij}\mathbf{M} \otimes (\mathbf{R}_{ij}\mathbf{Y}_n' - \mathbf{D}_n\mathbf{x}_n)\|_2^2$$

$$\text{subject to} \quad \|\mathbf{x}_n^{ij}\|_0 \leq T_n \tag{12}$$

where we denote $\otimes$ as the element-wise multiplication between two vectors.

Figure 7 shows that our algorithm can fix big holes and gaps but the KSVD can not. In this experiments we only compare with KSVD algorithm, because multiscale KSVD simply increases the dimension of atoms, which leads proportionally more atoms to form an overcomplete dictionary. At the same time, multiscale KSVD still fails to handle holes larger than the dimension of atoms.

5   Conclusion

We presented a dictionary learning and sparse coding method on cascaded residuals. Our cascade allows capturing both local and global information. Its coarse-to-fine structure prevent from reconstructing the regions that can be well represented by the coarser layers. Our sparse coding can be used to progressively improve the quality of the decoded image.

Our method provides significant improvement over the state-of-the-art solutions in terms of the quality of reconstructed image, reduction in the number of coefficients, and computational complexity. It generates much higher quality images using less number of coefficients. It produces superior results on image inpainting, in particular, in handling of very large ratios of missing pixels and large gaps.

Acknowledgment. This work was supported by the Australian Research Council's Discovery Projects funding scheme (project DP150104645).

References

1. Mairal, J., Bach, F., Ponce, J., Sapiro, G., Zisserman, A.: Non-local sparse models for image restoration. In: 2009 IEEE 12th International Conference on Computer Vision, pp. 2272–2279. IEEE (2009)
2. Aharon, M., Elad, M., Bruckstein, A.: K-SVD: an algorithm for designing overcomplete dictionaries for sparse representation. IEEE Trans. Sig. Process. **54**, 4311–4322 (2006)

3. Mairal, J., Ponce, J., Sapiro, G., Zisserman, A., Bach, F.R.: Supervised dictionary learning. In: Advances in Neural Information Processing Systems, pp. 1033–1040 (2009)
4. Yan, R., Shao, L., Liu, Y.: Nonlocal hierarchical dictionary learning using wavelets for image denoising. IEEE Trans. Image Process. **22**, 4689–4698 (2013)
5. Ophir, B., Lustig, M., Elad, M.: Multi-scale dictionary learning using wavelets. IEEE J. Sel. Top. Sig. Process. **5**, 1014–1024 (2011)
6. Sulam, J., Ophir, B., Elad, M.: Image denoising through multi-scale learnt dictionaries. In: 2014 IEEE International Conference on Image Processing (ICIP), pp. 808–812. IEEE (2014)
7. Ahmed, N., Natarajan, T., Rao, K.R.: Discrete cosine transform. IEEE Trans. Comput. **23**, 90–93 (1974)
8. Mallat, S.: A Wavelet Tour of Signal Processing. Academic Press (1999)
9. Candes, E.J., Donoho, D.L.: Curvelets: A surprisingly effective nonadaptive representation for objects with edges. Technical report, DTIC Document (2000)
10. Do, M.N., Vetterli, M.: The contourlet transform: an efficient directional multiresolution image representation. IEEE Trans. Image Process. **14**, 2091–2106 (2005)
11. Labate, D., Lim, W.Q., Kutyniok, G., Weiss, G.: Sparse multidimensional representation using shearlets. In: Optics & Photonics 2005, p. 59140U. International Society for Optics and Photonics (2005)
12. Engan, K., Aase, S.O., Husoy, J.H.: Method of optimal directions for frame design. In: Proceedings of the 1999 IEEE International Conference on Acoustics, Speech, and Signal Processing, ICASSP 1999, vol. 5, pp. 2443–2446 (1999)
13. Vidal, R., Ma, Y., Sastry, S.: Generalized principal component analysis (gpca). IEEE Trans. Pattern Anal. Mach. Intell. **27**, 1945–1959 (2005)
14. Mairal, J., Bach, F., Ponce, J., Sapiro, G.: Online dictionary learning for sparse coding. In: Proceedings of the 26th International Conference on Machine Learning, pp. 1–8 (2009)
15. Tarquino, J., Rueda, A., Romero, E.: A multiscale/sparse representation for diffusion weighted imaging (DWI) super-resolution. In: 2014 IEEE 11th International Symposium on Biomedical Imaging (ISBI), pp. 983–986. IEEE (2014)
16. Liu, Y., Liu, S., Wang, Z.: A general framework for image fusion based on multiscale transform and sparse representation. Inf. Fusion **24**, 147–164 (2015)
17. Yin, H.: Sparse representation with learned multiscale dictionary for image fusion. Neurocomputing **148**, 600–610 (2015)
18. Mairal, J., Sapiro, G., Elad, M.: Learning multiscale sparse representations for image and video restoration. Multiscale Model. Simul. **7**, 214–241 (2008)
19. Mallat, S.G., Zhang, Z.: Matching pursuits with time-frequency dictionaries. IEEE Trans. Sig. Process. **41**, 3397–3415 (1993)
20. Pati, Y.C., Rezaiifar, R., Krishnaprasad, P.: Orthogonal matching pursuit: recursive function approximation with applications to wavelet decomposition, pp. 40–44 (1993)
21. Chen, S.S., Donoho, D.L., Saunders, M.A.: Atomic decomposition by basis pursuit. SIAM Rev. **43**, 129–159 (2001)
22. Gorodnitsky, I.F., Rao, B.D.: Sparse signal reconstruction from limited data using FOCUSS: a re-weighted minimum norm algorithm. IEEE Trans. Sig. Process. **45**, 600–616 (1997)
23. Efron, B., Hastie, T., Johnstone, I., Tibshirani, R., et al.: Least angle regression. Ann. Stat. **32**, 407–499 (2004)
24. Le Pennec, E., Mallat, S.: Sparse geometric image representations with bandelets. IEEE Trans. Image Process. **14**, 423–438 (2005)

25. Mallat, S.G.: A theory for multiresolution signal decomposition: the wavelet representation. IEEE Trans. Pattern Anal. Mach. Intell. **11**, 674–693 (1989)
26. Simoncelli, E.P., Freeman, W.T.: The steerable pyramid: A flexible architecture for multi-scale derivative computation. In: ICIP, p. 3444. IEEE (1995)
27. Burt, P.J., Adelson, E.H.: The laplacian pyramid as a compact image code. IEEE Trans. Commun. **31**, 532–540 (1983)
28. Dong, W., Zhang, L., Lukac, R., Shi, G.: Sparse representation based image interpolation with nonlocal autoregressive modeling. IEEE Trans. Image Process. **22**, 1382–1394 (2013)
29. Rubinstein, R., Zibulevsky, M., Elad, M.: Efficient implementation of the K-SVD algorithm using batch orthogonal matching pursuit. CS Technion **40**, 1–15 (2008)
30. Tropp, J.A.: Greed is good: algorithmic results for sparse approximation. IEEE Trans. Inf. Theory **50**, 2231–2242 (2004)
31. Dabov, K., Foi, A., Katkovnik, V., Egiazarian, K.: Image denoising by sparse 3-d transform-domain collaborative filtering. IEEE Trans. Image Process. **16**, 2080–2095 (2007)

End-to-End Learning for Image Burst Deblurring

Patrick Wieschollek[1,2]([✉]), Bernhard Schölkopf[1], Hendrik P.A. Lensch[2], and Michael Hirsch[1]

[1] Max Planck Institute for Intelligent Systems, Stuttgart, Germany
patrick@wieschollek.info
[2] University of Tübingen, Tübingen, Germany

Abstract. We present a neural network model approach for multi-frame blind deconvolution. The discriminative approach adopts and combines two recent techniques for image deblurring into a single neural network architecture. Our proposed hybrid-architecture combines the explicit prediction of a deconvolution filter and non-trivial averaging of Fourier coefficients in the frequency domain. In order to make full use of the information contained in all images in one burst, the proposed network embeds smaller networks, which explicitly allow the model to transfer information between images in early layers. Our system is trained end-to-end using standard backpropagation on a set of artificially generated training examples, enabling competitive performance in multi-frame blind deconvolution, both with respect to quality and runtime.

1 Introduction

Nowadays, consumer cameras are able to capture an entire series of photographs in rapid succession. Hand-held acquisition of a burst of images is likely to cause blur due to unwanted camera shake during image capture. This is particularly true along with longer exposure times needed in low-light environments.

Motion blurring due to camera shake is commonly modeled as a spatially invariant convolution of a latent sharp image X with an unknown blur kernel k

$$Y = k * X + \varepsilon, \tag{1}$$

where $*$ denotes the convolution operator, Y the blurred observation and ε additive noise. Single image blind deconvolution (BD), i.e. recovering X from Y without knowing k, is a highly ill-posed problem for a variety of reasons. In contrast, multi-frame blind deconvolution or burst deblurring methods aim at recovering a single sharp high-quality image from a sequence of blurry and noisy observed images $Y_1, Y_2, \ldots, Y_N$. Accumulating information from several observations can help to solve the reconstruction problem associated with Eq. (1) more effectively.

Electronic supplementary material The online version of this chapter (doi:10. 1007/978-3-319-54190-7_3) contains supplementary material, which is available to authorized users.

S.-H. Lai et al. (Eds.): ACCV 2016, Part IV, LNCS 10114, pp. 35–51, 2017.
DOI: 10.1007/978-3-319-54190-7_3

Traditionally, generative models are used for blind image deconvolution. While they offer much flexibility they are often computationally demanding and time-consuming.

Discriminative approaches on the other hand keep the promise of fast processing times, and are particularly suited for situations where an exact modeling of the image formation process is not possible. A popular choice in this context are neural networks. They gained some momentum due to the great success of deep learning in many supervised computer vision tasks, but also for a number of low-level vision tasks state-of-the-art results have been reported [1,2]. Our proposed method lines up with the latter approaches and comes along with the following main contributions:

1. A robust state-of-the-art method for multi-image blind-deconvolution for both invariant and spatially-varying blur.
2. A hybrid neural network architecture as a discriminative approach for image deblurring supporting end-to-end learning in the fashion of deep learning.
3. A neural network layer version of Fourier-Burst-Accumulation [3] with learnable weights.
4. The proposed embedding of a small neural network allowing for information sharing across the image burst early in the processing stage.

2   Related Work

Blind image deconvolution (BD) has seen considerable progress in the last decade. A comprehensive review is provided in the recent overview article by Wang and Tao [4].

Single image blind deconvolution. Approaches for single image BD that report state-of-the-art results include the methods of Sun et al. [5], and Michaeli and Irani [6] that use powerful patch-based priors for image prediction. Following the success of deep learning methods in computer vision, also a number of neural network based methods have been proposed for image restoration tasks including *non-blind* deconvolution [2,7,8] which seeks to restore a blurred image when the blur kernel is known, but also for the more challenging task of *blind* deconvolution [9–14] where the blur kernel is not known a priori. Most relevant to our work is the recent work of Chakrabarti [11] which proposes a neural network that is trained to output the complex Fourier coefficients of a deconvolution filter. When applied to an input patch in the frequency domain, the network returns a prediction of the Fourier transform of the corresponding latent sharp image patch. For whole image restoration, the input image is cut into overlapping patches, each of which is independently processed by the network. The outputs are recomposed to yield an initial estimate of the latent sharp image, which is then used together with the blurry input image for the estimation of a single invariant blur kernel. The final result is obtained using the state-of-the-art nonblind deconvolution method of Zoran and Weiss [15].

Multi-frame blind deconvolution. Splitting an exposure budget across many photos can lead to a significant quality advantage [16]. Thus, it has been shown that multiple captured images can help alleviating the illposedness of the BD problem [17]. This has been exploited in several approaches [18–22] for *multi-frame* BD, i.e. combining multiple, differently blurred images into a single latent sharp image. Generative methods, that make explicit use of an image formation model, mainly differ in the prior and/or the optimization procedure they use. State-of-the-art methods use sparse priors with fast Bregman splitting techniques for optimization [20], or within a variational inference framework [18], cross-blur penalty functions between image pairs [21,22], also in combination with robust cost functions [19].

More recently proposed methods [23–26] also model the inter-frame motion and try to exploit the interrelation between camera motion, blur and image mis-alignment. Camera motion, when integrated during the exposure time of a single frame will produce intra-frame motion blur while leading to inter-frame mis-alignment during readout time between consecutive image capture.

All of the above-mentioned methods employ generative models and try to explicitly estimate one unknown blur kernel for each blurry input frame along with predicting the latent sharp image. A common shortcoming is the large computational burden with typical computation times in the order of tens of minutes, which hinders their wide-spread use in practice.

Recently, Delbracio and Sapiro have presented a fast method that aggregates a burst of images into a single image that is both sharper and less noisy than all the images in the burst [3]. The approach is inspired by a recently proposed Lucky Imaging method [27] targeted for astronomical imaging. Traditional Lucky Imaging approaches would select only a few "lucky" frames from a stack of hundreds to thousands recorded short-exposure images and combine them via non-rigid shift-and-add techniques into a single sharp image. In contrast, the authors of [27] propose to take all images into account (rather than a small subset of carefully chosen frames) taking and combining what is less blurred of each frame to form an improved image. It does so by computing a weighted average of the Fourier coefficients of the registered images in the burst. In [3,28] it has been demonstrated that this approach can be adapted successfully to remove camera shake originated from hand tremor vibrations. Their Fourier Burst Accumulation (FBA) approach allows for fast processing even for Megapixel images while at the same time yielding high-quality results provided that a "lucky", i.e. almost sharp frame is amongst the captured image burst.

In our work, we not only present a learning-based variant of FBA but also show how to alleviate the drawback of requiring a sharp frame amongst the input sequence of images. To this end we combine the single image BD method of [11] with FBA in a single network architecture which facilitates end-to-end learning. To the best of our knowledge this is the first time that a fully discriminative approach has been presented for the challenging problem of multi-frame BD.

3   Method

Let's assume we have given a burst of observed color images $Y_1, Y_2, \ldots, Y_N \in \mathcal{I}$ capturing the same scene $X \in \mathcal{I}$. Assuming each image in the captured sequence is blurred differently, our image formation model reads

$$Y_t = k_t * X + \varepsilon_t, \tag{2}$$

where $*$ denotes the convolution operator, k_t the blur kernel for observation Y_t and ε_t additive zero-mean Gaussian noise.

We aim at predicting a latent single sharp image $\hat{X}$ through a deep neural network architecture, i.e.

$$\pi^{(\theta)}: \mathcal{I}_p^N \to \mathcal{I}_p, \qquad (y_1, y_2, \ldots, y_N) \mapsto \hat{x} = \pi^{(\theta)}(y_1, y_2, \ldots, y_N).$$

The network operates on a patch-by-patch basis, here $y_t \in \mathcal{I}_p$ and $\hat{x} \in \mathcal{I}_p$ denote a patch in Y_t and X respectively. The patches are chosen to be overlapping. Our network predicts a single sharp patch $\hat{x} \in \mathcal{I}_p$ from multiple input patches $y_t \in \mathcal{I}_p$. All predicted patches are recomposed to form the final prediction $\hat{X}$ by averaging the predicted pixel values. During the training phase we optimize the learning parameters θ by directly minimizing the objective

$$\|\pi^{(\theta)}(y_1, y_2, \ldots, y_N) - x\|_2^2. \tag{3}$$

In the following we will describe the construction of $\pi^{(\theta)}(\cdot)$, the optimization of network parameters θ during the training of the neural network and the restoration of an entire sharp image.

3.1   Network Architecture

The architecture $\pi^{(\theta)}(\cdot)$ consists of several stages: (a) frequency band analysis with Fourier coefficient prediction, (b) a deconvolution part and (c) image fusion. Figure 1 illustrates the first two stages of our proposed system.

(a) **Frequency band analysis.** The frequency band analysis computes the discrete Fourier transform of the observed patch y_t according to the neural network approach in [11] at three different sizes ($17 \times 17, 33 \times 33, 65 \times 65$) using different sample sizes, which we will refer to bands b_1, b_2, b_3. In addition, band b_4 represents a low-pass band containing all coefficients with max $|z| \leq 4$ from band b_3. This is depicted in Fig. 1. To enable early information sharing within one burst of patches, we allow the neural network to spread the per band information extracted from one patch across all images of the burst using 1×1 convolution.

This essentially embeds a fully connected neural network for each Fourier coefficient $(f_{ij})_t$ with weight sharing. Since we use these operations in the image fusion stage again, we elaborate on this idea in more detail.

The values of one Fourier coefficient $(f_{ij})_t$ at frequency position (i, j) across the entire burst $t = 1, 2, \ldots, N$ can be considered as a single vector $(f_{ij})_{t=1,2,\ldots,N}$

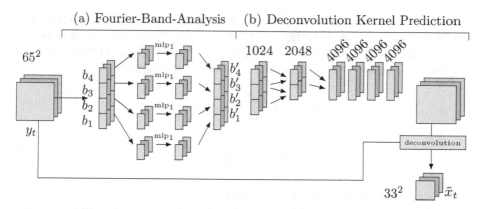

Fig. 1. Frequency band analysis and deconvolution for an image burst with 3 patches y_1, y_2, y_3. Following the work of Chakrabarti [11] we separate the Fourier spectrum in 4 different bands $b_1, \ldots, b_4$. In addition, we allow each band separately to interact across all images in one burst to support early information sharing. The predicted output of the deconvolution step are smaller patches $\tilde{x}_1, \tilde{x}_2, \tilde{x}_3$.

of dimension N (compare Fig. 2). Each of these vectors is fed through a small network of fully connected layers, labeled by mlp_1 in Fig. 1. This allows the neural network to adjust the extracted Fourier coefficients right before a dimensionality reduction occurs. These modified values $(f'_{tj})_{t=1,2,\ldots,N}$ give rise to adjusted Fourier bands b'_1, b'_2, b'_3, b'_4.

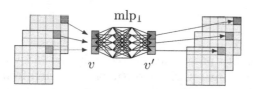

Fig. 2. For arbitrary inputs (bands b_1, b_2, b_3, b_4 or later FBA weights) we interpret each coefficient across one burst as a single vector. A transformed version of this excerpt will be placed at the same location in the output patch again. To reduced the number of learnable parameters, we employ weight sharing independent of the position.

(b) Deconvolution. Pairwise merging of the resulting bands b'_1, b'_2, b'_3, b'_4 with modified Fourier coefficients using fully connected layers with ReLU activation units entails a dimensionality reduction. The produced 4096 feature vector encoding is then fed through several fully connected layers producing a 4225 dimensional prediction of the filter coefficients of the deconvolution kernel. Applying the deconvolution kernel predicts a sharp patch $\hat{x}$ of size 33×33 from each input sequence of patches. This step is implemented as a multiplication of the predicted Wiener Filter with the Fourier transform of the input patch.

(c) **Image fusion.** In the last part of our pipeline we fuse all available sharp patches $y_1, y_2, \ldots, y_N$ by adopting the FBA approach described in [3] as a neural network component with learnable weights. The vanilla FBA algorithm applies the following weighted sum to a Fourier transform $\hat{\alpha}$ of a patch α:

$$u(\hat{\alpha}) = \mathcal{F}^{-1} \left(\sum_{i=1}^{N} w_i(\zeta) \hat{\alpha}_i(\zeta) \right) (x) \tag{4}$$

$$w_i(\zeta) = \frac{|\hat{\alpha}_i(\zeta)|^p}{\sum_{j=1}^{N} |\hat{\alpha}_j(\zeta)|^p}, \tag{5}$$

where w_i denotes the contribution of frequency ζ of a patch α_i. Note, that $u(\hat{\alpha})$ is differentiable in $\hat{\alpha}$ allowing to pass gradient information to previous layers through back-propagation. To incorporate this algorithm as a neural network layer into our pipeline, we replace Eq. (4) by a parametrized version

$$u(\hat{\alpha}) = \mathcal{F}^{-1} \left(\sum_{i=1}^{N} h_\phi(\zeta) \hat{\alpha}(\zeta) \right) (x). \tag{6}$$

Hence, instead of a hard-coded weight-averaging (using w_i) the network is able to learn a data-dependent weighted-averaging scheme. Again, the function $h_\phi(\cdot)$ represents two 1×1 convolutional layers with trainable parameters ϕ following the same idea of considering the Fourier coefficient across one burst as a single vector (compare Fig. 2).

3.2   Training

The network is trained on an artificially generated dataset obtained by applying synthetic blur kernels to patches extracted from the MS COCO dataset [29]. This dataset consists of real-world photographs collected from the internet. To increase the quality of ground-truth patches guiding the training process we reject patches with too small image gradients. This process gives us 542217 sharp patches. For a fair evaluation we use a splitting[1] in training and validation set. Optimizing the neural network parameters is done on the training set only. The input bursts of 14 blurry images are generated on-the-fly by applying synthetic blur kernels to the ground-truth patches. These synthetic blur kernels of sizes 17×17 and 7×7 pixels are generated using a Gaussian process with a Matérn covariance function following [9], a random subset of which is shown in Fig. 3. In addition, we apply standard data augmentation methods like rotating and mirroring to the ground-truth data. Hence, this approach gives nearly an infinite amount of training data. We also add zero-mean Gaussian noise with variance 0.1. The validation data is precomputed to ensure fair evaluation during training.

Unfortunately, sophisticated stepsize heuristics like Adam [30] or Adagrad [31] failed to guarantee a stable training. We suspect the large range of values in

---

[1] Provided by [29].

Fig. 3. Some of the synthetically generated PSFs using a Gaussian process for generating training examples on-the-fly.

the Fourier space to mislead those heuristics. Instead, we use stochastic gradient descent with momentum ($\beta = 0.9$), batchsize 32 and an initial learning rate of $\eta = 2$ which decreases every 5000 steps by a factor of 0.8. Training the neural network took 6 days using TensorFlow [32] on a NVIDIA Titan X.

The FBA approach [3] applies Gaussian smoothing to the weights w_i to account for the fact that small camera shakes are likely to vary the Fourier spectrum in a smooth way. While this removes strong artefacts in the restored recomposed image, it prevents the network to convergence during training. Following this idea we tried a fixed Gaussian blur with parameters set to the reported values of [3] as well as learning a blur kernel (initialized by a Gaussian) during training. In both cases we observed no convergence during training. Therefore, we apply this smoothing only for the final application of the neural network.

3.3 Deployment

During deployment we feed input patches of size 65×65 into our neural network with stride 5. Using overlapping patches helps to average multiple predictions. For recombination of overlapping patches we apply a 2-dimensional Hanning window to each patch to favour pixel values in the patch center and devaluate information at the border of the patch.

While the predicted images $\hat{X}$ generated by our neural network contain well-defined sharp edges we observed desaturation in color contrast. To correct the color of the predicted image we replace its ab-channel in the Lab color space by the ab-channel of the FBA results (compare Fig. 4).

Regarding runtime the most expensive step is the frequency band analysis. Given a burst of 14 images of size 1000×700 pixels the entire reconstruction process takes roughly 5 min per channel with our unoptimized implementation.

Fig. 4. Deblurring a burst of degraded images from a groundtruth image (left) results in a desaturated image (middle). Therefore we correct those colors (right image) using color transfer. (Color figure online)

4   Experiments

To evaluate and validate our approach we conduct several experiments including a comprehensive comparison with state-of-the-art techniques on a real-world dataset, and a performance evaluation on a synthetic dataset to test the robustness of our approach with varying image quality of the input sequence.

4.1   Comparison on Real-World Dataset

We compare the restored images with other state-of-the-art multi-image blind deconvolution algorithms. In particular, we compare with the multichannel blind deconvolution method from Šroubek et al. [21], the sparse-prior method of [33] and the FBA method proposed in [3]. We used the data provided by [3], which contains typical photographs captured with hand-held cameras (iPad back camera, Canon 400 D). As they are captured under various challenging lighting conditions they exhibit both noise and saturated pixels. As shown in [3] the FBA

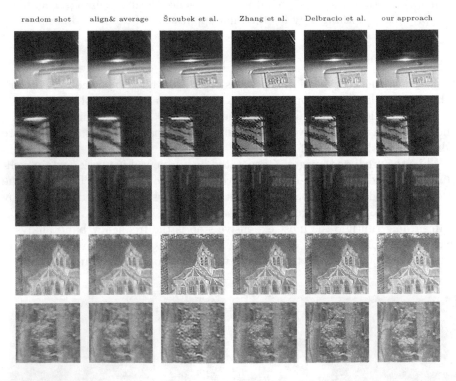

Fig. 5. Comparison to state-of-the-art multi-frame blind deconvolution algorithms on real-world data. See the supplementary material for high-resolution images. Note that our approach produces the sharpest results except for the last scene, which could be caused by the color transfer described in Sect. 3.3. (Color figure online)

algorithms demonstrated superior performance compared to previous state-of-the-art multi-image blind deconvolution algorithms [21,33] in both reconstruction quality and runtime. Figure 5 shows crops of the deblurred results on these images. The high-resolution images are enclosed in the supplemental material. Our trained neural network featuring the FBA-like averaging yields comparable if not superior results compared to previous approaches [3,21,33]. In direct comparison to the FBA results, our method is better removing blur due to our additional prepended deconvolution module.

4.2   Deblurring Bursts with Varying Number of Frames and Quality

Here, we analyse the performance of our approach depending on the burst "quality". Sorting all images provided by [3] within one burst according to their PSNR beginning with images of strong blur and consequently adding sharper shots to the burst gives a series of bursts starting with images of poor quality up to bursts with at least one close-to-sharp shot. Since our architecture is trained for deblurring bursts with exactly 14 input images, we duplicated images of bursts with fewer frames. Figure 6 clearly indicates good performance of our neural network even for a relative small number of input images with strong blur artifact.

4.3   Deblurring Image Bursts Without Reasonable Sharp Frames

To further challenge our neural network approach, we artificially sampled image bursts from unseen images taken from the MS COCO validation set and blurred them by applying synthetic blur kernels of size 14×14. The restored sharp images from the input bursts of 14 artificially blurred images under absence of a close-to-sharp frame (best shot) are depicted in Fig. 7. As the experiments indicate the explicit deconvolution step in our approach is absolutely necessary to handle these kind of snapshots and to remove blur artifacts. In contrast, while FBA [3] stands out in small memory footprint and fast processing times it clearly failed to recover sharp images for cases where no reasonably sharp frame is available amongst the input sequence.

4.4   Comparing to a Baseline Version

One might ask, how our trained neural network compares to an approach that applies the methods of Chakrabarti [11] and Delbracio and Sapiro [3] subsequently, each in a separate step. We fine-tuned the provided weights from [11] in combination with our FBA-layer. Figure 8 shows the training progress for an exemplar patch, where the improvement in sharpness is clearly visible.

In addition, we run the *entire* pipeline of Chakrabarti [11] including the costly non-blind deconvolution EPLL step and afterwards FBA. The approach is significantly slower and results in less sharp reconstructions (see Fig. 9).

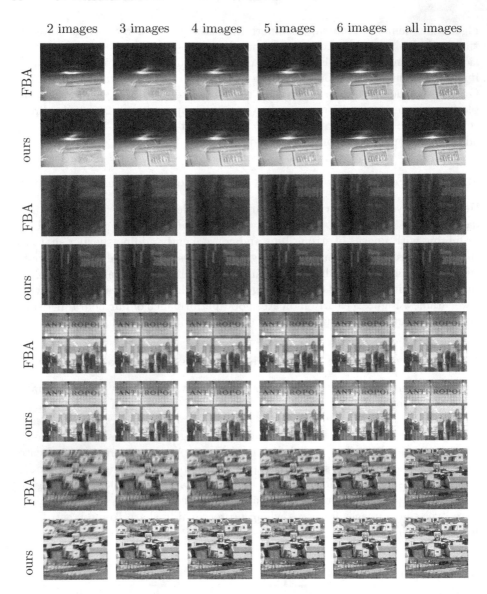

Fig. 6. FBA and our algorithm are compared on bursts with a growing number of images of increasing quality. The individual images are sorted according to their PSNR starting with the most blurry images. The input images were taken from [3].

4.5 Spatially-Varying Blur

To test whether our network is also able to deal with spatially-varying blur we generated a burst of images degraded by non-stationary blur. To this end, we took one of the recorded camera trajectories of [34] that are provided on

typical shot   best shot   FBA   ours

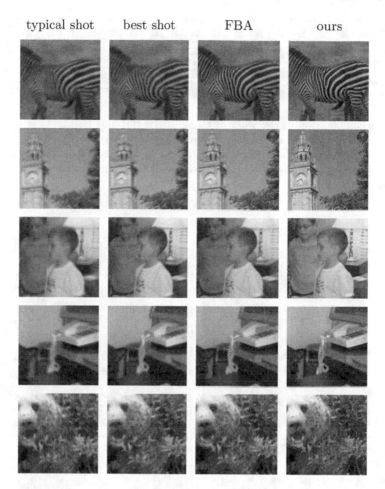

Fig. 7. Comparing FBA (third column) and our trained neural network (fourth column) against best shot and a typical shot. These images are taken from the validation set. For image bursts without a single sharp frame lucky imaging approaches fail due to a missing explicit deconvolution step, while our approach gives reasonable results.

the project webpage[2]. The camera trajectory has been recorded with a Vicon system at 500 fps and represents the camera motion during a slightly longer-exposed shot (1/30 s). The trajectory comprises a 6-dimensional time series with 167 time samples. We divided this time series into 8 fragments of approximately equal lengths.

With a Matlab script (see Supplemental material) 8 spatially-varying PSFs are generated as shown at the bottom of Fig. 10.

The spatially varying kernels of size 17×17 pixels are applied using the Efficient Filter Flow model of [35]. The results of our network along with results

---

[2] http://webdav.is.mpg.de/pixel/benchmark4camerashake/

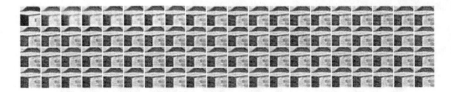

Fig. 8. The combination of the work of Chakrabarti [11] and Delbracio et al. [3] can be considered as a baseline version of our neural network. We fine-tuned the published weights from the work of Chakrabarti [11] in an end-to-end fashion in combination with our FBA-layer. The left-most patch is the ground-truth patch. Note how the sharpness continuously increases with training.

FBA   our network

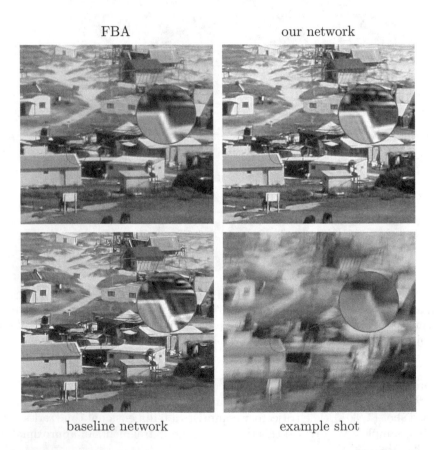

baseline network   example shot

Fig. 9. Comparison to a baseline approach of simply stacking [3,11]. Without end-to-end training ringing-artifacts are clearly visible on the blue roof. They are significantly dampened after training. (Color figure online)

of FBA for three example images are shown in Fig. 10. Our results are consistently sharper and demonstrate that our approach is also able to correct for spatially-varying blur.

FBA   ours

Fig. 10. Comparison to FBA on image sequences with spatially-varying blur. Our approach is able to reconstruct consistently sharper images.

5   Conclusion, Limitations and Future Work

We presented a discriminative approach for multi-frame blind deconvolution (BD) by posing it as a nonlinear regression problem. As a function approximator, we use a deep layered neural network, whose optimal parameters are learned from artificially generated data. Our proposed network architecture draws inspiration from two recent works as (a) a neural network approach to single image blind deconvolution of Chakrabarti [11], and (b) the Fourier Burst Accumulation (FBA) algorithm of Delbracio and Sapiro [3]. The latter takes a burst of images as input and combines them through a weighted average in the frequency domain to a single sharp image. We reformulated FBA as a learning method and casted it into a deep layered neural network. Instead of resorting to heuristics and hand-tuned parameters for weight computation, we learn optimal weights as network parameters through end-to-end training.

By prepending parts of the network of Chakrabarti to our FBA network we are able to extend its applicability by alleviating the necessity of a close-to-sharp frame being amongst the image burst. Our system is trained end-to-end on a set of artificially generated training examples, enabling competitive performance in multi-frame BD, both with respect to quality and runtime. Due to its novel information sharing in the frequency band analysis stage and its explicit deconvolution step, our network outperforms state-of-the-art techniques like FBA [3] especially for bursts with few severely degraded images.

Our contribution resides at the experimental level and despite competitive results with state-of-the-art, our proposed approach is subject to a number of limitations. However, at the same time it opens up several exciting directions for future research:

- Our proposed approach doesn't exploit the temporal structure of the input image sequence, which encodes valuable information about intra-frame blur and inter-frame image mis-alignment [24–26]. Embedding our described network into a network architecture akin to the spatio-temporal auto-encoder of Pătrăucean et al. [36] might enable such non-trivial inference.
- Our current model assumes a static scence and is not able to handle object motion. Inserting a Spatial Transformer Network Layer [37] which also facilitates optical flow estimation [36] could be an interesting avenue to capture and correct for object motion occurring between consecutive frames.

Acknowledgement. This work has been partially supported by the DFG Emmy Noether fellowship Le 1341/1-1 and an NVIDIA hardware grant.

References

1. Burger, H.C., Schuler, C.J., Harmeling, S.: Image denoising: can plain neural networks compete with BM3D? In: Proceedings of the IEEE Conference on Computer Vision and Pattern Recognition (CVPR). IEEE Computer Society (2012)

2. Schuler, C., Burger, H., Harmeling, S., Scholköpf, B.: A machine learning approach for non-blind image deconvolution. In: Proceedings of the IEEE Conference on Computer Vision and Pattern Recognition (CVPR). IEEE Computer Society (2013)
3. Delbracio, M., Sapiro, G.: Burst deblurring: removing camera shake through fourier burst accumulation. In: Proceedings of the IEEE Conference on Computer Vision and Pattern Recognition (CVPR). IEEE Computer Society (2015)
4. Wang, R., Tao, D.: Recent progress in image deblurring. arXiv preprint (2014). arXiv:1409.6838
5. Sun, L., Cho, S., Wang, J., Hays, J.: Edge-based blur kernel estimation using patch priors. In: IEEE International Conference in Computational Photography (ICCP), pp. 1–8. IEEE (2013)
6. Michaeli, T., Irani, M.: Blind deblurring using internal patch recurrence. In: Fleet, D., Pajdla, T., Schiele, B., Tuytelaars, T. (eds.) ECCV 2014. LNCS, vol. 8691, pp. 783–798. Springer, Heidelberg (2014). doi:10.1007/978-3-319-10578-9_51
7. Xu, L., Ren, J.S., Liu, C., Jia, J.: Deep convolutional neural network for image deconvolution. In: Ghahramani, Z., Welling, M., Cortes, C., Lawrence, N.D., Weinberger, K.Q., (eds.): Advances in Neural Information Processing Systems (NIPS), pp. 1790–1798. Curran Associates, Inc. (2014)
8. Rosenbaum, D., Weiss, Y.: The return of the gating network: combining generative models and discriminative training in natural image priors. In: Advances in Neural Information Processing Systems (NIPS), pp. 2665–2673 (2015)
9. Schuler, C.J., Hirsch, M., Harmeling, S., Schölkopf, B.: Learning to deblur. IEEE Transactions on Pattern Analysis and Machine Intelligence (2015)
10. Sun, J., Cao, W., Xu, Z., Ponce, J.: Learning a convolutional neural network for non-uniform motion blur removal. In: Proceedings of the IEEE Conference on Computer Vision and Pattern Recognition (CVPR). IEEE Computer Society (2015)
11. Chakrabarti, A.: A neural approach to blind motion deblurring. In: Leibe, B., Matas, J., Sebe, N., Welling, M. (eds.) ECCV 2016. LNCS, vol. 9907, pp. 221–235. Springer, Heidelberg (2016). doi:10.1007/978-3-319-46487-9_14
12. Hradiš, M., Kotera, J., Zemcík, P., Šroubek, F.: Convolutional neural networks for direct text deblurring. In: Proceedings of BMVC, vol. 10 (2015)
13. Svoboda, P., Hradi, M., Mark, L., Zemck, P.: CNN for license plate motion deblurring. In: IEEE International Conference on Image Processing (ICIP), pp. 3832–3836 (2016)
14. Loktyushin, A., Schuler, C., Scheffler, K., Schölkopf, B.: Retrospective motion correction of magnitude-input MR images. In: Bhatia, K.K., Lombaert, H. (eds.) MLMMI 2015. LNCS, vol. 9487, pp. 3–12. Springer, Cham (2015). doi:10.1007/978-3-319-27929-9_1
15. Zoran, D., Weiss, Y.: From learning models of natural image patches to whole image restoration. In: Proceedings of the IEEE International Conference on Computer Vision (ICCV), pp. 479–486. IEEE Computer Society (2011)
16. Hasinoff, S.W., Kutulakos, K.N., Durand, F., Freeman, W.T.: Time-constrained photography. In: Proceedings of the IEEE International Conference on Computer Vision (ICCV), pp. 333–340. IEEE (2009)
17. Rav-Acha, A., Peleg, S.: Two motion-blurred images are better than one. Pattern Recogn. Lett. **26**, 311–317 (2005)
18. Zhang, H., Wipf, D.P., Zhang, Y.: Multi-observation blind deconvolution with an adaptive sparse prior. IEEE Trans. Pattern Anal. Mach. Intell. **36**, 1628–1643 (2014)

19. Chen, J., Yuan, L., Tang, C.K., Quan, L.: Robust dual motion deblurring. In: IEEE Proceedings of the IEEE Conference on Computer Vision and Pattern Recognition (CVPR), pp. 1–8. IEEE Computer Society (2008)

20. Cai, J.F., Ji, H., Liu, C., Shen, Z.: Blind motion deblurring using multiple images. J. Comput. Phys. **228**, 5057–5071 (2009)

21. Šroubek, F., Milanfar, P.: Robust multichannel blind deconvolution via fast alternating minimization. IEEE Trans. Image Process. **21**, 1687–1700 (2012)

22. Zhu, X., Šroubek, F., Milanfar, P.: Deconvolving PSFs for a better motion deblurring using multiple images. In: Fitzgibbon, A., Lazebnik, S., Perona, P., Sato, Y., Schmid, C. (eds.) ECCV 2012. LNCS, vol. 7576, pp. 636–647. Springer, Heidelberg (2012). doi:10.1007/978-3-642-33715-4_46

23. Zhang, H., Carin, L.: Multi-shot imaging: joint alignment, deblurring and resolution-enhancement. In: Proceedings of the IEEE Conference on Computer Vision and Pattern Recognition (CVPR). IEEE Computer Society (2014)

24. Zhang, H., Yang, J.: Intra-frame deblurring by leveraging inter-frame camera motion. In: Proceedings of the IEEE Conference on Computer Vision and Pattern Recognition (CVPR). IEEE Computer Society (2015)

25. Kim, T.H., Nah, S., Lee, K.M.: Dynamic scene deblurring using a locally adaptive linear blur model. arXiv preprint (2016). arXiv:1603.04265

26. Ito, A., Sankaranarayanan, A.C., Veeraraghavan, A., Baraniuk, R.G.: BlurBurst: removing blur due to camera shake using multiple images. ACM Trans. Graph. **3** (2014). Submitted

27. Garrel, V., Guyon, O., Baudoz, P.: A highly efficient lucky imaging algorithm: image synthesis based on fourier amplitude selection. Publ. Astron. Soc. Pac. **124**, 861–867 (2012)

28. Delbracio, M., Sapiro, G.: Hand-held video deblurring via efficient fourier aggregation. IEEE Trans. Comput. Imaging **1**, 270–283 (2015)

29. Lin, T., Maire, M., Belongie, S.J., Bourdev, L.D., Girshick, R.B., Hays, J., Perona, P., Ramanan, D., Dollár, P., Zitnick, C.L.: Microsoft COCO: common objects in context. arXiv preprint (2014). arXiv:1405.0312

30. Kingma, D.P., Ba, J.: Adam: A method for stochastic optimization. arXiv preprint (2014). arXiv:1412.6980

31. Duchi, J., Hazan, E., Singer, Y.: Adaptive subgradient methods for online learning and stochastic optimization. J. Mach. Learn. Res. **12**, 2121–2159 (2011)

32. Abadi, M., Agarwal, A., Barham, P., Brevdo, E., Chen, Z., Citro, C., Corrado, G.S., Davis, A., Dean, J., Devin, M., Ghemawat, S., Goodfellow, I., Harp, A., Irving, G., Isard, M., Jia, Y., Jozefowicz, R., Kaiser, L., Kudlur, M., Levenberg, J., Mané, D., Monga, R., Moore, S., Murray, D., Olah, C., Schuster, M., Shlens, J., Steiner, B., Sutskever, I., Talwar, K., Tucker, P., Vanhoucke, V., Vasudevan, V., Viégas, F., Vinyals, O., Warden, P., Wattenberg, M., Wicke, M., Yu, Y., Zheng, X.: TensorFlow: Large-scale machine learning on heterogeneous systems (2015). Software. tensorflow.org

33. Zhang, H., Wipf, D., Zhang, Y.: Multi-image blind deblurring using a coupled adaptive sparse prior. In: Proceedings of the IEEE Conference on Computer Vision and Pattern Recognition (CVPR), pp. 1051–1058. IEEE Computer Society (2013)

34. Köhler, R., Hirsch, M., Mohler, B., Schölkopf, B., Harmeling, S.: Recording and playback of camera shake: benchmarking blind deconvolution with a real-world database. In: Fitzgibbon, A., Lazebnik, S., Perona, P., Sato, Y., Schmid, C. (eds.) ECCV 2012. LNCS, vol. 7578, pp. 27–40. Springer, Heidelberg (2012). doi:10.1007/978-3-642-33786-4_3

35. Hirsch, M., Sra, S., Schölkopf, B., Harmeling, S.: Efficient filter flow for space-variant multiframe blind deconvolution. In: Proceedings of the IEEE Conference on Computer Vision and Pattern Recognition (CVPR). IEEE (2010)
36. Pătrăucean, V., Handa, A., Cipolla, R.: Spatio-temporal video autoencoder with differentiable memory. arXiv preprint (2015). arXiv:1511.06309
37. Jaderberg, M., Simonyan, K., Zisserman, A., et al.: Spatial transformer networks. In: Advances in Neural Information Processing Systems (NIPS), pp. 2008–2016 (2015)

Spectral Reflectance Recovery with Interreflection Using a Hyperspectral Image

Hiroki Okawa[1], Yinqiang Zheng[2(✉)], Antony Lam[3], and Imari Sato[1,2]

[1] Tokyo Institute of Technology, Meguro, Japan
{okawa-h,imarik}@nii.ac.jp
[2] National Institute of Informatics, Chiyoda, Japan
yqzheng@nii.ac.jp
[3] Saitama University, Saitama, Japan
antonylam@cv.ics.saitama-u.ac.jp

Abstract. The capture of scene spectral reflectance (SR) provides a wealth of information about the material properties of objects, and has proven useful for applications including classification, synthetic relighting, medical imaging, and more. Thus many methods for SR capture have been proposed. While effective, past methods do not consider the effects of indirectly bounced light from within the scene, and the estimated SR from traditional techniques is largely affected by interreflection. For example, different lighting directions can cause different SR estimates. On the other hand, past work has shown that accurate interreflection separation in hyperspectral images is possible but the SR of all surface points needs to be known a priori. Thus we see that the estimation of SR and interreflection in its current form constitutes a chicken and egg dilemma. In this work, we propose the challenging and novel problem of simultaneously performing SR recovery and interreflection removal from a single hyperspectral image, and develop the first strategy to address it. Specifically, we model this problem using a compact sparsity regularized nonnegative matrix factorization (NMF) formulation, and introduce a scalable optimization algorithm on the basis of the alternating direction method of multipliers (ADMM). Our experiments have demonstrated its effectiveness on scenes with a single or two reflectance colors, containing possibly concave surfaces that lead to interreflection.

1 Introduction

The ability to capture the spectral reflectance (SR) of a scene allows us to see a wealth of information about the material properties of objects. This in turn, is useful for numerous applications including classification, synthetic relighting, medical imaging, and more. As a result, many methods for capturing scene SR have been proposed. These include, the brute force capture of narrowband images [1], computational techniques with conventional cameras [2], use of specialized optics or filters [3–7], and active lighting [8–12].

© Springer International Publishing AG 2017
S.-H. Lai et al. (Eds.): ACCV 2016, Part IV, LNCS 10114, pp. 52–67, 2017.
DOI: 10.1007/978-3-319-54190-7_4

Fig. 1. The effect of interreflection for a pink flower. (Color figure online)

While effective, the aforementioned approaches do not consider the effects of indirect illumination, which limits accuracy. Briefly, when an opaque scene is illuminated, the radiance at each point can be thought of as consisting of direct lighting and interreflection, whose light source comes from light bounced off of other surface points in the scene. Such interreflection is related to scene geometry including lighting directions. As a result, when only direct lighting is considered, the estimated SR from traditional techniques is largely affected by interreflection. For example, different lighting directions can cause different SR estimates. The situation is made even more complex when the scene's surfaces consist of different colors that could skew the recovered SR.

To illustrate the challenges more concretely, let us first demonstrate how both direct lighting and interreflection contribute to a captured image. Imagine a pink flower with its concave surfaces facing a white light, as shown in Fig. 1(left). In this case, light would bounce around inside the flower, and the surface points would be partially illuminated by pink interreflection in addition to the direct white light. Figure 1(middle) shows a rendered color image where red denotes direct lighting and green denotes interreflection. The spectral distributions of the inside and outside points of the flower are shown in Fig. 1(right). In the images, we clearly see both direct lighting and interrefelction inside the flower shown as a mixture of red (direct) and green (interreflection) colors and that its spectral distribution becomes more reddish due to interreflected pink light. Therefore, even in the case of a single material, a means of removing the effects of interreflection is needed for high accuracy SR recovery.

Interreflection analysis and removal have been extensively studied in existing computer vision literature [13–17]. However, all these methods do not take into account the hyperspectral imaging case. The only exception that we know of is Nam and Kim [18], in which they extended the basic approach of Liao et al. [16] to the hyperspectral case. They were able to separate direct and indirect illumination, which aided greatly in photometric stereo. Their approach is effective but requires either that the SR of the scene be known or that there be known convex parts of the object from which SR can be computed without needing to account for interreflection. Also, they only showed their solution for the monochromatic surface case.

Our discussion above indicates that the estimation of SR and interreflection in its current form constitutes a chicken and egg problem, where SR recovery requires interreflection removal from the scene, and interreflrection removal requires surface color information. In this paper, we propose solving the novel problem of simultaneously performing SR recovery and interreflection removal. We make the first contribution to solving this challenging problem, and show that it can be achieved using only a single hyperspectral image and considering up to the 2nd bounce interreflection. Specifically, we first model this problem using a compact sparsity regularized nonnegative matrix factorization (NMF) formulation. Considering that the number of pixels in a high resolution hyperspectral image is tremendous, we introduce a scalable algorithm for optimization on the basis of the alternating direction method of multipliers (ADMM). We demonstrate effective removal of color shift effects caused by interreflected light for single color, nonplanar objects. We also show that our proposed formulation and algorithm can be extended to handle objects with multiple colors in a straightforward way.

The remainder of this paper is organized as follows. In Sect. 2, we briefly review closely related works. We present our formulation for simultaneous SR recovery and interreflection removal in the presence of a single color in Sect. 3. Sections 4 and 5 include the optimization algorithm and experiment results, respectively. We show how to extend our formulation and algorithm to handle the case of two colors in Sect. 6. We briefly discuss the intrinsic difficulty of simultaneous SR recovery and interreflection removal in Sect. 7, and conclude this work in Sect. 8.

2   Related Work

This section provides related work on spectral reflectance imaging and interreflection analysis. The most straightforward way to obtain the spectral reflectance of a scene is to perform hyperspectral imaging under a controlled light source. For example, one approach uses a monochromator and CCD camera for narrowband imaging at multiple wavelengths [1]. Another common approach is to use a wideband light source and narrowband filters [19].

There have also been approximate approaches such as Tominaga [7], which used a six channel camera for spectral reflectance recovery. In addition, approaches using conventional three channel cameras with some constraints on scenes have been proposed [2,20].

Another interesting departure from the above approaches is to combine conventional cameras with active lighting in order to recover spectral reflectance [8–12]. In many of these methods, the types of active lighting were determined to improve the accuracy of recovered spectral reflectance.

Despite the effectiveness of all the aforementioned approaches, none of them address the issue of interreflection. As a result, the recovered spectral reflectance of a given surface point can include a combination of effects from direct illumination and indirect illumination, which was bounced off of different surfaces from the scene. These various surfaces in the scene may also be of different colors, which would skew the recovered spectral reflectance of the given surface point even more.

Interreflection and object shape go hand in hand. As a result, there are a number of methods that analyze the effects of interreflection and shape recovery. Koenderink and Doorn [21] devised a general model for diffuse interreflections. Forsyth et al. [13, 14] studied the effects of interreflections on shape recovery. Nayar et al. [22] explored the interplay between interreflections and shape through iterative refinement of surface reflectance and shape in scenes. Their method was then extended to multi-colored scenes [23].

Other work has also approached the problem of modeling and separating direct and indirect illumination. For example, Funt et al. [24, 25] separated direct and indirect illumination using the color different effect in interreflections. Later, Seitz et al. [26] devised an inverse light transport operator that could be used to separate m-bounced light from scenes with uniform albedo and Lambertian surfaces. However, their method requires scenes be captured with a laser, which can be time consuming. Nayar et al. [15] were able to separate direct and indirect lighting for complex scenes using spatially high-frequency lighting patterns. More recently, Liao et al. [16], removed interreflections using spectrum dependent albedo in scenes but they required multiple images under different illuminations. Fu et al. [17] extended this work to the fluorescent case and they were able to separate direct and indirect lighting using only one RGB image allowing for analysis of dynamic scenes, but the reflectance and fluorescence properties of the scenes need to be known a priori.

It should be noted that all these methods do not take into account the hyperspectral imaging case. An exception is Nam and Kim [18] where they extended the basic approach of Liao et al. [16] to the hyperspectral case. They were able to separate direct and indirect illumination, which aided greatly in photometric stereo. Their approach is effective but requires either that the spectral reflectance of the scene be known or that there be known convex parts of the object from which spectral reflectance can be computed without needing to account for interreflection. Also, they only showed their solution for the monochromatic surface case.

In this paper, we make the first contribution to solving the challenging and novel problem of simultaneously performing both tasks. Using only a single hyperspectral image, we show effective performance for single color objects, which could be nonplanar. We also pave the way for future extensions to the multicolor case with some separation results in the two-color case and discuss the difficulties in multicolor cases.

3   Spectral Reflectance Recovery with Interreflection

To highlight the problem itself, we will consider the simpler case with a single diffuse color here. A straightforward extension to the case with two colors will be shown in Sect. 6.

We denote the spectrum distribution of this diffuse color as $\mathbf{r}$, with a reflectance coefficient $r(\lambda_j)$ at wavelength λ_j. Assume that the object is illuminated by a spatially uniform light source with irradiance intensity $l(\lambda_j)$ at λ_j, and viewed by a hyperspectral sensor, whose response function is denoted by $a(\lambda_j)$. The intensity of the k-th pixel at λ_j can be described by

$$i_k(\lambda_j) = w_k * a(\lambda_j) * l(\lambda_j) * r(\lambda_j), \tag{1}$$

in which w_k denotes the shading factor dependent on the surface geometry and lighting direction.

In the presence of interreflection, the imaging equation becomes much more complicated, due to complex light interactions between different surface points that include multiple bounces of light. However, interreflections beyond the 2nd bounce are negligible in terms of intensity. Thus we will consider the 2nd bounce interreflection only. Given a single diffuse reflectance spectrum $r(\lambda_j)$, the spectrum of the 2nd bounce interreflection between any two surface points should be $r^2(\lambda_j)$. Therefore, the observed intensity of the k-th pixel at wavelength λ_j is

$$i_k(\lambda_j) = a(\lambda_j) * [w_k * l(\lambda_j) * r(\lambda_j) + v_k * l(\lambda_j) * r^2(\lambda_j)], \tag{2}$$

where v_k denotes the intensity factor of the 2nd bounce interreflection at the k-th pixel.

To simplify the notation, we assume that the illumination spectrum $l(\lambda_j)$ and the camera response spectrum $a(\lambda_j)$ have been precalibrated, and both can be normalized to 1. Then, the imaging equation can be simplified to

$$i_k(\lambda_j) = w_k * r(\lambda_j) + v_k * r^2(\lambda_j). \tag{3}$$

By stacking the measurements at all n pixels and m wavelength bands, we obtain the following matrix form

$$\begin{bmatrix} i_{11} & i_{12} & \cdots & i_{1n} \\ i_{21} & i_{22} & \cdots & i_{2n} \\ \cdots & \cdots & \cdots & \cdots \\ i_{m1} & i_{m2} & \cdots & i_{mn} \end{bmatrix} = \begin{bmatrix} r_1 & r_1^2 \\ r_2 & r_2^2 \\ \cdots & \cdots \\ r_m & r_m^2 \end{bmatrix} \begin{bmatrix} w_1 & w_2 & \cdots & w_n \\ v_1 & v_2 & \cdots & v_n \end{bmatrix}, \tag{4}$$

which can be cast into a nonnegative matrix factorization problem

$$I = [\mathbf{r} \ \mathbf{r} * \mathbf{r}] \ W = RW, R \geq 0, W \geq 0. \tag{5}$$

The nonnegative constraints on R and W are natural, which account for the physical nonnegative restrictions on all entities involved in the imaging process.

Considering that the strength of interreflection is usually much weaker than that of direct reflectance, it is reasonable to believe that $w_k \gg v_k$ for $1 \le k \le n$. In addition, interreflection only occurs at concave surface points, thus v_k is mostly close to zero, for an ordinary 3D object. These two observations indicate that the matrix W is sparse in general, which can be accounted for by using the L_1-norm sparsity prior. Therefore, the problem of spectral reflectance recovery with interreflection can be modeled using the following sparsity regularized NMF formulation

$$\min_{\mathbf{r},W} \|I - RW\|_F^2 + \alpha |W|_1^1, s.t., R \ge 0, W \ge 0, \tag{6}$$

in which α denotes the weighting factor. Throughout all the experiments, we fix α to be 1e-3.

Because of the nonnegative constraint on W, the L_1-norm sparsity regularizer is equivalent to the sum of all elements of W, and thus Eq. (6) can be rewritten into

$$\min_{\mathbf{r},W} \|I - RW\|_F^2 + \alpha \sum W, s.t., R \ge 0, W \ge 0. \tag{7}$$

NMF has been widely used for blind source separation, layer extraction and parts learning, for which there exists the popular multiplicative iterative updating scheme [27]. It has been generalized to handle the sparsity regularized NMF [28]. In contrast to the multiplicative scheme, the alternating direction method of multiplier (ADMM) has recently become quite popular in low-rank related optimization [29,30]. Xu et al. [31] adapted the ADMM method for NMF and found that it is more appropriate for large scale NMFs. Inspired by this observation, we decide to develop a scalable algorithm for Eq. (7) on the basis of ADMM.

4   Scalable Algorithm for Sparsity Regularized NMF

There are two sets of variables R and W in Eq. (7). To avoid the scale floating issue between R and W, we always fix the scale of R by enforcing that $\sum \mathbf{r} = 1$.

In order to circumvent the difficulty in ensuring the nonnegativity of the large matrix W, we introduce an auxiliary matrix U such that $U = W$, and rewrite Eq. (7) into

$$\min_{\mathbf{r},W,U} \quad \|I - R \cdot W\|_F^2 + \alpha \sum W$$
$$s.t., \quad \sum \mathbf{r} = 1, R \ge 0, U \ge 0, U = W. \tag{8}$$

Its corresponding augmented Lagrange function reads

$$f(\mathbf{r}, W, U, L, \mu) = \|I - R \cdot W\|_F^2 + \alpha \sum W + \langle L, U - W \rangle + \frac{\mu}{2} \|U - W\|_F^2, \tag{9}$$

in which L is the Lagrange multiplier and μ the penalty parameter. $\langle A, B \rangle$ is equivalent to the trace of $A^{\mathrm{T}}B$, where A^{T} denotes the transpose of A.

By following the inexact augmented Lagrange multiplier iterative method in [29], we need to solve the following constrained optimization problem at the i-th iteration with known L_i and μ_i

$$\min_{\mathbf{r},W,U} \quad \|I - R \cdot W\|_F^2 + \alpha \sum W + \langle L_i, U - W \rangle + \frac{\mu_i}{2} \|U - W\|_F^2$$
$$\text{s.t.,} \quad \sum \mathbf{r} = 1, R \geq 0, U \geq 0. \tag{10}$$

Rather than optimizing Eq. (10) over all variables simultaneously, we choose to minimize the cost function in an alternating manner as follows.

Solving W. Given $\mathbf{r}$, U and L, Eq. (10) can be simplified into the following unconstrained quadratic minimization problem

$$\min_{W} \quad \|I - R \cdot W\|_F^2 + \alpha \sum W + \langle L_i, U - W \rangle + \frac{\mu_i}{2} \|U - W\|_F^2, \tag{11}$$

This leads to a simple updating rule for W

$$W \leftarrow \left(R^{\mathrm{T}} R + \frac{\mu_i}{2} E \right)^{-1} \left(R^{\mathrm{T}} I - \frac{\alpha}{2} E_H + \frac{L_i}{2} + \frac{\mu_i}{2} U \right), \tag{12}$$

where E is the identity matrix of the matrix product, and E_H is the identity matrix of the Hadamard product.

Solving r. Given U, L and W, we solve the following constrained nonlinear least squares problem

$$\min_{\mathbf{r}} \quad \|I - RW\|_F^2$$
$$\text{s.t.,} \quad \sum \mathbf{r} = 1, R \geq 0, \tag{13}$$

which can be expanded using the trace-norm operation

$$\min_{\mathbf{r}} \quad \mathrm{tr}\left(RWW^{\mathrm{T}}R^{\mathrm{T}} - 2RWI^{\mathrm{T}} \right)$$
$$\text{s.t.,} \quad \sum \mathbf{r} = 1, R \geq 0. \tag{14}$$

We use a standard damped Gauss-Newton method to solve the small nonlinear least squares problem above, while keeping $\mathbf{r}$ to be nonnegative.

Solving U. Given L, $\mathbf{r}$ and W, Eq. (10) reduces to

$$\min_{U} \quad \langle L_i, U - W \rangle + \frac{\mu_i}{2} \|U - W\|_F^2$$
$$\text{s.t.,} \quad U \geq 0, \tag{15}$$

It is a convex quadratic problem, whose global optimum can be found in closed form as follows

$$U \leftarrow \left(W - \frac{1}{\mu_i} L_i \right)_+ , \tag{16}$$

in which the projection operation $(A)_+$ is to clamp all negative elements of A to be zero.

The complete ADMM based optimization algorithm is summarized in Algorithm 1. We introduce a parameter ρ to control the increasing rate of the penalty parameter μ, which is fixed to be 1.05. As for the initialization parameters, we use random initial values for all variables, except the initial μ, which is set to 1e-6. We terminate the iteration when the difference of the objective value at two consecutive iterations is small than 1e-8.

5   Experimental Results

We capture hyperspectral images by using an EBA Japan NH-7 camera, whose spatial resolution is 1280×1024 pixels and spectral resolution is 5 nm in the range from 420 nm to 700 nm. A wideband white LED is used for illumination. We use a standard white reference to calibrate the response function of the camera and the spectral distribution of the illumination.

We simultaneously estimate the spectral reflectance and remove interreflection by using Algorithm 1. For comparison, we average the spectra of different pixels of the same color, and refer to it as the conventional baseline method. As for the ground truth reflectance spectrum, we use a spectrometer to measure a planar surface without any interreflection. We measure the relative error of the estimated reflectance spectrum $\mathbf{r}$ with respect to its ground truth $\hat{\mathbf{r}}$ by $\|\hat{\mathbf{r}} - \mathbf{r}\| / \|\hat{\mathbf{r}}\|$. We show all hyperspectral images in RGB for visualization.

We first use three foldable paper fans as the test targets, which are best suited for our experiments, since we can unfold them to capture the ground

---

Algorithm 1. Spectral Reflectance Recovery and Interreflection Removal via ADMM

Input: $I \in \mathbb{R}^{m \times n}$, $\alpha = 1e^{-3}$, $\rho = 1.05$,
Output: $\mathbf{r} \in \mathbb{R}^m$, $W \in \mathbb{R}^{2 \times n}$,
   Random initialization for $\mathbf{r}$, W, U and L; $\mu = 1e^{-6}$; $i \leftarrow 1$,
  while not converged **do**
    Update W via Eq. (12),
    Update $\mathbf{r}$ via Eq. (14),
    Update U via Eq. (16),
    $L_{i+1} \leftarrow L_i + \mu_i (U - W)$,
    $\mu_{i+1} \leftarrow \min\{\rho\mu_i, 10^{20}\}$
    $i \leftarrow i + 1$
  end while

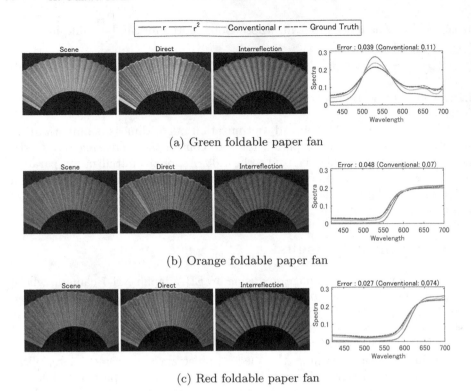

(a) Green foldable paper fan

(b) Orange foldable paper fan

(c) Red foldable paper fan

Fig. 2. Spectral reflectance recovery and interreflection removal of a green (a), orange (b) and red (c) foldable paper fan. In each row, the original scene, the estimated direct reflectance component, the interreflection component and the spectral comparison are shown from the left to right sequentially. (Color figure online)

truth reflectance spectra, without any interference from interreflections. Figure 2 shows the separation results of a green (a), orange (b) and red (c) paper fan, from which we can clearly observe that our estimated reflectance spectrum is much more accurate than the conventional method without considering interreflection. For example, for the green fan, the relative error of our method is 0.039, which is much lower than that of the conventional method's relative error of 0.11.

To verify that our proposed method will benefit a great variety of colors, we test on 24 different colored sheets paper as shown in Fig. 3(a). The estimation errors for these sheets of paper are shown in Fig. 3(b), from which we can see that our method improves over the conventional method by a nontrivial margin for all colors. As a representative, we show the separation results for the first color in Fig. 3(c, f). The spectral distributions for the best and worst cases are shown in Fig. 3(d) and (e), respectively.

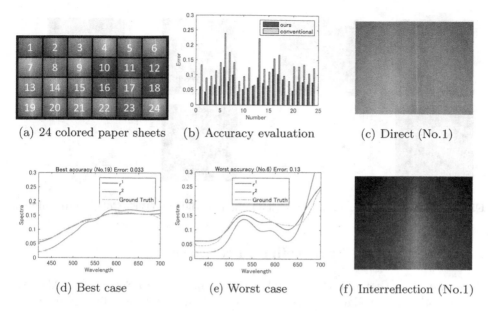

(a) 24 colored paper sheets   (b) Accuracy evaluation   (c) Direct (No.1)

(d) Best case   (e) Worst case   (f) Interreflection (No.1)

Fig. 3. Spectral reflectance recovery accuracy for a variety of colors. (a) shows the 24 sheets of colored paper. The recovery accuracy is shown in (b). As a representative, (c) and (f) show the direct and interreflection components for the scene containing paper sheet No.1. (d) and (e) illustrate the estimated spectra in the best and worst case, respectively.

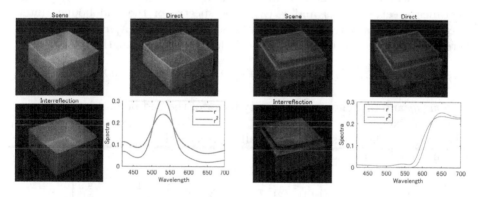

Fig. 4. Experiment results for a green (left) and red (right) box. (Color figure online)

We also use some real objects with more complex surface geometry, including green and red boxes shown in Fig. 4, as well as pink and red flowers in Fig. 5. We can also intuitively see that our separation results are reasonable. For example, the interreflection around the inner bottom of the box or the central part of the flower becomes much stronger.

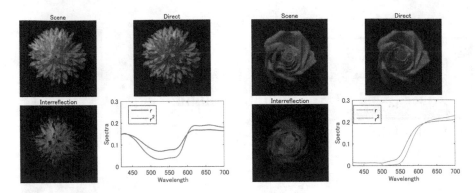

Fig. 5. Experiment results for a pink (left) and red (right) flower. (Color figure online)

6   Extension to Two Reflectance Colors

Here, we show that our regularized NMF formulation in Sect. 3 and ADMM based optimization algorithm in Sect. 4 can be straightforwardly extended to handle an object with two reflectance colors.

6.1   Mathematical Formulations

Again, we will consider 2nd bounce interreflections only. Given two diffuse reflectance spectra $r_1(\lambda_j)$ and $r_2(\lambda_j)$, the spectra of the 2nd bounce inter-reflection between any two surface points can be $r_1^2(\lambda_j)$, $r_1(\lambda_j)r_2(\lambda_j)$ or $r_2^2(\lambda_j)$. Therefore, the observed intensity of the k-th pixel at wavelength λ_j should be

$$i_k(\lambda_j) = w_k * r_1(\lambda_j) + p_k * r_2(\lambda_j) + v_k * r_1^2(\lambda_j) + q_k * r_1(\lambda_j) * r_2(\lambda_j) + u_k * r_2^2(\lambda_j), \tag{17}$$

where v_k, q_k and u_k denote the intensity factor of the 2nd bounce interreflection at the k-th pixel.

By stacking the measurements at all n pixels and m wavelength bands, the following matrix form can be obtained

$$
\begin{bmatrix}
i_{11} & i_{12} & \cdots & i_{1n} \\
i_{21} & i_{22} & \cdots & i_{2n} \\
\cdots & \cdots & \cdots & \cdots \\
i_{m1} & i_{m2} & \cdots & i_{mn}
\end{bmatrix}
=
\begin{bmatrix}
r_{11} & r_{21} & r_{11}^2 & r_{11}r_{21} & r_{21}^2 \\
r_{12} & r_{22} & r_{12}^2 & r_{12}r_{22} & r_{22}^2 \\
\cdots & \cdots & \cdots & \cdots & \cdots \\
r_{1m} & r_{2m} & r_{1m}^2 & r_{1m}r_{2m} & r_{2m}^2
\end{bmatrix}
\begin{bmatrix}
w_1 & w_2 & \cdots & w_n \\
p_1 & p_2 & \cdots & p_n \\
v_1 & v_2 & \cdots & v_n \\
q_1 & q_2 & \cdots & q_n \\
u_1 & u_2 & \cdots & u_n
\end{bmatrix}, \tag{18}
$$

which again fits into the nonnegative matrix factorization formulation

$$I = \begin{bmatrix} \mathbf{r}_1 \ \mathbf{r}_2 \ \mathbf{r}_1 * \mathbf{r}_1 \ \mathbf{r}_1 * \mathbf{r}_2 \ \mathbf{r}_2 * \mathbf{r}_2 \end{bmatrix} \tilde{W} = \tilde{R}\tilde{W}, \tilde{R} \geq 0, \tilde{W} \geq 0. \tag{19}$$

Similarly, the matrix $\tilde{W}$ should be sparse in general, because of the facts that interreflection is usually much weaker than direct reflectance and that each scene point does not contain all interreflection permutations. Similar to Eq. (7), after introducing the L_1-norm sparsity term, the spectral reflectance recovery and interreflection removal problem in the presence of two colors can be modeled as

$$\min_{\mathbf{r_1},\mathbf{r_2},\tilde{W}} \|I - \tilde{R}\tilde{W}\|_F^2 + \alpha \sum \tilde{W}, \, s.t., \, \tilde{R} \geq 0, \tilde{W} \geq 0. \tag{20}$$

6.2   Optimization Algorithm

By comparing Eq. (20) with Eq. (7), we recognize that Algorithm 1 can be directly used here. The only modification needed is to update $\mathbf{r_1}$ and $\mathbf{r_2}$ by solving

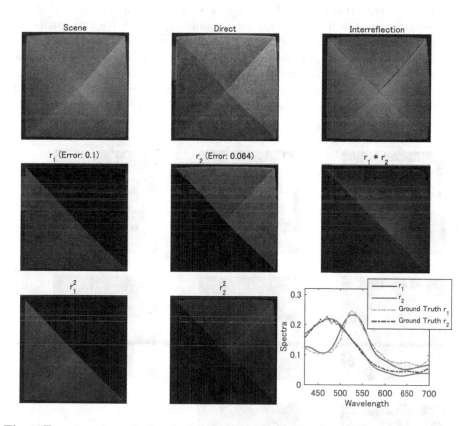

Fig. 6. Experiment results for the interior surface of a blue/green pyramid. In addition to correctly separating interreflection from direct reflectance, it is important to note that our proposed method further decomposes the total interreflection component into different interreflection permutations of the two reflectance colors. (Color figure online)

the following nonlinear least squares problem with nonnegative constraints

$$\min_{\mathbf{r_1},\mathbf{r_2}} \quad \left\| I - \tilde{R}\,\tilde{W} \right\|_F^2$$
$$\text{s.t.,} \quad \sum \mathbf{r_1} = 1, \sum \mathbf{r_2} = 1, \tilde{R} \geq 0. \tag{21}$$

6.3   Experiment Results

We also conduct some experiments using real images of scenes with two colors. We make a pyramid by folding two sheets of paper with different colors, and capture a hyperspectral image of its interior surface as input. Figure 6 shows the experiment results for a blue/green pyramid. Quantitative comparison of the estimated spectra with their respective ground truth verifies the accuracy of our proposed algorithm. Qualitatively, we can see that the separation of the direct and interreflection components is quite visually pleasing. In particular, our method correctly decomposes the total interreflection component into different interreflection permutations of the two reflectance colors. The experiment results

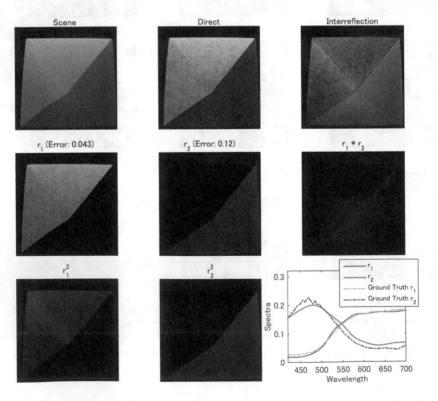

Fig. 7. Experiment results for the interior surface of a yellow/blue pyramid. (Color figure online)

in Fig. 7 for a yellow/blue pyramid demonstrate the outstanding performance of our method again.

7 Discussion

Here, we roughly discuss the difficulty inherent to the task of simultaneous spectral reflectance recovery and interreflection removal, even in the presence of one or two reflectance colors only.

It is straightforward to recognize that direct spectral reflectance and interreflection are indistinguishable up to a scale for a white surface. Therefore, any methods including ours will hardly work for a white or gray scene.

In the case of two colors, the estimation tends to be unstable, as the two reflectance spectra become similar to each other. This intrinsic degeneracy can be observed in Fig. 8, in which the spectra of the green and light yellow paper sheets are very similar, and the estimation results from our method are less accurate.

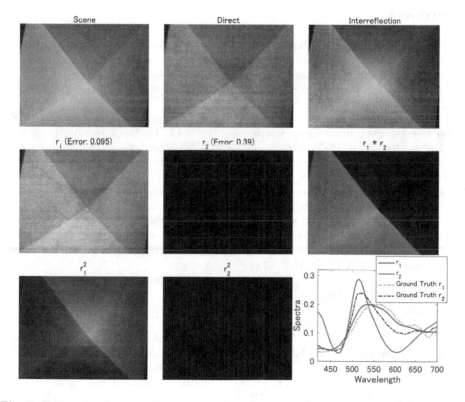

Fig. 8. Estimation becomes inaccurate when the two colors are similar. (Color figure online)

8   Conclusion

In this paper, we have introduced the novel problem of simultaneous spectral reflectance recovery and interreflection removal by using a single hyperspectral image, which often arises in practical scenarios in acquiring the spectral information of full scenes with concavity. We have shown our first endeavors in addressing this challenging problem, by modeling it using a sparsity regularized nonnegative matrix factorization formulation and developing a scalable optimization algorithm. Experiment results have demonstrated the effectiveness of our formulation and algorithm for scenes with a single or two reflectance colors.

As future work, we plan to address more general scenes with multiple colors. We envision that more constraints to fully disambiguate various permutations of interreflection components are needed.

Acknowledgements. This work was supported in part by Grant-in-Aid for Scientific Research on Innovative Areas (No.15H05918) from MEXT, Japan.

References

1. Balas, C., Papadakis, V., Papadakis, N., Papadakis, A., Vazgiouraki, E., Themelis, G.: A novel hyper-spectral imaging apparatus for the non-destructive analysis of objects of artistic and historic value. J. Cult. Herit. **4**, 330–337 (2003)
2. Maloney, L.T., Wandell, B.A.: Color constancy: a method for recovering surface spectral reflectance. JOSA A **3**, 29–33 (1986)
3. Gao, L., Kester, R.T., Tkaczyk, T.S.: Compact image slicing spectrometer (ISS) for hyperspectral fluorescence microscopy. OpEx **17**, 12293–12308 (2009)
4. Gao, L., Kester, R.T., Hagen, N., Tkaczyk, T.S.: Snapshot image mapping spectrometer (IMS) with high sampling density for hyperspectral microscopy. OpEx **18**, 14330–14344 (2010)
5. Gorman, A., Fletcher-Holmes, D.W., Harvey, A.R.: Generalization of the lyot filter and its application to snapshot spectral imaging. OpEx **18**, 5602–5608 (2010)
6. Gorman, A., Muyo, G., Harvey, A.R.: Snapshot spectral imaging using birefringent interferometry and image replication. The Optical Society (2010)
7. Tominaga, S.: Multichannel vision system for estimating surface and illumination functions. JOSA A **13**, 2163–2173 (1996)
8. Chi, C., Yoo, H., Ben-Ezra, M.: Multi-spectral imaging by optimized wide band illumination. IJCV **86**, 140–151 (2010)
9. DiCarlo, J.M., Xiao, F., Wandell, B.A.: Illuminating illumination. In: CIC, IS&T/SID, pp. 27–34 (2001)
10. Han, S., Sato, I., Okabe, T., Sato, Y.: Fast spectral reflectance recovery using DLP projector. In: Kimmel, R., Klette, R., Sugimoto, A. (eds.) ACCV 2010. LNCS, vol. 6492, pp. 323–335. Springer, Heidelberg (2011). doi:10.1007/978-3-642-19315-6_25
11. Lam, A., Subpa-Asa, A., Sato, I., Okabe, T., Sato, Y.: Spectral imaging using basis lights. In: BMVC. BMVA Press (2013)
12. Park, J.I., Lee, M.H., Grossberg, M.D., Nayar, S.K.: Multispectral imaging using multiplexed illumination. In: ICCV, pp. 1–8 (2007)
13. Forsyth, D., Zisserman, A.: Mutual illumination. In: CVPR, pp. 466–473 (1989)

14. Forsyth, D., Zisserman, A.: Shape from shading in the light of mutual illumination. Image Vis. Comput. **8**, 42–49 (1990)
15. Nayar, S.K., Krishnan, G., Grossberg, M.D., Raskar, R.: Fast separation of direct and global components of a scene using high frequency illumination. In: ACM SIGGRAPH, pp. 935–944 (2006)
16. Liao, M., Huang, X., Yang, R.: Interreflection removal for photometric stereo by using spectrum-dependent albedo. In: CVPR, pp. 689–696 (2011)
17. Fu, Y., Lam, A., Matsushita, Y., Sato, I., Sato, Y.: Interreflection removal using fluorescence. In: Fleet, D., Pajdla, T., Schiele, B., Tuytelaars, T. (eds.) ECCV 2014. LNCS, vol. 8693, pp. 203–217. Springer, Heidelberg (2014). doi:10.1007/978-3-319-10602-1_14
18. Nam, G., Kim, M.: Multispectral photometric stereo for acquiring high-fidelity surface normals. Comput. Graph. Appl. **34**, 57–68 (2014)
19. Imai, F.H., Rosen, M.R., Berns, R.S.: Comparison of spectrally narrow-band capture versus wide-band with a priori sample analysis for spectral reflectance estimation. In: Proceedings of Eighth Color Imaging Conference: Color Science and Engineering, Systems, Technologies and Applications, IS&T, pp. 234–241 (2000)
20. Jiang, J., Gu, J.: Recovering spectral reflectance under commonly available lighting conditions. In: 2012 IEEE Computer Society Conference on Computer Vision and Pattern Recognition Workshops (CVPRW), pp. 1–8 (2012)
21. Koenderink, J.J., van Doorn, A.J.: Geometrical modes as a general method to treat diffuse interreflections in radiometry. JOSA **73**, 843–850 (1983)
22. Nayar, S.K., Ikeuchi, K., Kanade, T.: Shape from interreflections. IJCV **6**, 173–195 (1991)
23. Nayar, S.K., Gao, Y.: Colored interreflections and shape recovery. In: Proceedings of the Image Understanding Workshop (1992)
24. Funt, B.V., Drew, M.S.: Color space analysis of mutual illumination. PAMI **15**, 1319–1326 (1993)
25. Funt, B.V., Drew, M.S., Ho, J.: Color constancy from mutual reflection. IJCV **6**, 5–24 (1991)
26. Seitz, S.M., Matsushita, Y., Kutulakos, K.N.: A theory of inverse light transport. In: ICCV, pp. 1440–1447 (2005)
27. Lee, D., Seung, H.: Learning the parts of objects by non-negative matrix factorization. Nature **401**, 788–791 (1999)
28. Eggert, J., Korner, E.: Sparse coding and NMF. In: IJCNN, pp. 2529–2533 (2004)
29. Lin, Z., Chen, M., Wu, L., Ma, Y.: The augmented lagrange multiplier method for exact recovery of corrupted low-rank matrices. ArXiv e-prints (2010)
30. Zheng, Y., Liu, G., Sugimoto, S., Yan, S., Okutomi, M.: Practical low-rank matrix approximation under robust l1-norm. In: CVPR, pp. 1410–1417 (2012)
31. Xu, Y., Yin, W., Wen, Z., Zhang, Y.: An alternating direction algorithm for matrix completion with nonnegative factors. Front. Math. China **7**, 365–384 (2012)

Learning Contextual Dependencies for Optical Flow with Recurrent Neural Networks

Minlong Lu[1,2(✉)], Zhiwei Deng[2], and Ze-Nian Li[2]

[1] College of Computer Science, Zhejiang University, Hangzhou, China
minlongl@sfu.ca
[2] School of Computing Science, Simon Fraser University, Burnaby, BC, Canada

Abstract. Pixel-level prediction tasks, such as optical flow estimation, play an important role in computer vision. Recent approaches have attempted to use the feature learning capability of Convolutional Neural Networks (CNNs) to tackle dense per-pixel predictions. However, CNNs have not been as successful in optical flow estimation as they are in many other vision tasks, such as image classification and object detection. It is challenging to adapt CNNs designated for high-level vision tasks to handle pixel-level predictions. First, CNNs do not have a mechanism to explicitly model contextual dependencies among image units. Second, the convolutional filters and pooling operations result in reduced feature maps and hence produce coarse outputs when upsampled to the original resolution. These two aspects render CNNs limited ability to delineate object details, which often result in inconsistent predictions. In this paper, we propose a recurrent neural network to alleviate this issue. Specifically, a row convolutional long short-term memory (RC-LSTM) network is introduced to model contextual dependencies of local image features. This recurrent network can be integrated with CNNs, giving rise to an end-to-end trainable network. The experimental results demonstrate that our model can learn context-aware features for optical flow estimation and achieve competitive accuracy with the state-of-the-art algorithms at a frame rate of 5 to 10 fps.

1 Introduction

Convolutional Neural Networks (CNNs) [1] have brought a revolution in computer vision community with its powerful feature learning capability based on large-scale datasets. They have been immensely successful in high-level computer vision tasks, such as image classification [2,3] and object detection [4,5]. CNNs are good at extracting abstract image features by using convolution and pooling layers to progressively shrink the feature maps, which produces translation invariant local features and allows the aggregation of information over large areas of the input images.

Recently, researchers have been attempting to employ CNNs to tackle pixel-level prediction tasks, such as semantic segmentation [6,7] and optical flow estimation [8,9]. These tasks differ from the previous high-level tasks in that they not only require precise single pixel prediction, but also require semantically

© Springer International Publishing AG 2017
S.-H. Lai et al. (Eds.): ACCV 2016, Part IV, LNCS 10114, pp. 68–83, 2017.
DOI: 10.1007/978-3-319-54190-7_5

meaningful and contextually consistent predictions among a set of pixels within objects. Optical flow estimation is even more difficult because it requires finding the x-y flow field between a pair of images, which involves a very large continuous labeling space.

There are significant challenges in adapting CNNs to handle dense per-pixel optical flow estimation. First, CNNs do not have a mechanism to explicitly model contextual dependencies among image pixels. Although the local features learned with CNNs play an important role in classifying individual pixels, it is similarly important to consider factors such as appearance and spatial consistency while assigning labels in order to obtain precise and consistent results. Besides, the convolution and pooling operations result in reduced feature maps, and hence produce coarse outputs when upsampled to the original resolution to produce pixel-level labels. These two aspects render CNNs limited ability to delineate object details, and can result in blob-like shapes, non-sharp borders and inconsistent labeling within objects.

In this paper, Recurrent Neural Networks (RNNs) are incorporated to alleviate this problem. RNNs have achieved great success in modeling temporal dependencies for sequential data, and have been widely used in natural language processing [10], image captioning [11], etc. Long short-term memory (LSTM) [12] is a special RNN structure that is stable and powerful for modeling long-range dependencies without suffering from the vanishing gradient problem of vanilla RNN models [13]. We propose a row convolutional LSTM (RC-LSTM), which has convolution operators in both the input-to-state and state-to-state transitions to handle structure inputs. We treat an image as a sequence of rows and use our RC-LSTMs to explicitly model the spatial dependencies among the rows of pixels, which encodes the neighborhood contexture into local image representation.

The proposed RC-LSTM structure can be integrated with CNNs to enhance the learned feature representations and produce context-aware features. In our experiments, we integrate our RC-LSTM with FlowNet [8], the state-of-the-art CNN-based model for optical flow estimation, to form an end-to-end trainable network. We test the integrated network on several datasets, the experimental results demonstrate that our RC-LSTM structure can enhance the CNN features and produce more accurate and consistent optical flow maps. Our model achieves competitive accuracy with the state-of-the-art methods and has the best performance among the real-time ones.

2 Related Work

2.1 Optical Flow

Optical flow estimation has been one of the key problems in computer vision. Starting from the original approaches of Horn and Schunck [14] as well as Lucas and Kanade [15], many improvements have been introduced to deal with the shortcomings of previous models. Most of those methods model the optical flow problem as an energy minimization framework, and are usually carried out in a coarse-to-fine scheme [16–18]. Due to the complexity of the energy minimization, such methods have the problem of local minima and may not be able to estimate large displacements accurately. The methods in [19, 20] integrate descriptor matching into a variational approach to deal with large displacements problem. The method in [21] emphasizes on sparse matching and uses edge-preserving interpolation to obtain dense flow fields, which achieves state-of-the-art optical flow estimation performance.

2.2 CNNs for Pixel-Level Prediction

CNNs have become the method of choice and achieved state-of-the-art performance in many high-level vision tasks, e.g. image classification. Recently, researchers start attempting to use CNNs for pixel-level labeling problems, including semantic segmentation [6, 7, 22, 23], depth prediction [24], and optical flow estimation [8]. The traditional CNNs require fixed-size inputs, and the fully connected layers transform the feature maps into vector representations which are difficult to reconstruct for 2D predictions. Therefore, it is not straightforward in adapting CNNs designated to produce a high-level label for an image to tackle the tasks of pixel-level predictions.

One simple way to use CNNs in such applications is to apply a CNN in a "sliding window" way and predict a single label for the current image patch [22]. This works well in many situations, but the main drawback is the high computational cost. Fully Convolutional Networks (FCN) is proposed in [6] which can take input of arbitrary size and produce output in the same size with efficient inference. The key insight is to transform the fully connected layers into convolution layers which produce coarse predictions. Then deconvolution layers are incorporated to iteratively refine the prediction to the original size. This method achieves significant improvement on segmentation performance on several datasets.

A similar scheme is utilized in FlowNet [8] to predict an optical flow field with convolution layers to extract compact feature maps and deconvolution layers to upsample them to the desired resolution. The difference is that not only the coarse predictions, but the whole coarse feature maps are "de-convolved", allowing the transfer of more information to the final prediction. Two FlowNet architectures are proposed and compared in [8]. The FlowNetS architecture simply stacks the image pair together and feeds them through a generic CNN. The FlowNetC architecture includes a layer that correlates feature vectors at different image locations. Although the performance of FlowNet is not as good as the state-of-the-art traditional non-CNN methods, it opens the direction of learning optical flow with CNNs and can be considered as the state-of-the-art CNN-based model.

The central issue in the methodology of [6, 8] is that the convolution and pooling operations result in reduced feature maps and hence produce coarse outputs when upsampled to the original resolution. Besides, CNNs lack smoothness constraints to model label consistency between pixels. To solve this problem, Zheng et al. [7] combines CNNs with a Conditional Random Fields (CRFs)-based probabilistic graphical model. Mean-field approximate inference for the CRFs is formulated as Recurrent Neural Networks, which enables training CRFs end-to-end together with CNNs. This CRF-RNN network is integrated with FCN [6] and achieves more precise segmentation results. However, the inference of mean-field approximation in [7] is for discrete labeling, making it not applicable to refine optical flow maps which is a continuous labeling task with very large label space.

2.3   RNNs and LSTMs

Recurrent Neural Networks (RNNs) have achieved great success in modeling temporal dependency for chain-structured data, such as natural language and speeches [10]. Long short-term memory (LSTM) [12] is a special RNN structure that is stable and powerful for modeling long-range temporal dependencies without the vanishing gradient problem.

RNNs have been extended to model spatial and contextual dependencies among image pixels [25–27]. The key insight is to define different connection structures among pixels within an image and build spatial sequences of pixels. Multi-dimensional RNNs are proposed in [28] and are applied to offline arabic handwriting recognition. 2D-RNNs [29], tree-structured RNNs [30], and directed acyclic graph RNNs [23] are proposed to model different connections between image pixels for different vision tasks. Applying RNNs to specifically defined graph structures on images are different with the idea of CRF-RNN [7]. Instead of implicitly encoding neighborhood information with a pairwise term in an energy minimization framework, the RNNs enables explicit information propagation via the recurrent connections.

Our approach intergrates ideas from these methods. We propose to model the contextual dependencies among each row of pixels with a row convolutional LSTM(RC-LSTM). Similar as [31], our RC-LSTM utilize convolution operators in both the input-to-state and state-to-state transitions instead of the full matrix multiplication in LSTM to better handle structure inputs. The feature vectors of the pixels in each row form one input to the RC-LSTM, and the rows are processed in sequence. The RC-LSTM enables the information propagation/message passing among the rows of pixels, which encodes the contextual dependencies into the local feature representations. This RC-LSTM structure is integrated with convolution and deconvolution layers, giving rise to an end-to-end trainable network. To the best of our knowledge, our work is the first attempt to integrate CNNs with RNNs for optical flow estimation.

3   Approach

To predict dense pixel-level optical flow from a pair of images, the images are processed by three types of network components: convolution layers, deconvolution layers, and our RC-LSTM network. Functionally, the convolution layers transform raw image pixels to compact and discriminative representations.

The deconvolution layers then upsample the feature maps to the desired output resolution. Based on them, the proposed RC-LSTM models the contextual dependencies of local features and produce context-aware representations, which are used to predict the final optical flow map.

In this section, we will first briefly review the LSTM, and then introduce our RC-LSTM for modeling contextual dependencies among image rows. After that, we will explain how our RC-LSTM model can be integrated with CNNs to form an end-to-end trainable network.

3.1 LSTM Revisited

Long short-term memory (LSTM) [12] is a special RNN structure that is stable and powerful for modeling long-range temporal dependencies. Its innovation is the introduction of the "memory cell" c_t to accumulate the state information. The cell is accessed, written and cleared by several controlling gates, which enables LSTM to selectively forget its previous memory states and learn long-term dynamics without the vanishing gradient problem of simple RNNs. Given x_t as the input of an LSTM cell at time t, the cell activation can be formulated as:

$$
\begin{aligned}
i_t &= \sigma(W_{xi}x_t + W_{hi}h_{t-1} + b_i) \\
f_t &= \sigma(W_{xf}x_t + W_{hf}h_{t-1} + b_f) \\
o_t &= \sigma(W_{xo}x_t + W_{ho}h_{t-1} + b_o) \\
g_t &= \phi(W_{xc}x_t + W_{hc}h_{t-1} + b_c) \\
c_t &= f_t \odot c_{t-1} + i_t \odot g_t \\
h_t &= o_t \odot \phi(c_t)
\end{aligned}
\tag{1}
$$

where σ stands for the sigmoid function, ϕ stands for the tanh function, and $\odot$ denotes the element-wise multiplication. In addition to the hidden state h_t and memory cell c_t, LSTM has four controlling gates: i_t, f_t, o_t, and g_t, which are the input, forget, output, and input modulation gate respectively.

The input gate i_t controls what information in g_t to be accumulated into the cell c_t. While the forget gate f_t helps the c_t to maintain and selectively forget information in previous state c_{t-1}. Whether the updated cell state c_t will be propagated to the final hidden state h_t representation is controlled by the output gate o_t. Multiple LSTMs can be temporally concatenated to form more complex structures to solve many real-life sequence modeling problems.

3.2 RC-LSTM

The aforementioned LSTMs are designed to model temporal dependency for a data sequence. To applying LSTMs to one image, a spatial order of the image pixels needs to be defined. The simplest way is to consider each pixel as an individual input data, and the image is reshaped to a sequence of pixels to feed into the LSTM model. However, the interactions among pixels are beyond this

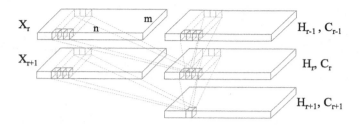

Fig. 1. The structure of the RC-LSTM.

chain-structured sequence. This simple way will loss the spatial relationships, because adjacent pixels may not be neighbors in this sequence.

In order to model the spatial relationships among pixels, we consider each row of the image as one input data, and consider the image as a sequence of rows. This results in structure inputs which LSTM has difficulty to handle. We propose a row convolutional long short-term memory (RC-LSTM) to cope with the structure inputs and learn the contextual dependencies among the rows of pixels. Our RC-LSTM is illustrated in Fig. 1. The feature vectors of the pixels in the rth row form one input matrix $X_r \in \mathcal{R}^{m \times n}$ to the RC-LSTM, where n is the number of pixels in each row and m is the feature dimension. The RC-LSTM determines the hidden state H_r by the input X_r and the states H_{r-1}, C_{r-1} of the previous row. Convolution operations are used in both the input-to-state and state-to-state transitions, and the RC-LSTM can be formulated as:

$$
\begin{aligned}
i_r &= \sigma(w_{xi} \otimes X_r + w_{hi} \otimes H_{r-1} + b_i) \\
f_r &= \sigma(w_{xf} \otimes X_r + w_{hf} \otimes H_{r-1} + b_f) \\
o_r &= \sigma(w_{xo} \otimes X_r + w_{ho} \otimes H_{r-1} + b_o) \\
g_r &= \phi(w_{xc} \otimes X_r + w_{hc} \otimes H_{r-1} + b_c) \\
C_r &= f_r \odot C_{r-1} + i_r \odot g_r \\
H_r &= o_r \odot \phi(C_r)
\end{aligned}
\tag{2}
$$

where the w-s are convolution kernels with size $1 \times k$ and $\otimes$ denotes convolution. The distinguishing feature of our RC-LSTM is that the inputs X_r, cell states C_r, hidden states H_r, and the gates i_r, f_r, g_r, o_r are all matrices. In this sense, our RC-LSTM can be considered as a generalization of the traditional vector-based LSTM to handle structure input. Due to the convolution operations, our model also has fewer parameters and less redundancy, in comparison to the matrix multiplications in Eq. 1.

The RC-LSTM model enables the message passing in one direction: from the top to the bottom of the image. The pixel $v_{(r,c)}$ at location (r, c) will get the information propagated from its ancestors in the previous row, which are a small neighborhood of k pixels near the pixel $v_{(r-1,c)}$. Figure 2 illustrated the message passing among the pixels. Our RC-LSTM explicitly models the contextual dependencies among the pixels using this 1-direction message passing.

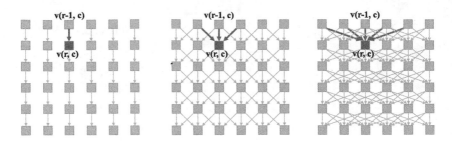

Fig. 2. Illustration of top-to-bottom message passing among the image pixels with k equals 1, 3, and 5.

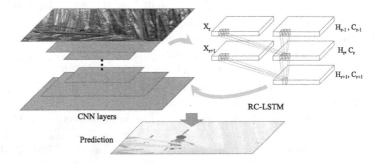

Fig. 3. The overall framework of an end-to-end trainable network. RC-LSTMs can be used in any point to refine the feature maps, by modeling the spatial dependencies of local features and produce context-aware representations.

Spatial dependencies of a pixel come from surrounding pixels in all directions within an image, and the 1-direction message passing might not be enough to model all these dependencies. This can be addressed by using the RC-LSTM 4 times: row by row from top to bottom, from bottom to top, column by column from left to right, and from right to left. We call this method 4-direction message passing RC-LSTM, which is able to model more complete contextual dependencies.

3.3   Integration with CNNs

Our RC-LSTMs can be integrated with any CNNs structures, other networks (e.g. auto-encoder [32]), and even hand-crafted features, to model the spatial dependencies of the local features and produce context-aware feature representations. Figure 3 shows a generic framework that integrates the RC-LSTM model with a general CNN. The RC-LSTM model gets input from the CNN feature maps, refine the local features with message passing, and output context-aware feature maps, which can be further processed by CNN layers or directly used to predict output flows.

In our experiment, we integrate our RC-LSTM with FlowNetS [8], which has a generic network structure and is shown to have better generalization abilities.

The image pair is stacked as the input of the FlowNetS, and the network has 10 convolution layers and 4 deconvolution layers. Detailed structure of the integrated network can be found in Sect. 4.1.

4 Experiments

4.1 Datesets and Experiment Setup

Optical flow evaluation requires dense per-pixel ground truth, which is very difficult to obtain from real world images. Therefore, the amount of real images with dense ground truth optical flow field is very small. In Middlebury dataset [33], optical flow ground truth of 6 image pairs are generated by tracking the hidden fluorescent paint applied to the scene surfaces, with computer-controlled lighting and motion stages for camera and scene. The KITTI dataset [34] is larger (389 image pairs) and the semi-dense optical flow ground truth is obtained by the use of 4 cameras and a laser scanner.

In order to facilitate the evaluation of optical flow algorithms, some larger synthetic datasets with a variant of motion types are generated [8,35]. In our experiments, we attempt to use both real world images and synthetic images to evaluate our method. The datasets used in our experiments are introduces below, and example images from some of these datasets are shown in Fig. 4.

- **Middlebury dataset** [33] contains 8 training image pairs and 8 testing image pairs with dense optical flow ground truth. Six image pairs are real world images and others are synthetic ones. The image resolution ranges from

(a)   (b)

(c)   (d)

Fig. 4. Sample images from (a) Middlebury, (b) Sintel Clean, (c) Flying Chairs, and (d) Sintel Final dataset. The Final version of Sintel (d) includes motion blur and atmospheric effects to the Clean version (b).

Fig. 5. Optical flow color coding scheme: the vector from the center to a pixel is encoded using the color of that pixel.

316×252 to 640×480, and the displacements are small, usually less than 10 pixels. The ground truth flows of the training images are publicly available, while the ground truth of the testing images is hidden from the public, researchers can upload their testing results to an evaluation server.

- **KITTI dataset** [34] contains 194 training image pairs and 195 testing image pairs. An average of 50% pixels in the images have ground truth optical flow. The images are captured using cameras mounted on an autonomous car in real world scenes, and the image resolution is about 1240×376. This dataset contains strong projective transformations and special types of motions. Similar with Middlebury dataset, the ground truth flows of the training images are publicly available, while the testing results are evaluated on a server.
- **Sintel dataset** [35] contains computer rendered artificial scenes of a 3D movie with dense per-pixel ground truth. It includes large displacements, and pays special attention to achieve realistic image properties. The dataset provides "Clean" and "Final" versions. The Final version includes atmospheric effects (e.g. fog), reflections, and motion blur, while the Clean version does not includes these effects. Each version contains 1041 training image pairs, and 552 testing image pairs. The image resolution is 1024×436. The ground truth flows of the training images are publicly available, while the testing results are evaluated on a server.
- **Flying Chairs dataset** [8] is a large synthetic dataset which is built to provide sufficient data for training CNNs for optical flow estimation. It contains 22232 training image pairs and 640 testing image pairs. The images have resolution 512×386 and are generated by rendering 3D chair models on background images from Flickr. The ground truth flows of the whole dataset are publicly available.

We train our network on the training set of Flying Chairs dataset, and test the network on the Sintel, KITTI, Middlebury, and the testing set of Flying Chairs datasets. Note that we only train our model on the training set of Flying Chairs dataset, and we do not train or fine-tune our model on the other datasets. Therefore, both the "train" and "test" data of other datasets can serve as testing

Table 1. A summary of the datasets used to TEST our method.

Dataset	Image pairs #	Evaluation method
Sintel Clean Train	1041	Compare results with GT
Sintel Clean Test	552	Upload to evaluation server
Sintel Final Train	1041	Compare results with GT
Sintel Final Test	552	Upload to evaluation server
KITTI Train	194	Compare results with GT
Middlebury Train	8	Compare results with GT
Flying Chairs Test	640	Compare results with GT

Table 2. The detailed structures of the Proposed-1/4dir networks with RC-LSTM + FlowNetS. The illustrated input/output resolutions are based on the input image with size 512×384. pr stands for prediction, and pr' stands for the upsampled pr.

Layer name	Kernel sz	Str	I/O Ch#	InputRes	OutputRes	Input
conv1	7×7	2	6/64	512×384	256×192	Stacked image pair
conv2	5×5	2	64/128	256×192	128×96	conv1
conv3	5×5	2	128/256	128×96	64×48	conv2
conv3_1	3×3	1	256/256	64×48	64×48	conv3
conv4	3×3	2	256/512	64×48	32×24	conv3_1
conv4_1	3×3	1	512/512	32×24	32×24	conv4
conv5	3×3	2	512/512	32×24	16×12	conv4_1
conv5_1	3×3	1	512/512	16×12	16×12	conv5
conv6	3×3	2	512/1024	16×12	8×6	conv5_1
conv6_1	3×3	1	1024/1024	8×6	8×6	conv6
pr6+loss6	3×3	1	1024/2	8×6	8×6	conv6_1
deconv5	4×4	2	1024/512	8×6	16×12	conv6_1
pr5+loss5	3×3	1	1026/2	16×12	16×12	conv5_1+deconv5+pr6'
deconv4	4×4	2	1026/256	16×12	32×24	conv5_1+deconv5+pr6'
pr4+loss4	3×3	1	770/2	32×24	32×24	conv4_1+deconv4+pr5'
deconv3	4×4	2	770/128	32×24	64×48	conv4_1+deconv4+pr5'
pr3+loss3	3×3	1	386/2	64×48	64×48	conv3_1+deconv3+pr4'
deconv2	4×4	2	386/64	64×48	128×96	conv3_1+deconv3+pr4'
RC-LSTM 1/4dir	1×3	1	194/200	128×96	128×96	conv2+deconv2+pr3'
pr2+loss2	3×3	1	200/2	128×96	128×96	RC-LSTM 1/4dir
upsample	-	-	2/2	128×96	512×384	pr2

data in our experiments, the same scheme is used in [8]. The datasets used to test our method are summarized in Table 1.

Two architectures of our RC-LSTM as described in Sect. 3.2, i.e. 1-direction and 4-direction message passing, are integrated with FlowNetS to build an end-to-end trainable network. We do not include the variational refinement because it is essentially using CNN results to initialize a traditional flow estimation [19] and is not end-to-end trainable. Specifically, our RC-LSTMs are plugged into FlowNetS after the last deonvolution layer and before the final prediction layer. The detailed network structure can be found in Table 2 and our implementation is based on Caffe [36]. The kernel size is set to be 1×3. For the sake of convenience, we call the integrated networks Proposed-1dir and Proposed-4dir. The networks are evaluated both qualitatively and quantitatively described as follows.

The endpoint error (EPE) is used to evaluate the performance of different methods, which is define as:

$$EPE = \frac{1}{N} \sum \sqrt{(u_i - u_{GTi})^2 + (v_i - v_{GTi})^2} \tag{3}$$

where N is the total number of image pixels, (u_i, v_i) and (u_{GTi}, v_{GTi}) are the predicted flow vector and the ground truth flow vector for pixel i, respectively.

The optical flow field of an image pair is also visualized as an image by color-coding the flow field as in [33,35]. Flow direction is encoded with color and flow magnitude is encoded with color intensity. Figure 5 shows the color coding scheme: the vector from the center to a pixel is encoded using the color of that pixel. We use the open source tool provided in [33] to generate color-coded flow maps. The estimated flow maps on Flying Chairs test, Sintel train, and Middlebury train are visualized and compared with the visualized ground truth.

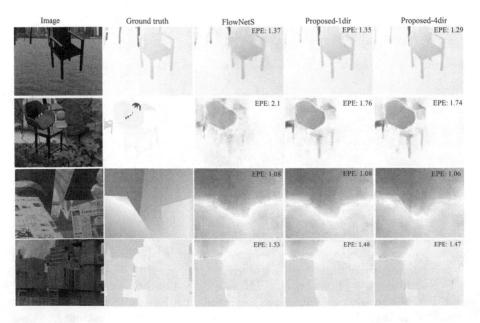

Fig. 6. The estimated optical flow maps from Flying Chairs dataset(top two) and Middlebury dataset(bottom two).

4.2   Results

Figure 6 compares the Proposed-1dir and Proposed-4dir networks to the FlowNetS method on the Flying Chairs and Middlebury datasets. The results produced by FlowNetS are often blurry and inconsistent, which lose some of the object details (e.g. chair legs). Both our proposed networks are able to produce better visualized flow maps, which are more accurate and contains more consistent objects with finer details. This demonstrates that our RC-LSTM is able to enhance the CNN features of FlowNetS, by explicitly modeling the spatial dependencies among pixels and producing context-aware features for optical flow

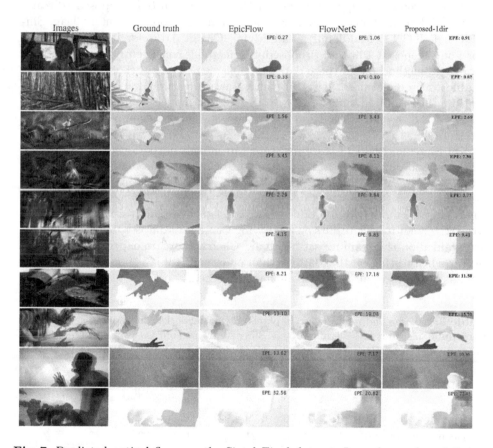

Fig. 7. Predicted optical flows on the Sintel Final dataset. In each row from left to right: overlaid image pair, ground truth flow, the predicted results of EpicFlow [21], FlowNetS [8], and Proposed-1dir RC-LSTM. Endpoint error (EPE) is shown at the top-right corner of each prediction. In most cases (Row 1–8), the proposed method produces visually better results with lower EPE than the FlowNetS. Although the EPEs of the proposed method are somewhat worse than that of EpicFlow, our model often produces better object details, see Row 3–5, 7–8, 10.

estimation. It can be found the Proposed-4dir network produces lower endpoint errors (EPE) than the Proposed-1dir network.

In Fig. 7, we qualitatively compare our Proposed-1dir to FlowNetS as well as the state-of-the-art method, EpicFlow [21], on more examples from Sintel Final dataset. The images of EpicFlow and FlowNetS are from the results in [8]. It can be seen from the figure that in most cases (Row 1–8), the Proposed-1dir method produces visually better results and lower endpoint errors (EPE) than the FlowNetS. Although the EPEs of the proposed method are somewhat worse than that of EpicFlow, our model often produces better object details, see Row 3–5, 7–8, and 10 in Fig. 7.

Table 3 shows the quantitative comparison between our Proposed-1dir and Proposed-4dir methods to many well-performing methods on Sintel Clean, Sintel Final, KITTI, Middlebury, and Flying Chairs datasets. It is shown that both our Proposed-1dir and Proposed-4dir methods consistently outperform the FlowNetS method on all these datasets. This demonstrates that our RC-LSTM is able to produce more powerful context-aware features for optical flow estimation. The running time of FlowNetS and our models is measured using an *NVIDIA TITAN* GPU on Middlebury dataset, and the time of other methods are from [8].

On Sintel and KITTI train datasets, our models outperform the LDOF [19] method. And our models are comparable to the real-time method EPPM [37], while our Proposed-1dir is two times faster. Although the proposed methods performs not as well as the state-of-the-art EpicFlow [21], it has been shown in Fig. 7 that our model often produce more consistent results. It is more interesting to see the quantitative results on the Flying Chairs test dataset. Since our models are trained on the training set of Flying Chairs, they are expected to perform better when testing on this dataset than on others. Table 3 shows that both our models outperform all the state-of-the-art methods.

Table 3. Average endpoint errors (in pixels) of our networks compared to several well-performing methods on several datasets. Since we trained our network on Flying Chairs dataset, we can test our model on both the train and test images on other datasets.

Method	Sintel Clean		Sintel Final		KITTI	Middle	Chairs	Time(s)	
	Train	Test	Train	Test	Train	Train	Test	CPU	GPU
EpicFlow [21]	2.40	4.12	3.70	6.29	3.47	0.31	2.94	16	-
DeepFlow [20]	3.31	5.38	4.56	7.21	4.58	0.21	3.53	17	-
EPPM [37]	-	6.49	-	8.38	-	-	-	-	0.2
LDOF [19]	4.29	7.56	6.42	9.12	13.73	0.45	3.47	65	2.5
FlowNetS [8]	4.50	7.42	5.45	8.43	8.26	1.09	2.71	-	0.05
Proposed-1dir	3.77	6.72	4.93	7.94	7.55	1.02	2.64	-	0.08
Proposed-4dir	3.77	6.69	4.90	7.91	7.54	1.01	2.56	-	0.2

The overall experimental results show that our models achieve best performance on the Flying Chairs dataset, and generalize well to other existing datasets. Note the fact that training data is essential to the performance of the CNN models, while the Flying Chairs dataset contains unrealistic images with 3D chair models rendered on background images. These results indicate that our method is very promising and may perform even better, if sufficiently large datasets with more realistic images are available.

5   Conclusion

In this paper, we proposed a row convolutional long short-term memory (RC-LSTM) network to model contextual dependencies of image pixels. The RC-LSTM is integrated with FlowNetS to enhance its learned feature representations and produce context-aware features for optical flow estimation. The experimental results demonstrated that our model can produce more accurate and consistent optical flows than the comparing CNN-based models. Our model also achieved competitive accuracy with state-of-the-art methods on several datasets.

Acknowledgement. This work was supported in part by the Natural Sciences and Engineering Research Council of Canada under the Grant RGP36726. We gratefully acknowledge the support of NVIDIA Corporation with the donation of the Titan GPU for this research.

References

1. LeCun, Y., Bottou, L., Bengio, Y., Haffner, P.: Gradient-based learning applied to document recognition. Proc. IEEE **86**, 2278–2324 (1998)
2. Krizhevsky, A., Sutskever, I., Hinton, G.E.: Imagenet classification with deep convolutional neural networks. In: Advances in Neural Information Processing Systems (NIPS), pp. 1097–1105 (2012)
3. Szegedy, C., Liu, W., Jia, Y., Sermanet, P., Reed, S., Anguelov, D., Erhan, D., Vanhoucke, V., Rabinovich, A.: Going deeper with convolutions. In: IEEE Conference on Computer Vision and Pattern Recognition (CVPR), pp. 1–9 (2015)
4. Girshick, R., Donahue, J., Darrell, T., Malik, J.: Rich feature hierarchies for accurate object detection and semantic segmentation. In: IEEE Conference on Computer Vision and Pattern Recognition (CVPR), pp. 580–587 (2014)
5. Ouyang, W., Wang, X., Zeng, X., Qiu, S., Luo, P., Tian, Y., Li, H., Yang, S., Wang, Z., Loy, C.C., et al.: DeepID-Net: deformable deep convolutional neural networks for object detection. In: IEEE Conference on Computer Vision and Pattern Recognition (CVPR), pp. 2403–2412 (2015)
6. Long, J., Shelhamer, E., Darrell, T.: Fully convolutional networks for semantic segmentation. In: IEEE Conference on Computer Vision and Pattern Recognition (CVPR), pp. 3431–3440 (2015)
7. Zheng, S., Jayasumana, S., Romera-Paredes, B., Vineet, V., Su, Z., Du, D., Huang, C., Torr, P.H.: Conditional random fields as recurrent neural networks. In: IEEE International Conference on Computer Vision (ICCV), pp. 1529–1537 (2015)

8. Dosovitskiy, A., Fischer, P., Ilg, E., Hausser, P., Hazirbas, C., Golkov, V., van der Smagt, P., Cremers, D., Brox, T.: Flownet: Learning optical flow with convolutional networks. In: IEEE International Conference on Computer Vision (ICCV), pp. 2758–2766 (2015)
9. Teney, D., Hebert, M.: Learning to extract motion from videos in convolutional neural networks. arXiv preprint arXiv:1601.07532 (2016)
10. Graves, A., Jaitly, N.: Towards end-to-end speech recognition with recurrent neural networks. In: International Conference on Machine Learning (ICML), pp. 1764–1772 (2014)
11. Karpathy, A., Fei-Fei, L.: Deep visual-semantic alignments for generating image descriptions. In: IEEE Conference on Computer Vision and Pattern Recognition (CVPR), pp. 3128–3137 (2015)
12. Hochreiter, S., Schmidhuber, J.: Long short-term memory. Neural Comput. **9**, 1735–1780 (1997)
13. Pascanu, R., Mikolov, T., Bengio, Y.: On the difficulty of training recurrent neural networks. In: International Conference on Machine Learning (ICML), pp. 1310–1318 (2013)
14. Horn, B.K., Schunck, B.G.: Determining optical flow. In: Technical Symposium East, pp. 319–331. International Society for Optics and Photonics (1981)
15. Lucas, B.D., Kanade, T., et al.: An iterative image registration technique with an application to stereo vision. In: International Joint Conference on Artificial Intelligence (IJCAI), vol. 81, pp. 674–679 (1981)
16. Brox, T., Bruhn, A., Papenberg, N., Weickert, J.: High accuracy optical flow estimation based on a theory for warping. In: Pajdla, T., Matas, J. (eds.) ECCV 2004. LNCS, vol. 3024, pp. 25–36. Springer, Heidelberg (2004). doi:10.1007/978-3-540-24673-2_3
17. Wedel, A., Cremers, D., Pock, T., Bischof, H.: Structure-and motion-adaptive regularization for high accuracy optic flow. In: IEEE International Conference on Computer Vision (ICCV), pp. 1663–1668 (2009)
18. Sun, D., Roth, S., Black, M.J.: A quantitative analysis of current practices in optical flow estimation and the principles behind them. Int. J. Comput. Vis. **106**, 115–137 (2014)
19. Brox, T., Malik, J.: Large displacement optical flow: descriptor matching in variational motion estimation. IEEE Trans. Pattern Anal. Mach. Intell. **33**, 500–513 (2011)
20. Weinzaepfel, P., Revaud, J., Harchaoui, Z., Schmid, C.: Deepflow: Large displacement optical flow with deep matching. In: IEEE International Conference on Computer Vision (ICCV), pp. 1385–1392 (2013)
21. Revaud, J., Weinzaepfel, P., Harchaoui, Z., Schmid, C.: Epicflow: edge-preserving interpolation of correspondences for optical flow. In: IEEE Conference on Computer Vision and Pattern Recognition (CVPR), pp. 1164–1172 (2015)
22. Farabet, C., Couprie, C., Najman, L., LeCun, Y.: Learning hierarchical features for scene labeling. IEEE Trans. Pattern Anal. Mach. Intell. **35**, 1915–1929 (2013)
23. Shuai, B., Zuo, Z., Wang, G., Wang, B.: DAG-Recurrent neural networks for scene labeling. arXiv preprint arXiv:1509.00552 (2015)
24. Eigen, D., Puhrsch, C., Fergus, R.: Depth map prediction from a single image using a multi-scale deep network. In: Advances in Neural Information Processing Systems (NIPS), pp. 2366–2374 (2014)
25. Liang, M., Hu, X.: Recurrent convolutional neural network for object recognition. In: IEEE Conference on Computer Vision and Pattern Recognition (CVPR), pp. 3367–3375 (2015)

26. van den Oord, A., Kalchbrenner, N., Kavukcuoglu, K.: Pixel recurrent neural networks. In: International Conference on Machine Learning (ICML) (2016)
27. Visin, F., Kastner, K., Cho, K., Matteucci, M., Courville, A., Bengio, Y.: Renet: A recurrent neural network based alternative to convolutional networks. arXiv preprint arXiv:1505.00393 (2015)
28. Graves, A., Schmidhuber, J.: Offline handwriting recognition with multidimensional recurrent neural networks. In: Advances in Neural Information Processing Systems (NIPS), pp. 545–552 (2009)
29. Shuai, B., Zuo, Z., Wang, G.: Quaddirectional 2D-recurrent neural networks for image labeling. IEEE Sig. Process. Lett. **22**, 1990–1994 (2015)
30. Tai, K.S., Socher, R., Manning, C.D.: Improved semantic representations from tree-structured long short-term memory networks. arXiv preprint arXiv:1503.00075 (2015)
31. Shi, X., Chen, Z., Wang, H., Yeung, D.Y., Wong, W.K., Woo, W.c.: Convolutional lstm network: A machine learning approach for precipitation nowcasting. Advances in Neural Information Processing Systems (NIPS) (2015)
32. Rolfe, J.T., LeCun, Y.: Discriminative recurrent sparse auto-encoders. arXiv preprint arXiv:1301.3775 (2013)
33. Baker, S., Scharstein, D., Lewis, J., Roth, S., Black, M.J., Szeliski, R.: A database and evaluation methodology for optical flow. Int. J. Comput. Vis. **92**, 1–31 (2011)
34. Geiger, A., Lenz, P., Urtasun, R.: Are we ready for autonomous driving? the KITTI vision benchmark suite. In: IEEE Conference on Computer Vision and Pattern Recognition (CVPR) (2012)
35. Butler, D.J., Wulff, J., Stanley, G.B., Black, M.J.: A naturalistic open source movie for optical flow evaluation. In: Fitzgibbon, A., Lazebnik, S., Perona, P., Sato, Y., Schmid, C. (eds.) ECCV 2012. LNCS, vol. 7577, pp. 611–625. Springer, Heidelberg (2012). doi:10.1007/978-3-642-33783-3_44
36. Jia, Y., Shelhamer, E., Donahue, J., Karayev, S., Long, J., Girshick, R., Guadarrama, S., Darrell, T.: Caffe: Convolutional architecture for fast feature embedding. arXiv preprint arXiv:1408.5093 (2014)
37. Bao, L., Yang, Q., Jin, H.: Fast edge-preserving patchmatch for large displacement optical flow. In: IEEE Conference on Computer Vision and Pattern Recognition (CVPR), pp. 3534–3541 (2014)

Language and Video

Auto-Illustrating Poems and Songs with Style

Katharina Schwarz[1]([✉]), Tamara L. Berg[2], and Hendrik P.A. Lensch[1]

[1] University of Tübingen, Tübingen, Germany
katharina.schwarz@uni-tuebingen.de
[2] University of North Carolina, Chapel Hill, USA

Abstract. We develop an optimization based framework to automatically illustrate poems and songs. Our method is able to produce both semantically relevant and visually coherent illustrations, all while matching a particular user selected visual style. We demonstrate our method on a selection of 200 popular poems and songs collected from the internet and operate on around 14M Flickr images. A user study evaluates variations on our optimization procedure. Finally, we present two applications, identifying textual style, and automatic music video generation.

1 Introduction

When an artist creates a poem or song they weave their story carefully, selecting words that produce a vivid visual story in our minds. In this work, we explore the goal of automatically illustrating such creative pieces of art with images. As artists compositions are intended to be highly emotional and lyrical, we aim to select images that are aesthetically pleasing and highly stylistic according to the style of the artwork, e.g., we might illustrate a poem about love with images predicted to be "romantic" but a heavy metal song in "horror" style. Although those kind of texts are quite challenging due to their high level of abstraction, they often display beautiful language, lending themselves well to our goal of auto-illustration with style. Additionally, their nice repetitive structure perfectly suits our approach. As such texts often deal with a certain theme, we incorporate the global idea as well as allowing for visual adaptions to content changes.

An overview of our processing pipeline is illustrated in Fig. 2. Given a large collection of annotated internet photos, we would like to auto-illustrate poems and songs with images reflecting a user-specified style. These illustrations should tell a visual story of the text, matching both the specified image style and demonstrating visual coherence between selected images. This is achieved by first selecting images from the collection that match important words from the input story, depicting the specified style. Suitable candidate images are selected in this manner for each text line. Next, a coherent image sequence is generated to illustrate the story by optimizing style scores and consistency between successive images along the text lines. Consistency is measured using a combination of textual and visual coherence scores related to image content and color.

Electronic supplementary material The online version of this chapter (doi:10.1007/978-3-319-54190-7_6) contains supplementary material, which is available to authorized users.

© Springer International Publishing AG 2017
S.-H. Lai et al. (Eds.): ACCV 2016, Part IV, LNCS 10114, pp. 87–103, 2017.
DOI: 10.1007/978-3-319-54190-7_6

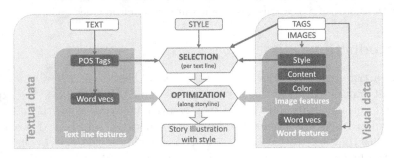

Fig. 1. Automatically generated illustrations of the first text lines (bottom) of the song "The Mamas The Papas - CALIFORNIA DREAMIN" in three different styles (left).

Fig. 2. Overview. Given a large collection of annotated images, we automatically illustrate texts in two steps. First, for each text line, linguistically relevant images are selected that match important words and depict a style. Then, based on this collection, we optimize along the storyline to match style and coherence between selected photos.

In order to select good images for illustration, we extract a number of visual and textual features for comparing text to images. In particular, image tags are used in combination with parsed words and word2vec embedding representations [1] to better match not only the syntactic, but also the semantic meaning of images and text. To encourage semantic similarity of image content between subsequent images, the response vector of a deep neural network pre-trained for image classification is utilized, judging content similarity between two images as the distance between the corresponding representations. Finally, the style of each input picture is predicted based on the work of [2]. We use the predictions for 20 different image styles to tell a picture story in a particular illustration style. Each of these criteria can now be used to both assemble a selection of candidate images per text line and then to optimize for consistency along the story using global discrete energy optimization.

The novel combination of considering textual semantic search, content similarity, style classification, and discrete optimization allows us to generate picture story illustrations with controllable style, even for challenging abstract text types such as poems and song lyrics. Some example outputs of our pipeline for different styles are shown in Fig. 1. The resulting sequences can easily be synchronized with a song to generate a music video.

2   Related Work

Web-scale Photo Collections: Large community photo collections have been successfully exploited for tasks such as scene reconstruction [3,4] or scene completion [5]. However, utilizing search engines to query for content usually leads to noisy results due to the weak nature of associated text on the internet. Methods to identify and distill relevant images from these large unstructured data collections are required [6]. Other work has investigated related problems on large web collections, such as extracting storylines depicted in Flickr images of an event like the fourth of July [7], or using YouTube videos to enhance finding an ordering of a photo stream [8]. The latter also uses the images of a photostream to identify keyframes in a user video.

Images to Text: Recently, many researchers have started to explore the relationship between images and the natural language used to describe this imagery. Automatically creating natural language descriptions from images has been presented in several lines of research [9–15]. [9] exploit statistics from parsing large text corpora as well as visual recognition algorithms and output relevant sentences for images. [10] assemble a large filtered set of Flickr images associated with relevant captions to approach the challenge of generating simple image descriptions. Most recent approaches take advantage of deep learning based models and features [11–14], some with models of attention mechanisms [15].

Text to Images: In the opposite direction, namely starting from a textual description to generate or retrieve a visual representation of language has also been investigated in a number of works. [16,17] provide a graphics engine to render static 3D scenes from natural language descriptions. Whereas [16] allows for user interaction to generate "natural" looking scenes including colors and textures, [17] focuses on automatically resolving spatial relations between 3D objects correctly. An approach for 2D abstract scene generation was presented in [18,19], modeling scenes using clip-art images produced by workers on Amazon's Mechanical Turk. They are able to learn visual interpretations of simple sentences and automatically illustrate short texts. [20] tackles the Text-to-Image co-reference problem, identifying which visual objects a text refers to, exploiting natural sentential descriptions of RGB-D scenes to improve 3D semantic parsing.

Auto-Illustration: Most relevant to our work, several approaches have been proposed to automatically illustrate text using images. Joshi et al. [21] incorporate an unsupervised ranking scheme to select pictures to illustrate a given story. The Text-to-Video pipeline introduced by [22] assembles a relevant image set using a hierarchical algorithm to retrieve Flickr images with high precision to a given text snippet. Images along the text are chosen by considering RGB color information between candidates of neighboring text parts. The method proposed by [23] learns an association between image sequences and multiple sentences.

Style: Similarly to previous auto-illustration approaches, we also retrieve images that match an input text. However, we also optimize for two additional storyline features: (1) we select images for illustration according to a particular story style, and (2) we attempt to select a visually coherent set of images for illustration. In order to illustrate stories according to a particular style, we rely on previous work for style recognition in images [2]. This approach predicts style using features from a pre-trained multi-layer deep network, fine-tuned to predict image style on 80K Flickr photographs depicting 20 different styles.

3   Feature Extraction

In order to support our algorithm we extract a set of features from images and their associated tags as well as from lines of text. The features allow matching both content and style between individual lines of text and images, and between pairs of images selected to illustrate a sequence of text lines.

3.1   Text Features

Language features are extracted by applying **parsing** to determine relevant words from the input text while **image tags** are used directly. **Word-vector representations** are generated for both extracted words and image tags.

Parsing. A line l of text T is first analyzed by tokenizing and parsing to determine part of speech (POS) labels for each word [24] and a lemmatizer based on WordNet [24,25] improves performance. The extracted nouns, verbs, adjectives, and adverbs are gathered in a set $\tau(l)$. POS parsing enables us to select the most relevant words for each matching task. For candidate retrieval, we consider the subset $\tau_{NV}(l) = \{w_1...w_A\} \subseteq \tau(l)$ of extracted nouns and verbs to obtain a broad and large enough image set. Later, the word-vector representations for text lines are computed based on the entire set, $\tau(l)$, to capture additional meaning.

Image Tags. All images considered for illustration have user associated tags. Thus, for each image I we store its associated list of tags $\kappa_I = \{w_1...w_K\}$ in an inverted file table, making it efficient to access all images with tags matching words $\tau_{NV}(l)$ from a text line l.

Word-vector Representation. In addition to directly matching between words, we exploit recent work that maps words to vectors (word2vec) based on a continuous skip-gram model, providing a mapping of phrases into a 300d vector space [1,26]. The mapping keeps and expresses a large number of precise syntactic and semantic word relationships while compressing semantic similarity. For a single word w, we obtain its word2vec representation $V(w)$ and, for a

Table 1. Style in YFCC14M. For all 20 provided styles, the table indicates the amount of images within the YFCC14M subset with $I_{sty} > 50\%$ for a certain style. "Detailed" contains the highest amount and half of the styles hold at least more than 300K images.

Style	> 50%	Style	> 50%	Style	> 50%	Style	> 50%	Style	> 50%
Detailed	674K	Noir	352K	Serene	324K	Melancholy	234K	Sunny	201K
Bright	519K	Hazy	335K	Minimal	303K	Long Exposure	227K	Pastel	184K
Horror	498K	Geom. Comp.	334K	HDR	296K	Vintage	224K	Macro	135K
Depth of field	408K	Bokeh	327K	Romantic	259K	Texture	219K	Ethereal	87K

set of words, we average the vector representations of all words in the set. We calculate word vectors for tags of a text line $\tau(l) = \tau_l$ as V_{τ_l} and for image tags κ_I as V_{κ_I}.

3.2 Image Features

Features are extracted from the images to identify **style**, **content**, and for ensuring **color** consistency between selected images.

Style Feature. Estimates for 20 style-classes are extracted using the method of Karayev et al. [2] which classifies image style using a convolutional-neural-network approach. This estimate is used to consider only images that match the specified style, and to ensure consistency between images selected for illustration. We assume that an image I matches a certain style, sty, if its prediction score for this style is greater than 0.5, defining the style constraint as $I_{sty} > 50\%$.

Image Content Feature. To compare the visual content of two images we make use of deep learning results, in particular, the VGG 16-layer model [27]. This deep convolutional network has demonstrated high accuracy on image classification. We use a pre-trained model which has been trained to recognize 1000 classes from the ImageNet Challenge. For our image content representation we use the 4096d features extracted from the first fully-connected network layer.

Color Feature. As color significantly influences our visual impressions of images and is not well represented using pre-trained CNN features, we incorporate a simple RGB color histogram feature for evaluating image similarity. We extract 256 bins per color channel for each image.

4   Corpora

To demonstrate our approach, we make use of publicly available data, crawling a set of 200 famous poems and song lyrics, and use the YFCC100M dataset.

Creative Text Corpus. We assemble a set of creative texts, namely songs and poems to demonstrate our approach. We obtained lyrics to 100 songs from http://www.songlyrics.com. In order to cover different music styles, we crawled the lyrics from the "Top 100 Songs of All Time" category. For poems, the website http://100.best-poems.net provides best poem texts for various categories. To retrieve a broad set across categories, we downloaded the poems in their "Top 100 Poems" category which covers famous poems of all time.

Image Corpus. For generating illustrations, we make use of the "Yahoo Flickr Creative Commons 100 Million Dataset (YFCC100M)" [28] recently published by [29]. It contains about 100 million Flickr images and videos with associated meta-data. We consider the image portion, around 70% of which have textual tags. We select images that are potentially relevant to our text corpus by filtering out images that are not tagged with relevant words, i.e., nouns and verbs appearing in our creative text corpus. Text pre-processing is used to lemmatize, remove stopwords, and select words that occur in both the poems and songs, leaving us with a representative set of about 400 words. The frequency of the resulting words mentioned in the tags reached up to 820079 for the word "music". For around 60 words more than 100000 images do match. Only Flickr images whose tags match at least one of the words in this representative list are selected, leaving us with a subset of 14 million images. Table 1 indicates the amount in our YFCC14M subset with a prediction greater than 50% for each style, typically yielding several hundred thousand images that are likely to depict that style.

5   Selection and Optimization

Given the feature vectors computed for each database image and the similarity measures defined in Sect. 3, the selection process for auto-illustration first computes a suitable candidate set of images matching to each text line. Then, based on those pre-selected candidate image lists, an optimization step estimates the best sequence of images for illustration, maximizing text and style match scores, as well as cohesion in content, color, and style along the illustration.

5.1   Selecting Candidate Images

For every line in a text, a set of candidate images is selected that semantically matches the text line and corresponds to the specified illustration style. Specifically, given the POS analysis, candidate selection is performed for each text line l by including each image $I \in$ YFCC14M into the set of candidate images I_{cand_l}, if the following condition is fulfilled:

$$(\kappa_I \cap \tau_{NV}(l) \neq \emptyset) \wedge (I_{sty} > 50\%) \Rightarrow (I \in I_{cand_l}) \qquad (1)$$

This means we select all images whose tags match at least one noun or verb present in the text line and, at the same time, depict the requested illustration style with a style prediction score greater than 50%. Sorting this list according to style score, the top 1000 candidate images per text line form a suitable basis for our optimization phase.

5.2 Storyline Optimization

Given the candidate image sets for each text line, we would like to select a final set of images to illustrate our story that are: (1) semantic relevant to the story, (2) good representatives of the selected style, and (3) visually coherent along the story line. Figure 3 visualizes the underlying structure for two successive text parts, each connected with a small subset of potential images for selection. The task of choosing the best image from each subset, while preserving semantic relatedness and visual consistency can be described as a discrete optimization problem with pairwise variables and formulated as an energy with unary and pairwise terms.

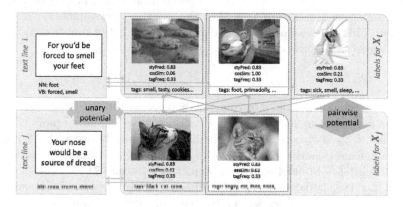

Fig. 3. Optimization structure for two text lines. The images in the middle are finally picked. ("J. Prelutsky - BE GLAD YOUR NOSE YOUR FACE", "depth of field")

Thus, from an assembled and presorted set of images per text line, the goal is to select one image for each position, X_i, of the corresponding text line $i \in \nu$. We formulate this as minimization of an energy function E with image labels for a text line i along all lines ν (nodes) and over consecutive text pairs ε (edges):

$$E(X) = \sum_{i \in \nu} U(X_i) + \sum_{i,j \in \varepsilon} P(X_i, X_j) \tag{2}$$

The unary potential function U measures semantic and stylistic relatedness between a text line and a potential image for illustrating that line. The pairwise consistency terms P describe the interaction potential between pairs of images.

Unary terms: Two types of unary terms measure semantic (text) relatedness between the text lines and images, s_{freq} and s_{sem}, combined in a weighted sum (Eq. 3). In order to turn the similarity into a cost for the minimization, we calculate $1 - \sum weightedUnaryTerms$.

$$U(X_i) = 1 - (\lambda_1 s_{freq}(x_i) + \lambda_2 s_{sem}(x_i)) \tag{3}$$

- **Tag frequency.** s_{freq} computes the overlap between all nouns and verbs extracted from the text line and the tags associated with an image:

$$s_{freq} = \frac{1}{A} \sum_{a \in A} \xi_a, \text{ with } \xi_a = \begin{cases} 1, & w_a \in \tau_{NV} \text{ occurs in } \kappa_I, \forall a \in A \\ 0, & else \end{cases}$$

- **Semantic.** s_{sem} calculates the word2vec similarity between the text line words τ and the image tags κ using cosine similarity between the average representation vectors V_τ and V_κ as $s_{sem} = \text{cossim}(V_\tau, V_\kappa) = \frac{V_\tau \cdot V_\kappa}{\|V_\tau\|\|V_\kappa\|}$. This allows for similarity comparisons beyond exact word matches.

Pairwise terms: Between all possible candidate image pairs for each successive text line, a pairwise energy term is computed that is minimized to obtain a globally consistent illustration such that the storyline flows smoothly along the illustration. This pairwise potential is defined as a weighted sum (Eq. 4) of three types of consistency: style d_{sty}, color d_{col}, and content d_{cont}.

$$P(X_i, X_j) = \mu_1 d_{sty}(x_i, x_j) + \mu_2 d_{cont}(x_i, x_j) + \mu_3 d_{col}(x_i, x_j) \qquad (4)$$

- **Style.** d_{sty} computes image to image coherence as the normalized Euclidean distance between the 20d style vectors of successive candidate pairs.
- **Content.** d_{cont} is obtained by the Euclidean distance between the 12 normalized CNN feature activation vectors to encourage smoothness between what successive images in our illustration depict (Figs. 4 and 7).
- **Color.** d_{col} is calculated as the Euclidean distance between RGB color histograms computed between successive pairs of candidate images.

To minimize E, an NP-hard problem, we use the "sequential tree-reweighted message passing algorithm" (TRW-S) proposed by [30] whose main property is that the value of the bound is guaranteed not to decrease and, thus, at least a "local" maximum of the bound is retrieved. The weights of the parameter sets $w_U = (\lambda_1|\lambda_2)$, $w_P = (\mu_1|\mu_2|\mu_3)$ will be discussed in the following section.

6   Human Evaluation

Aligning a story with appropriate images in a pleasant style is a subjective task, especially for abstract texts such as poems and songs. Thus, in order to obtain more general ratings from a wide variety of people, we performed experiments on Amazon Mechanical Turk (AMT) to measure the quality of our method. First, we tease apart the relative contributions from various pieces of our system. We perform an experiment on the semantic connection between text and images, the unary terms, and evaluate the pairwise terms which regulate visual consistency along image sequences. Finally, we validate the quality of resulting illustrations.

6.1   Experiment Data Set

Our experiments are designed based on the assembled data described in Sect. 4, consisting of about 200 creative texts. In total, 20 styles can be used for illustration. We randomly select 110 text-style combinations consisting of half poems and half song lyrics and ensure that every style is represented. Due to the style constraint, some text lines may result in only a few or even no image responses for a requested style. Thus, we only use text lines l with enough image candidates ($\#I_{cand_l} > 1000$) to optimize over 1000 image labels and at least one word present in the pre-trained Google word2vec representation ($\exists V_{\tau_l}$), ensuring that we can perform a proper parameter set evaluation. Finally, to evaluate visual consistency along an image sequence, the number of image responses for each of the succeeding text lines should also be large enough. Thus, we only accept consecutive text parts T_q with M succeeding lines $T_q = \{l_{q_1}...l_{q_M}\}$ such that all consecutive lines l_{q_m} fulfill the word2vec and candidate set requirements.

Table 2. Examples from AMT user studies. Exp. 1 (left): Evaluation of semantic text-to-image relation. Exp. 2 (right): Two example pairs of image sequences to evaluate visual consistency. Exp. 3 (bottom row): Evaluation of text stream illustrations.

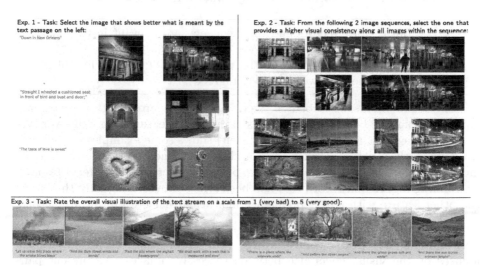

6.2   User Study

People may have different internal rating systems, especially for subjective tasks. Thus, to measure relative contributions, we formulate our first experiments (Exp. 1, Exp. 2) as binary forced-choice tasks. Each pairwise preference test is designed as a data-pair selected by two parameter sets controlling different portions of the features contributing to the optimization and is presented to 5 Turkers. Depending on the type of study, data within a pair either consists of two images compared to a text line or two image sequences for evaluating visual

Table 3. User study results. Different style groups S are identified due similar impact. S_{cont}: higher impact of content than color (e.g. "sunny", "hazy"), S_{col}: styles largely connected to color (e.g. "noir", "vintage"), S_{abst}: abstract styles (e.g. "minimal").

	Exp.1: Unary contributions			Exp.2: Pairwise contributions			
	tagFreq < wordVec	wordVec < all	tagFreq < all	col < cont	cont < sty	col < sty	noPE < allPE
S_{cont}	68%	60%	78%	85%	58%	73%	71%
S_{col}	54%	63%	60%	26%	71%	55%	74%
S_{abst}	69%	60%	80%	84%	58%	60%	47%

consistency. Randomized ordering and positioning are used to negate click biases. Tasks resulting in rather unclear (2–3)-decisions out of 5 Turkers are filtered out afterwards as they are not suitable to detect trends. Based on the derived best parameter settings, we let Turkers rate the quality of final illustrations along the according text streams in a third experiment (Exp. 3) on a 5 pt Likert scale.

Experiment 1: Text-to-image semantic. Our first experiment evaluates the semantic connection between text and images, represented by the unary terms in the optimization. We formulate our hypotheses $H_{T \leftrightarrow I}$ as:

$H_{T \leftrightarrow I}$: – Both, word vectors and tag frequency are relevant in the unary terms. (tagFreq > 0, wordVec > 0)
  – The positive influence of the word vectors is higher than of the tag frequency (wordVec >= tagFreq).

The requirements described in Sect. 6.1 result in around 2110 text-image pairs. We compare binary contributions of the unary features, e.g. only wordVec $w_U = (0|1)$ against all in $(1|1)$. Table 2 (left) shows some examples of the tasks we gave to the Turkers consisting of a short text line and 2 images. Overall, including only the wordVecs has been preferred over only tagFreq (Table 3). Combining tagFreq and wordVecs has been selected over using only the one or the other feature with a 67% preference indicating that both terms are needed.

Experiment 2: Consistency along storyline. Our second experiment focuses on the visual coherence between successive images. We present pairs of sequences containing 4 images per stream to provide a reasonable evaluation set, and, without the underlying text lines to focus on the visual coherence. We evaluate the contribution of our visual features, the pairwise terms, to the optimization, formulating the hypotheses $H_{I_{seq}}$ as:

$H_{I_{seq}}$: – All three features are necessary. (style, content, color > 0)
  – Highest preference results are retrieved for relation:
    color $\lessgtr$ content < style

| The sea is calm tonight. | The tide is full, the moon lies fair | Upon the straits; on the French coast the light | Gleams and is gone; the cliffs of England stand, | Glimmering and vast, out in the tranquil bay. | Come to the window, sweet is the night-air! |

Fig. 4. Poem "M. Arnold - DOVER BEACH" in: "sunny" (top), "minimal" (bottom).

The constraints described in Sect. 6.1 lead to a dataset of about 1000 image sequence pairs. Based on the outcome of Exp. 1, we set $w_U = (1|1)$ and compare binary contributions between the pairwise feature terms to retrieve relations between them. Table 2 (right) shows some tested sequences. Results are shown in Table 3. Style was always preferred over the other features to ensure coherence. In this experiment, we identified different groups S of styles. Styles like "sunny" profit from a higher contribution of content than color (S_{cont}, 85% preference). Other styles, e.g. "noir" largely depend on color being preferred 74% over content (S_{col}). Rather abstract styles like "minimal" are not as suitable for auto-illustration as $(0|0|0)$ was preferred 53% over $(1|1|1)$ (S_{abst}).

The results of the study demonstrate the importance of distinguishing between certain styles. Thus, based on the performed binary experiments and the obtained relations, we experimentally obtained different parameter sets relating the proportions of the features for different groups of styles and, similar to experiment 2, tested them against all weights set to 1. Most of the styles worked best for a weight set of $w_U = (.8|1), w_P = (1|.5|.2)$, e.g. "hazy" 80%. Contrarily, for "horror" $(1|1|1)$ was preferred in average 92% over partial combinations. However, very color depending styles worked better combining the features with $w_U = (.8|1), w_P = (1|.2|.5)$, thus setting color > content, e.g. "noir" was preferred 90% over all set to 1 and "vintage" 80%.

Experiment 3: Text illustration. Based on the previously derived parameter settings, we let Turkers rate the quality on a subset of 45 illustrations along text lines on a 5 pt Likert scale from 1 (very bad) to 5 (very good). The subjective outcome of our system makes it challenging to obtain scores rating the overall quality. However, for many styles the resulting mean μ was around 4 indicating good quality, e.g., "long exposure" $\mu = 4.0$ ($\sigma = .91$), "noir" $\mu = 3.9$ ($\sigma = 1.07$). We tended to find that results in the style group G_{cont} had stronger decisions (smaller σ-values). "Minimal" only obtained a top-two box acceptance of 37%. However, non-abstract styles obtained top-two box acceptance rates between 70% and 80%, indicating high acceptance of our results, e.g., "sunny" 75%.

7   Results and Discussion

The main challenge of our approach is to balance semantic relevance with producing an illustration that both depicts the requested style and demonstrates strong visual coherence along the illustration. Figure 5 provides a consistent visual appearance of the style "noir" while preserving the meaning of the underlying text lines, even distinguishing between the raven "croaking" and "quoting".

The weak nature of tags and polysemy makes this problem highly challenging, e.g., in Fig. 6, all images are tagged with "cloud" although the last image does not show a cloud. However, sometimes our method works surprisingly well, e.g., in Table 2 (left) the text "The taste of love is sweet" shows an amazing result with our features, providing an idea of taste. Overall, the nice repetitive structure of creative texts allows us to search an image for each text line instead of forcing text splits that can lead to wrongly combined words. Additionally, restricting the candidates per text line by the style prediction attribute makes it even more challenging to create a semantically relevant illustration. In Fig. 1, the line "Well, I get down on my knees and pretend to pray" works nicely for the first and last row with the styles "hdr" and "serene", but provides a rather strange X-ray image of a knee for "noir" as there was no better candidate available. Overall, as shown in Sect. 6.2, the way we combine our textual features generally supports selecting suitable images illustrating the meaning of the text.

Further, recurring lines in poems as well as refrains in songs often serve as stylistic device to strengthen the main message by repetition. Thus, if similar content is described in text lines, the images should provide similar or even identical visualizations, e.g., Fig. 7 visualizes recurring images for lines with similar descriptions. Often, such creative texts deal with a main theme. We capture its global idea by first providing pre-selection of similar candidate lists by recur-

| "Surely," said I, "surely that is something at my window lattice" | Ghastly grim and ancient raven wandering from the Nightly shore | Respite - respite and nepenthe, from thy memories of Lenore! | What this grim, ungainly, ghastly, gaunt and ominous bird of yore | Meant in croaking "Nevermore." | It shall clasp a sainted maiden whom the angels name Lenore | Quoth the Raven, "Nevermore." |

Fig. 5. Poem "E.A. Poe - THE RAVEN" illustrated in style "noir". Note "nevermore" presented in form of a raven and the different selections for "croaking" and "quoting".

Fig. 6. Poem "W. - I WANDERED LONELY CLOUD". The text line "I wandered lonely as a cloud" is presented in different illustration styles (horror, sunny, noir, bright, depth of field, geom. composition, texture). All image tags include the word "cloud".

Fig. 7. Song "J. Cash - RING OF FIRE" illustrated with the styles "long exposure" (top) and "sunny" (bottom). Recurrence is provided for similar texts parts.

Table 4. Example for candidate lists along part of song "The doors - LIGHT MY FIRE" in illustration style "long exposure". Left: succeeding text lines. Second column: selected optimization result. Right: candidate images per text line.

rence, and, then, the content feature selects images with highest content similarity along the storyline. Table 4 shows an example of candidate lists sorted by highest style prediction (right) along succeeding lines (left) and estimated optimization results (middle). We can observe that the images are selected properly with high coherence in their visual appearance and style along the sequence.

Overall, not every style is similarly suitable to illustrate text. In our user study (Sect. 6.2) we identified different style groups and retrieved lowest acceptance rate for abstract styles. However, even for such styles we were able to retrieve nice results, e.g., in Fig. 4 for the abstract style "minimal". Additional examples for different styles are shown in Fig. 7 or 9. Please see the supplemental material for longer and more extensive illustration results.

Finally, rather than restricting our framework to creative texts only, we enable the input of new text data of arbitrary type and length. For a selected style, the text is then parsed and illustrated visually consistent by our system with one image per text line out of the YFCC14M images.

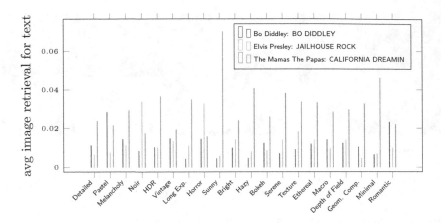

Fig. 8. Examples for average style-image-responses for some song lyrics for all 20 styles. Highest peaks indicate main moods of the text.

Fig. 9. Song "B. Diddley: BO DIDDLEY". Highest average style-image-responses are obtained for the styles "pastel" (top) and "romantic" (bottom).

8   Applications

Finally, our system can enable applications such as identifying the style of a text using our candidate selection process or automatically generating a music video by illustrating complete songs in different styles.

Text Style from Candidate Selection. Restricting the number of retrieved images for text lines by the style constraint $I_{sty} > 50\%$ (Sect. 5.1) leads to interesting insights about the connection between text and styles. We calculated the average number of retrieved images for the text lines in a story relative to the available amount of images for a certain style $\#I_{sty}$ in the YFCC14M subset. Thus, for a story text T and a style sty, the average style-image-response T_{sty} is calculated as shown in Eq. 5. Note, that only text lines $l \in L$ with number of candidate images $\#I_{cand_l} > 0$ are considered.

$$T_{sty} = I_{ret}(sty) = \frac{\sum_{l \in L} I_{cand_l}(sty)}{\#l} \cdot \frac{1}{\#I_{sty}}, \quad \text{with } \forall l \in L, \#I_{cand_l} > 0 \quad (5)$$

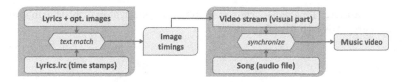

Fig. 10. Music video generation pipeline.

Figure 8 demonstrates this idea, showing the resulting average style-image-response of stories for all provided styles. Some texts have a peak in one or two styles, e.g., "The Mamas The Papas: CALIFORNIA DREAMIN" in "sunny". Others even have peaks in connected styles, e.g., "E. Presley JAILHOUSE ROCK" in "noir" + "horror" or "B. Diddley: BO DIDDLEY" (Fig. 9) in "pastel" + "romantic". Interestingly, these styles do seem to indicate the main moods of the texts as they often work better as styles with a lower image-style-response.

Music Video Generation. Further, we enable the generation of simple music videos. An overview is given in Fig. 10. For that purpose, our system outputs a file listing the song lyrics and selected images links. Additionally, the audio version of the song and its ".lrc"-file are needed. The LRC format [31], is usually used in karaoke to align song lyrics with the music. The structure simply consists of the text lines of a song and its associated time-stamps. Optionally, additional ID tags indicating artist and song meta information might be attached.

For the music video generation, we start by matching the text lines between the ".lrc"-file and our "lyrics2images"-file. To be robust against spelling errors, we compare the stems of the words text line to obtain the time-stamps for the images. Our tool converts them into duration timings for each image. A video stream is then simply generated by displaying the images in their proposed ordering and duration timings similar to a slide show. This video stream is joined together with the audio file resulting in a music video. The synchronization is already provided by the image timings. However, beat detection could improve synchronization in future work.

9   Conclusion

Given a preferred style the presented pipeline automatically generates an illustration for a poem or song. Our framework optimizes both semantic relevance and visual coherence while selecting images, exploiting recent advances in convolutional neural networks for image and style classification. The generated sequences have been evaluated in a user study, indicating that combining multiple features improves over simpler image selection processes and obtaining highest acceptance rates for non-abstract styles. We also demonstrate applications to story style classification and music video generation.

Acknowledgement. This work originated from a research stay of Katharina Schwarz at the University of North Carolina (UNC).

References

1. Mikolov, T., Sutskever, I., Chen, K., Corrado, G.S., Dean, J.: Distributed representations of words and phrases and their compositionality. In: NIPS, pp. 3111–3119 (2013)
2. Karayev, S., Trentacoste, M., Han, H., Agarwala, A., Darrell, T., Hertzmann, A., Winnemoeller, H.: Recognizing image style. In: BMVC (2014)
3. Snavely, K.N.: Scene reconstruction and visualization from internet photo collections. PhD thesis, University of Washington (2009)
4. Frahm, J.-M., et al.: Building Rome on a cloudless day. In: Daniilidis, K., Maragos, P., Paragios, N. (eds.) ECCV 2010. LNCS, vol. 6314, pp. 368–381. Springer, Heidelberg (2010). doi:10.1007/978-3-642-15561-1_27
5. Hays, J., Efros, A.A.: Scene completion using millions of photographs. In: ACM SIGGRAPH (2007)
6. Averbuch-Elor, H., Wang, Y., Qian, Y., Gong, M., Kopf, J., Zhang, H., Cohen-Or, D.: Distilled collections from textual image queries. Comput. Graph. Forum **34**(2), 131–142 (2015)
7. Kim, G., Xing, E.P.: Reconstructing storyline graphs for image recommendation from web community photos. In: CVPR, pp. 3882–3889 (2014)
8. Kim, G., Sigal, L., Xing, E.P.: Joint summarization of large-scale collections of web images and videos for storyline reconstruction. In: CVPR, pp. 4225–4232 (2014)
9. Kulkarni, G., Premraj, V., Dhar, S., Li, S., Choi, Y., Berg, A.C., Berg, T.L.: Baby talk: understanding and generating image descriptions. In: CVPR, pp. 1601–1608 (2011)
10. Ordonez, V., Kulkarni, G., Berg, T.L.: Im2Text: describing images using 1 million captioned photographs. In: NIPS, pp. 1143–1151 (2011)
11. Karpathy, A., Li, F.F.: Deep visual-semantic alignments for generating image descriptions. In: CVPR, pp. 3128–3137 (2015)
12. Vinyals, O., Toshev, A., Bengio, S., Erhan, D.: Show and tell: a neural image caption generator. In: CVPR, pp. 3156–3164 (2015)
13. Devlin, J., Cheng, H., Fang, H., Gupta, S., Deng, L., He, X., Zweig, G., Mitchell, M.: Language models for image captioning: the quirks and what works. In: ACL, pp. 100–105 (2015)
14. Fang, H., Gupta, S., Iandola, F., Srivastava, R.K., Deng, L., Dollar, P., Gao, J., He, X., Mitchell, M., Platt, J.C., Zitnick, C.L., Zweig, G.: From captions to visual concepts and back. In: CVPR, pp. 1473–1482 (2015)
15. Xu, K., Ba, J., Kiros, R., Cho, K., Courville, A.C., Salakhutdinov, R., Zemel, R.S., Bengio, Y.: Show, attend and tell: neural image caption generation with visual attention. In: ICML, pp. 2048–2057 (2015)
16. Coyne, B., Sproat, R.: WordsEye: an automatic text-to-scene conversion system. In: SIGGRAPH, pp. 487–496. ACM (2001)
17. Spika, C., Schwarz, K., Dammertz, H., Lensch, H.P.A.: AVDT - automatic visualization of descriptive texts. In: VMV, pp. 129–136 (2011)
18. Zitnick, C.L., Parikh, D.: Bringing semantics into focus using visual abstraction. In: CVPR, pp. 3009–3016 (2013)
19. Zitnick, C.L., Parikh, D., Vanderwende, L.: Learning the visual interpretation of sentences. In: ICCV, pp. 1681–1688 (2013)

20. Kong, C., Lin, D., Bansal, M., Urtasun, R., Fidler, S.: What are you talking about? Text-to-image coreference. In: CVPR, pp. 3558–3565 (2014)
21. Joshi, D., Wang, J.Z., Li, J.: The story picturing engine–a system for automatic text illustration. TOMCCAP **2**(1), 68–89 (2006)
22. Schwarz, K., Rojtberg, P., Caspar, J., Gurevych, I., Goesele, M., Lensch, H.P.A.: Text-to-video: story illustration from online photo collections. In: Setchi, R., Jordanov, I., Howlett, R.J., Jain, L.C. (eds.) KES 2010. LNCS (LNAI), vol. 6279, pp. 402–409. Springer, Heidelberg (2010). doi:10.1007/978-3-642-15384-6_43
23. Kim, G., Moon, S., Sigal, L.: Ranking and retrieval of image sequences from multiple paragraph queries. In: CVPR, pp. 1993–2001 (2015)
24. Bird, S., Klein, E., Loper, E.: Natural Language Processing with Python, 1st edn. O'Reilly Media, Inc., Sebastopol (2009)
25. Fellbaum, C. (ed.): WordNet: An Electronic Lexical Database. MIT Press, Cambridge (1998)
26. Mikolov, T., Chen, K., Corrado, G., Dean, J.: Efficient Estimation of Word Representations in Vector Space. CoRR (2013)
27. Simonyan, K., Zisserman, A.: Very Deep Convolutional Networks for Large-Scale Image Recognition. CoRR (2014)
28. Thomee, B.: Yahoo! Webscope dataset YFCC-100M (2014). http://labs.yahoo.com/Academic_Relations
29. Thomee, B., Shamma, D.A., Friedland, G., Elizalde, B., Ni, K., Poland, D., Borth, D., Li, L.: The New Data and New Challenges in Multimedia Research. CoRR (2015)
30. Kolmogorov, V.: Convergent tree-reweighted message passing for energy minimization. IEEE Trans. Pattern Anal. Mach. Intell. **28**(10), 1568–1583 (2006)
31. Shiang-shiang, K.D.: Information about LRC (2012). http://www.mobile-mir.com/en/HowToLRC.php

Spatio-Temporal Attention Models
for Grounded Video Captioning

Mihai Zanfir[2], Elisabeta Marinoiu[2], and Cristian Sminchisescu[1,2(✉)]

[1] Department of Mathematics, Faculty of Engineering, Lund University,
Lund, Sweden
cristian.sminchisescu@math.lth.se
[2] Institute of Mathematics of the Romanian Academy, Bucharest, Romania
{mihai.zanfir,elisabeta.marinoiu}@imar.ro

Abstract. Automatic video captioning is challenging due to the complex interactions in dynamic real scenes. A comprehensive system would ultimately localize and track the objects, actions and interactions present in a video and generate a description that relies on temporal localization in order to ground the visual concepts. However, most existing automatic video captioning systems map from raw video data to high level textual description, bypassing localization and recognition, thus discarding potentially valuable information for content localization and generalization. In this work we present an automatic video captioning model that combines spatio-temporal attention and image classification by means of deep neural network structures based on long short-term memory. The resulting system is demonstrated to produce state-of-the-art results in the standard YouTube captioning benchmark while also offering the advantage of localizing the visual concepts (subjects, verbs, objects), with no grounding supervision, over space and time.

1   Introduction

In this work, we consider the problem of automatic video captioning, where given an input video, a learned model should describe its content with one or more sentences. This is important considering the increasing rate at which multimedia content is uploaded on the Internet, which in turn requires automatic understanding and description for the retrieval of meaningful content. Automatic video captioning would also be beneficial to human-computer interaction, surveillance and monitoring, and as an aid to the blind and visually-impaired.

However, the problem of translating from the visual domain to a textual one is challenging, as it ideally involves understanding the key actors, objects and their interaction in the scene, followed by the construction of a both semantically and grammatically correct natural language description. The first part is made difficult by the large number of semantic categories, which exhibit a high interclass variability. Objects may be of different sizes, shapes and colors, or only partially visible. The lack of available, rich annotated data, and the absence

---

M. Zanfir and E. Marinoiu—Authors contributed equally.

© Springer International Publishing AG 2017
S.-H. Lai et al. (Eds.): ACCV 2016, Part IV, LNCS 10114, pp. 104–119, 2017.
DOI: 10.1007/978-3-319-54190-7_7

of localization information for the key elements present in a video makes the problem even harder. Data is difficult to collect due to the tedious and time-consuming process of annotating individual video frames. Progress has been made in the related problem of automatic image captioning, where annotations are plentiful [1,2]. Even with a complete understanding of the video content, there still remains the problem of delivering a sufficiently relevant digest, at different levels of abstraction, required by specific tasks. A bird enthusiast may require a specific video to be described as "Sudan golden sparrows are bathing in water", whereas a regular person may be satisfied with a simpler description like "Birds playing in water".

Very recent work has focused on attention mechanisms that ground textual elements into the video timeline [3,4]. However, the visual domain is usually represented only at a coarse frame level without explicitly revealing the spatio-temporal structure pertaining to a textual element. We believe that a video captioning model can benefit from the work in spatio-temporal segmentation in video [5–7], that could offer plausible proposals for localization of the textual elements. Our contributions can be listed as follows: (1) an attention mechanism that links textual elements to spatio-temporal object proposals and is able to provide localized visual support of the words in the generated sentence, with no grounding supervision, (2) integration of high-level semantic representations obtained both from classifiers learned on YouTube dataset [8] (subjects, verbs, objects) and from pre-trained models with state-of-the-art recurrent neural networks, and (3) competitive or better than state-of-the-art results on three different metrics on the challenging YouTube video description dataset. An overview of our model is given in Fig. 1. Illustrations of the detailed textual and visual output produced by our model appear in Fig. 5.

2   Related Work

Previous approaches to video and image captioning follow broadly two main lines of work: (1) intermediate concept prediction in the form of subject, verb, object or place (S,V,O,P) followed by a template sentence generation step, or (2) full sentence generation using recurrent neural networks, mainly long short-term memory units (LSTMs).

Concept Discovery. Earlier work on video captioning has attempted to first detect a subject, a verb and an object for each video and then form a sentence using a template model and a learned language model. In [9] the authors first mine (S,V,O) triplets from the video descriptions. Then, they learn a semantic hierarchy over subjects, verbs and objects and use a multi-channel SVM to predict an (S,V,O) triplet over the learned hierarchies by trading-off specificity and semantic similarity. Once an (S,V,O) triplet is obtained for each video, a sentence is formed using a template-based approach. In [10] a factor graph model is proposed to combine visual detections with language statistics in order to learn an (S,V,O,P) tuple for each video. The sentence generation step is similar to that of [9]. In [11] the authors propose a framework to jointly model language and vision. The language model is a compositional one, learned over (S,V,O) triplets while the vision model is a two-layer neural network built on top of deep features. Treating concept discovery and sentence generation in separate steps has the advantage of solving two potentially easier and better specified problems instead of a harder, less constrained one. However, the downside is that the resulting sentences can be rigid and may fail to capture the richness of human descriptions.

Recurrent Networks for Image and Video Captioning. Inspired by the recent success of recurrent neural networks in automatic language translation [12,13], a series of papers made use of similar models in "translating" from a visual input to a textual output where the visual information is usually encoded using convolutional neural networks (CNN). In the case of image captioning, significant progress has been made in recent work [14–17]. The authors in [15] use an attention mechanism on top of a CNN and extract features from a lower convolutional layer in order to obtain correspondences between the feature vectors and regions of the 2D image. [16] use external region proposals and learn an alignment model between image regions and sentences. In [17], using a rich annotated dataset [2] of image regions and corresponding textual descriptions, the problem of localization and description is addressed jointly. They propose a fully convolutional localization network (FCLN) for dense captioning.

In video captioning, the authors of [18] use a stack of two recurrent sequence models (LSTMs) to tackle the problems of activity recognition, image and video description. For image description, features extracted from a pre-trained CNN are fed directly into the LSTMs, while in the case of video description, first, a CRF model is used to obtain a distribution over subjects, verbs and objects. Then, the CRF responses are fed to the LSTMs to form a full sentence. Similarly, [19] uses a two stack LSTMs model for video description, but the visual information is encoded as mean-pooled CNN feature over the video frames. They also show improvement by transfer learning from the image domain (where more training data is available) to video. Approaches have also been pursued using semantic classifiers responses for subjects, verbs and places in combination with LSTMs [20]. Despite the fact that such approaches have achieved a significant improvement (under BLEU and METEOR metrics) compared to previous template based approaches, they do not fully exploit the underlying structure of the video, nor do they attempt to explicitly identify (localize) the main actors or objects that correspond to the textual descriptions produced. [4] incorporate a spatio-temporal 3D CNN trained on video action recognition and use a temporal attention mechanism to select the most relevant temporal segments. [3] are interested in generating multiple sentences and discuss temporal and spatial attention mechanisms. The spatial elements are obtained by sampling image patches around a central actor, on datasets where this assumption holds. They use deep convolutional features like VGG [21] and C3D [22] to represent an image frame. The purpose of these frame-level attention approaches is to guide the model towards different frames of the video at each time step (when a word is produced). Unlike these, our attention model focuses on a pool of spatio-temporal proposals and learns to choose the best spatial-temporal support for every word of the sentence.

3   Methodology

Our approach to video captioning has two main components: first revealing the spatio-temporal visual support of words in video and then guiding the sentence generation process by including semantic information in the learning process. We integrate a soft-attention mechanism, operating over a pool of spatio-temporal proposals, into a state-of-the-art recurrent network. The joint model learns to produce semantically meaningful sentences while attending to different parts of the video. The semantic information is obtained in two ways: (a) by learning to predict subjects, verbs and objects (S,V,O) and (b) by using pre-trained state-of-the-art image classification and object detection models. An overview of our modeling and computational pipeline is shown in Fig. 1. Section 3.1 briefly introduces the recurrent model while Sect. 3.2 describes the attention mechanism. Learning the semantic concepts is explained in Sect. 3.3 and the experimental details and results are given in Sect. 4.

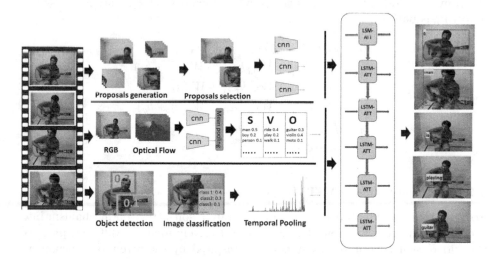

Fig. 1. Overview of our approach. We build an attention mechanism on top of spatio-temporal object proposals and integrate it with state-of-the-art image classifiers, object detectors and recurrent neural networks (LSTM). The image classifiers together with learned high-level semantic features in the form of (Subject, Verb, Object) are provided as contextual features to the Attention-LSTM. Our model is able to visually ground each of the words from the sentence it generates, spatially and temporally, with no additional supervision.

3.1   Recurrent Networks for Video Captioning

Recurrent Neural Networks (RNNs) make use of sequential information and learn temporal dynamics by mapping a sequence of inputs to hidden states and then learn to decode the hidden states into a series of outputs. Their major drawback, however, is the *vanishing gradient* which makes it difficult to learn long-range dependencies that exist in the input sequences [23]. A solution to this issue is to incorporate explicit unit memories, controlled by gates deciding at each step which information should be passed on and which one should be forgotten. Those units, known as long short-term memory (LSTM) units [24], have proven to perform well for machine translation [12] and have recently been used for both image [15,18] and video captioning [19,25,26]. A schema of the LSTM unit (introduced in [27]) used in our experiments is shown in Fig. 2. The LSTM unit consists of a memory cell, c_t, that encodes the information transmitted from previous units up to current step and *gates* deciding how the information in the memory cell is updated and what the output should be. The input i_t, forget f_t, and output o_t gates are sigmoid functions that decide how much to consider from the current input(u_t), what to retain from the previous cell memory (c_{t-1}) and how much information from the memory cell to be transferred to the hidden state (h_t). The updates at time step t given textual input u_t, visual input z_t, the previous hidden state h_t and the previous memory cell c_{t-1} are given by the following equations:

$$
\begin{aligned}
i_t &= \sigma(W_{xi}u_t + W_{hi}h_{t-1} + W_{zi}z_t + b_i) \\
f_t &= \sigma(W_{xf}u_t + W_{hf}h_{t-1} + W_{zf}z_t + b_f) \\
o_t &= \sigma(W_{xo}u_t + W_{ho}h_{t-1} + W_{zo}z_t + b_o) \\
g_t &= \phi(W_{xg}u_t + W_{hg}h_{t-1} + W_{zg}z_t + b_g) \\
c_t &= f_t \odot c_{t-1} + i_t \odot g_t \\
h_t &= o_t \odot \phi(c_t)
\end{aligned}
\tag{1}
$$

3.2   Attention-Based LSTM

Soft-Attention Mechanism. We incorporate a soft-attention mechanism into the LSTM in order to allow the model to selectively focus on different parts of the video each time it produces a word. Inspired by the attention mechanism that exploits the spatial layout of an image [15], a few recent methods have attempted to exploit the temporal structure of the video by learning how to assign different weights to frames in a video sequence [3,4]. These methods, however, do not localize objects in images, and thus the same frame can offer support to very different words in the output. This approach can work when a video selectively focuses on individual objects at a time and thus in a single frame very few objects of interest are present. However, in most videos, there are multiple actors and objects present in a frame. To address this problem, we allow the sequence model (LSTM) to choose where to focus among a pool of coherent spatio-temporal proposals at each time step t. Thus, the model is able to indicate what is the *localized* visual support used to produce a particular word from the video description.

Considering $P = [p_1, p_2, \ldots, p_m]$ the temporal feature vector where m is the number of proposals and p_i the descriptor for the i-th proposal, the LSTM learns at each time step a series of m weights β_{ti} such that the final encoding of the visual input is

$$z_t = \sum_{i=1}^{m} \beta_{ti} p_i, \quad \text{with} \quad \sum_{i=1}^{m} \beta_{ti} = 1 \qquad (2)$$

The weights β_{ti} represent the importance of the i-th proposal for generating the word at the current time step, given the previously generated words. First, for each proposal we learn a score ϵ_{ti} based on its visual feature p_i and the previous hidden state h_{t-1}, which is given by

$$\epsilon_{ti} = W_{ph}\phi(W_p p_i + W_h h_{t-1} + b_{ph}) \qquad (3)$$

where ϕ is the hyperbolic tangent and W_{ph}, W_p, W_h, b_{ph} are parameters to be learned. Those scores are then normalized to obtain the β_{ti} weights:

$$\beta_{ti} = \frac{e^{\epsilon_{ti}}}{\sum_{i=1}^{m} e^{\epsilon_{ti}}} \qquad (4)$$

A schematic view of the Attention-LSTM model is shown in Fig. 2. At testing time, by inspecting the weights in decreasing order, we can interpret the visual information preferred (selected) by the model when producing a particular word in the textual description of the video.

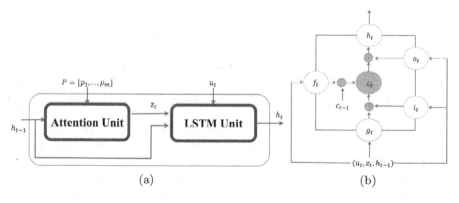

Fig. 2. (a) Integration of an attention mechanism with LSTM in our model: at each timestep t, given the previous hidden state h_{t-1} and m spatio-temporal proposals, the Attention Unit outputs a weighted mean-pooled visual feature z_t. The LSTM Unit receives the visual feature z_t, the previous hidden state h_{t-1} and the previously generated word u_t, and outputs the current hidden state h_t. (b) Schema of LSTM Unit described in Eq. 1.

3.3 High-Level Semantic Description

Generating a textual description of a video requires identifying the actors and their interactions and then constructing a grammatically well-formed sentence. For this purpose, in order to generate a human-like textual description of a video, we first represent a video in the form of a Subject(S), a Verb(V) and an Object(O) (similarly to earlier works [9]). We then integrate this representation with state-of-the-art recurrent models, along with spatio-temporal localization processes and object detection and classification information.

SVO Representation and Vocabulary Construction. In order to learn a semantic high-level representation for each video, we represent a sentence in a compact and simplified manner that preserves its main idea by extracting a (S,V,O) tuple - e.g. the sentence *A cat plays with a toy* is represented as (cat, play, toy)). We initially used the SVO vocabulary proposed in [9], but found it to be too small (only 45 subjects, 218 verbs and 241 objects) and too semantically restrictive (e.g. no different words for *man* and *woman*, as it only contains *person*). We mine the intermediate concepts differently from [9], such that our vocabulary is richer and with fewer constraints. The important changes we made in the way we build the vocabulary are: (1) considering both the indirect and direct objects when parsing the sentences as opposed to only the direct objects, (2) not grouping words into very general classes, and (3) an S, V or O is added to the final vocabulary if it is mentioned in at least two different sentences in any given video. We use the parser available from [28] to extract from each sentence a subject, verb and an object. Our final vocabulary set is a tuple $\mathcal{D} = \{\mathcal{S}, \mathcal{V}, \mathcal{O}\}$ of the corresponding vocabularies for each sentence part and it has a considerably larger size: 246 subjects, 459 verbs and 801 objects.

SVO Classification. We treat the three vocabularies separately and use Least Squares Support Vector Machine (LS-SVM) as a classifier in a one-vs-all approach. Note that an input video can have multiple labels from each vocabulary (e.g. a video can have labels 'cat', 'animal', 'kitten' for the subject class). We use LS-SVM because it provides a closed form solution both for the leave-one-out prediction and the prediction error via the block inversion lemma [29]. We represent a video as a classifier response vector for all the classes in the combined vocabulary $\mathcal{D}$. Training videos use the leave-one-out prediction and testing videos use the prediction based on the classifiers learned on the whole training set. This is different from [20], where classifiers scores are learned in the same way for training and testing. The dataset used in [20] has around 56k training examples, with roughly the same vocabulary size as ours. Given that our dataset has a much smaller number of training videos ($\approx 1,300$), we argue that LS-SVM is a better option; we can tune parameters without relying on a separate validation set (further decreasing the amount of label data) and we can better simulate the testing conditions. Our choice is also supported by the increase in classification accuracy when compared to other methods using the same vocabulary (see Sect. 4.3).

4 Experimental Details

Dataset Description. We perform our experiments on the YouTube dataset [8] which consists of 1,967 short videos (between 10 s and 25 s length) collected from YouTube that usually depict only one main activity. Each video has approximately 40 human-generated English descriptions collected through Amazon Mechanical Turk. We use the same splitting into train (1,197 videos), validation (100 videos) and test (670 videos) subsets as previous methods, so that our results are directly comparable to them.

Evaluation Measures. We report our results under BLEU [30] and METEOR [31] metrics which were originally proposed for the evaluation of automatic translation approaches and have also been adopted by previous works in video and image captioning. BLEU@n computes the geometric average of the n-gram precision between generated and reference sentences. METEOR computes an alignment score between sentences by taking into account the exact tokens, the stemmed ones and semantic similarities between them. We use the evaluation software provided by [1] which we adapt to our dataset.

4.1 Spatio-Temporal Object Proposals

We use the method from [6] to gather a pool of spatio-temporal object proposals. We split each video into parts using a shot boundary detection method [32]. Around 1,000 spatio-temporal proposals are extracted separately for each sub-video and together they form the pool of proposals for the whole video. We filter out the proposals that have a small spatial or temporal extent. To diversify the pool and to eliminate very similar proposals, we keep only those that have low IoU scores with each other. From the pool of proposals, we are interested mainly in those that could be attached a semantic meaning. Thus, we sort the proposals according to a semantic measure based on two scores and retain the top m. In our experiments we set $m = 20$. The first score is obtained by running the image classification CNN VGG-19 from [21](trained on 1.3M images from ImageNet Large-Scale Visual Recognition Challenge (ILSVRC) [33]) on every bounding box of each proposal, retaining the maximum activation in each frame among the 1,000 classes and averaging across all frames. The second score is obtained by running the 20-class object detector from [34] on every frame of each video. For each frame of the proposal, we compute the maximum detection score (multiplied by the IoU between the bounding box of the detection and the spatial extent of the proposal) and then we average the scores across all proposal frames. The final proposal score is the average of the image classification and detection scores.

Given a proposal, for each of its bounding boxes in the video frames, we extract the output of the fc_7 layer of the VGG-19. The feature descriptor for a proposal is obtained by mean-pooling over all the bounding boxes. We represent a video by m such descriptors corresponding to best scoring m spatio-temporal proposals and we refer to this $m \times D$ descriptor as the *temporal visual feature*.

Fig. 3. Examples of selected object proposals. For each video we generated a large pool of spatio-temporal object proposals and then learned to automatically select those that are most likely to overlap with easily recognisable semantic categories.

In practice, for some videos, the number of duplicate proposals is very large, and the final number of proposals can be less than m. Since the network takes as input a fixed sized array, for the videos that do not have at least m spatio-temporal proposals we pad the feature matrix to obtain a fixed size descriptor. We mark the padding proposals so that they will be ignored in the learning and testing processes. Examples of the selected proposals can be seen in Fig. 3.

4.2   Attention LSTM with Spatio-Temporal Object Proposals

Vocabulary. We use all the words in the training sentences without performing any pre-processing step. This results in a vocabulary of size 9,070. Similarly to previous works, we represent the words as one-hot vectors, set the maximum sentence length to 20 and mark with special characters the beginning and end of a sentence.

Training. We implemented our attention mechanism using the Caffe [35] framework, integrating it on top of the LSTM provided by [18]. We refer to our proposed LSTM model which uses an attention mechanism over spatio-temporal object proposals as LSTM-ATT. The training phase in LSTM-ATT is a sequential one: at each time step t the unit is given the temporal feature P (representing the m proposals), the embedded vector u_t, corresponding to the previous ground-truth word w_{t-1}, and the previous hidden state h_{t-1}. The output hidden-state h_t (see Fig. 2) is then used to predict a distribution $P(w_t)$ over the words in the vocabulary. We use the softmax loss and a dropout of 0.5 to avoid over-fitting. We train our models for a maximum of 128 epochs and use the validation set to choose the best iteration for each of the two metrics, BLEU and METEOR.

Inference. Inference is also performed in a sequential manner: given m proposals and the previous emitted word at time $t - 1$, sampled from $P(w_{t-1})$, the model generates the current word until the special character for end of sentence is met or the maximum length of a sentence is reached. We perform the sampling using beam search with beam size 20. Because we noticed that standard beam search implementations sometimes tends to end sentences too early, we modified it to force longer sentences to be produced (at least 4 words).

4.3   Integrating Contextual Semantic Features

Semantic SVO Representation. In order to obtain the SVO responses, we use the LS-SVM described in Sect. 3.3 and consider 3 different classification problems, one for S, one for V and one for O. Each video can then be described by concatenating the responses to these three classification problems. We use different features, depending on the part of sentence we want to classify. For the subject and object classes, we use the VGG-19 of [21], extract feature responses from the fc7 layer for each frame of the video and then perform mean pooling. For the verb class we use two types of motion features: the trajectory features of [36] and the motion-CNN features of [37], again followed by mean pooling. For S, V and O we obtained the following classification accuracies, respectively: 62.5%, 40.9% and 28.30%. Among the 3 classification problems, the results on O are the lowest since the number of classes in the object vocabulary is the largest: 801 compared with 264 for S and 459 for V. Also, the objects have a smaller spatio-temporal extent in video as they usually represent the objects manipulated by a person or animal (e.g. *onion*, *ball*, etc.). Since we considerably augmented the vocabulary, our classification results on this vocabulary do not compare directly with previous methods. However, we run our method on the initial proposed vocabulary [9] and show results in Table 1, against the most common (S,V,O) triple found in human annotations for each video. We show state-of-the-art results for S and O and a slightly lower accuracy than state-of-the-art for V. Notice that the methods marked with (*) generate a full sentence using a recurrent neural network and then extract the S, V and O using a dependency parser. Our aim is to use these intermediate semantic concepts as features to guide and ground our LSTM attention model (LSTM-ATT model) in the sentence generation process.

Semantic Representations. Apart from the SVO responses obtained by training using only the YouTube dataset, we also extract high level semantic features using state-of-the-art image classification and detection models. More precisely, we run the VGG CNN from [21] in each frame of the video and obtain the 1,000 dimensional score vector representing the classification responses over the 1,000 classes from the ImageNet classification dataset. To obtain a semantic representation of the video, we experimented with both average and max pooling over the frames and noticed that average pooling performs better in our experiments. We also run the 20 class object detector of [34] in each frame of every video and compute a 20-dimensional descriptor. For each class, we retain the detection

Table 1. Binary SVO accuracy computed against the most common (S,V,O) triple provided by humans. Entries marked with (*) first obtain a sentence describing the whole video and then mine the (S,V,O), whereas the others perform a classification over the S, V and O vocabularies.

Model	S%	V%	O%
HVC [10]	76.5	22.2	11.9
FGM [10]	76.4	21.3	12.3
JointEmbed [11]	78.2	24.4	11.9
(*) LSTM-E (VGG+C3D) [38]	80.4	**29.8**	13.8
(*) LSTM-YT-coco [19]	76.0	23.3	14.0
(*) LSTM-YT-coco+flicker [19]	75.61	25.3	12.4
LS-SVM(ours)	**83.6**	28.1	**23.1**

response scores in every frame then perform temporal pooling across a window of 25 frames. The final score for a class is the maximum of the temporal pooled scores for that class. The temporal pooling ensures that the detection observed is stable and lasts for at least 1 second. The maximum over such detection scores represents the confidence in having seen a particular object in video.

Integration with LSTM-ATT. We have experimented with two methods to integrate the high-level semantic features - SVO classification, object detection (DET) and image classification scores (CLS) - with the LSTM-ATT. In the first method, both the temporal visual features P and semantic features s are provided as input to the LSTM-ATT. In the second one, we stack a LSTM, that processes only the semantic features s, on top of LSTM-ATT which receives the temporal visual feature P as input. We refer to these methods as LSTM-ATT(SEM) and LSTM2-ATT(SEM), respectively, where (SEM) stands for different subsets of semantic features. A schematic view of the two models we use is shown in Fig. 4.

4.4 Experimental Results

Quantitative Results. Results obtained with the proposed models are shown in Table 2. We first check whether the attention mechanism provides an advantage over mean pooling the temporal visual feature, as quantified by the currently most used metrics, BLEU and METEOR. Using a simple LSTM that receives as input the mean pooled temporal visual feature, we obtain a score of 45.4% on BLEU@4 and 31.2% on METEOR. With LSTM-ATT, the results are considerably higher: 48.7% on BLEU@4 and 31.9% on METEOR, which demonstrates that the attention mechanism not only provides a way to focus selectively on the input video but also improves results. This is also true when adding semantic features both to the standard LSTM (with mean pooled temporal visual feature) and to the LSTM-ATT.

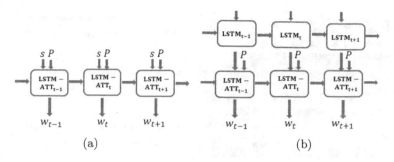

(a)   (b)

Fig. 4. Integration of semantic features with LSTM-ATT. (a) LSTM-ATT(SEM): both semantic features s and temporal features P are processed by the LSTM-ATT unit. (b) LSTM2-ATT(SEM): we stack a LSTM unit that processes only the semantic input s on top of the LSTM-ATT.

Table 2. Comparison with previous works on BLEU@1 - BLEU@4 and METEOR metrics. Values are reported as percentage %.

Method	BLEU@1	BLEU@2	BLEU@3	BLEU@4	METEOR
FGM [10]	-	-	-	13.68	23.9
S2VT [25]	-	-	-	-	29.8
MM-VDN [26]	-	-	-	37.64	29.00
LSTM-YT-coco [19]	-	-	-	33.29	29.07
LSTM-YT-coco+flicker [19]	-	-	-	33.29	28.88
Temporal attention [4]	-	-	-	41.92	29.60
LSTM-E (VGG+C3D) [38]	78.8	66.0	55.4	45.3	31.0
h-RNN [3]	81.5	70.4	60.4	49.9	32.6
HRNE with attention [39]	79.2	66.3	55.1	43.8	**33.1**
GRU-RCNN [40]	-	-	-	49.63	31.70
LSTM	78.0	66.4	56.7	45.4	31.2
LSTM(SVO)	80.1	68.1	57.5	45.8	31.2
LSTM(DET,CLS)	81.2	68.9	57.9	46.2	31.1
LSTM(SVO,DET,CLS)	80.8	69.3	59.3	48.3	30.7
LSTM-ATT	80.1	68.9	59.4	48.7	31.9
LSTM-ATT(SVO)	81.0	70.5	61.2	50.5	32.3
LSTM-ATT(DET,CLS)	81.9	70.9	60.9	50.5	31.6
LSTM-ATT(SVO,DET,CLS)	82.0	71.6	62.4	51.5	32.0
LSTM2-ATT(SVO)	**82.4**	**71.8**	**62.5**	**52.0**	32.3
LSTM2-ATT(DET,CLS)	80.6	68.1	57.4	46.0	31.8
LSTM2-ATT(SVO,DET,CLS)	81.5	70.8	61.5	50.6	32.4

Fig. 5. Highest scoring proposals of our model (according to β Eq. 4) for each emitted word in the sentence. We only illustrate the grounding of the main words in the sentence and ignore linking words. The complete sentence is shown in the right column together with the closest reference from the human annotations. For each proposal we show a single, randomly selected frame (Color figure online).

Our LSTM-ATT model achieves competitive results compared to other methods. Adding semantic features on top of this model improves the state-of-the-art results on the BLEU@n metric, while also performing well on METEOR. We show results with both LSTM-ATT(SEM) and LSTM2-ATT(SEM). The contributions of the SVO semantic features alone and in conjunction with DET and CLS features are also presented. In the case of SVO features alone, the best results are obtained with LSTM2-ATT(SVO) method for both evaluation metrics (BLEU@4 52.0%, METEOR 32.3%), while when using the full semantic features, our best performing method under BLEU is LSTM-ATT(SVO,DET,CLS) (50.6%) and under METEOR is LSTM2-ATT(SVO,DET,CLS) (32.4%).

Qualitative Results. Our attention mechanism, built on top of spatio-temporal object proposals, allows for a *visual explanation* of what the model ranked as the most relevant visual support for emitting a particular word. This can be done by inspecting the learned weights β (see Eq. 4) and their associated proposals. In Fig. 5 we show the proposal with the highest associated β weight (a random frame from it) that was used in generating a particular word. For display purposes, we ignore linking words and articles that do not have a visual grounding in video. Our model correctly indicates the localization of the key video description components and the words it emitted, even for those having a very small spatial extent such as *pepper, ball, toy, gun*. There are cases when a single spatio-temporal proposal is chosen as the best visual explanation for multiple words, as it is with (*girl, riding, horse*). We also show examples when the obtained sentences are wrong (marked with red in Fig. 5). In some of the cases, our algorithm correctly identifies parts of the sentences - especially subjects and verbs such as (*man-cutting, dog-playing*) - but fails to find the correct object. This is due to the large variability in objects appearance and size and also depends on the quality of the spatio-temporal proposals pool.

5   Conclusions

In this paper we have addressed some of the challenges in automatic video captioning by aiming to spatio-temporally ground the semantic video concepts, as an intermediate step, without grounding supervision. In contrast to most existing automatic video captioning systems that map from the raw video to the high level textual description, bypassing localization, we aim at aggregating additional, potentially valuable information, by relying on spatio-temporal video proposals and image classification responses for content localization and improved generalization, fused using deep neural network attention models, based on long short-term memory. Our resulting system produces competitive, state-of-the-art results in the standard YouTube captioning benchmark and offers the additional advantage of localizing the concepts (subjects, verbs, objects), with no grounding supervision, over space and time.

Acknowledgement. This work was supported in part by CNCS-UEFISCDI under PCE-2011-3-0438, JRP-RO-FR-2014-16 and NVIDIA through a GPU donation.

References

1. Chen, X., Fang, H., Lin, T., Vedantam, R., Gupta, S., Dollár, P., Zitnick, C.L.: Microsoft COCO captions: data collection and evaluation server. arXiv preprint arXiv:1504.00325 (2015)
2. Krishna, R., Zhu, Y., Groth, O., Johnson, J., Hata, K., Kravitz, J., Chen, S., Kalantidis, Y., Li, L., Shamma, D.A., Bernstein, M.S., Li, F.: Visual genome: connecting language and vision using crowdsourced dense image annotations. arXiv preprint arXiv:1602.07332 (2016)
3. Yu, H., Wang, J., Huang, Z., Yang, Y., Xu, W.: Video paragraph captioning using hierarchical recurrent neural networks. In: CVPR (2016)
4. Yao, L., Torabi, A., Cho, K., Ballas, N., Pal, C., Larochelle, H., Courville, A.: Describing videos by exploiting temporal structure. In: ICCV (2015)
5. Taralova, E.H., Torre, F., Hebert, M.: Motion words for videos. In: Fleet, D., Pajdla, T., Schiele, B., Tuytelaars, T. (eds.) ECCV 2014. LNCS, vol. 8689, pp. 725–740. Springer, Cham (2014). doi:10.1007/978-3-319-10590-1_47
6. Oneata, D., Revaud, J., Verbeek, J., Schmid, C.: Spatio-temporal object detection proposals. In: Fleet, D., Pajdla, T., Schiele, B., Tuytelaars, T. (eds.) ECCV 2014. LNCS, vol. 8691, pp. 737–752. Springer, Cham (2014). doi:10.1007/978-3-319-10578-9_48
7. Fragkiadaki, K., Arbelaez, P., Felsen, P., Malik, J.: Learning to segment moving objects in videos. In: CVPR (2015)
8. Chen, D.L., Dolan, W.B.: Collecting highly parallel data for paraphrase evaluation. In: ACL (2011)
9. Guadarrama, S., Krishnamoorthy, N., Malkarnenkar, G., Venugopalan, S., Mooney, R., Darrell, T., Saenko, K.: Youtube2text: recognizing and describing arbitrary activities using semantic hierarchies and zero-shot recognition. In: ICCV (2013)
10. Thomason, J., Venugopalan, S., Guadarrama, S., Saenko, K., Mooney, R.: Integrating language and vision to generate natural language descriptions of videos in the wild. In: COLING (2014)
11. Xu, R., Xiong, C., Chen, W., Corso, J.J.: Jointly modeling deep video and compositional text to bridge vision and language in a unified framework. In: AAAI Conference on Artificial Intelligence (2015)
12. Sutskever, I., Vinyals, O., Le, Q.V.: Sequence to sequence learning with neural networks. In: NIPS (2014)
13. Bahdanau, D., Cho, K., Bengio, Y.: Neural machine translation by jointly learning to align and translate. In: ICLR (2015)
14. Vinyals, O., Toshev, A., Bengio, S., Erhan, D.: Show and tell: a neural image caption generator. In: CVPR (2015)
15. Xu, K., Ba, J., Kiros, R., Cho, K., Courville, A.C., Salakhutdinov, R., Zemel, R.S., Bengio, Y.: Show, attend and tell: neural image caption generation with visual attention. In: ICML (2015)
16. Karpathy, A., Fei-Fei, L.: Deep visual-semantic alignments for generating image descriptions. In: CVPR (2015)
17. Johnson, J., Karpathy, A., Fei-Fei, L.: Densecap: Fully convolutional localization networks for dense captioning. arXiv preprint arXiv:1511.07571 (2015)
18. Donahue, J., Hendricks, L.A., Guadarrama, S., Rohrbach, M., Venugopalan, S., Saenko, K., Darrell, T.: Long-term recurrent convolutional networks for visual recognition and description. In: CVPR (2015)

19. Venugopalan, S., Xu, H., Donahue, J., Rohrbach, M., Mooney, R., Saenko, K.: Translating videos to natural language using deep recurrent neural networks. In: NAACL HLT (2015)

20. Rohrbach, A., Rohrbach, M., Schiele, B.: The long-short story of movie description. In: Gall, J., Gehler, P., Leibe, B. (eds.) GCPR 2015. LNCS, vol. 9358, pp. 209–221. Springer, Cham (2015). doi:10.1007/978-3-319-24947-6_17

21. Simonyan, K., Zisserman, A.: Very deep convolutional networks for large-scale image recognition. In: ICLR (2014)

22. Tran, D., Bourdev, L.D., Fergus, R., Torresani, L., Paluri, M.: Learning spatiotemporal features with 3D convolutional networks. In: ICCV (2015)

23. Hochreiter, S., Bengio, Y., Frasconi, P., Schmidhuber, J.: Gradient flow in recurrent nets: the difficulty of learning long-term dependencies (2001)

24. Hochreiter, S., Schmidhuber, J.: Long short-term memory. Neural Comput. **9**, 1735–1780 (1997)

25. Venugopalan, S., Rohrbach, M., Donahue, J., Mooney, R., Darrell, T., Saenko, K.: Sequence to sequence - video to text. In: ICCV (2015)

26. Xu, H., Venugopalan, S., Ramanishka, V., Rohrbach, M., Saenko, K.: A multi-scale multiple instance video description network. In: arXiv preprint arXiv:1505.05914 (2015)

27. Zaremba, W., Sutskever, I.: Learning to execute. arXiv preprint arXiv:1410.4615 (2014)

28. Manning, C.D., Surdeanu, M., Bauer, J., Finkel, J., Bethard, S.J., McClosky, D.: The Stanford CoreNLP natural language processing toolkit. In: ACL (2014)

29. Cawley, G.C.: Leave-one-out cross-validation based model selection criteria for weighted ls-svms. In: IJCNN (2006)

30. Papineni, K., Roukos, S., Ward, T., Zhu, W.J.: Bleu: a method for automatic evaluation of machine translation. In: ACL (2002)

31. Lavie, A., Agarwal, A.: Meteor: an automatic metric for MT evaluation with improved correlation with human judgments, pp. 65–72 (2005)

32. Lienhart, R.W.: Comparison of automatic shot boundary detection algorithms. In: International Society for Optics and Photonics on Electronic Imaging 1999, pp. 290–301 (1998)

33. Russakovsky, O., Deng, J., Su, H., Krause, J., Satheesh, S., Ma, S., Huang, Z., Karpathy, A., Khosla, A., Bernstein, M., Berg, A.C., Fei-Fei, L.: ImageNet large scale visual recognition challenge. IJCV **115**, 211–252 (2015)

34. Ren, S., He, K., Girshick, R., Sun, J.: Faster R-CNN: towards real-time object detection with region proposal networks. In: NIPS (2015)

35. Jia, Y., Shelhamer, E., Donahue, J., Karayev, S., Long, J., Girshick, R., Guadarrama, S., Darrell, T.: Caffe: convolutional architecture for fast feature embedding. In: ACMMM (2014)

36. Wang, H., Kläser, A., Schmid, C., Liu, C.L.: Action recognition by dense trajectories. In: CVPR. IEEE (2011)

37. Gkioxari, G., Malik, J.: Finding action tubes. In: CVPR (2015)

38. Pan, Y., T.M., Yao, T., Li, H., Rui, Y.: Jointly modeling embedding and translation to bridge video and language. In: arXiv preprint arXiv:1505.01861. (2015)

39. Pan, P., Xu, Z., Yang, Y., Wu, F., Zhuang, Y.: Hierarchical recurrent neural encoder for video representation with application to captioning. arXiv preprint arXiv:1511.03476 (2015)

40. Ballas, N., Yao, L., Pal, C., Courville, A.C.: Delving deeper into convolutional networks for learning video representations. arXiv preprint arXiv:1511.06432 (2015)

Variational Convolutional Networks
for Human-Centric Annotations

Tsung-Wei Ke[1], Che-Wei Lin[1], Tyng-Luh Liu[1(✉)], and Davi Geiger[2]

[1] Institute of Information Science, Academia Sinica, Taipei, Taiwan
liutyng@iis.sinica.edu.tw
[2] Courant Institute of Mathematical Sciences,
New York University, New York City, USA

Abstract. To model how a human would annotate an image is an important and interesting task relevant to image captioning. Its main challenge is that a same visual concept may be important in some images but becomes less salient in other situations. Further, the subjective viewpoints of a human annotator also play a crucial role in finalizing the annotations. To deal with such high variability, we introduce a new deep net model that integrates a CNN with a variational auto-encoder (VAE). With the latent features embedded in a VAE, it becomes more flexible to tackle the uncertainly of human-centric annotations. On the other hand, the supervised generalization further enables the discriminative power of the generative VAE model. The resulting model can be end-to-end fine-tuned to further improve the performance on predicting visual concepts. The provided experimental results show that our method is state-of-the-art over two benchmark datasets: MS COCO and Flickr30K, producing mAP of 36.6 and 23.49, and PHR (Precision at Human Recall) of 49.9 and 32.04, respectively.

1 Introduction

Exploring the intriguing relationships between language and vision models has recently become an active research topic in computer vision community. Notable efforts include generating text descriptions for images, *e.g.*, [1–4] or videos [5,6], while their main idea is to discover important spatial or spatial-temporal visual information and express it with appropriate wording. Another interesting development has been centered on the problem of *image question answering* [7]. The task often results in a more complex and challenging vision-language computational model, which would require learning different levels/types of semantics to address the various combinations of questions and underlying scenes. Yet, in contrast to dealing with *image captioning*, there are also techniques, *e.g.*, [8], aiming at solving language-to-image problems to generate images according to the given descriptions.

We instead focus on the problem of *human-centric annotations* [9] for images, which can be considered a subtask of image captioning. From popular image caption collections such as MS COCO [10] and Flickr30K [11], one can conclude

© Springer International Publishing AG 2017
S.-H. Lai et al. (Eds.): ACCV 2016, Part IV, LNCS 10114, pp. 120–135, 2017.
DOI: 10.1007/978-3-319-54190-7_8

Visual Concepts	
air	musical
band	performing
concert	performs
crowd	plays
fans	put
hands	stage

Fig. 1. An image example with 12 visual concepts as the ground truth.

that it is inappropriate and also impossible to use a caption to name every content in the image. For example, when describing a basketball in a scene, a sensible caption would not state "a round basketball" but simply "a basketball" instead. However, the same concept of "round" would become meaningful if the shape of a target object such as a building or a church is to be emphasized. The example pinpoints that image annotations are highly correlated to *important* properties of the image, and are inherently linked to the annotator's viewpoints. Following [12], we consider the image annotations termed as *visual concepts*, whose labeling depends on the *subjective* judgment of a human annotator. To construct the set of visual concepts from an image caption dataset, we single out those words with the top most appearances in the captions. The ground truth of visual concepts of an image can then be formed by intersecting all its captions with the set of visual concepts. Figure 1 shows an image and the corresponding visual concepts, with which the task of human-centric annotations aims to predict.

Recent studies have shown that learning to predict human-centric annotations could improve the performances of image captioning [13] and image question answering [4]. Misra *et al.* [12] consider human-centric annotations as visual concepts. Their method can predict both the visual concepts and their presence in an image whether a human would annotate the concept or not. Motivated by the promising progress, we aim to more satisfactorily address the problem of human-centric annotations. In particular, to model the subtlety of how annotations are achieved, we decompose the process into two stages. We first predict the presences of all the available concepts in an image, and then *simulate* how a human would decide their relevance in the final annotations. The reasoning can be realized by fusing a Convolutional Neural Network (CNN) with a Variational Auto-Encoder (VAE) [14], where the resulting network architecture will be termed as a Variational CNN (VCNN). The annotation process by the proposed VCNN proceeds as follows. It starts by using a deep CNN to output the probabilities of all the concepts, and then passes the visual features and the information (or more precisely, the probabilities) of concept presence to a (stacked) VAE model to generate the annotation predictions. The proposed two-stage processing can be seamlessly coupled to form an end-to-end VCNN model, as illustrated in Fig. 2. One crucial difference between our method and [12] is that with the proposed VCNN model, the probability of annotating a particular visual concept is conditioned on the presence information of all the concepts, rather than the concept alone.

2   Related Work

Methods dealing with image captioning can be divided into two categories, namely, *caption retrieval* and *caption generation*. For caption retrieval, Devlin *et al.* [15] propose to search for a set of the nearest neighbor images, and gather from them the candidate captions. The description that is most similar to the other candidates is chosen from the set to represent the query image. In [16], Klein *et al.* exploit the alignment between linguistic descriptors, derived from the Gaussian-Laplacian Mixture Model, with CNN-based visual features for caption retrieval. For caption generation, most techniques rely on using deep net models. A popular formulation is to use two subnetworks, which typically consists of a CNN as the vision model and a Recurrent Neural Network (RNN) as the language model [1–4,17]. And the variants of RNN include the Long Short-Term Memory (LSTM) network [2,17], the bidirectional RNN [1], etc. Furthermore, in [17], Jia *et al.* extend the input to the LSTM with the extracted semantic information to improve the performance of image caption generation. Xu *et al.* [3] introduce an attention model that aims to help LSTM to emphasize salient objects while generating descriptions. In [4,13], the CNN module is fine-tuned to detect possible attributes/words in the image, and the resulting prediction is then taken as the input to the language model.

Apart from dealing with a single image, video description generation has also gained increasing attention and interest. Rohrbach *et al.* [18] formulate the task as a machine translation problem by learning a CRF to yield the semantic representation and translating it into the video description. In [19], a factor graph is constructed to combine visual detections on subject, verb, object and scene elements with linguistic statistics to infer the most likely tuple for sentence generation. Yao *et al.* [5] propose to capture spatio-temporal dynamics and build an attention model. With the temporal attention, the most relevant video subsequences are selected for RNN to describe. Venugopalan *et al.* [6] divide text generation into two subtasks: a stacked LSTM network is used to first encode a video sequence and then decode it into a sentence.

Understating the underlying factors behind human-centric annotations has been an interesting topic in computer vision. The analysis conducted by Berg *et al.* [9] investigates three types of factors, including composition, semantics, and context, which are all closely related to how people evaluate the importance of a content in the image. In [20], Turakhia *et al.* model the attribute dominance and argue that more dominant attributes would be described first when seeing an image. Yun et al. [21] explore the relationships among images, eye movements and descriptions, and use a gaze-enabled model for detection and annotation. In addition, there are several techniques aiming at directly predicting user-supplied tags. Chen et al. [22] propose to pre-train a CNN on easy images to learn an initial visual representation. The weights are then transferred and fine-tuned on realistic images. When testing with image-tag pairs, the resulting two-stage learning approach is shown to outperform schemes with only fine-tuning. In [23], Izadinia et al. have focused on predicting 5400 tags over a dataset with 5M Flickr images. Besides recognizing the user-supplied tags, [12,13,24] are to predict words filtered from the image captions. Taking these words as noisy labels, Misra et al. [12] propose a factor-decoupling model to implicitly predict visual labels, where the classifier is trained essentially with the human-centric annotations. In [24], Joulin et al. have attempted to predict 100,000 words over an extremely large-scale dataset with approximately 100M images.

The VAE model by Kingma et al. [14] is established by integrating a top-down deep generative network with a bottom-up recognition network. The recognition model is optimized with respect to a variational lower bound to achieve approximate posterior inference. Its extension to semi-supervised applications is proposed in [25]. Another generalization can be found in the so-called Importance Weighted Auto-Encoder (IWAE) [26], which employs a similar network as the VAE, but is learned with a tighter log-likelihood lower bound. Besides these efforts, a popular application of VAE is to include the model to enable variational inference with an RNN, e.g., [27–29]. In [27], Fabius et al. generalize the encoding-decoding procedure to the temporal domain. While the distribution over the latent variable is decided from the last state of the recurrent recognition model, the recurrent generative model outputs data with the initial state computed from the updated latent representation. Recently, Chung et al. [29] introduce a high-level latent variable into an RNN to model the variability in rich-structured sequential data. The VAE-based models are also used in tackling image generation [28,30].

3   Our Method

We begin by casting the problem of how a human would annotate an image as follows. Let $\mathcal{V} = \{v_k\}_{k=1}^K$ be a set of K visual concepts. Then, the human-centric annotations for a given image $\mathbf{x}$ form a subset of $\mathcal{V}$, denoted as

$$\mathcal{A}_{\mathbf{x}} = \{v_k \,|\, y_k = 1,\ 1 \le k \le K\} \subseteq \mathcal{V} \tag{1}$$

where $y_k \in \{0, 1\}$ is a binary random variable specifying whether visual concept v_k is mentioned in the annotations. Analogous to the formulation in [12], we define a latent random variable $c_k \in \{0, 1\}$ as the *visual label* of v_k and use it to indicate whether the visual concept v_k is present in the image. For convenience, we write $\mathbf{c} = (c_1, \ldots, c_K)^\top$ and marginalizing over $\mathbf{c}$ would yield

$$p(y_k|\mathbf{x}) = \sum_{\mathbf{c} \in \{0,1\}^K} p(y_k|\mathbf{c}, \mathbf{x})\, p(\mathbf{c}|\mathbf{x}) \approx p(y_k|\mathbf{c}^*, \mathbf{x})\, p(\mathbf{c}^*|\mathbf{x}) \tag{2}$$

where the approximation is the result of assuming that the probability distribution $p(\mathbf{c}|\mathbf{x})$ peaks very sharply at $\mathbf{c}^*$. Indeed, the approximation in (2) is *exact* if we do have the factual information about the presence of each concept v_k. That is, the closer $\mathbf{c}^*$ is to the (unavailable) ground truth of visual labels, the more valid the approximation will be. With (2), we carry out our method in two sequential stages.

1. Construct a convolutional neural network (CNN) to yield $p(\mathbf{c}^*|\mathbf{x})$.
2. Learn a variational auto-encoder (VAE) to output $p(y_k|\mathbf{c}^*, \mathbf{x})$ for each concept v_k.

Details about how we sequentially learn the two types of neural networks and fine-tune them as an end-to-end system will be described in the next two subsections. We now remark that unlike the formulation in [12], we estimate $p(y_k|\mathbf{x})$ by marginalizing over $\mathbf{c}$ rather than just c_k. The distinction is crucial, as in many practical situations, the mentioning of a visual concept v_k depends on not only c_k but also the presence of other relevant visual concepts.

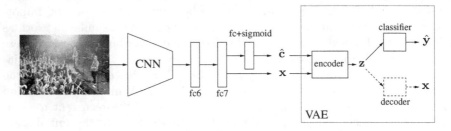

Fig. 2. We couple CNN and VAE to form a variational CNN for human-centric annotations.

3.1   On $p(\mathbf{c}^*|\mathbf{x})$

To model the multi-label learning for $p(\mathbf{c}^*|\mathbf{x})$, we assume the independence of visual labels in an image. That is,

$$p(\mathbf{c}^*|\mathbf{x}) = \prod_{k=1}^{K} p(c_k^*|\mathbf{x}). \tag{3}$$

We employ the VGG net [31] pre-trained on ImageNet as the adopted CNN, and modify the network by adding on top of the fc7 a discriminative classifier composed of a fully-connected layer and a sigmoid function. (See Fig. 2.) Due to the lack of visual-label ground truth in the training dataset, we use the information of visual concepts as the *noisy* ground truth and fine-tune the VGG net with the human-centric annotations to yield the probabilities of visual labels.

3.2   On $p(y_k|\mathbf{c}^*, \mathbf{x})$

With the CNN learned in the first stage, we extract features from fc7 and represent each image with $\mathbf{x} \in \mathbb{R}^L$. ($L = 4096$ for VGG.) On the other hand, simply using $\mathbf{c}^* \in \{0,1\}^K$ does not fully utilize the visual-label information. We instead consider their probabilities, and denote them by $\hat{\mathbf{c}} \in \mathbb{R}^K$, whose kth component is the probability $p(c_k^*|\mathbf{x})$ yielded by the CNN. To simplify the notation, we write $\mathbf{w} = \hat{\mathbf{c}} \oplus \mathbf{x} \in \mathbb{R}^{K+L}$ where $\oplus$ denotes vector concatenation. Further, we use $\hat{y}_k$ to denote $p(y_k|\mathbf{w})$, the probability of mentioning concept v_k in the annotations and let $\hat{\mathbf{y}} = (\hat{y}_1, \ldots, \hat{y}_K)^\top \in \mathbb{R}^K$. Before we explain the proposed VAE formulation, we first describe a naïve approach to predicting the probabilities of visual concepts. Assume that the training dataset has N images, represented by $\{(\mathbf{x}_i, \mathbf{y}_i^*)\}_{i=1}^N$, where $\mathbf{y}_i^* \in \{0,1\}^K$ is the visual-concept ground truth of image $\mathbf{x}_i$. We can construct a neural network (detailed in Subsect. 4.4) to directly model $p(y_k|\hat{\mathbf{c}}, \mathbf{x}) = p(y_k|\mathbf{w})$ with a cross-entropy objective function:

$$\mathcal{E}_{\text{naive}} = -\sum_{i=1}^N \sum_{k=1}^K \mathbb{I}(\mathbf{y}_i^*(k) = 1) \, \log p(y_k|\mathbf{w}_i) \tag{4}$$

where $\mathbb{I}(\cdot)$ is the indicator function and $\mathbb{I}(\mathbf{y}_i^*(k) = 1)$ verifies that visual concept v_k is mentioned in the ground truth $\mathbf{y}_i^*$.

We next describe the proposed VAE model. Our method is inspired by [25], but we extend it to a combined generative and supervised learning. To begin with, we hypothesize the following data generative process:

$$p_{\boldsymbol{\theta}}(\mathbf{z}) = \mathcal{N}(\mathbf{z}|\mathbf{0}, \mathbf{I}) \quad \text{and} \quad p_{\boldsymbol{\theta}}(\mathbf{x}|\mathbf{z}) = f(\mathbf{x}; \mathbf{z}, \boldsymbol{\theta}) \tag{5}$$

where the prior of the latent variable $\mathbf{z} \in \mathbb{R}^D$ is assumed to be the centered isotropic multivariate Gaussian and $f(\cdot)$ is a suitable likelihood function, while $\boldsymbol{\theta}$ are VAE generative parameters. We then introduce a distribution $q_{\boldsymbol{\phi}}(\mathbf{z}|\mathbf{w})$ to approximate the true posterior distribution $p_{\boldsymbol{\theta}}(\mathbf{z}|\mathbf{w})$ where $\boldsymbol{\phi}$ are variational parameters. More specifically, we have

$$q_{\boldsymbol{\phi}}(\mathbf{z}|\mathbf{w}) = \mathcal{N}(\mathbf{z}|\boldsymbol{\mu}_{\boldsymbol{\phi}}(\mathbf{w}), \text{diag}(\boldsymbol{\sigma}_{\boldsymbol{\phi}}^2(\mathbf{w}))) \tag{6}$$

where $\boldsymbol{\mu}_{\boldsymbol{\phi}}(\mathbf{w})$ and $\boldsymbol{\sigma}_{\boldsymbol{\phi}}(\mathbf{w})$ respectively denote a vector of means and a vector of standard deviations. In our formulation, both are represented by the neural network. Then we can derive the *variational lower bound*, $\mathcal{L}(\boldsymbol{\theta}, \boldsymbol{\phi}; \mathbf{w})$:

$$\log p_{\boldsymbol{\theta}}(\mathbf{w}) \geq \mathcal{L}(\boldsymbol{\theta}, \boldsymbol{\phi}; \mathbf{w}) = \mathbb{E}_{q_{\boldsymbol{\phi}}(\mathbf{z}|\mathbf{w})}[\log p_{\boldsymbol{\theta}}(\mathbf{w}|\mathbf{z}) + \log p_{\boldsymbol{\theta}}(\mathbf{z}) - \log q_{\boldsymbol{\phi}}(\mathbf{z}|\mathbf{w})]$$
$$= -D_{KL}(q_{\boldsymbol{\phi}}(\mathbf{z}|\mathbf{w})\|p_{\boldsymbol{\theta}}(\mathbf{z})) + \mathbb{E}_{q_{\boldsymbol{\phi}}(\mathbf{z}|\mathbf{w})}[\log p_{\boldsymbol{\theta}}(\mathbf{w}|\mathbf{z})]. \tag{7}$$

The derivation so far follows the standard analysis of variational approximation. To incorporate the ground-truth information of visual concepts and to boost the discriminative power to our model, the last term in (7) is approximated by

$$\mathbb{E}_{q_{\boldsymbol{\phi}}(\mathbf{z}|\mathbf{w})}[\log p_{\boldsymbol{\theta}}(\mathbf{w}|\mathbf{z})] \approx \mathbb{E}_{q_{\boldsymbol{\phi}}(\mathbf{z}|\mathbf{w})}[\log p_{\boldsymbol{\theta}}(\mathbf{x}|\mathbf{z})] + \mathbb{E}_{q_{\boldsymbol{\phi}}(\mathbf{z}|\mathbf{w})}[\log p_{\boldsymbol{\theta}}(\mathbf{y}|\mathbf{z})] \tag{8}$$

where the approximation decouples the joint generative process into unsupervised decoding and classification, respectively. (See Fig. 2.) Thus, the objective function to be minimized in learning the supervised VAE is defined by

$$\mathcal{E}_{\text{VAE}}(\boldsymbol{\theta}, \boldsymbol{\phi}, \mathbf{w}) = \sum_{i=1}^{N} D_{KL}(q_{\boldsymbol{\phi}}(\mathbf{z}_i|\mathbf{w}_i)\|p_{\boldsymbol{\theta}}(\mathbf{z}_i)) - \mathbb{E}_{q_{\boldsymbol{\phi}}(\mathbf{z}|\mathbf{w})}[\log p_{\boldsymbol{\theta}}(\mathbf{x}_i|\mathbf{z}_i)]$$

$$- \sum_{i=1}^{N} \sum_{k=1}^{K} \mathbb{I}(\mathbf{y}_i^*(k) = 1) \log p_{\boldsymbol{\theta}}(y_k|\mathbf{z}_i). \tag{9}$$

Using the reparameterization trick for $\mathbb{E}_{q_{\boldsymbol{\phi}}(\mathbf{z}|\mathbf{w})}[\log p_{\boldsymbol{\theta}}(\mathbf{x}|\mathbf{z})]$ and the KL divergence closed-form:

$$D_{KL}(q_{\boldsymbol{\phi}}(\mathbf{z}|\mathbf{w})\|p_{\boldsymbol{\theta}}(\mathbf{z})) = -\frac{1}{2} \sum_{j=1}^{D} (1 + \log(\sigma_{\phi,j}^2) - \mu_{\phi,j}^2 - \sigma_{\phi,j}^2) \tag{10}$$

where $\sigma_{\phi,j}^2$, $\mu_{\phi,j}^2$ are respectively the jth elements of $\boldsymbol{\sigma}_{\phi}^2(\mathbf{w})$ and $\boldsymbol{\mu}_{\phi}^2(\mathbf{w})$, the supervised VAE can be learned with the Stochastic Gradient Variational Bayes (SGVB) [14]. Having sequentially trained the CNN and the VAE, we link the two models and remove the decoder module (shown as the dotted rectangle in Fig. 2) from the architecture. This way we can enhance the discriminative power of the VCNN by end-to-end fine-tuning with only the classification loss function $\mathcal{E}_{naive}$ defined in (4).

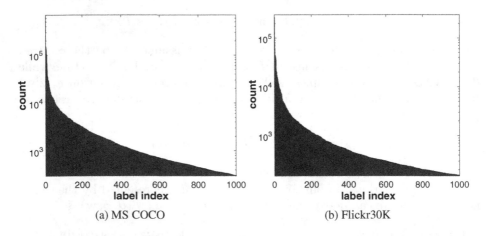

(a) MS COCO   (b) Flickr30K

Fig. 3. Total number of presences for each *visual concept* in a dataset.

4   Experimental Results

We evaluate the proposed VCNN model on two image caption datasets: MS COCO [10] and Flickr30K [11]. Numbers, punctuation symbols, accents and special characters are removed from the captions. Every caption is then lower-cased and tokenized into words. For each dataset, we select $K = 1000$ most common words, including nouns, verbs, adjectives and other parts of speech, to form the set of visual concepts for human-centric annotations.

4.1 Datasets

MS COCO [10] includes 82,783 training images and 40,504 validation images. Each image is provided with five human-annotated captions. Following [12], we split the collection of validation images into equally-sized validation and test set, where the split is the same as that in [12]. Flickr30K [11] is composed of 158,915 crowd-sourced captions describing 31,783 images. As in [1], we divide them into training, validation and test sets, each of which contains 29,783, 1000, and 1000 images, respectively.

To generate the ground-truth annotations of visual concepts for each training image, we use a 1000-dimensional binary vector to indicate which of the selected 1000 most common words appeared in any of the 5 corresponding captions. Based on these binary vectors of ground truth, our models and [12,13] are all learned with the same setting in the experiments. Unless otherwise mentioned, we report the results on the test sets of MS COCO and Flickr30K.

4.2 Cost-Sensitive Criterion

Because the visual concepts are derived from image captions annotated by humans, some words are mentioned much more frequently than the others. For example, in the two datasets, boy, girl would be used more often than lion or elephant. We have counted the total number of each visual concept present in the images over MS COCO and Flickr30K. The results are plotted in Fig. 3. Such an imbalanced distribution of word labels could cause biases on learning the VAE model. To address this issue, we separate the set of visual concepts into a common set and a rare set, denoted by $\mathcal{V} = \mathcal{V}_c \sqcup \mathcal{V}_r$. We extend the classification loss term in (9) into a cost-sensitive one by

$$\mathcal{E}_{\mathrm{cs}}(\mathbf{y}) = \left(\sum_{\mathbf{x}_i \in \mathcal{V}_c} \lambda_c + \sum_{\mathbf{x}_i \in \mathcal{V}_r} \lambda_r \right) \sum_{k=1}^{K} \mathbb{I}(\mathbf{y}_i^*(k) = 1) \log p_\theta(y_k|\mathbf{z}_i)) \qquad (11)$$

where λ_c, λ_r are the cost-sensitive weighting parameters. In the experiments, we set $\lambda_r > \lambda_c$ to avoid the penalty dominance from misclassifying common words.

4.3 Stacked VAE

We also try stacking two latent variables to discover more effective architecture of the supervised VAE. The architecture of our stacked VAE is shown in Fig. 4. Specifically, we first learn a latent variable $\mathbf{z}_1$ based on Sect. 3.2 and subsequently learn $\mathbf{z}_2$ using $\mathbf{z}_1$. The deep generative model can be described by

$$p(\mathbf{w}, \mathbf{z}_1, \mathbf{z}_2) = p(\mathbf{x}, \hat{\mathbf{c}}, \mathbf{z}_1, \mathbf{z}_2) = p(\hat{\mathbf{c}})p(\mathbf{z}_1)p(\mathbf{z}_2|\mathbf{z}_1)p(\mathbf{x}|\mathbf{z}_2). \qquad (12)$$

Analogously, we can derive the variational lower bound as

$$\mathcal{L}_{\text{stacked}}(\boldsymbol{\theta}, \boldsymbol{\phi}; \mathbf{w}) \approx -D_{KL}(q_{\boldsymbol{\phi}}(\mathbf{z}_1|\mathbf{w})\|p_{\boldsymbol{\theta}}(\mathbf{z}_1)) - D_{KL}(q_{\boldsymbol{\phi}}(\mathbf{z}_2|\mathbf{z}_1)\|p_{\boldsymbol{\theta}}(\mathbf{z}_2|\mathbf{z}_1))$$
$$+ \mathbb{E}_{q_{\boldsymbol{\phi}}(\mathbf{z}_2|\mathbf{w})}[\log p_{\boldsymbol{\theta}}(\mathbf{x}|\mathbf{z}_2) + \log p_{\boldsymbol{\theta}}(\hat{\mathbf{c}})].$$

$$(13)$$

Using the holistically-nested structure proposed by Xie [32], we add two side-output classifiers to the latent variables. The summation (before applying the activation) and the probability of the side-output of latent variable $\mathbf{z}_k, k \in \{1, 2\}$ are denoted as $\mathbf{S}^{(k)}, p^{(k)}(\mathbf{y})$, where $p^{(k)}(\mathbf{y}) = \psi(\mathbf{S}^{(k)})$, and ψ is the nonlinear activation function of the classifiers. Then we can construct another classifier by fusing these side-output layers:

$$\mathbf{S}^{(3)} = \boldsymbol{\alpha}_1 \mathbf{S}^{(1)} + \boldsymbol{\alpha}_2 \mathbf{S}^{(2)} \quad \text{and} \quad \mathbf{y}^{(3)} = \psi(\mathbf{S}^{(3)}) \tag{14}$$

where α_1 and α_2 are learnable weights. The objective function of the stacked VAE can be obtained by replacing the classification loss term in (9) with $\sum_{k=1}^{3} \mathcal{E}_{\text{cs}}(\mathbf{y}^{(k)})$.

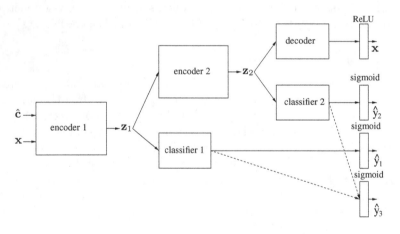

Fig. 4. The architecture of the stacked VAE network.

4.4   Implementation Details

As the scales of the two datasets are significantly different, the adopted VAE architecture differs in depth for MS COCO and Flickr30K datasets. In testing with MS COCO, we set the layer-wise number of neurons in the encoder as 5096-2500-2500/2500. We also construct the generative decoder with the size of 2500-2000-4096 and the label classifier with the size of 2500-2000-1000. For Flickr30K, the sizes of encoder, decoder and label classifier are set as 5096-2500/2500, 2500-4096 and 2500-1000, respectively. The VAE model is first pre-trained to optimize (9), learn to generate visual features and predict the visual concepts. We use the sigmoid function and ReLU [33] as the respective activation function for label classifier and for the generative decoder. ReLU is also used as the activation function of every hidden unit in the VAE. The values of hyperparameters used in training are stated as follows: batch size of 256, learning rate of 0.001 and weight decay of 0.0005. The network is trained for 20 epochs. For cost-sensitive learning, we separate $\mathcal{V}$ into two subsets, $\mathcal{V}_c$ and $\mathcal{V}_r$, that $\mathcal{V}_c$ is composed of the 100 most common words and the rest of the words belong to $\mathcal{V}_r$. We set the cost-sensitive weight λ_c to 0.001 and λ_r to 1 for balancing penalty.

To construct the stacked-VAE, we first remove the generative decoder from the pre-trained network and keep the encoder and the classifier, denoted as enc_1 and $class_1$. We initialize a new encoder (enc_2), generative decoder (rec_2) and label classifier ($class_2$) that are later attached on top of enc_1. enc_2 shares the same latent variable $\mathbf{z}_1$ with $class_1$. In MS COCO, we set the enc_2 to the size of 2500-2500-2500, rec_2 to 2500-2000-4096 and $class_2$ to 2500-2000-1000. For Flickr30K, these layer-wise sizes are set to 2500-2500, 2500-4096 and 2500-1000. We follow the same learning procedure and strategy to train the stacked-VAE and start learning the value fusion weights α_1 and α_2 from 0.5. For end-to-end fine-tuning the proposed VCNN, we set batch size to 1, learning rate to 0.000015 and weight decay to 0.0005. Stochastic Gradient Descent (SGD) is used to optimize the objective function and update parameter weights in our model. It takes seven days to train the complete model on one Titan-X GPU.

4.5 SGVB Estimator Evaluation

The VAE is learned to optimize the objective function (9), consisting of a SGVB estimator and a supervised term. We argue that such a criterion can help the recognition model (encoder) to capture global information of the contents in an image. To demonstrate the advantage of the proposed scheme, we compare our model with the naïve model mentioned in Sect. 3.2. We construct a network which has the same architecture with the VAE, except that the generative decoder is removed. Then, the network is optimized with (4) by adopting the same learning strategy described in Sect. 4.4 and using SGD to update neuron weights. We report the results on Mean Average Precision of the visual concepts over MS COCO and Flickr30K with 224×224 input images. Table 1 shows that owing to the use of supervised VAE, our model can improve the mAP from 32.90 to 33.07, 22.67 to 22.90 for MS COCO and Flickr30K, respectively. We also perform similar comparisons for the stacked-VAE. The mAP performance has increased by 0.08 and 0.09 for the respective datasets.

4.6 Cost-Sensitive Evaluation

As discussed in Sect. 4.2, the imbalanced distribution of various labels of visual concepts could cause biases on learning. Fitting common labels will be too dominant such that the penalty on error prediction of uncommon words may be ignored. We have proposed a cost-sensitive criterion to address this issue and

Table 1. Ablation evaluations on SGVB (mAP).

Method	MS COCO	Flickr30K
VCNN w/o SGVB	32.90	22.67
VCNN with SGVB	33.07	22.90
stacked-VCNN w/o SGVB	33.67	22.95
stacked-VCNN with SGVB	33.75	23.04

Table 2. Cost-sensitive learning (mAP).

Method	MS COCO	Flickr30K
VCNN w/o c.s	33.07	22.90
VCNN with c.s	33.37	23.01
stacked-VCNN w/o c.s	33.75	23.04
stacked-VCNN with c.s	34.32	23.49

help learn the model more efficiently. Table 2 reports that mAP of VCNN and that of stacked-VCNN are originally 33.07 and 33.75, and are boosted to 33.37 and 34.32 if the cost-sensitive criterion is considered in testing with the MS COCO dataset. As for Flickr30K, the criterion also improves mAP from 22.90 to 23.01 and 23.04 to 23.49 for both networks.

4.7   Stacked-VAE Evaluation

The proposed stacked architecture can help VAE to learn richer representations in both low-level and high-level latent variables, namely, z_1 and z_2. We compare the performance of non-stacked and stacked architecture on both the naïve model and VAE. In Table 1, we show that stacked VAE can improve the performance and increase mAP by 0.68 in MS COCO and 0.14 in Flickr30K. Even with the naïve network (*i.e.*, without SGVB estimator), the mAP has been boosted from 32.90 to 33.67 in MS COCO, and 22.67 to 22.95 in Flickr30K. We also evaluate such architecture jointly learned with the cost-sensitive criterion. In Table 2, the resulting models indeed benefit from stacked structure that mAP is respectively improved from 33.37 to 34.32 and 23.01 to 23.49 that both are state-of-the-art for the MS COCO and Flickr30K datasets.

4.8   More on MS COCO and Flickr30K

Besides focusing on mAP, we follow [12,13] and conduct further evaluations based on Precision at Human Recall (PHR). Chen *et al.* [34] propose an evaluation metric which takes human agreement into consideration for the task of word prediction. A "human recall" value is an estimated probability of human using the word when viewing the image. The metric computes human recall given multiple references per image and retrieves precision at this human recall value as

PHR. As pointed out in [34], it is more stable to evaluate with PHR than with mAP for the prediction of human annotations.

Two training schemes are both considered in [12,13] to learn the network models, including (1) using only fine-tuning and (2) employing weakly-supervised approach of Multiple Instance Learning (MIL) [35]. With the first scheme, the models are end-to-end fine-tuned with 224×224 input images. In the MIL formulation, a noisy-OR version of MIL [35] is adopted. That is, the probability of a word is computed by scrutinizing the instance probabilities from individual patches of the image:

$$1 - \prod_{u,v}(1 - p_{u,v}^{w_i}) \tag{15}$$

where $p_{u,v}^{w_i}$ denotes the probability of word w_i at region (u, v). For the noisy-OR MIL learning, we transform the fully-connected layers in our model to 1×1 convolutional layers and also resize the input image to 565×565. The convolutional network then performs sliding over the image with a 224×224 window and a stride of 32, which would produce a 12×12 map at both fc7 and fc8. The probability for each label is then computed ranging over this 12×12 spatial grid. Unless otherwise stated, models learned with noisy-OR MIL are marked with MIL in our reported results.

To evaluate the various models by their predictions of visual concepts, we sort the annotations into the following categories of part of speech (POS): Nouns (NN), Verbs (VB), Adjectives (JJ), Pronouns (PRP) and Prepositions (IN). We report results based on these POS tags and also compute overall mAP and PHR, which are respectively denoted as All in Tables 3, 4, 5, and 6. We take VCNN with stacked-VAE, which is learned with the cost-sensitive criterion, as our final model and compare it with [12,13]. The experimental results we obtained are state-of-the-art in both MS COCO and Flicrk30K. Table 3 shows that our model yields better mAP results than the other two. We achieve at mAP of 34.3 (direct classification) and 36.6 (MIL). When the performance is evaluated with the PHR criterion, the overall precisions by our method are respectively 47.1 and 49.9, as in Table 4. For the Flickr30K dataset, we only conduct experiments with 224×224 input images in that [12] has not yet released the code and also does not report results on Flickr30K. We implement the method of [12] on our own and follow their training strategy to obtain the experimental results. It can be inferred from both Tables 5 and 6 that the proposed VCNN still yields better results, 23.49 for mAP and 32.04 for PHR, while the techniques of [12,13] achieve similar performances.

4.9   Qualitative Results

Fig. 5 shows six examples of how VCNN correctly predicts *visual concepts* by inferring from the distribution of relevant *visual labels*. In the left image of the top row, our model predicts *shovel* should be mentioned with the knowledge of presence of boy, playing, holding, yellow and sand. In most situations,

Table 3. Mean Average Precision of MS COCO (mAP).

Method	NN	VB	JJ	DT	PRP	IN	Others	All
Classification [13]	34.9	18.1	20.5	32.8	19.2	21.8	16.3	29.0
Classification + Latent [12]	38.7	20.1	22.6	33.8	21.2	23.0	17.5	32.0
VCNN (Ours)	**41.6**	**21.5**	**24.3**	33.6	**22.2**	**23.5**	17.3	**34.3**
MILVC [13]	41.6	20.7	23.9	33.4	20.4	22.5	16.3	34.0
MILVC + Latent [12]	44.3	22.3	25.8	34.4	21.8	23.6	17.3	36.3
MIL-VCNN (Ours)	**44.6**	**22.7**	**26.1**	33.9	**22.5**	23.4	17.2	**36.6**

Table 4. Precision at Human Recall of MS COCO (PHR).

Method	NN	VB	JJ	DT	PRP	IN	Others	All
Classification [13]	42.5	30.4	33.9	40.5	30.4	30.7	23.8	38.2
Classification + Latent [12]	47.8	33.7	37.9	42.5	34.2	34.4	29.0	42.9
VCNN (Ours)	**52.7**	**36.3**	**44.1**	41.0	**36.8**	**35.9**	26.9	**47.1**
MILVC [13]	52.7	32.8	40.5	40.3	32.2	33.0	24.6	45.8
MILVC + Latent [12]	55.5	36.3	44.7	42.9	32.1	37.3	26.4	48.9
MIL-VCNN (Ours)	**56.8**	**37.2**	**44.9**	**43.1**	36.3	**37.4**	**26.7**	**49.9**

boy
playing
holding
yellow
sand
shovel

playing
ball
grass
soccer
kick

large
wall
colorful
mural
art

mountain
climbing
hill
rocky
climbs
gray

woman
white
dressed
standing
tennis
holding
hand

street
food
cart
walking
selling
vendor
restaurant

Fig. 5. Visualization results. The proposed VAE predicts that the visual concepts marked in blue should be additionally mentioned, while those marked in red should be removed, given the information of the presence of the visual labels (marked in black) (Color figure online).

humans tend not to mention the typical color of the object. For example, Rocks are commonly `gray`. Likewise, our VCNN is able to lower the probability of mentioning the specific visual concept in such a condition. In the left image of

Table 5. Mean Average Precision of Flickr30K (mAP).

Method	NN	VB	JJ	DT	PRP	IN	Others	All
Classification [13]	24.80	17.20	17.38	28.50	20.38	23.40	15.72	21.75
Classification + Latent [12]	24.61	16.52	16.79	28.42	20.30	23.40	16.43	21.44
VCNN (Ours)	**26.99**	**18.66**	**18.58**	**28.81**	**20.55**	**24.36**	15.84	**23.49**

Table 6. Precision at Human Recall of Flickr30K (PHR).

Method	NN	VB	JJ	DT	PRP	IN	Others	All
Classification [13]	31.12	27.61	27.33	32.43	31.24	32.80	18.47	29.49
Classification + Latent [12]	31.80	25.73	25.64	35.98	30.15	32.67	21.89	29.36
VCNN (Ours)	**33.82**	**28.41**	**31.72**	34.61	**35.90**	**37.78**	21.18	**32.04**

the second row, the visual concept gray is removed when mountains, climbing, hill and rocky are already detected.

5 Discussions

We have proposed a new deep net model to address the problem of human-centric annotations. Our method relies on decomposing the annotation probability that results in two relevant subtasks, where we have used a CNN and a VAE to tackle them, respectively. The integrated architecture is a variational convolutional network that can be end-to-end fine-tuned to improve predicting visual labels. Our main contribution is to introduce an effective supervised learning formulation to enable the discriminative power of a VAE, while maintaining its generative property. The experimental results we have obtained are state-of-the-art over two benchmark datasets: MS COCO and Flickr30K, under two different evaluation metrics. Two promising directions for future work are to include attention mechanisms to our model to help capture salient patches in the image, and to integrate techniques in natural language processing to better address the linguistic issues in human-centric annotations.

Acknowledgement. We thank the reviewers for their valuable comments. This work was supported in part by MOST grants 102-2221-E-001-021-MY3, 105-2221-E-001-027-MY2 and an NSF grant 1422021.

References

1. Karpathy, A., Fei-Fei, L.: Deep visual-semantic alignments for generating image descriptions. In: Proceedings of the IEEE Conference on Computer Vision and Pattern Recognition, pp. 3128–3137 (2015)

2. Donahue, J., Anne Hendricks, L., Guadarrama, S., Rohrbach, M., Venugopalan, S., Saenko, K., Darrell, T.: Long-term recurrent convolutional networks for visual recognition and description. In: Proceedings of the IEEE Conference on Computer Vision and Pattern Recognition, pp. 2625–2634 (2015)
3. Xu, K., Ba, J., Kiros, R., Courville, A., Salakhutdinov, R., Zemel, R., Bengio, Y.: Show, attend and tell: Neural image caption generation with visual attention. arXiv preprint arXiv:1502.03044 (2015)
4. Wu, Q., Shen, C., Liu, L., Dick, A., van den Hengel, A.: What value do explicit high level concepts have in vision to language problems? In: Proceedings IEEE Conference Computer Vision and Pattern Recognition, vol. 2(4) (2016)
5. Yao, L., Torabi, A., Cho, K., Ballas, N., Pal, C., Larochelle, H., Courville, A.: Describing videos by exploiting temporal structure. In: Proceedings of the IEEE International Conference on Computer Vision, pp. 4507–4515 (2015)
6. Venugopalan, S., Rohrbach, M., Donahue, J., Mooney, R., Darrell, T., Saenko, K.: Sequence to sequence-video to text. In: Proceedings of the IEEE International Conference on Computer Vision, pp. 4534–4542 (2015)
7. Antol, S., Agrawal, A., Lu, J., Mitchell, M., Batra, D., Zitnick, C.L., Parikh, D.: Vqa: Visual question answering. In: International Conference on Computer Vision (ICCV) (2015)
8. Mansimov, E., Parisotto, E., Ba, J., Salakhutdinov, R.: Generating images from captions with attention. In: ICLR (2016)
9. Berg, A.C., Berg, T.L., Daume III., H., Dodge, J., Goyal, A., Han, X., Mensch, A., Mitchell, M., Sood, A., Stratos, K., et al.: Understanding and predicting importance in images. In: 2012 IEEE Conference on Computer Vision and Pattern Recognition (CVPR), pp. 3562–3569. IEEE (2012)
10. Lin, T.-Y., Maire, M., Belongie, S., Hays, J., Perona, P., Ramanan, D., Dollár, P., Zitnick, C.L.: Microsoft COCO: common objects in context. In: Fleet, D., Pajdla, T., Schiele, B., Tuytelaars, T. (eds.) ECCV 2014. LNCS, vol. 8693, pp. 740–755. Springer, Cham (2014). doi:10.1007/978-3-319-10602-1_48
11. Plummer, B.A., Wang, L., Cervantes, C.M., Caicedo, J.C., Hockenmaier, J., Lazebnik, S.: Flickr30k entities: collecting region-to-phrase correspondences for richer image-to-sentence models. In: Proceedings of the IEEE International Conference on Computer Vision, pp. 2641–2649 (2015)
12. Misra, I., Zitnick, C.L., Mitchell, M., Girshick, R.: Seeing through the human reporting bias: visual classifiers from noisy human-centric labels. In: CVPR (2016)
13. Fang, H., Gupta, S., Iandola, F., Srivastava, R.K., Deng, L., Dollár, P., Gao, J., He, X., Mitchell, M., Platt, J.C., et al.: From captions to visual concepts and back. In: Proceedings of the IEEE Conference on Computer Vision and Pattern Recognition, pp. 1473–1482 (2015)
14. Kingma, D.P., Welling, M.: Auto-encoding variational bayes. In: Proceedings of the 2nd International Conference on Learning Representations (ICLR), vol. 2014 (2013)
15. Devlin, J., Gupta, S., Girshick, R., Mitchell, M., Zitnick, C.L.: Exploring nearest neighbor approaches for image captioning. arXiv preprint arXiv:1505.04467 (2015)
16. Klein, B., Lev, G., Sadeh, G., Wolf, L.: Associating neural word embeddings with deep image representations using fisher vectors. In: Proceedings of the IEEE Conference on Computer Vision and Pattern Recognition, pp. 4437–4446 (2015)
17. Jia, X., Gavves, E., Fernando, B., Tuytelaars, T.: Guiding long-short term memory for image caption generation. arXiv preprint arXiv:1509.04942 (2015)

18. Rohrbach, M., Qiu, W., Titov, I., Thater, S., Pinkal, M., Schiele, B.: Translating video content to natural language descriptions. In: Proceedings of the IEEE International Conference on Computer Vision, pp. 433–440 (2013)

19. Thomason, J., Venugopalan, S., Guadarrama, S., Saenko, K., Mooney, R.J.: Integrating language and vision to generate natural language descriptions of videos in the wild. In: COLING, vol. 9 (2014)

20. Turakhia, N., Parikh, D.: Attribute dominance: What pops out? In: Proceedings of the IEEE International Conference on Computer Vision, pp. 1225–1232 (2013)

21. Yun, K., Peng, Y., Samaras, D., Zelinsky, G., Berg, T.: Studying relationships between human gaze, description, and computer vision. In: Proceedings of the IEEE Conference on Computer Vision and Pattern Recognition, pp. 739–746 (2013)

22. Chen, X., Gupta, A.: Webly supervised learning of convolutional networks In: Proceedings of the IEEE International Conference on Computer Vision, pp. 1431–1439 (2015)

23. Izadinia, H., Russell, B.C., Farhadi, A., Hoffman, M.D., Hertzmann, A.: Deep classifiers from image tags in the wild. In: Proceedings of the 2015 Workshop on Community-Organized Multimodal Mining: Opportunities for Novel Solutions, pp. 13–18. ACM (2015)

24. Joulin, A., van der Maaten, L., Jabri, A., Vasilache, N.: Learning visual features from large weakly supervised data. arXiv preprint arXiv:1511.02251 (2015)

25. Kingma, D.P., Mohamed, S., Rezende, D.J., Welling, M.: Semi-supervised learning with deep generative models. In: Advances in Neural Information Processing Systems, pp. 3581–3589 (2014)

26. Burda, Y., Grosse, R., Salakhutdinov, R.: Importance weighted autoencoders. arXiv preprint arXiv:1509.00519 (2015)

27. Fabius, O., van Amersfoort, J.R.: Variational recurrent auto-encoders. arXiv preprint arXiv:1412.6581 (2014)

28. Gregor, K., Danihelka, I., Graves, A., Wierstra, D.: Draw: A recurrent neural network for image generation. arXiv preprint arXiv:1502.04623 (2015)

29. Chung, J., Kastner, K., Dinh, L., Goel, K., Courville, A.C., Bengio, Y.: A recurrent latent variable model for sequential data. In: Advances in Neural Information Processing Systems, pp. 2962–2970 (2015)

30. Kulkarni, T.D., Whitney, W.F., Kohli, P., Tenenbaum, J.: Deep convolutional inverse graphics network. In: Cortes, C., Lawrence, N.D., Lee, D.D., Sugiyama, M., Garnett, R. (eds.): Advances in Neural Information Processing Systems, vol. 28, pp. 2539–2547. Curran Associates, Inc., New York (2015)

31. Simonyan, K., Zisserman, A.: Very deep convolutional networks for large-scale image recognition. In: International Conference on Learning Representations (2015)

32. Xie, S., Tu, Z.: Holistically-nested edge detection. In: Proceedings of the IEEE International Conference on Computer Vision, pp. 1395–1403 (2015)

33. Nair, V., Hinton, G.E.: Rectified linear units improve restricted boltzmann machines. In: Proceedings of the 27th International Conference on Machine Learning (ICML 2010), pp. 807–814 (2010)

34. Chen, X., Fang, H., Lin, T.Y., Vedantam, R., Gupta, S., Dollár, P., Zitnick, C.L.: Microsoft COCO captions: Data collection and evaluation server. arXiv preprint arXiv:1504.00325 (2015)

35. Zhang, C., Platt, J.C., Viola, P.A.: Multiple instance boosting for object detection. In: Advances in Neural Information Processing Systems, pp. 1417–1424 (2005)

Anticipating Accidents in Dashcam Videos

Fu-Hsiang Chan[1]([⊠]), Yu-Ting Chen[1], Yu Xiang[2], and Min Sun[1]

[1] Department of Electrical Engineering, National Tsing Hua University,
Hsinchu, Taiwan
corgi1205@gmail.com, s728039@gmail.com, sunmin@ee.nthu.edu.tw
[2] Department of Computer Science and Engineering,
University of Washington, Seattle, USA
yuxiang@cs.washington.edu

Abstract. We propose a Dynamic-Spatial-Attention (DSA) Recurrent Neural Network (RNN) for anticipating accidents in dashcam videos (Fig. 1). Our DSA-RNN learns to (1) distribute soft-attention to candidate objects dynamically to gather subtle cues and (2) model the temporal dependencies of all cues to robustly anticipate an accident. Anticipating accidents is much less addressed than anticipating events such as changing a lane, making a turn, etc., since accidents are rare to be observed and can happen in many different ways mostly in a sudden. To overcome these challenges, we (1) utilize state-of-the-art object detector [3] to detect candidate objects, and (2) incorporate full-frame and object-based appearance and motion features in our model. We also harvest a diverse dataset of 678 dashcam accident videos on the web (Fig. 3). The dataset is unique, since various accidents (e.g., a motorbike hits a car, a car hits another car, etc.) occur in all videos. We manually mark the time-location of accidents and use them as supervision to train and evaluate our method. We show that our method anticipates accidents about 2 s before they occur with 80% recall and 56.14% precision. Most importantly, it achieves the highest mean average precision (74.35%) outperforming other baselines without attention or RNN.

1 Introduction

Driving a car by an Artificial Intelligent (AI) agent has been one of the greatest dream in AI for decades. In the past 10 years, significant advances have been achieved. Since 2009, Google's self-driving car has accumulated 1,011,338 autonomous driving miles on highways and busy urban streets [4]. Recently, Tesla's Autopilot can drive on highway-like environment primarily relying on cheap vision sensors. Despite these great advances, there are two major challenges. The first challenge is how to drive safely with other "human drivers". Google's self-driving car is involved in 12 minor accidents [4] mostly caused by

Electronic supplementary material The online version of this chapter (doi:10.1007/978-3-319-54190-7_9) contains supplementary material, which is available to authorized users.

© Springer International Publishing AG 2017
S.-H. Lai et al. (Eds.): ACCV 2016, Part IV, LNCS 10114, pp. 136–153, 2017.
DOI: 10.1007/978-3-319-54190-7_9

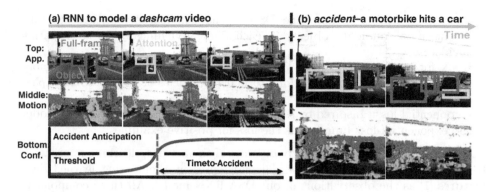

Fig. 1. Illustration of Accident Anticipation. Given a dashcam video (playing from left to right in panel (a)), we extract both appearance (top-row: App.) [1] and motion (middle-row) [2] features. For appearance feature, we consider both full-frame and object-level (red-boxes) features. For motion feature, keypoints with large motion are shown in green dots. Our proposed dynamic-spatial-attention Recurrent Neural Network (RNN) distributes attention to objects (yellow-boxes indicate strong attention) and predicts the confidence of accident anticipation at each frame (bottom-row: Conf.). Once the confidence reaches a threshold (brown-dash-line), our system will trigger an accident alert "time-to-accident" seconds before the true accident (Panel (b)). (Color figure online)

other human drivers. This suggests that a self-driving car should learn to anticipate others' behaviors in order to avoid these accidents. The other important challenge is how to scale-up the learning process. In particular, how to learn from as many corner cases as possible? We propose to take advantage of the cheap and widely available dashboard cameras (later referred to as dashcam) to observe corner cases at scale.

Dashcam is very popular in places such as Russia, Taiwan and Korean. For instance, dashcams are equipped on almost all new cars in the last three years in Taiwan. Its most common use case is to record how accidents occur in order to clarify responsibilities. As a result, many dashcam videos involving accidents have been recorded. Moreover, according to statistics, ~ 90 people died per day due to road accidents in the US [5]. In order to avoid these casualty, we propose a method to learn from dashcam videos for anticipating various accidents and providing warnings a few seconds before the accidents occur (see Fig. 1).

Learning to anticipate accidents is an extremely challenging task, since accidents are very diverse and they typically happen in a sudden. Human drivers learn from experiences to pay attention on subtle cues including scene semantic, object appearance and motion. We propose a Dynamic-Spatial-Attention (DSA) Recurrent Neural Network (RNN) to anticipate accidents before they occur. Our method consists of three important model designs:

- Dynamic-spatial-attention: The DSA mechanism learns to distribute soft-attention to candidate objects in each frame dynamically for anticipating accidents.
- RNN for sequence modeling: We use RNN with Long Short-Term Memory (LSTM) [6] cells to model the long-term dependencies of all cues to anticipate an accident.
- Exponential-loss: Inspired by [7] on anticipating drivers' maneuvers, we adopt the exponential-loss function as the loss for positive examples.

To effectively extract cues, we rely on state-of-the-art deep learning approaches to reliably detect moving objects [3] and represent them using learned deep features [1] as the observations of our DSA-RNN model. All these components together enable accident anticipation using a cheap vision-based sensor.

In order to evaluate our proposed method, we download 678 dashcam videos with high video quality (720p in resolution) from the web. The dataset is unique, since various accidents (e.g., a motorbike hits a car, a car hits another car, etc.) occur in all videos. Moreover, most videos are captured across six cities in Taiwan. Due to the crowded road with many moving objects and complicated street signs/billboards, it is a challenging dataset for vision-based method (Fig. 3-Right). For each video, we manually annotated the bounding boxes of car, motorbike, bicycle, human, and the time when the accident occurs. 58 out of 678 videos are used only for training the object detector. Among the remaining 620 videos, we manually select 620 positive clips and 1130 negative clips, where each clip consists of 100 frames. A positive clip contains the moment of accident at the last 10 frames, and a negative clip contains no accident. We split the dataset into training and testing, where the number of training clips is roughly three times the number of testing clips. We show in experiments that all our model designs help improve the anticipation accuracy. Our method can anticipate accident 1.8559 s ahead with high recall (80%) and reasonable precision (56.14%).

In summary, our method has the following contributions:

- We show that, using deep learning, a vision-based sensor (dashcam) can provide subtle cues to anticipate accidents in complex urban driving scenarios.
- We propose a dynamic-spatial-attention RNN to achieve accident anticipation 1.8559 s ahead with 80% recall and 56.14% precision.
- We show that potentially a vast amount of dashcam videos can be used to improve self-driving car ability.
- We introduce the first crowd-sourced dashcam video dataset for accident anticipation available online at http://aliensunmin.github.io/project/dashcam/.

In the following sections, we will first describe the related work. Then, our method is introduced in Sect. 3. Finally, experiments are discussed in Sect. 4.

2   Related Work

We first discuss related work of anticipation in computer vision, robotics and intelligent vehicle. Then, we mention recent works incorporating RNN with attention mechanism in computer vision. Finally, we compare our dashcam accident anticipation dataset with two large-scale dashcam video datasets.

2.1 Anticipation

A few works have been proposed to anticipate events — classify "future" event given "current" observation. Ryoo [8] proposes a feature matching techniques for early recognition of unfinished activities. Hoai and Torre [9] propose a max-margin-based classifier to predict subtle facial expressions before they occur. Lan et al. [10] propose a new representation to predict the future actions of people in unconstrained "in-the-wild" footages. Our method is related to event anticipation, since accident can be consider as a special event. Moreover, our dashcam videos are very challenging, since these videos are captured by different moving cameras observing static stuff (e.g., building, road signs/billboards, etc.) and moving objects (e.g., motorbikes, cars, etc.) on the road. Therefore, we propose a dynamic-spatial-attention mechanism to discovery subtle cues for anticipating accidents.

Anticipation has been applied in tasks other than event prediction. Kitani et al. [11] propose to forecast human trajectory by surrounding physical environment (e.g., road, pavement, etc.). The paper also shows that the forecasted trajectory can be used to improve object tracking accuracy. Yuen and Torralba [12] propose to predict motion from still images. Julian et al. [13] propose a novel visual appearance prediction method based on mid-level visual elements with temporal modeling methods.

Event anticipation is also popular in the robotic community [14–17]. Wang et al. [14] propose a latent variable model for inferring unknown human intentions. Koppula and Saxena [15] address the problem of anticipating future activities from RGB-D data. A real robotic system has executed the proposed method to assist humans in daily tasks. [16,17] also propose to anticipate human activities for improving human-robot collaboration.

There are also many works for predicting drivers' intention in the intelligent vehicle community. [18–21] have used vehicle trajectories to predict the intent for lane change or turn maneuver. A few works [22–24] address maneuver anticipation through sensory-fusion from multiple cameras, GPS, and vehicle dynamics. However, most methods assume that informative cues always appear at a fixed time before the maneuver. [7,25] are two exceptions which use an input-output HMM and a RNN, respectively, to model the temporal order of cues. On one hand, our proposed method is very relevant to [7], since they also use RNN to model temporal order of cues. On the other hand, anticipating accidents is mush less addressed than anticipating specific-maneuvers such as lane change or turn, since accidents are rare to be observed and can happen in many different ways mostly in a sudden. In order to address the challenges in accident anticipation, our method incorporates a RNN with spatial attention mechanism to focus on object-specific cues at each frame dynamically. In summary, all these previous methods focus on anticipating specific-maneuvers such as lane change or turn.

In contrast, we aim at anticipating various accidents observed in naturally captured dashcam videos.

2.2 RNN with Attention

Recently, RNN with attention has been applied on a few core computer vision tasks: video/image captioning and object recognition. On one hand, RNN with soft-attention has been used to jointly model a visual observation and a sentence for video/image caption generation [26,27]. Yao et al. [26] incorporate a "temporal" soft-attention mechanism to select critical frames to generate a video caption. Xu et al. [27] demonstrate the power of spatial-attention mechanism for generating an image caption. Compared to [27], our proposed dynamic-spatial-attention RNN has two main differences: (1) their spatial-attention is for a single frame, whereas our spatial-attention changes dynamically at each frame in a sequence; (2) rather than applying spatial-attention on a regular grid, we apply a state-of-the-art object detector [3] to extract object candidates for assigning the spatial-attention. On the other hand, a few RNN with hard-attention models have been proposed. Mnih et al. [28] propose to train a RNN with hard-attention model with a reinforcement learning method for image classification tasks. Similarly, Ba et al. [29] propose a RNN with hard-attention model to both localize and recognize multiple objects in an image. They apply their method on transcribing house number sequences from Google Street View images.

2.3 Dashcam Video Dataset

Dashcam videos or videos captured by cameras on vehicles have been collected for studying many recognition and localization tasks. For instance, CamVid [30], Leuven [31], and Daimler Urban Segmentation [32] have been introduced to study semantic understanding of urban scenes. There are also two recently collected large-scale datasets [33,34]. KITTI [33] is a well-known vision benchmark dataset to study vision-based self-driving tasks including object detection, multiple-objects tracking, road/lane detection, semantic segmentation, and visual odometry. KITTI consists of videos captured around the mid-size city of Karlsruhe, in rural areas and on highways. Moreover, all videos are captured by vehicles with the same equipment under normal driving situation (i.e., no accident), whereas our dataset consists of accident videos harvested from many online users across six cities. Recently a large-scale dashcam dataset [34] is released for evaluating semantic segmentation task. It consists of frames captured in 50 cities. Among them, 5 k frames and 30 k frames are labeled with detail and coarse semantic labels, respectively. Despite the diverse observation in this new dataset, most frames are still captured under normal driving situation. We believe our dashcam accident anticipation dataset is one of the first crowd-sourced datasets for anticipating accidents.

3   Our System

We formally define accident anticipation and then present our proposed Dynamic-spatial-attention (DSA) Recurrent Neural Network (RNN). The goal of accident anticipation is to use observations in a dashcam video to predict an accident before it occurs. We define our observations and accident label for the jth video as $((\mathbf{x}_1, \mathbf{x}_2, \ldots, \mathbf{x}_T)_j, y_j)$, where $\mathbf{x}_t$ is the observation at frame t, T is the total number of frames in the video, and y_j is the accident label to specify at which frame the accident started (defined below). For instance, if $y = \hat{t}$, any $t < \hat{t}$ is a frame before the accident. With a bit abuse of notation, we use $y = \infty$ to specify the video as free from accident. During training, all the observations and accident labels are given to train a model for anticipation. While in testing, our system are given an observation of $\mathbf{x}_t$ one at a time following the order of the frames. The goal is to predict the future accident as early as possible given the observations $(\mathbf{x}_1, \mathbf{x}_2, \ldots, \mathbf{x}_t) | t < y$ before accident occurs at frame y.

Our proposed dynamic-spatial-attention RNN is built upon standard RNN based on Long Short-Term Memory (LSTM) cells. We first give preliminaries of the standard RNN and LSTM before we describe the dynamic-spatial-attention mechanism (Sect. 3.2) and training procedure for anticipation (Sect. 3.3).

3.1   Preliminaries

RNN. Standard RNN is a special type of network which takes a sequence of observations $(\mathbf{x}_1, \mathbf{x}_2, \ldots, \mathbf{x}_T)$ as input and outputs a sequence of learned hidden representations $(\mathbf{h}_1, \mathbf{h}_2, \ldots, \mathbf{h}_T)$, where $\mathbf{h}_t$ encodes the sequence observations $(\mathbf{x}_1, \mathbf{x}_2, \ldots, \mathbf{x}_t)$ up to frame t. The hidden representation is generated by a recursive equation below,

$$\mathbf{h}_t = g(\mathbf{W}\mathbf{x}_t + \mathbf{H}\mathbf{h}_{t-1} + \mathbf{b}),\qquad(1)$$

where $g(\cdot)$ is a non-linear function applied element-wise (e.g., sigmoid), $\mathbf{W}, \mathbf{H}, \mathbf{b}$ are the model parameters to be learned. The hidden representation $\mathbf{h}_t$ is used to predict a target output. In our case, the target output is the probability of discrete event a_t,

$$\mathbf{a}_t = \mathrm{softmax}(\mathbf{W}_a \mathbf{h}_t + \mathbf{b}_a),\qquad(2)$$

where $\mathbf{a}_t = [\ldots, a_t^i, \ldots]$. The softmax function computes the probability of events (i.e., $\sum_i a_t^i = 1$), and $\mathbf{W}_a, \mathbf{b}_a$ are the model parameters to be learned. For accident anticipation, accident and non-accident are the discrete events and their probabilities are denoted by a_t^0 and a_t^1, respectively. In this work, we denote matrices with bold, capital letters, and vectors with bold, lower-case letters. The recursive design of RNN is clear and easy to understand. However, it suffers from a well-known problem of vanishing gradients [35] such that it is hard to train a RNN to capture long-term dependencies. A common way to address this issue is to replace function $g(\cdot)$ with a complicated Long Short-Term (LSTM) Memory cell [6]. We now give an overview of the LSTM cell and then define our dynamic-spatial-attention RNN based on LSTM cells.

Long-Short Term Memory Cells. LSTM introduces a memory cell **c** to maintain information over time. It can be considered as the state of the recurrent system. LSTM extends the standard RNN by replacing the recursive equation in Eq. 1 with

$$(\mathbf{h}_t, \mathbf{c}_t) = \text{LSTM}(\mathbf{x}_t, \mathbf{h}_{t-1}, \mathbf{c}_{t-1}),\qquad(3)$$

where the memory cell **c** allows RNN to model long-term contextual dependencies. To control the interaction among the input, memory cell, and output, three gates are designed: input gate **i**, forget gate **f**, and output gate **o** (see Fig. 3.1). Each gate is designed to either block or non-block information. At each frame t, LSTM first computes gate activations: $\mathbf{i}_t, \mathbf{f}_t$ (Eq. 4, 5) and updates its memory cell from $\mathbf{c}_{t-1}$ to $\mathbf{c}_t$ (Eq. 6). Then it computes the output gate activation $\mathbf{o}_t$ (Eq. 7), and outputs a hidden representation $\mathbf{h}_t$ (Eq. 8).

Workflow of LSTM [6].

$$\mathbf{i}_t = \sigma(\mathbf{W}_i\mathbf{x}_t + \mathbf{U}_i\mathbf{h}_{t-1} + \mathbf{V}_i\mathbf{c}_{t-1} + \mathbf{b}_i)\qquad(4)$$

$$\mathbf{f}_t = \sigma(\mathbf{W}_f\mathbf{x}_t + \mathbf{U}_f\mathbf{h}_{t-1} + \mathbf{V}_f\mathbf{c}_{t-1} + \mathbf{b}_f)\qquad(5)$$

$$\mathbf{c}_t = \mathbf{f}_t \odot \mathbf{c}_{t-1} + \mathbf{i}_t \odot \rho(\mathbf{W}_c\mathbf{x}_t + \mathbf{U}_c\mathbf{h}_{t-1} + \mathbf{b}_c)\qquad(6)$$

$$\mathbf{o}_t = \sigma(\mathbf{W}_o\mathbf{x}_t + \mathbf{U}_o\mathbf{h}_{t-1} + \mathbf{V}_o\mathbf{c}_t + \mathbf{b}_o)\qquad(7)$$

$$\mathbf{h}_t = \mathbf{o}_t \odot \rho(\mathbf{c}_t)\qquad(8)$$

We now define the common notations in Eq. 4–8. $\odot$ is an element-wise product, and the logistic function σ and the hyperbolic tangent function ρ are both applied element-wise. $\mathbf{W}_*, \mathbf{V}_*, \mathbf{U}_*, \mathbf{b}_*$, and $\mathbf{V}_*{}^1$ are the parameters. Note that the input and forget gates participate in updating the memory cell (Eq. 6). More specifically, forget gate controls the part of memory to forget, and the input gate allows newly computed values (based on the current observation and previous hidden representation) to add to the memory cell. The output gate together with the memory cell computes the hidden representation (Eq. 8). Since the current memory cell only goes through a binary operation (i.e., forget gate) and a summation (Eq. 6), the gradient with respect to the memory cell does not vanish as fast as standard RNN. We now describe our dynamic-spatial-attention RNN architecture based on LSTMs for anticipation.

3.2   Dynamic Spatial Attention

For accident anticipation, we would like our RNN to focus on spatial-specific observations corresponding to vehicles, pedestrian, or other objects in the scene. We propose to learn a dynamic spatial attention model to focus on candidate objects on specific spatial locations at each frame (Fig. 2). We assume that there

---

[1] The subscript $*$ denotes any symbol.

are J spatial-specific object observations $\mathbf{X}_t = \{\hat{\mathbf{x}}_t^j\}_{j \in \{1,...,J\}}$ and their corresponding locations $\mathcal{L}_t = \{\mathbf{l}_t^j\}_{j \in \{1,...,J\}}$. We propose to adapt the recently proposed soft-attention mechanism [27] to take dynamic weighted-sum of spatially-specific object observations $\mathbf{X}_t$ as below,

$$\phi(\mathbf{X}_t, \boldsymbol{\alpha}_t) = \sum_{j=1}^{J} \alpha_t^j \hat{\mathbf{x}}_t^j, \qquad (9)$$

where $\phi(\mathbf{X}_t, \boldsymbol{\alpha}_t)$ is the dynamic weighted-sum function[2], $\sum_{j=1}^{J} \alpha_t^j = 1$ and α_t^j is computed at each frame t along with the LSTM. We refer $\boldsymbol{\alpha}_t = \{\alpha_t^j\}_j$ as the attention weights. They are computed from unnormalized attention weights $\mathbf{e}_t = \{e_t^j\}_j$ as below,

$$\alpha_t^j = \frac{\exp(e_t^j)}{\sum_j \exp(e_t^j)}. \qquad (10)$$

We design the unnormalized attention weights to measure the relevance between the previous hidden representation $\mathbf{h}_{t-1}$ and each spatial-specific observation $\hat{\mathbf{x}}_t^j$ as below,

$$e_t^j - \mathbf{w}^T \rho(\mathbf{W}_e \mathbf{h}_{t-1} + \mathbf{U}_e \hat{\mathbf{x}}_t^j + \mathbf{b}_e), \qquad (11)$$

where $\mathbf{w}, \mathbf{W}_e, \mathbf{U}_e$, and $\mathbf{b}_e$ are model parameters. Then, we replace all $\mathbf{x}_t$ in Eq. 4, 5, 6, 7, 8 with $\phi(\mathbf{X}_t)$, which is a shorthand notation of $\phi(\mathbf{X}_t, \boldsymbol{\alpha}_t)$ in Eq. 9. Note that Xu et al. [27] apply spatial-attention on a regular grid in a single frame, whereas our method applies spatial-attention on candidate object regions detected by state-of-the-art deep-learning-based detector [3]. Moreover, rather than applying spatial-attention on a single frame for caption generation, we apply spatial-attention on a "sequence" of frames dynamically which are jointly modeled using RNN.

Combining with Full-Frame Feature. Spatial-specific object observations incorporate detail cues of moving objects which might involve in the accident. However, full-frame feature can capture important cues related to the scene or motion of the camera, etc. We propose two ways to combine the full-frame feature with spatial-specific object feature.

- Concatenation. We can simply concatenate the full-frame feature $\mathbf{x}^F$ with the weighted-summed object feature $\phi(\mathbf{X})$ as $\mathbf{x} = [\mathbf{x}^F; \phi(\mathbf{X})]$.
- Weighted-sum. We can treat the full-frame as an object as large as the whole frame. Then, the attention model will assign a soft-weight for the full-frame feature using the mechanism described above. Note that this way of combining reduces the combined feature dimension by two.

---

[2] α_t is often omitted for conciseness.

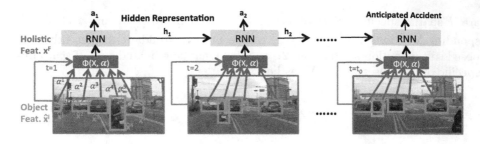

Fig. 2. The model visualization of our dynamic-spatial-attention RNN which takes weighted sum of the full-frame feature $\mathbf{x}^F$ and object features $\mathbf{X} = \{\hat{\mathbf{x}}^i\}_i$ as observation (one variant in Sect. 3.2). This example shows that the accident is anticipated at time t_0, which is $y-t_0$ seconds before true accident. $\hat{\mathbf{x}}^i$ denotes observation of the ith object, The function $\phi(\mathbf{X}, \boldsymbol{\alpha})$ in Eq. 9 computes the weighted-sum of all features. $\mathbf{a}_t$ is the probability of a future accident defined in Eq. 2. $\mathbf{h}_t$ is the learned hidden representation which propagates to the next RNN (see Eq. 7, 8). Feat. stands for feature. Note that the subscript t is omitted when it is clear from the context that a variable is time-specific.

3.3   Training Procedure

Accident probability $\mathbf{a}_t$ is the targeted output of our DSA-RNN. We describe its corresponding loss function.

Anticipation Loss. Intuitively, the penalty of failing to anticipate an accident at a frame very close to the accident should be higher than the penalty at a frame far away from the accident. Hence, we use the following exponential loss [7] for positive accident training videos,

$$L_p(\{\mathbf{a}_t\}) = \sum_t -e^{-\max(0, y-t)} \log(a_t^0), \tag{12}$$

where the accident happens at frame y, and a_t^0 is the anticipated probability of accident at frame t. For negative training videos, we use the standard cross-entropy loss,

$$L_n(\{\mathbf{a}_t\}) = \sum_t -\log(a_t^1), \tag{13}$$

where a_t^1 is the anticipated probability of non-accident at frame t.
The final loss is the sum of all these losses across all training videos,

$$\sum_{j \in P} L_p(\{\mathbf{a}_t^j\}) + \sum_{j \in N} L_n(\{\mathbf{a}_t^j\}), \tag{14}$$

where j is the video index, $P = \{j; y_j \neq \infty\}$ is the set of positive videos, and $N = \{j; y_j = \infty\}$ is the set of negative videos. Since all loss functions are differentiable, we use stochastic gradient with the standard back-propagation through time (BPTT) algorithm [36] to train our model. Detail training parameters are described in Sect. 4.2.

4   Experiments

In this section, we first introduce our novel dashcam accident dataset and describe the implementation details. Finally, we describe all the baseline methods for comparison and report the experimental results.

4.1   Dashcam Accident Dataset

A dashcam is a cheap aftermarket camera, which can be mounted inside a vehicle to record street-level visual observation from the driver's point-of-view (see Fig. 3-Top-Right-Corner). In certain places such as Russia and Taiwan, dashcams are equipped on almost all new cars in the last three years. Hence, a large number of dashcam videos have been shared on video sharing websites such as YouTube[3]. Instead of recording dashcam videos ourselves similar to other datasets [33,34], we harvest dashcam videos shared online from many users. In particular, we target at accident videos with human annotated address information or GPS locations. In this way, we have collected various accident videos with high video quality (720p in resolution). The dataset consists of 678 videos captured in six major cities in Taiwan (Fig. 3-Right). Our diverse accidents include: 42.6% motorbike hits car, 19.7% car hits car, 15.6% motorbike hits motorbike, and 20% other types. Figure 3 shows a few sample videos and their corresponding locations on Google map. We can see that almost all big cities on the west coast of Taiwan are covered. Our videos are more challenging than videos in the KITTI [33] dataset due to the following reasons,

- Complicated road scene: The street signs and billboards in Taiwan are significantly more complex than those in Europe.
- Crowded streets: The number of moving cars, motorbikes, and pedestrians per frame are typically larger than other datasets [33,34].
- Diverse accidents: Accidents involving cars, motorbikes, etc. are all included in our dataset.

We manually annotate the temporal locations of accidents and the moving objects in each video. 58 videos are used only for training the object detector. Among the remaining 620 videos, we sample 1750 clips, where each clip consists of 100 frames (5 s). These clips contain 620 positive clips containing the moment of accident at the last 10 frames[4], and 1130 negative clips containing no accidents. We randomly split the dataset into training and testing, where the number of training clips is about three times the number of testing clips: 1284 training clips 455 positive and 829 negative clips) and 466 testing clips (165 positive and 301 negative clips). We will make all the original videos, their annotated accident locations, and our sampled clips publicly available.

---

[3] https://www.youtube.com/watch?v=YHFvSCAg4DE.
[4] Hence, we use the first 90 frames to anticipate accidents.

Fig. 3. Our dashcam accident dataset consists of a large number of diverse accident dashcam videos (Right-panel). It typically contains more moving objects and complicated street signs/billboards than the KITTI [33] dataset (Left-panel).

4.2   Implementation Details

Features. Both appearance and motion cues are intuitively important for accident anticipation. We extract both single-frame-based appearance and clip-based local motion features. For capturing appearance, we use pre-trained VGG [1] network to extract a fixed 4096 dimension feature for each frame at 20fps. For motion feature, we extract improved dense trajectory (IDT) feature [2][5] for a clip consisting of 5 consecutive frames. Then, we first use PCA to reduce the trajectory feature dimension to 100, and train a Gaussian-Mixture-Model (GMM) with 64 clusters. Finally, we use the 1st order statistic of fisher vector encoding to compute a fixed 6400 dimension feature. For VGG, we extract features both on a full-frame and on each candidate object, and we combine them following the methods described in Sect. 3.2. For IDT, we only extract features on a full-frame, since many candidate object regions do not contain enough trajectories to compute a robust IDT feature. In addition, we design a Relativity-Motion (RM) features using relative 2D motion among nearby objects ˙(5 × 5 median motion encoding).

Candidate Objects. As mentioned in Sect. 3.2, we assume our model observes J spatial-specific object regions. Our proposed dynamic-spatial-attention model will learn to distribute its soft-attention to these regions. Given an image, there are a huge number of possible object regions, when considering all locations and scales. To limit the number of object regions, we use a state-of-the-art object

---

[5] IDT also includes Histogram of Oriented Gradient (HOG) [37] (an appearance feature) on the motion boundary.

detector [3] to generate less than 20 candidate object regions for each frame. Since the object detector pre-trained on MSCOCO dataset [38] is not trained to detect objects in street scenes, we finetune the last three fully connected layers on street scenes data including KITTI dataset [33], our collected 58 videos, and randomly sampled 10 frames in 455 positive training clips. Our finetuned detector achieves 52.3% mean Average Precision (mAP) across five categories[6], which significantly outperforms the pre-trained detector (41.53%) (see more detail in supplementary material).

Model Learning. All experiments use 0.0001 learning rate, 40 maximum epoch, 10 batch size. We implement our method on TensorFlow [39].

4.3   Evaluation Metric

We evaluate every method based on the correctness of anticipating a future accident. Given a video, a method needs to generate the confidence/probability of future accident a_t^0 at each frame. At frame t when the confidence is higher than or equal to a threshold q, the method claims that there will be an accident in the future. If the video is an accident video, this is a True Positive (TP) anticipation. The accident is correctly anticipated at frame t, which is $y - t$ frames before it occurs at frame y. We define $y - t$ as time-to-accident. If the video is a non-accident video, this is a False Positive (FP) anticipation. On the other hand, if all the confidence $\{a_t^0\}_{t < y}$ are smaller than the threshold q, the method claims that there will not be an accident in the future. If the video is an accident video, this is a False Negative (FN) prediction. If the video is a non-accident video, this is a True Negative (TN) prediction. For each threshold q, we can compute the precision $= \frac{TP}{TP+FP}$ and recall $= \frac{TP}{TP+FN}$. By changing the threshold q, we can compute many pairs of precision and recall and plot the precision v.s. recall curve (see Fig. 4-Left). Given a sequence of precision and recall pairs, we can compute the average precision, which is used to show the system's overall accuracy. For each threshold q, we can also collect all the Time-to-accident (ToA) of the true positive anticipation, and compute the average ToA as the expected anticipation time.

4.4   Baseline Methods

We compare different variants of our method using RNN and a few baseline methods without modeling the temporal relation between frames. Here we present these variants and baselines as a series of simplifications on our proposed method.

– Dynamic-Spatial-Attention RNN. This is our proposed method. Our method has three variants (see Sect. 3.2): (1) no full-frame features, only attention on object candidates (D); (2) weighted-summing full-frame feature with object-specific features (F+D-sum); (3) concatenating full-frame features with object features (F+D-con.).

---

[6] Human, bicycle, motorbike, car and bus.

- Average-Attention RNN. We replace the inferred spatial-attention with a average attention (no dynamic attention), where all candidate object observations are average-pooled to a fixed dimension feature. Then, we either use only the average attention feature (avg.-D), or concatenate the full-frame feature with the average attention feature (F+avg.-D-con.). These baselines highlight the effect of using dynamic spatial-attention.
- Frame-based RNN. We remove all candidate object observations and use only full frame observation (F). This baseline highlights the effect of using candidate object observations.
- Average-Attention Single-frame Classifier (SFC). We start from Average-Attention RNN (avg.-D and F+avg.-D-con.) and replace RNN with a Single-frame Classifier (SFC). Then, the same loss function in our method is used to train the single-frame classifier using standard back-propagation. These baselines highlight the importance of RNN.
- Maximum-Probability Single-frame Classifier (SFC). We replace the average-attention with the maximum accident anticipation probability over all objects as the accident anticipation probability at each frame. We either use only the object feature (max.-D). These baselines highlight the effect of using RM vs. VGG.
- Frame-based Single-frame Classifier (SFC). We start from Frame-based RNN (F) and replace RNN with a single-frame classifier. Then, the same loss functions in our method are used to train the single-frame classifier using standard back-propagation. This baseline also highlights the importance of RNN.

We first evaluate all methods using VGG appearance feature and IDT motion feature separately to compare the effectiveness of both features. Next, we combine the best VGG variant with the best IDT variant using late-fusion to take advantage of both appearance and motion features.

4.5   Results

We report the Average Precision (AP) of all methods in Table 1, and discuss our results below.

- For VGG feature,
    - RNN consistently outperforms SFC. Without using dynamic attention, VGG+RNN (the first row in Table 1) consistently outperform VGG+SFC (the second row in Table 1) by at most 23.80% in AP (see avg.-D).
    - Object observation improves over full-frame observation. Both VGG+ RNN+ avg.-D and VGG+RNN+F+avg,-D-con. outperform VGG+ RNN+F.
    - Dynamic Spatial-attention further improves over RNN. Both dynamic attention F+D-sum and F+D-con. outperform average attention (VGG+RNN+ F+avg.-D-con.) by at most 21.02% in AP. Object only dynamic attention (VGG+RNN+D) also outperforms object only average attention (VGG+RNN+ avg.-D) by 3.28% in AP.

Table 1. Accident anticipation accuracy in Average Precision (AP). avg. stands for average. con. stands for concatenate. All methods are defined in Sect. 4.4

Type	No dynamic attention			Dynamic attention		
	F	avg.-D	F+avg.-D-con	D	F+D-sum	F+D-con
VGG+RNN	51.89%	**64.88%**	**52.51%**	68.16%	68.21%	**73.53%**
VGG+SFC	46.61%	41.08%	49.01%	—	—	—
IDT+RNN	49.73%	—	—	—	—	—
IDT+SFC	**54.15%**	—	—	—	—	—

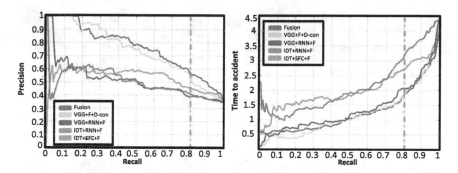

Fig. 4. Left panel shows Precision v.s. Recall (PR) curves. Right panel shows average Time-to-Accident v.s. Recall (ToAR) curves. As indicated by the dash-vertical-lines in both panels, our fused method on average anticipates the accident 1.8559 s before it occurs with 56.14% precision and 80% recall. Note that, compared to other methods, IDT+RNN+F has longer ToA but much worse precision. This implies IDT+RNN+F has a much higher false alarm rate.

- For IDT feature,
    - IDT is a powerful full-frame feature. IDT's frame-based SFC outperforms VGG+SFC+F and VGG+RNN+F by at least 2.26% in AP.
    - RNN is worse than SFC. This is different from our finding using VGG feature. We believe that when the long IDT feature (6400 dimensions) is forced to embedded into 512 dimensions for RNN encoding, some discriminative information might be lost.
- For RM feature,
    - RM+SFC+max.-D (49.36%) is worse than VGG+SFC+max.-D.(66.05%) It shows just detecting objects and estimating their motion direction can not compare with VGG.
- We combine the best IDT method (IDT+SFC+F) with the best VGG method (VGG+RNN+F+D-con.) into Fused-F+D-con. In particular, we fuse the anticipation probability outputs of both methods using equal-weight-summation. This fused method achieves the best 74.35% AP.

We plot the precision v.s. recall curves of the combined method (Fused-F+D-con.), the best VGG method (VGG+RNN+F+D-con.), and many full-frame baselines in Fig. 4-Left.

Average Time-to-accident (ToA). We report the average time-to-accident v.s. recall curves of the combined method (Fused-F+D-con.), the best VGG method (VGG+RNN+ F+D-con.), and many full-frame baselines in

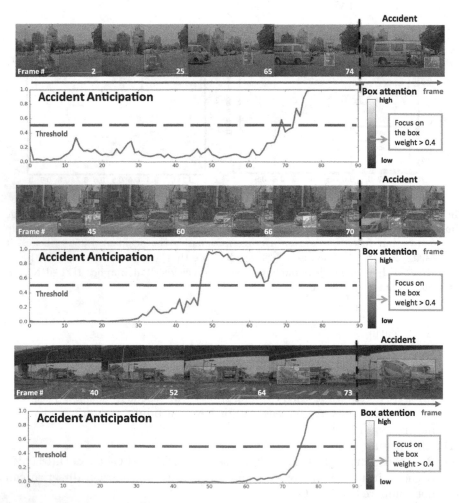

Fig. 5. Typical accident anticipation examples. In each example, we show the sampled frames overlaid with the attention weights (i.e., a value between zero and one) on the object bounding boxes, where yellow, red, and dark indicate high, medium, and low attention, respectively. When the outline of a box turns green, this indicates that its attention is higher than 0.4. On the bottom row, we visualize the predicted confidence of anticipated accident. The threshold is set to 0.5 for visualization. (Color figure online)

Fig. 4-Right. Our fused method on average anticipate the accident 1.8559 s before it occurs with 56.14% precision and 80% recall. We report the performance at 80% recall, since our system aims at detecting most true accidents. Note that, compared to other methods, IDT+RNN+F and IDT+SFC+F has longer ToA but much worse precision. This implies that they have much higher false alarm rates (Fig. 5).

5 Conclusion

We propose a Dynamic-Spatial-Attention RNN model to anticipate accidents in dashcam videos. A large number of dashcam videos containing accidents have been harvested from the web. In this challenging dataset, our proposed method consistently outperforms other baselines without attention or RNN. Finally, our method fusing VGG appearance and IDT motion features can achieve accident anticipation about 2 s before it occurs with 80% recall and 56.14% precision. We believe the accuracy can be further improved if other sensory information such as GPS or map information can be utilized.

Acknowledgements. We thank Industrial Technology Research Institute for their support.

References

1. Simonyan, K., Zisserman, A.: Very deep convolutional networks for large-scale image recognition. In: ICLR (2015)
2. Wang, H., Schmid, C.: Action recognition with improved trajectories. In: ICCV (2013)
3. Ren, S., He, K., Girshick, R., Sun, J.: Faster R-CNN: towards real-time object detection with region proposal networks. In: NIPS (2015)
4. Google Inc.: Google self-driving car project monthly report (2015)
5. National highway traffic safety administration: 2012 motor vehicle crashes: overview (2013)
6. Hochreiter, S., Schmidhuber, J.: Long short-term memory. Neural Comput. **9**, 1735–1780 (1997)
7. Jain, A., Singh, A., Koppula, H.S., Soh, S., Saxena, A.: Recurrent neural networks for driver activity anticipation via sensory-fusion architecture. In: ICRA (2016)
8. Ryoo, M.S.: Human activity prediction: early recognition of ongoing activities from streaming videos. In: ICCV (2011)
9. Hoai, M., De la Torre, F.: Max-margin early event detectors. In: CVPR (2012)
10. Lan, T., Chen, T.-C., Savarese, S.: A hierarchical representation for future action prediction. In: Fleet, D., Pajdla, T., Schiele, B., Tuytelaars, T. (eds.) ECCV 2014. LNCS, vol. 8691, pp. 689–704. Springer, Cham (2014). doi:10.1007/978-3-319-10578-9_45
11. Kitani, K.M., Ziebart, B.D., Bagnell, J.A., Hebert, M.: Activity forecasting. In: Fitzgibbon, A., Lazebnik, S., Perona, P., Sato, Y., Schmid, C. (eds.) ECCV 2012. LNCS, vol. 7575, pp. 201–214. Springer, Heidelberg (2012). doi:10.1007/978-3-642-33765-9_15

12. Yuen, J., Torralba, A.: A data-driven approach for event prediction. In: Daniilidis, K., Maragos, P., Paragios, N. (eds.) ECCV 2010. LNCS, vol. 6312, pp. 707–720. Springer, Heidelberg (2010). doi:10.1007/978-3-642-15552-9_51

13. Walker, J., Gupta, A., Hebert, M.: Patch to the future: unsupervised visual prediction. In: CVPR (2014)

14. Wang, Z., Deisenroth, M., Ben Amor, H., Vogt, D., Schölkopf, B., Peters, J.: Probabilistic modeling of human movements for intention inference. In: RSS (2012)

15. Koppula, H.S., Saxena, A.: Anticipating human activities using object affordances for reactive robotic response. PAMI **38**, 14–29 (2016)

16. Koppula, H.S., Jain, A., Saxena, A.: Anticipatory planning for human-robot teams. In: ISER (2014)

17. Mainprice, J., Berenson, D.: Human-robot collaborative manipulation planning using early prediction of human motion. In: IROS (2013)

18. Berndt, H., Emmert, J., Dietmayer, K.: Continuous driver intention recognition with hidden markov models. In: Intelligent Transportation Systems (2008)

19. Frohlich, B., Enzweiler, M., Franke, U.: Will this car change the lane? - Turn signal recognition in the frequency domain. In: Intelligent Vehicles Symposium (IV) (2014)

20. Kumar, P., Perrollaz, M., Lefévre, S., Laugier, C.: Learning-based approach for online lane change intention prediction. In: Intelligent Vehicles Symposium (IV) (2013)

21. Liebner, M., Baumann, M., Klanner, F., Stiller, C.: Driver intent inference at urban intersections using the intelligent driver model. In: Intelligent Vehicles Symposium (IV) (2012)

22. Morris, B., Doshi, A., Trivedi, M.: Lane change intent prediction for driver assistance: on-road design and evaluation. In: Intelligent Vehicles Symposium (IV) (2011)

23. Doshi, A., Morris, B., Trivedi, M.: On-road prediction of driver's intent with multimodal sensory cues. IEEE Pervasive Comput. **10**, 22–34 (2011)

24. Trivedi, M.M., Gandhi, T., McCall, J.: Looking-in and looking-out of a vehicle: computer-vision-based enhanced vehicle safety. IEEE Trans. Intell. Transp. Syst. **8**, 108–120 (2007)

25. Jain, A., Koppula, H.S., Raghavan, B., Soh, S., Saxena, A.: Car that knows before you do: anticipating maneuvers via learning temporal driving models. In: ICCV (2015)

26. Yao, L., Torabi, A., Cho, K., Ballas, N., Pal, C., Larochelle, H., Courville., A.: Describing videos by exploiting temporal structure. In: ICCV (2015)

27. Xu, K., Ba, J., Kiros, R., Courville, A., Salakhutdinov, R., Zemel, R., Bengio, Y.: Show, attend and tell: neural image caption generation with visual attention. arXiv preprint (2015). arXiv:1502.03044

28. Mnih, V., Heess, N., Graves, A., kavukcuoglu, k.: Recurrent models of visual attention. In: NIPS (2014)

29. Ba, J., Mnih, V., Kavukcuoglu, K.: Multiple object recognition with visual attention. In: ICLR (2015)

30. Brostow, G.J., Shotton, J., Fauqueur, J., Cipolla, R.: Segmentation and recognition using structure from motion point clouds. In: Forsyth, D., Torr, P., Zisserman, A. (eds.) ECCV 2008. LNCS, vol. 5302, pp. 44–57. Springer, Heidelberg (2008). doi:10. 1007/978-3-540-88682-2_5

31. Leibe, B., Cornelis, N., Cornelis, K., Gool, L.V.: Dynamic 3D scene analysis from a moving vehicle. In: CVPR (2007)

32. Scharwächter, T., Enzweiler, M., Franke, U., Roth, S.: Efficient multi-cue scene segmentation. In: Weickert, J., Hein, M., Schiele, B. (eds.) GCPR 2013. LNCS, vol. 8142, pp. 435–445. Springer, Heidelberg (2013). doi:10.1007/978-3-642-40602-7_46
33. Geiger, A., Lenz, P., Urtasun, R.: Are we ready for autonomous driving? The KITTI vision benchmark suite. In: CVPR (2012)
34. Cordts, M., Omran, M., Ramos, S., Scharwächter, T., Enzweiler, M., Benenson, R., Franke, U., Roth, S., Schiele, B.: The cityscapes dataset. In: CVPR Workshop on the Future of Datasets in Vision (2015)
35. Pascanu, R., Mikolov, T., Bengio, Y.: On the difficulty of training recurrent neural networks. arXiv preprint (2012). arXiv:1211.5063
36. Werbos, P.J.: Backpropagation through time: what it does and how to do it. Proc. IEEE **78**, 1550–1560 (1990)
37. Dalal, N., Triggs, B.: Histograms of oriented gradients for human detection. In: CVPR (2005)
38. Lin, T.-Y., Maire, M., Belongie, S., Hays, J., Perona, P., Ramanan, D., Dollár, P., Zitnick, C.L.: Microsoft COCO: common objects in context. In: Fleet, D., Pajdla, T., Schiele, B., Tuytelaars, T. (eds.) ECCV 2014. LNCS, vol. 8693, pp. 740–755. Springer, Cham (2014). doi:10.1007/978-3-319-10602-1_48
39. Abadi, M., Agarwal, A., Barham, P., Brevdo, E., Chen, Z., Citro, C., Corrado, G.S., Davis, A., Dean, J., Devin, M., Ghemawat, S., Goodfellow, I., Harp, A., Irving, G., Isard, M., Jia, Y., Jozefowicz, R., Kaiser, L., Kudlur, M., Levenberg, J., Mané, D., Monga, R., Moore, S., Murray, D., Olah, C., Schuster, M., Shlens, J., Steiner, B., Sutskever, I., Talwar, K., Tucker, P., Vanhoucke, V., Vasudevan, V., Viégas, F., Vinyals, O., Warden, P., Wattenberg, M., Wicke, M., Yu, Y., Zheng, X.: TensorFlow: large-scale machine learning on heterogeneous systems (2015). Software. tensorflow.org

Pano2Vid: Automatic Cinematography for Watching 360° Videos

Yu-Chuan Su$^{(\boxtimes)}$, Dinesh Jayaraman, and Kristen Grauman

The University of Texas at Austin, Austin, USA
`ycsu@cs.utexas.edu`

Abstract. We introduce the novel task of Pano2Vid — *automatic cinematography* in panoramic 360° videos. Given a 360° video, the goal is to direct an imaginary camera to *virtually* capture natural-looking normal field-of-view (NFOV) video. By selecting "where to look" within the panorama at each time step, Pano2Vid aims to free both the videographer and the end viewer from the task of determining what to watch. Towards this goal, we first compile a dataset of 360° videos downloaded from the web, together with human-edited NFOV camera trajectories to facilitate evaluation. Next, we propose AUTOCAM, a data-driven approach to solve the Pano2Vid task. AUTOCAM leverages NFOV web video to discriminatively identify space-time "glimpses" of interest at each time instant, and then uses dynamic programming to select optimal human-like camera trajectories. Through experimental evaluation on multiple newly defined Pano2Vid performance measures against several baselines, we show that our method successfully produces informative videos that could conceivably have been captured by human videographers.

1 Introduction

A 360° video camera captures the entire visual world as observable from its optical center. This is a dramatic improvement over standard *normal field-of-view (NFOV)* video, which is usually limited to 65°. This increase in field of view affords exciting new ways to record and experience visual content. For example, imagine a scientist in a shark cage studying the behavior of a pack of sharks that are swimming in circles all around her. It would be impossible for her to observe each shark closely in the moment. If, however, she had a 360° camera continuously recording spherical panoramic video of the scene all around her, she could later replay her entire visual experience hundreds of times, "choosing her own adventure" each time, focusing on a different shark etc. Similarly, a footballer wearing a 360° camera could review a game from his in-game positions at all times, studying passes which were open to him that he had not noticed in the heat of the game.

Electronic supplementary material The online version of this chapter (doi:10.1007/978-3-319-54190-7_10) contains supplementary material, which is available to authorized users.

S.-H. Lai et al. (Eds.): ACCV 2016, Part IV, LNCS 10114, pp. 154–171, 2017.
DOI: 10.1007/978-3-319-54190-7_10

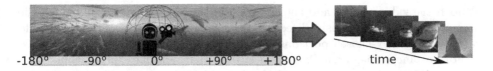

Fig. 1. The "Pano2Vid" task: convert input 360° video to output NFOV video.

In such cases and many others, 360° video offers (1) a richer way to experience the visual world unrestricted by a limited field of view, attention, and cognitive capacity, even compared to actually being present *in situ*, while (2) partially freeing the videographer of camera control. Indeed, 360° videos are growing increasingly popular as consumer- and production-grade 360° cameras (e.g., Ricoh, Bublcam, 360Fly, GoPro) enter the market, and websites such as YouTube and Facebook begin to support 360° content with viewers.

This new medium brings with it some new challenges. Foremost, it largely transfers the choice of "where to look" from the videographer to the viewer. This makes 360° video hard to view effectively, since a human viewer must now somehow make the "where to look" choice and convey it to a video player in real time. Currently, there are three main approaches. In the first approach, the user navigates a 360° video manually. A standard viewer displays a small portion of the 360° video corresponding to a normal field-of-view (NFOV) camera[1]. The user must either drag using a mouse, or click on up-down-left-right cursors, to adjust the virtual camera pose continuously, for the full duration of the video. A second approach is to show the user the entire spherical panoramic video unwrapped into its warped, equirectangular projection[2]. While less effort for the user, the distortions in this projected view make watching such video difficult and unintuitive. The third approach is to wear a virtual reality headset, a natural way to view 360° video that permits rich visual experiences. However, a user must usually be standing and moving about, with a headset obscuring all real-world visual input, for the full duration of the video. This can be uncomfortable and/or impractical over long durations. Plus, similar to the click-based navigation, the user remains "in the dark" about what is happening elsewhere in the scene, and may find it difficult to decide where to look to find interesting content in real time. In short, all three existing paradigms have interaction bottlenecks, and viewing 360° video remains cumbersome.

To address this difficulty, we define "Pano2Vid", a new computer vision problem (see Fig. 1). The task is to design an algorithm to automatically control the pose and motion of a virtual NFOV camera within an input 360° video. The output of the system is the NFOV video captured by this virtual camera. Camera control must be optimized to produce video that could conceivably have been captured by a human observer equipped with a *real* NFOV camera. A successful Pano2Vid system would therefore take the burden of choosing "where to look"

---

[1] See, e.g., https://www.youtube.com/watch?v=HNOT_feL27Y,
[2] See, e.g., https://www.youtube.com/watch?v=Nv6MCkaR5mc.

off both the videographer and the end viewer: the videographer could enjoy the moment without consciously directing her camera, while the end viewer could watch intelligently-chosen portions of the video in the familiar NFOV format.

For instance, imagine a Pano2Vid system that automatically outputs hundreds of NFOV videos for the 360° shark cage video, e.g., focusing on different sharks/subgroups of sharks in turn. This would make analysis much easier for the scientist, compared to manually selecting and watching hundreds of different camera trajectories through the original 360° video. A machine-selected camera trajectory could also serve as a useful default initialization for viewing 360° content, where a user is not forced to interact with the video player continuously, but could *opt* to do so when they desire. Such a system could even be useful as an editing aid for cinema. 360° cameras could partially offload camera control from the cinematographer to the editor, who might start by selecting from machine-recommended camera trajectory proposals.

This work both formulates the Pano2Vid problem and introduces the first approach to address it. The proposed "AUTOCAM" approach first learns a discriminative model of human-captured NFOV web video. It then uses this model to identify candidate viewpoints and events of interest to capture in 360° video, before finally stitching them together through optimal camera motions using a dynamic programming formulation for presentation to human viewers. Unlike prior attempts at automatic cinematography, which focus on virtual 3D worlds and employ heuristics to encode popular idioms from cinematography [1–6], AUTOCAM is (a) the first to tackle real video from dynamic cameras and (b) the first to consider directly *learning* cinematographic tendencies from data.

The contributions of this work are four-fold: (1) we formulate the computer vision problem of automatic cinematography in 360° video (Pano2Vid), (2) we propose a novel Pano2Vid algorithm (AUTOCAM), (3) we compile a dataset of 360° web video, annotated with ground truth human-directed NFOV camera trajectories[3] and (4) we propose a comprehensive suite of objective and subjective evaluation protocols to benchmark Pano2Vid task performance. We benchmark AUTOCAM against several baselines and show that it is the most successful at virtually capturing natural-looking NFOV video.

2   Related Work

Video Summarization. Video summarization methods condense videos in *time* by identifying important events [7]. A summary can take the form of a keyframe sequence [8–13], a sequence of video highlight clips [14–19], or montages of frames [20] or video clip excerpts [21]. Among these, our proposed AUTOCAM shares with [10–12, 18, 19] the idea of using user-generated visual content from the web as exemplars for informative content. However, whereas existing methods address *temporal* summarization of NFOV video, we consider a novel form of *spatial* summarization of 360° video. While existing methods decide *which frames to keep* to shorten a video, our problem is instead to choose *where to look* at each

---

[3] http://vision.cs.utexas.edu/projects/Pano2Vid.

time instant. Moreover, existing summarization work assumes video captured intentionally by a camera person (or, at least, a well-placed surveillance camera). In contrast, our input videos largely lack this deliberate control. Moreover, we aim not only to capture all *important* content in the original 360° video, but to do so in a *natural, human-like* way so that the final output video resembles video shot by human videographers with standard NFOV cameras.

Camera Selection for Multi-video Summarization. Some efforts address *multi-video* summarization [22–24], where the objective is to select, at each time instant, video feed from one camera among many to include in a summary video. The input cameras are human-directed, whether stationary or dynamic [24]. In contrast, we deal with a single hand-held 360° camera, which is not intentionally directed to point anywhere.

Video Retargeting. Video retargeting aims to adapt a video to better suit the aspect ratio of a target display, with minimal loss of content that has already been purposefully selected by an editor or videographer [25–29]. In our setting, 360° video is captured *without* human-directed content selection; instead, the system must automatically select the content to capture. Furthermore, the spatial downsampling demanded by Pano2Vid will typically be much greater than that required in retargeting.

Visual Saliency. Salient regions are usually defined as those that capture the visual attention of a human observer, e.g., as measured by gaze tracking. While saliency detectors most often deal with static images [30–33], some are developed for video [34–38], including work that models temporal continuity in saliency [37]. Whereas saliency methods aim to capture where human eyes move subconsciously during a free-viewing task, our Pano2Vid task is instead to capture where human videographers would *consciously point their cameras*, for the specific purpose of capturing a video that is *presentable to other human viewers*. In our experiments, we empirically verify that saliency is not adequate for automatic cinematography.

Virtual Cinematography. Ours is the first attempt to automate cinematography in complex real-world settings. Existing virtual cinematography work focuses on camera manipulation within much simpler *virtual* environments/video games [1–4], where the perception problem is bypassed (3-D positions and poses of all entities are knowable, sometimes even controllable), and there is full freedom to position and manipulate the camera. Some prior work [5,6] attempts virtual camera control within restricted *static* wide field-of-view video of classroom and video conference settings, by tracking the centroid of optical flow in the scene. In contrast, we deal with unrestricted 360° web video of complex real-world scenes, captured by moving amateur videographers with shaky hand-held devices, where such simple heuristics are insufficient. Importantly, our approach is also the first to *learn content-based camera control from data*, rather than relying on hand-crafted guidelines/heuristics as all prior attempts do.

3   Approach

We first define the Pano2Vid problem in more detail (Sect. 3.1) and describe our data collection process (Sect. 3.2). Then we introduce our AUTOCAM approach (Sect. 3.3). Finally, we introduce several evaluation methodologies for quantifying performance on this complex task (Sect. 3.4), including an annotation collection procedure to gather human-edited videos for evaluation (Sect. 3.5).

3.1   Pano2Vid Problem Definition

First, we define the Pano2Vid task of automatic videography for 360° videos. Given a dynamic panoramic 360° video, the goal is to produce "natural-looking" normal-field-of-view (NFOV) video. For this work, we define NFOV as spanning a horizontal angle of 65.5° (corresponding to a typical 28 mm focal length full-frame Single Lens Reflex Camera [39]) with a 4:3 aspect ratio.

Broadly, a natural-looking NFOV video is one which is indistinguishable from human-captured NFOV video (henceforth "HumanCam"). Our ideal video output should be such that it could conceivably have been captured by a human videographer equipped with an NFOV camera whose optical center coincides exactly with that of the 360° video camera, with the objective of best presenting the event(s) in the scene. In this work, we do not allow skips in time nor camera zoom, so the NFOV video is defined completely by the camera trajectory, i.e., the time sequence of the camera's principal axis directions. To solve the Pano2Vid problem, a system must determine a NFOV camera trajectory through the 360° video to carve it into a HumanCam-like NFOV video.

3.2   Data Collection: 360° Test Videos and NFOV Training Videos

Human-directed camera trajectories are content-based and often present scenes in *idiomatic* ways that are specific to the situations, and with specific intentions such as to tell a story [40]. Rather than hand-code such characteristics through cinematographic rules/heuristics [1–4], we propose to *learn* to capture NFOV videos, by observing HumanCam videos from the web. The following overviews our data collection procedure.

360° videos. We collect 360° videos from YouTube using the keywords "Soccer," "Mountain Climbing," "Parade," and "Hiking." These terms were selected to have (i) a large number of relevant 360° video results, (ii) dynamic activity, i.e., spatio-temporal *events*, rather than just static scenes, and (iii) possibly multiple regions/events of interest at the same time. For each query term, we download the top 100 videos sorted by relevance and filter out any that are not truly 360° videos (e.g., animations, slide shows of panoramas, restricted FOV) or have poor lighting, resolution, or stitching quality. This yields a Pano2Vid test set of 86 total 360° videos with a combined length of 7.3 h. See the project webpage (See footnote 3) for example videos.

HumanCam NFOV Videos. In both the learning stage of AUTOCAM (Sect. 3.3) and the proposed evaluation methods (Sect. 3.4), we need a model for HumanCam. We collect a large diverse set of HumanCam NFOV videos from YouTube using the same query terms as above and imposing a per-video max length of 4 min. For each query term, we collect about 2,000 videos, yielding a final HumanCam set of 9,171 videos totalling 343 h. See Supp. for details.

3.3   AutoCam: Proposed Solution for the Pano2Vid Task

We now present AUTOCAM, our approach to solve the Pano2Vid task. The input to the system is an arbitrary 360° video, and the output is a natural looking NFOV video extracted from it.

AUTOCAM works in two steps. First, it evaluates all virtual NFOV spatio-temporal "glimpses" (ST-glimpses) sampled from the 360° video for their "capture-worthiness"—their likelihood of appearing in HumanCam NFOV video. Next, it selects virtual NFOV camera trajectories, prioritizing both (i) high-scoring ST-glimpses from the first step, and (ii) smooth human-like camera movements. AUTOCAM is fully automatic and does not require any human input. Furthermore, as we will see next, the proposed learning approach is unsupervised—it learns a model of human-captured NFOV video simply by watching clips people upload to YouTube.

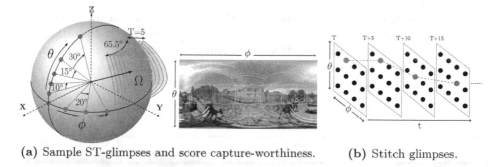

(a) Sample ST-glimpses and score capture-worthiness.   (b) Stitch glimpses.

Fig. 2. AUTOCAM first samples and scores the capture-worthiness of ST-glimpses. It then jointly selects a glimpse for each time step and stitches them together to form the output NFOV video. Best viewed in color.

Capture-Worthiness of Spatio-Temporal Glimpses. The first stage aims to find content that is likely to be captured by human videographers. We achieve this by scoring the capture-worthiness of candidate ST-glimpses sampled from the 360° video. An ST-glimpse is a five-second NFOV video clip recorded from the 360° video by directing the camera to a fixed direction in the 360° camera axes. One such glimpse is depicted as the blue stack of frame excerpts on the surface of the sphere in Fig. 2a. These are not rectangular regions in the equirectangular projection (Fig. 2a, right) so they are projected into

NFOV videos before processing. We sample candidate ST-glimpses at longitudes $\phi \in \Phi = \{0, 20, 40, \ldots, 340\}$ and latitudes $\theta \in \Theta = \{0, \pm10, \pm20, \pm30, \pm45, \pm75\}$ and intervals of 5 s. Each candidate ST-glimpse is defined by the camera principal axis direction (θ, ϕ) and time t: $\Omega_{t,\theta,\phi} \equiv (\theta_t, \phi_t) \in \Theta \times \Phi$. See Supp.

Our approach learns to score capture-worthiness from HumanCam data. We expect capture-worthiness to rely on two main facets: content and composition. The *content* captured by human videographers is naturally very diverse. For example, in a mountain climbing video, people may consider capturing the recorder and his companion as well as a beautiful scene such as the sunrise as being equally important. Similarly, in a soccer video, a player dribbling and a goalkeeper blocking the ball may both be capture-worthy. Our approach accounts for this diversity both by learning from a wide array of NFOV HumanCam clips and by targeting related domains via the keyword query data collection described above. The *composition* in HumanCam data is a meta-cue, largely independent of semantic content, that involves the framing effects chosen by a human videographer. For example, an ST-glimpse that captures only the bottom half of a human face is not capture-worthy, while a framing that captures the full face is; a composition for outdoor scenes may tend to put the horizon towards the middle of the frame, etc.

Rather than attempt to characterize capture-worthiness through rules, AUTOCAM *learns* a data-driven model. We make the following hypotheses: (i) the majority of content in HumanCam NFOV videos were considered capture-worthy by their respective videographers (ii) most random ST-glimpses would *not* be capture-worthy. Based on these hypotheses, we train a capture-worthiness classifier. Specifically, we divide each HumanCam video into non-overlapping 5-second clips, to be used as positives, following (i) above. Next, *all* candidate ST-glimpses extracted from (disjoint) 360° videos are treated as negatives, per hypothesis (ii) above. Due to the weak nature of this supervision, both positives and negatives may have some label noise.

Fig. 3. Example glimpses scored by AUTOCAM. Left 4 columns are glimpses considered capture-worthy by our method; each column is from the same time step in the same video. Right column shows non-capture-worthy glimpses.

To represent each ST-glimpse and each 5 s HumanCam clip, we use off-the-shelf convolutional 3D features (C3D) [41]. C3D is a generic video feature based on 3D (spatial+temporal) convolution that captures appearance and motion information in a single vector representation, and is known to be useful for recognition tasks. We use a leave-one-video-out protocol to train one capture-worthiness classifier for each 360° video. Both the positive and negative training samples are from videos returned by the same keyword query term as the test video, and we sub-sample the 360° videos so that the total number of negatives is twice that of positives. We use logistic regression classifiers; positive class probability estimates of ST-glimpses from the left-out video are now treated as their capture-worthiness scores.

Figure 3 shows examples of "capture-worthy" and "non-capture-worthy" glimpses as predicted by our system. We see that there may be multiple capture-worthy glimpses at the same moment, and both the content and composition are important for capture-worthiness. Please see the Supp. file for further analysis, including a study of how our predictions correlate with the viewpoint angles.

Camera Trajectory Selection. After obtaining the capture-worthiness score of each candidate ST-glimpse, we construct a camera trajectory by finding a path over the ST-glimpses that maximizes the *aggregate* capture-worthiness score, while simultaneously producing human-like smooth camera motions. A naive solution would be to choose the glimpse with the maximum score at each step. This trajectory would capture the maximum aggregate capture-worthiness, but the resultant NFOV video may have large/shaky unnatural camera motions. For example, when two ST-glimpses located in opposite directions on the viewing sphere have high capture-worthiness scores, such a naive solution would end up switching between these two directions at every time-step, producing unpleasant and even physically impossible camera movements.

Instead, to construct a trajectory with more human-like camera operation, we introduce a *smooth motion* prior when selecting the ST-glimpse at each time step. Our prior prefers trajectories that are stable over those that jump abruptly between directions. For the example described above, the smooth prior would suppress trajectories that switch between the two directions constantly and promote those that focus on one direction for a longer amount of time. In practice, we realize the smooth motion prior by restricting the trajectory from choosing an ST-glimpse that is displaced from the previous ST-glimpse by more than $\epsilon = 30°$ in both longitude and latitude, i.e.

$$|\Delta\Omega|_\theta = |\theta_t - \theta_{t-1}| \leq \epsilon, \quad |\Delta\Omega|_\phi = |\phi_t - \phi_{t-1}| \leq \epsilon. \tag{1}$$

Given (i) the capture-worthiness scores of all candidate ST-glimpses and (ii) the smooth motion constraint for trajectories, the problem of finding the trajectories with maximum aggregate capture-worthiness scores can be reduced to a shortest path problem. Let $C(\Omega_{t,\theta,\phi})$ be the capture-worthiness score of the ST-glimpse at time t and viewpoint (θ, ϕ). We construct a 2D lattice per time slice, where each node corresponds to an ST-glimpse at a given angle pair.

The edges in the lattice connect ST-glimpses from time step t to $t + 1$, and the weight for an edge is defined by:

$$E\left(\Omega_{t,\theta,\phi}, \Omega_{t+1,\theta',\phi'}\right) = \begin{cases} -C(\Omega_{t+1,\theta',\phi'}), & |\Omega_{t,\theta,\phi} - \Omega_{t+1,\theta',\phi'}| \leq \epsilon \\ \infty, & \text{otherwise,} \end{cases} \quad (2)$$

where the difference above is shorthand for the two angle requirements in Eq. 1. See Fig. 2b, middle and right.

Algorithm 1. Camera trajectory selection

$C \ \leftarrow$ Capture-worthiness scores
$\epsilon \leftarrow$ Valid camera motion
for all θ, ϕ **do**
    $Accum[\Omega_{1,\theta,\phi}] \ \leftarrow \ C[\Omega_{1,\theta,\phi}]$
end for
for $t \leftarrow 2, T$ **do**
    for all θ, ϕ **do**
        $\Omega_{t-1,\theta',\phi'} \ \leftarrow \ \arg\max_{\theta',\phi'} Accum[\Omega_{t-1,\theta',\phi'}]$
                $s.t. \ |\Omega_{t,\theta,\phi} - \Omega_{t-1,\theta',\phi'}| \leq \epsilon$
        $Accum[\Omega_{t,\theta,\phi}] \ \leftarrow \ Accum[\Omega_{t-1,\theta',\phi'}] + C[\Omega_{t,\theta,\phi}]$
        $TraceBack[\Omega_{t,\theta,\phi}] \ \leftarrow \ \Omega_{t-1,\theta',\phi'}$
    end for
end for
$\Omega \ \leftarrow \ \arg\max_{\theta,\phi} Accum[\Omega_{T,\theta,\phi}]$
for $t \leftarrow T, 1$ **do**
    $Traj[t] \ \leftarrow \ \Omega$
    $\Omega \ \leftarrow \ TraceBack[\Omega]$
end for

The solution to the shortest path problem over this graph now corresponds to camera trajectories with maximum aggregate capture-worthiness. This solution can be efficiently computed using dynamic programming. See pseudocode in Algorithm 1. At this point, the optimal trajectory indicated by this solution is "discrete" in the sense that it makes jumps between discrete directions after each 5-second time-step. To smooth over these jumps, we linearly interpolate the trajectories between the discrete time instants, so that the final camera motion trajectories output by AutoCam are continuous. In practice, we generate K NFOV outputs from each 360° input by (i) computing the best trajectory ending at each ST-glimpse location (of 198 possible), and (ii) picking the top K of these.

Note that AutoCam is an offline batch processing algorithm that watches the entire video before positioning the virtual NFOV camera at each frame. This matches the target setting of a human editing a pre-recorded 360° video to capture a virtual NFOV video, as the human is free to watch the video in full. In fact, we use human-selected edits to help evaluate AutoCam (Sects. 3.5 and 4.2).

3.4   Quantitative Evaluation of Pano2Vid Methods

Next we present evaluation metrics for the Pano2Vid problem. A good metric must measure how close a Pano2Vid algorithm's output videos are to human-generated NFOV video, while simultaneously being reproducible for easy benchmarking in future work. We devise two main criteria:

– **HumanCam-based metrics**: Algorithm outputs should look like Human-Cam videos—the more indistinguishable the algorithm outputs are from real manually captured NFOV videos, the better the algorithm.
– **HumanEdit-based metrics**: Algorithms should select camera trajectories close to human-selected trajectories ("HumanEdit")—The closer algorithmically selected camera motions are to those selected by humans editing the same 360° video, the better the algorithm.

The following fleshes out a family of metrics capturing these two criteria. All of which can easily be reproduced and compared to easily, given the same training/testing protocol is applied.

HumanCam-Based Metrics. We devise three HumanCam-based metrics:

Distinguishability: Is it possible to distinguish Pano2Vid and HumanCam outputs? Our first metric quantifies *distinguishability* between algorithmically generated and HumanCam videos. For a fully successful Pano2Vid algorithm, these sets would be entirely indistinguishable. This method can be considered as an automatic Turing test that is based on feature statistics instead of human perception; it is also motivated by the adversarial network framework [42] where the objective of the generative model is to disguise the discriminative model. We measure distinguishability using 5-fold cross validation performance of a discriminative classifier trained with HumanCam videos as positives, and algorithmically generated videos as negatives. Training and testing negatives in each split are generated from disjoint sets of 360° video.

HumanCam-likeness: Which Pano2Vid method gets closer to HumanCam? This metric directly compares outputs of multiple Pano2Vid methods using their relative distances from HumanCam videos in a semantic feature space (e.g., C3D space). Once again a classifier is trained on HumanCam videos as positives, but this time with *all* algorithm-generated videos as negatives. Similar to exemplar SVM [43], each algorithm-generated video is assigned a ranking based on its distance from the decision boundary (i.e. HumanCam-likeness), using a leave-one-360°-video-out training and testing scheme. We rank all Pano2Vid algorithms for each 360° video and compare their normalized mean rank; lower is better. We use classification score rather than raw feature distance because we are only interested in the factors that distinguish Pano2Vid and HumanCam. Since this metric depends on the relative comparison of all methods, it requires the output of all methods to be available during evaluation.

Transferability: Do semantic classifiers transfer between Pano2Vid and HumanCam video? This metric tries to answer the question: if we learn to distinguish between the 4 classes (based on search keywords) on HumanCam videos, would the classifier perform well on Pano2Vid videos (Human → Auto), and vice versa (Auto → Human)? Intuitively, the more similar the domains, the better the transferability. A similar method is used to evaluate automatic image colorization in [44]. To quantify transferability, we train a multi-class classifier on Auto(/Human) videos generated by a given Pano2Vid method and test it on Human(/Auto) videos. This test accuracy is the method's transferability score.

HumanEdit-Based Metrics. Our metrics thus far compare Pano2Vid outputs with generic NFOV videos. We now devise a set of HumanEdit-based metrics that compare algorithm outputs to human-selected NFOV camera trajectories, given the *same input 360° video*. Sect. 3.5 will explain how we obtain HumanEdit trajectories. Note that a single 360° video may have several equally valid HumanEdit camera trajectory annotations, e.g. from different annotators.

Mean cosine similarity: How closely do the camera trajectories match? To compute this metric, we first measure the frame-wise cosine distance (in the 360° camera axes) between the virtual NFOV camera principal axes selected by Pano2Vid and HumanEdit. These frame-wise distances are then pooled into one score in two different ways: (1) **Trajectory pooling:** Each Pano2Vid trajectory is compared to its best-matched HumanEdit trajectory. Frame-wise cosine distances to each human trajectory are first averaged. Each Pano2Vid output is then assigned a score corresponding to the minimum of its average distance to HumanEdit trajectories. Trajectory pooling rewards Pano2Vid outputs that are similar to at least one HumanEdit trajectory *over the whole video*, and (2) **Frame pooling:** This pooling method rewards Pano2Vid outputs that are similar to different HumanEdit tracks in different portions of the video. First, each frame is assigned a score based on its minimum frame-wise cosine distance to a HumanEdit trajectory. Now, we simply average this over all frames to produce the "frame distance" score for that trajectory. Frame pooling rewards Pano2Vid outputs that are similar to any HumanEdit trajectory at each frame.

Mean overlap: How much do the fields of view overlap? The cosine distance between principal axes ignores the fact that cameras have limited FOV. To account for this, we compute "overlap" metrics on Pano2Vid and HumanEdit camera FOVs on the unit sphere. Specifically, we approximate the overlap using $\max(1 - \frac{\Delta\Omega}{FOV}, 0)$, which is 1 when the principal axes coincide, and 0 for all $\Delta\Omega >$ FOV. We apply both trajectory and frame pooling as for the cosine distance.

Fig. 4. HumanEdit interface. We display the 360° video in equirectangular projection and ask annotators to direct the camera using the mouse. The NFOV video is rendered and displayed to the annotator offline. Best viewed in color.

3.5   HumanEdit Annotation Collection

To collect human editors' annotations, we ask multiple annotators to watch the 360° test videos and generate the camera trajectories from them. We next describe the annotation collection process. We then analyze the consistency of collected HumanEdit trajectories.

Annotation Interface and Collection Process. Figure 4 shows the HumanEdit annotation interface. We display the entire 360° video in equirectangular projection. Annotators are instructed to move a cursor to direct a virtual NFOV camera. Virtual NFOV frame boundaries are backprojected onto the display (shown in cyan) in real time as the camera is moved.

We design the interface to mitigate problems due to discontinuities at the edges. First, we extend the panoramic strip by 90° on the left and right as shown in Fig. 4. The cursor may now smoothly move over the 360° boundaries to mimic camera motion in the real world. Second, when passing over these boundaries, content is duplicated, and so is the cursor position and frame boundary rendering. When passing over an edge of this extended strip, the cursor is repositioned to the duplicated position that is already on-screen by this time. Finally, before each annotation, our interface allows the annotator to pan the panoramic strip to a chosen longitude to position important content centrally if they so choose. Please refer to Supp. for more visual examples and project webpage for the interface in action.

For each 360° video, annotators watch the full video first to familiarize themselves with its content. Thus, they have the same information as our AUTOCAM approach. Each annotator provides *two* camera trajectories per 360° video, to account for the possibility of multiple good trajectories. Each of 20 360° videos is annotated by 3 annotators, resulting in a final database with 120 human-annotated trajectories adding up to 202 min of NFOV video. Our annotators were 8 students aged between 24–30.

HumanEdit Consistency Analysis. After collecting HumanEdit, we measure the consistency between trajectories annotated by different annotators using the metrics described in Sect. 3.4.

Table 1 shows the results. The average cosine distance between human trajectories is 0.520, which translates to 59° difference in camera direction at every moment. The difference is significant, considering the NFOV is 65°. Frame differences, however, are much smaller—37° on average, and overlap of >50% across annotators at every frame. These differences indicate that there is more than one natural trajectory for each 360° video, and different annotators may pick different trajectories. Still, with >50% overlap at any given moment, we see that there is often something in the 360° video that catches everyone's eye; different trajectories arise because people place different priority on the content and choose to navigate through them in different manner. Overall, this analysis

Table 1. HumanEdit consistency.

Cosine	Trajectory	0.520
Similarity	Frame	0.803
Overlap	Trajectory	0.462
	Frame	0.643

Table 2. Baseline illustration. CENTER generates random trajectories biased toward the 360° video center. EYE-LEVEL generates static trajectories on the equator.

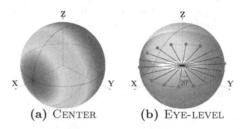

(a) CENTER   (b) EYE-LEVEL

justifies our design to ask each annotator to annotate twice and underscores the need for metrics that take the multiple trajectories into account, as we propose.

4   Experiment

Baseline. We compare our method to the following baselines.

- CENTER + Smooth motion prior—This baseline biases camera trajectories to lie close to the "center" of the standard 360° video axes, accounting for the fact that user-generated 360° videos often have regions of interest close to the centers of the standard equirectangular projection. We sample from a Gaussian starting at the center, then centered at the current direction for subsequent time-steps. See Table 2a.
- EYE-LEVEL prior—This baseline points the NFOV camera to some preselected direction on the equator (i.e. at 0° latitude) throughout the video. 0° latitude regions often correspond to eye-level in our dataset. We sample at 20° longitudinal intervals along the equator. See Table 2b.
- SALIENCY trajectory—This baseline uses our pipeline, but replaces our discriminative capture-worthiness scores with the saliency score from a popular saliency method [30].
- AUTOCAM W/O STITCHING—This is an ablated variant of our method that has no camera motion constraint. We generate multiple outputs by sampling ST-glimpses at each time step based on their capture-worthiness scores.

For each method we generate $K = 20$ outputs per 360° video, and average their results. See Supp. for details.

Implementation Details. Following [41], we split the input video into 16 frame clips then extract the $fc6$ activation for each clip and average them as the C3D video features (whether on glimpses or full HumanCam clips). We use temporally non-overlapping clips to reduce computation cost. We use the Sport1M model provided by the authors without fine-tuning. We use logistic regression with $C = 1$ in all experiments involving a discriminative classifier.

Table 3. Pano2Vid performance: HumanCam-based metrics. The arrows in column 3 indicate whether lower scores are better (⇓), or higher scores (⇑).

			CENTER	EYE-LEVEL	SALIENCY	AUTOCAM W/O STITCHING (ours)	AUTOCAM (ours)
Distinguishability	Error rate (%)	⇑	1.3	2.8	5.2	4.0	**7.0**
HumanCam-Likeness	Mean Rank	⇓	0.659	0.571	0.505	0.410	**0.388**
Transferability	Human → Auto	⇑	0.574	0.609	0.595	**0.637**	0.523
	Auto → Human		0.526	0.559	0.550	0.561	**0.588**

4.1   HumanCam-Based Evaluation

We first evaluate our method using the HumanCam-based metrics (defined in Sect. 3.4). Table 3 shows the results. Our full method (AUTOCAM) performs the best in nearly all metrics. It improves the Distinguishability and HumanCam-Likeness by 35% and 23% respectively, compared to the best baseline. The advantage in Transferability is not as significant, but is still 5% better in Auto → Human transfer. AUTOCAM W/O STITCHING is second-best overall, better than all other 3 baselines. These results establish that both components of our method—(i) capture-worthy ST-glimpse selection, and (ii) smooth camera motion selection—capture important aspects of human-like NFOV videos.

Fig. 5. Example AUTOCAM outputs. We show the result for two 360° videos, and two trajectories for each.

Among the remaining baselines, SALIENCY, which is content-based and also uses our smooth motion selection pipeline, performs significantly better than CENTER and EYE-LEVEL, which are uniformly poor throughout. However, SALIENCY falls well short of even AUTOCAM W/O STITCHING, establishing that saliency is a poor proxy for capture-worthiness.

We observe that the transferability metric results are asymmetric, and AUTO-CAM only does best on transferring semantic classifiers in the Auto → Human direction. Interestingly, AUTOCAM W/O STITCHING is best on Human → Auto, but the smooth motion constraint adversely affects this score for AUTOCAM. This performance drop may be caused by the content introduced when trying to stitch two spatially disjoint capture-worthy ST-glimpses. While AUTOCAM W/O STITCHING can jump directly between such glimpses, AUTOCAM is constrained to move through less capture-worthy content connecting them. Moreover, intuitively, scene recognition relies more on content selection than on camera motion, so incoherent motion might not disadvantage AUTOCAM W/O STITCHING.

Figure 5 shows example output NFOV videos of our algorithm for two 360° videos. For each video, we show two different generated trajectories. Our method is able to find multiple natural NFOV videos from each input. See project webpage for video examples and comparisons of different methods.

4.2   HumanEdit-Based Evaluation

Next, we evaluate all methods using the HumanEdit-based metrics (Sect. 3.4). Table 4 shows the results. Once again, our method performs best on all but the frame-pooling overlap metrics.

On the cosine distance metric in particular, AUTOCAM W/O STITCHING suffers significantly from having incoherent camera motion. EYE-LEVEL is second best on these metrics. It does better on frame-wise metrics, suggesting that humans rarely choose static eye-level trajectories. Further, EYE-LEVEL does better on overlap metrics, even outperforms AUTOCAM on average per-frame overlap, suggesting a tendency to make large mistakes which are penalized by cosine metrics but not by overlap metrics. SALIENCY scores poorly throughout; even though saliency may do well at predicting human gaze fixations, as discussed above, this is not equivalent to predicting plausible NFOV excerpts.

To sum up, our method performs consistently strongly across a wide range of metrics based on both resemblance to generic YouTube NFOV videos, and on closeness to human-created edits of 360° video. This serves as strong evidence that our approach succeeds in capturing human-like virtual NFOV videos.

Table 4. Pano2Vid performance: HumanEdit-based metrics. Higher is better.

		CENTER	EYE-LEVEL	SALIENCY	AUTOCAM W/O STITCHING (ours)	AUTOCAM (ours)
Cosine	Trajectory	0.257	0.268	0.063	0.184	**0.304**
Similarity	Frame	0.572	0.575	0.387	0.541	**0.581**
Overlap	Trajectory	0.194	0.243	0.094	0.202	**0.255**
	Frame	0.336	**0.392**	0.188	0.354	0.389

5   Conclusion

We formulate Pano2Vid: a new computer vision problem that aims to produce a natural-looking NFOV video from a dynamic panoramic 360° video. We collect a new dataset for the task, with an accompanying suite of Pano2Vid performance metrics. We further propose AUTOCAM, an approach to learn to generate camera trajectories from human-generated web video. We hope that this work will provide the foundation for a new line of research that requires both scene understanding and active decision making. In the future, we plan to explore supervised approaches to leverage HumanEdit data for learning the properties of good camera trajectories and incorporate more task specific features such as human detector.

Acknowledgement. This research is supported in part by NSF IIS -1514118 and a gift from Intel. We also gratefully acknowledge the support of Texas Advanced Computing Center (TACC).

References

1. Christianson, D.B., Anderson, S.E., He, L.W., Salesin, D.H., Weld, D.S., Cohen, M.F.: Declarative camera control for automatic cinematography. In: AAAI/IAAI, vol. 1 (1996)
2. He, L.w., Cohen, M.F., Salesin, D.H.: The virtual cinematographer: a paradigm for automatic real-time camera control and directing. In: ACM CGI (1996)
3. Elson, D.K., Riedl, M.O.: A lightweight intelligent virtual cinematography system for machinima production. In: AIIDE (2007)
4. Mindek, P., Čmolík, L., Viola, I., Gröller, E., Bruckner, S.: Automatized summarization of multiplayer games. In: ACM CCG (2015)
5. Foote, J., Kimber, D.: Flycam: practical panoramic video and automatic camera control. In: ICME (2000)
6. Sun, X., Foote, J., Kimber, D., Manjunath, B.: Region of interest extraction and virtual camera control based on panoramic video capturing. In: IEEE TOM (2005)
7. Truong, B.T., Venkatesh, S.: Video abstraction: a systematic review and classification. In: ACM TOMM (2007)
8. Goldman, D.B., Curless, B., Salesin, D., Seitz, S.M.: Schematic storyboarding for video visualization and editing. In: ACM TOG (2006)
9. Lee, Y.J., Ghosh, J., Grauman, K.: Discovering important people and objects for egocentric video summarization. In: CVPR (2012)
10. Kim, G., Xing, E.: Jointly aligning and segmenting multiple web photo streams for the inference of collective photo storylines. In: CVPR (2013)
11. Khosla, A., Hamid, R., Lin, C.J., Sundaresan, N.: Large-scale video summarization using web-image priors. In: CVPR (2013)
12. Xiong, B., Grauman, K.: Detecting snap points in egocentric video with a web photo prior. In: Fleet, D., Pajdla, T., Schiele, B., Tuytelaars, T. (eds.) ECCV 2014. LNCS, vol. 8693, pp. 282–298. Springer, Cham (2014). doi:10.1007/978-3-319-10602-1_19
13. Gong, B., Chao, W.L., Grauman, K., Sha, F.: Diverse sequential subset selection for supervised video summarization. In: NIPS (2014)
14. Gygli, M., Grabner, H., Riemenschneider, H., Gool, L.: Creating summaries from user videos. In: Fleet, D., Pajdla, T., Schiele, B., Tuytelaars, T. (eds.) ECCV 2014. LNCS, vol. 8695, pp. 505–520. Springer, Cham (2014). doi:10.1007/978-3-319-10584-0_33
15. Gygli, M., Grabner, H., Van Gool, L.: Video summarization by learning submodular mixtures of objectives. In: CVPR (2015)

16. Potapov, D., Douze, M., Harchaoui, Z., Schmid, C.: Category-specific video summarization. In: Fleet, D., Pajdla, T., Schiele, B., Tuytelaars, T. (eds.) ECCV 2014. LNCS, vol. 8694, pp. 540–555. Springer, Cham (2014). doi:10.1007/978-3-319-10599-4_35

17. Zhao, B., Xing, E.: Quasi real-time summarization for consumer videos. In: CVPR (2014)

18. Sun, M., Farhadi, A., Seitz, S.: Ranking domain-specific highlights by analyzing edited videos. In: Fleet, D., Pajdla, T., Schiele, B., Tuytelaars, T. (eds.) ECCV 2014. LNCS, vol. 8689, pp. 787–802. Springer, Cham (2014). doi:10.1007/978-3-319-10590-1_51

19. Song, Y., Vallmitjana, J., Stent, A., Jaimes, A.: Tvsum: Summarizing web videos using titles. In: CVPR (2015)

20. Sun, M., Farhadi, A., Taskar, B., Seitz, S.: Salient montages from unconstrained videos. In: Fleet, D., Pajdla, T., Schiele, B., Tuytelaars, T. (eds.) ECCV 2014. LNCS, vol. 8695, pp. 472–488. Springer, Cham (2014). doi:10.1007/978-3-319-10584-0_31

21. Pritch, Y., Rav-Acha, A., Gutman, A., Peleg, S.: Webcam synopsis: Peeking around the world. In: ICCV (2007)

22. Fu, Y., Guo, Y., Zhu, Y., Liu, F., Song, C., Zhou, Z.H.: Multi-view video summarization. In: IEEE TOM (2010)

23. Dale, K., Shechtman, E., Avidan, S., Pfister, H.: Multi-video browsing and summarization. In: CVPR (2012)

24. Arev, I., Park, H.S., Sheikh, Y., Hodgins, J., Shamir, A.: Automatic editing of footage from multiple social cameras. In: ACM TOG. ACM (2014)

25. Liu, F., Gleicher, M.: Video retargeting: automating pan and scan. In: ACM MM (2006)

26. Avidan, S., Shamir, A.: Seam carving for content-aware image resizing. In: ACM TOG (2007)

27. Rubinstein, M., Shamir, A., Avidan, S.: Improved seam carving for video retargeting. In: ACM TOG (2008)

28. Krähenbühl, P., Lang, M., Hornung, A., Gross, M.: A system for retargeting of streaming video. In: ACM TOG (2009)

29. Khoenkaw, P., Piamsa-Nga, P.: Automatic pan-and-scan algorithm for heterogeneous displays. MTA 74(4), 11837–11865 (2015)

30. Harel, J., Koch, C., Perona, P.: Graph-based visual saliency. In: NIPS (2006)

31. Liu, T., Yuan, Z., Sun, J., Wang, J., Zheng, N., Tang, X., Shum, H.Y.: Learning to detect a salient object. In: PAMI (2011)

32. Achanta, R., Hemami, S., Estrada, F., Susstrunk, S.: Frequency-tuned salient region detection. In: CVPR (2009)

33. Perazzi, F., Krähenbühl, P., Pritch, Y., Hornung, A.: Saliency filters: Contrast based filtering for salient region detection. In: CVPR (2012)

34. Itti, L., Baldi, P.F.: Bayesian surprise attracts human attention. In: NIPS (2005)

35. Zhai, Y., Shah, M.: Visual attention detection in video sequences using spatiotemporal cues. In: ACM MM (2006)

36. Guo, C., Zhang, L.: A novel multiresolution spatiotemporal saliency detection model and its applications in image and video compression. IEEE TIP 19, 185–198 (2010)

37. Rudoy, D., Goldman, D., Shechtman, E., Zelnik-Manor, L.: Learning video saliency from human gaze using candidate selection. In: CVPR (2013)

38. Wang, J., Borji, A., Kuo, C.C., Itti, L.: Learning a combined model of visual saliency for fixation prediction. In: IEEE TIP (2016)

39. Xiao, J., Ehinger, K.A., Oliva, A., Torralba, A.: Recognizing scene viewpoint using panoramic place representation. In: CVPR (2012)
40. Mascelli, J.V.: The Five C's of Cinematography: Motion Picture Filming Techniques. Silman-James Press, Los Angeles (1998)
41. Tran, D., Bourdev, L., Fergus, R., Torresani, L., Paluri, M.: Learning spatiotemporal features with 3d convolutional networks. In: ICCV (2015)
42. Goodfellow, I., Pouget-Abadie, J., Mirza, M., Xu, B., Warde-Farley, D., Ozair, S., Courville, A., Bengio, Y.: Generative adversarial nets. In: NIPS (2014)
43. Malisiewicz, T., Gupta, A., Efros, A.A.: Ensemble of exemplar-SVMs for object detection and beyond. In: ICCV (2011)
44. Zhang, R., Isola, P., Efros, A.A.: Colorful image colorization. arXiv preprint arXiv:1603.08511 (2016)

PicMarker: Data-Driven Image Categorization Based on Iterative Clustering

Jiagao Hu, Zhengxing Sun[✉], Bo Li, and Shuang Wang

State Key Laboratory for Novel Software Technology, Nanjing University,
Nanjing, People's Republic of China
szx@nju.edu.cn

Abstract. Facing the explosive growth of personal photos, an effective classification tool is becoming an urgent need for users to categorize images efficiently with personal preferences. As previous researches mainly focus on the accuracy of automatic classification within the predefined label space, they cannot be used directly for the personalized categorization. In this paper, we propose a data-driven classification method for personalized image classification tasks which can categorize images group by group. Firstly, we describe images from both the view of appearance and the view of semantic. Then, an iterative framework which incorporates spectral clustering with user intervention is utilized to categorize images group by group. To improve the quality of clustering, we propose an online multi-view metric learning algorithm to learn the similarity metrics in accordance with user's criterion, and constraint propagation is integrated to adjust the similarity matrix. In addition, we build a system named *PicMarker* based on the proposed method. Experimental results demonstrate the effectiveness of the proposed method.

1 Introduction

With the rapid development of portable camera and the increasing rise of self-shooting as the order of the day, personal photos have faced an explosive growth. This put forward an urgent requirement to categorize our own images with personal preferences. Though image classification methods have achieved great success [1,4] as it's always a hot research topic in the field of computer vision, they are still difficult to meet the requirements of personalized categorization. In practice, the contents and themes of personal photos could be diverse from each other, and different people could have different taxonomies. On one hand, as the training data utterly determines the scope of categories, almost all of these model-driven methods [15,26,29] cannot be used directly for image categorization with personal preference. On the other hand, to label personalized training data for classifiers to get personal taxonomies is impractical as well, since the training process needs a large number of labeled images and image labeling is a time-consuming and boring work, which is not so friendly to users. Recently, considering the challenge of gathering labeled training data, zero-shot classification, which aims to classify samples without any training data via knowledge transfer

© Springer International Publishing AG 2017
S.-H. Lai et al. (Eds.): ACCV 2016, Part IV, LNCS 10114, pp. 172–187, 2017.
DOI: 10.1007/978-3-319-54190-7_11

between seen and unseen classes, has become increasingly popular in computer vision community. But essentially, these methods need mid-level semantic representation defined by human experts [16,22,30] or extracted from auxiliary text sources [9,27] as prior knowledge to establish inter-class connections. Thus, they are also not suited for personalized categorization.

As the model-driven method has limitations in the personalized categorization, classifying image in a data-driven manner could be a nice alternative. Data-driven method has been widely used to discover the structure of data which can be used to assist user manipulation, such as image segmentation [25], region annotation [13], image and video tagging [2], image extrapolation [35]. For image categorization, users are more accustomed to categorize images group by group. With discovered structure by data-driven method, images could be clustered into groups better, which is more consistent with user's habit.

In this paper, we propose a data-driven image classification method based on iterative clustering which can get personalized result. Firstly, we propose to describe images from both the view of appearance and the view of semantic. Then, an iterative framework which incorporates spectral clustering with user intervention is utilized to categorize images group by group. The clustering can put similar images into groups, and the user intervention is essential for personalized classification. In each iteration, user can categorize images by marking the positive or negative samples in the primary cluster and labelling the positive ones. To improve the quality of clustering, we propose an online multi-view metric learning algorithm to learn the similarity metrics in accordance with user's criterion during each iteration, and constraint propagation is integrated to adjust the similarity between images. In addition, we build a system named *PicMarker* which can be used to categorize images based on the proposed method. The advantages of our method lies in three aspects:

(1) Users can categorize images group by group effectively with very little burden in an iterative manner.
(2) Users can categorize scalable and large-scale image collections with growing efficiency in an incremental manner.
(3) Users can categorize images flexibly and freely as they wish.

2   Related Work

Image Classification. A plenty of works on image classification follows the supervised framework, which first train classifiers offline with lots of pre-labeled images, then use classifiers to classify images automatically. These methods have made a lot of great achievements, for example, some good practices have been put forward [1,8], some new learning strategies have been proposed [26,28], the impressive deep Convolutional Neural Networks (CNNs) [15,33] have emerged. However, the drawback of these methods is that the training data utterly determines the scope of categories, which means that the taxonomies is fixed by the trained classifier. As a consequence, they can hardly be applied to the categorization of personal categorization tasks. Some other researchers focus on the image classification with user in the loop, for example, taking user feedback into consideration to tune the classifier which makes it more fit to the user's preference [37]. In Lu et al. [31], the most interesting image is selected and the user is asked to label it. With the help of user intervention, collections can be easily categorized with personal preference. But all these methods need a number of pre-labeled images as the training data before the categorization since they are still model-driven methods, which will bring extra burden for users to categorize their personal images.

Recently, zero-shot classification is getting popular considering the challenge of gathering labeled training data. It can transfers information from observed classes to recognize unseen classes, which offers a compelling solution where unseen classes that do not have any labeled instances. For the prior knowledge for model training, most existing zero-shot models use attributes [16,30]. They first predict attributes of an image, and then predict its class label as a function of attributes. In addition to attributes, other external knowledge sources can also be used. For example, Mensink et al. [22] exploits the class relationships based on the *ImageNet* hierarchy, Elhoseiny et al. [9] uses *Wikipedia* articles as prior knowledge, Rohrbach et al. [27] uses the semantic hierarchy of *WordNet* to mine the parts of object categories. Nevertheless, as all the prior knowledge are pre-defined, they can hardly meet the personalized classification criterions.

Clustering. One of the best practiced ways to classify images group by group is to combine the categorization with clustering analysis [7]. The clustering result depends heavily on the feature vectors and the similarity metric, thus many researchers focus on these fields. For example, Ciocca et al. [7] investigated the impacts of the supervised features for unsupervised clustering. Lee et al. [19] proposed to use contextual descriptors in the clustering. And a plenty of research [5,10,36] try to learn the similarity metric for clustering. In addition, the number of categories is another key element in the clustering. But in practice, the set of potential categories is effectively unbounded, and may grow over time. Thus, several works [13,18] try to discover the potential categories in an iterative manner by interleaving unsupervised clustering and metric learning with user in the loop. Our work is inspired by them, but our method is in a fully online manner which is more efficient without re-training all the labeled samples from scratch in each iteration, and it allows the categorization of incremental collection with growing efficiency. Furthermore, we have mined more associations to improve the quality of clustering for the online process.

3   Overview

As illustrated in Fig. 1, for an uncategorized image collection with appropriate scale, the proposed method categories it following these steps. (1) *Similarity Measurement*: the similarity between all images are measured by the online learned metrics. (2) *(Constrained) Clustering*: images will be clustered into several groups. (3) *Selecting*: the most prominent cluster is selected as primary

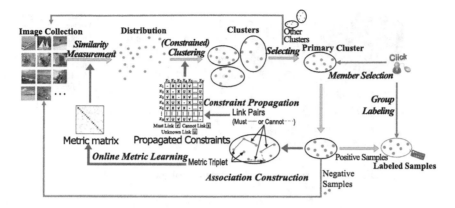

Fig. 1. Overview of the proposed method.

cluster which will be shown to the user for member selection, and other clusters will be send back for next iteration. (4) *Member Selection*: the user is asked to select member from the primary cluster, after that each image in the primary cluster will be marked as positive or negative. (5) *Group Labeling*: the positive samples will be gathered together so that the user can label them with a novel or an existing category, and the negative ones will be send back for next iteration. The whole iterative categorization process will continue until there are not enough uncategorized images for clustering.

To improve efficiency and lighten user's burden as much as possible, *online metric learning* is used to optimize the similarities between images, and *constraint propagation* is used to add constraints to uncategorized images. After the member selection in each iteration, *associations are first constructed*, which include triplets for metric learning and link pairs for constraint propagation. The learned metric will be used to measure the similarities between uncategorized images, and the constraints will be used to constrain the spectral clustering.

After the categorization of the current collection, the learned metrics can be used to categorize following collections directly. Therefore, for large-scale or incremental collection, we can partition it into several small ones and categorize them one by one with growing efficiency.

In the following sections, we will describe our method in detail.

4   Image Representation

Image representation plays an important role in classification tasks. In practice, different people may categorize images from different views. For example, some people prefer to categorize images according to what the image looks like, and others may prefer to categorize images by what the image contains. To cover both the appearance view and the semantic view, we propose to extract the visual appearance attribute together with the semantic object attribute to describe images.

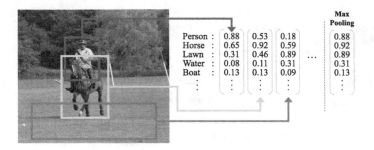

Fig. 2. Illustration of the semantic object attribute.

Visual Appearance Attribute. The visual appearance attributes of an image, such as the texture, color, shape, etc. are basic properties which could show distinctiveness of image. Thus, describing the appearance of images is helpful for the categorization. The most general way to describe the appearance is to gather the statistics of particular visual characters, such as the famous BoW methods [12,17] which build the descriptor with the statistical histogram. In recent years, there are some higher-level methods [3,32] which use classifiers to get the visual attribute of images. As these methods utilize classifiers instead of statistical histogram, their descriptors are typically more compact.

In this paper, we adopt the PiCoDes [3] which trained 2048 classifiers to extract the visual attributes. For a given image, we can get a 2048-dimensional vector to describe its appearance, such as shape, texture, and color patterns.

Semantic Object Attribute. The semantic object attribute could describe the semantic content of an image by perceiving the objects it contains, which is also important to the categorization of images. The intuitive manner to describe what an image contains is to use classifiers or detectors to estimate the existence of specific objects. For the sake of simplicity, we pre-define a list of objects and train a set of corresponding binary classifiers to extract the semantic attribute of images. The pre-defined objects should cover most of the nature to keep the descriptor's generalization. But in fact, not all types of object need to be detected because most of them overlap each other. Here, we define the object class list the same as Torresani et al. [34] does, *i.e.*, choose object classes from concepts in the Large Scale Concept Ontology for Multimedia [23]. To get more accurate judgment of the existence of objects by simple classifiers, we propose to classify subimages instead of the whole image.

A subimage is a region of the image which contains an entire object without too many irrelevant elements. As in Fig. 2, the three rectangles indicate three subimages in which the person, the horse and the lawn dominate respectively. Obviously, the simple classifiers can get more accurate judgment for subimages than the entire image. There are a lot of research about generating region proposals which can be regarded as subimages here. In this paper, we use BING [6] as it can extract accurate region proposals very fast.

To describe the semantic object attribute of a given image, a set of subimages (15 in our experiment) are generated at first. Then the object classifiers are used to classify the subimages. Finally, a maximum pooling is applied to get the final descriptor with each item indicate the probability that the given image contains the particular object as shown in Fig. 2.

5   Algorithm

In each iteration, unlabeled images are clustered, and the primary cluster is selected. Then the user is asked to mark the positive or negative instances of the primary cluster. The positive images will be categorized, while others will keep unlabeled. The primary cluster with positive and negative marks will be used to online learn the similarity metric to cluster unlabeled images in next iteration. In this section, we will describe these in detail.

5.1   Association Construction

Suppose the selected primary cluster is $\mathcal{P}$. After the member selection, it will be divided into two parts, *i.e.*, the positive set $\mathcal{P}^+$ whose members belong to a same category (the majority category), and the negative set $\mathcal{P}^-$ whose members do not belong to the majority category. With the two sets, some helpful associations can be built and discovered.

The first association we could discover is that the similarities between images in $\mathcal{P}^+$ should be greater than the similarity between an instance in $\mathcal{P}^+$ and an instance in $\mathcal{P}^-$. Therefore, we could construct a set of triplets as

$$\mathcal{T} = \left\{ \left(x, x^+, x^-\right) \middle| x, x^+ \in \mathcal{P}^+, x^- \in \mathcal{P}^- \right\}. \tag{1}$$

This triplet set can be used to learn a new similarity metric to ensure that the similarity between x and x^+ is greater than that between x and x^-, which could make similar images closer.

The second association we could mine is that instances similar to those in $\mathcal{P}^+$ should be in a same category, and should not be in a same category with instances similar to those in $\mathcal{P}^-$. That is to say, given unlabeled images x_a, x_b and x_c. If x_a is similar to $x_i \in \mathcal{P}^+$, and x_b is similar to $x_j \in \mathcal{P}^+$, then x_a and x_b are likely to be in a same category. And if x_c is similar to $x_k \in \mathcal{P}^-$, then x_a and x_c might not be in a same category. We formulate this type of association as must-link set $\mathcal{M}$ and cannot-link set $\mathcal{C}$ as

$$\mathcal{M} = \left\{ (x_i, x_j) \middle| x_i, x_j \in \mathcal{P}^+ \wedge x_i \neq x_j \right\}, \tag{2}$$

$$\mathcal{C} = \left\{ (x_i, x_j), (x_j, x_i) \middle| x_i \in \mathcal{P}^+ \wedge x_j \in \mathcal{P}^- \right\}. \tag{3}$$

In each iteration, we use these link sets to propagate constraints to unlabeled images, which will make similar images closer and dissimilar images farther.

5.2 Online Multi-view Similarity Metric Learning

As some people may concern more about the semantic view, while others may concern more about the appearance view when categorizing their image collections, it is necessary to optimally combine the multiple views. The general idea of our multi-view similarity metric learning is to learn a separate metric for each view, and meanwhile learn an optimal combination from different views.

The multi-view similarity between two images x_i and x_j is defined as a weighted sum of all M views:

$$S(x_i, x_j) = \sum_{m=1}^{M} \theta_m \mathcal{S}_{\mathbf{W}_m}(\phi_m(x_i), \phi_m(x_j)), \qquad (4)$$

where θ_m is the combinational weight for the m-th view, $\phi_m(x_i)$ is the feature vector of image x_i in the m-th view, and $\mathcal{S}_{\mathbf{W}_m}(\phi_m(x_i), \phi_m(x_j))$ denotes the similarity between x_i and x_j from the m-th view. In an individual view, the similarity is measured by the following bilinear function:

$$\mathcal{S}_{\mathbf{W}}(\phi(x_i), \phi(x_j)) = \phi(x_i)^{\mathrm{T}} \mathbf{W} \phi(x_j), \qquad (5)$$

where $\mathbf{W}$ is the similarity metric matrix. During each iteration, the multiple metric matrixes and combinational weights will be online updated using the triplet set $\mathcal{T}$. Initially, all similarity metric matrixes $\{\mathbf{W}_m \,|\, m = 1, ..., M\}$ are identity matrix, and combinational weights $\{\theta_m \,|\, m = 1, ..., M\}$ are set as $1/M$.

To optimize both the similarity metric matrix and the combinational weights for each individual view, we resort to the efficient OASIS algorithm [5] for online similarity metric learning in each view, and apply the Hedge algorithm [11] to learn the optimal weights.

OASIS optimizes the metric matrix $\mathbf{W}$ to make the similarity of each triplet (x, x^+, x^-) satisfies $\mathcal{S}_{\mathbf{W}}(\phi(x), \phi(x^+)) > \mathcal{S}_{\mathbf{W}}(\phi(x), \phi(x^-)) + 1$, where $+1$ means a safety margin to ensure a sufficiently large difference. Then the hinge loss function is defined as follows:

$$l_{\mathbf{W}}(x, x^+, x^-) = \max(0, 1 - \mathcal{S}_{\mathbf{W}}(\phi(x), \phi(x^+)) + \mathcal{S}_{\mathbf{W}}(\phi(x), \phi(x^-))). \quad (6)$$

The goal is to minimize a global loss that accumulates hinge losses over all triplets in $\mathcal{T}$. Passive-Aggressive algorithm is applied to iteratively optimize $\mathbf{W}$ over triplets. It is not difficult to derive that for the t-th triplet $(x^{(t)}, x^{(t)+}, x^{(t)-})$, the $\mathbf{W}^{(t)}$ should be online updated by the following closed-form solution:

$$\mathbf{W}^{(t)} = \mathbf{W}^{(t-1)} + \tau^{(t)} V^{(t)}, \qquad (7)$$

where $\tau^{(t)} = \min(C, l_{\mathbf{W}^{(t-1)}}(x^{(t)}, x^{(t)+}, x^{(t)-}) / \|V^{(t)}\|^2)$, C is a trade-off parameter, and $V^{(t)} = x^{(t)}(x^{(t)+} - x^{(t)-})^{\mathrm{T}}$. For more detail about the OASIS optimization, we refer the reader to [5].

To online learn the combinational weights $\{\theta_m \,|\, m = 1, ..., M\}$, we apply the well-known Hedge algorithm [11] which is simple and effective for online learning.

Algorithm 1. Online Multi-view Similarity Metric Learning (OMSML)

Input:

  – Triplet set: $\mathcal{T}$;
  – Number of pre-learned triplets: N_{Pre}.

1: **for** each (x, x^+, x^-) in $\mathcal{T}$ **do**
2:    **if** $S(x, x^+) < S(x, x^-) + 1$ **then**
3:       **for** $m = 1, ..., M$ **do**
4:          Update $\mathbf{W}_m$ by Eq. 7
5:       **end for**
6:    **end if**
7: **end for**
8: Set β by Eq. 9
9: **for** each $(x, x^!, x^-)$ in $\mathcal{T}$ **do**
10:    **for** $m = 1, ..., M$ **do**
11:       **if** $\mathcal{S}_{\mathbf{W}_m}(\phi_m(x), \phi_m(x^+)) \leq \mathcal{S}_{\mathbf{W}_m}(\phi_m(x), \phi_m(x^-))$ **then**
12:          Update $\theta_m - \theta_m \beta$
13:       **end if**
14:    **end for**
15:    $\theta_m = \frac{\theta_m}{\sum_{k=1}^{M} \theta_k}, \forall m = 1, ... M$
16: **end for**

In particular, after the metric learning of all views, for the t-th triplet, the weight is updated as follows:

$$\theta_m^{(t)} = \frac{\theta_m^{(t-1)} \beta^{z_m^{(t)}}}{\sum_{k=1}^{M} \theta_k^{(t-1)} \beta^{z_k^{(t)}}}, \tag{8}$$

where $\beta \in (0, 1)$ is a discounting parameter to penalize the poor view, and $z_m^{(t)}$ equals 1 when $\mathcal{S}_{\mathbf{W}_m}(\phi_m(x), \phi_m(x^+)) \leq \mathcal{S}_{\mathbf{W}_m}(\phi_m(x), \phi_m(x^-))$, and 0 otherwise. We follow the principles in [14] to set β in each iteration as:

$$\beta = \frac{\sqrt{(N_{\text{Pre}} + \text{size}(\mathcal{T}))}}{\sqrt{\ln M} + \sqrt{(N_{\text{Pre}} + \text{size}(\mathcal{T}))}}, \tag{9}$$

where N_{Pre} denotes the number of triplets in all previous iterations.

Algorithm 1 summarizes the proposed algorithm for online multi-view similarity metric learning (OMSML). After the learning of similarity metrics, we can calculate a new similarity matrix $\mathbf{S} = \{S_{ij}\}_{N \times N}$ with the learned metric and combinational weights, and in it, similar images will be closer. This could make the cluster larger, tighter and purer, which means the user effort would be lower.

5.3   Constraint Propagation

Although the learned metric makes similar images closer, the effect could be limited since only triplets of a specific category can be used to learn the metric in each iteration. To exploit the user intervention further, we use the link pairs to propagate constraints to unlabeled images. Here we apply the constraint propagation algorithm proposed by Lu and Ip [21] which provides an exhaustive and efficient solution using k-nearest neighbor graphs.

In [21], the algorithm to propagate constraint could be summarized as follows:

(1) Initialize the pairwise constraints matrix $\mathbf{Y} = \{Y_{ij}\}_{N \times N}$ with the link set $\mathcal{M}$ and $\mathcal{C}$: $Y_{ij} = 1$ where $(x_i, x_j) \in \mathcal{M}$, $Y_{ij} = -1$ where $(x_i, x_j) \in \mathcal{C}$, and $Y_{ij} = 0$ otherwise.

(2) For a given similarity matrix $\mathbf{S} = \{S_{ij}\}_{N \times N}$, Form the weight matrix $\mathbf{Z}$ by $Z_{ij} = S_{ij}/\sqrt{S_{ii}S_{jj}}$ if $x_j (j \neq i)$ is among the k-nearest neighbors of x_i, and $Z_{ij} = 0$ otherwise.

(3) Construct the matrix $\mathcal{L} = D^{-1/2}\mathbf{Z}D^{-1/2}$, where D is a diagonal matrix with $D_{ii} = \sum_{j=1}^{N} Z_{ij}$. Then the propagate constraints is

$$F = (1 - \alpha)^2 (I - \alpha\mathcal{L})^{-1}\mathbf{Y}(I - \alpha\mathcal{L})^{-1}, \tag{10}$$

where α is a parameter in range $(0,1)$, and I is an identity matrix.

(4) Adjust the similarity matrix with the propagated constraints F and the weight matrix $\mathbf{Z}$ as follows:

$$\mathbb{S}_{ij} = \begin{cases} 1 - (1 - F_{ij})(1 - \mathbf{Z}_{ij}), & F_{ij} \geqslant 0 \\ (1 + F_{ij})\mathbf{Z}_{ij}, & F_{ij} < 0 \end{cases}. \tag{11}$$

In each iteration, we will use the link set $\mathcal{M}$ and $\mathcal{C}$ to adjust the similarity matrix after the similarity metric learning.

5.4   Image Categorization Based on Iterative Clustering

With the adjusted similarity matrix, we use spectral clustering to cluster images into several groups. After clustering, the average intra-cluster distance of each cluster c is calculated by:

$$d_{\text{intra}}(c) = \frac{1}{|c| \times (|c| - 1)} \sum_{\substack{i,j \in c \\ i \neq j}} d(x_i, x_j), \tag{12}$$

where $d(x_i, x_j)$ is the distance between x_i and x_j in the spectral clustering space. And the cluster c^* with minimum intra-cluster distance is selected as primary cluster:

$$c^* = \arg\min_c d_{\text{intra}}(c). \tag{13}$$

The primary cluster will be shown to the user for member selection and labeling.

Algorithm 2 summarizes the proposed image categorization based on iterative clustering. Initially, the similarity metric matrixes are identity matrixes, and combinational weights are set as $1/M$. Then the similarity matrix is measured, and images are clustered for the first time. After user's member selection on the primary cluster, Algorithm 2 could repeat until all images classified.

Algorithm 2. Image Categorization Based on Iterative Clustering

---

Input:

- Primary cluster in the initialization: $\mathcal{P} = \{\mathcal{P}^+, \mathcal{P}^-\}$;
- Uncategorized images: $\mathcal{U} = \{x_1, ..., x_n\}$;
- Number of clusters: NC.

1: **repeat**
2:    Construct triplet set $\mathcal{T}$ using $\mathcal{P}$ by Eq. 1;
3:    Online learn the multi-view similarity metric (**Algorithm 1**);
4:    Calculate the similarity matrix of $\mathcal{U}$ by Eq. 4;
5:    Construct must-link set $\mathcal{M}$ and cannot-link set $\mathcal{C}$ using $\mathcal{P}$ by Eq. 2 and Eq. 3;
6:    Propagate pairwise constraints by Eq. 10;
7:    Adjust the similarity matrix by Eq. 11;
8:    Cluster $\mathcal{U}$ into NC clusters with adjusted similarity matrix;
9:    Select the primary cluster $\mathcal{P}$ by Eq. 13;
10:    Mark images in $\mathcal{P}$ as $\{\mathcal{P}^+, \mathcal{P}^-\}$, label $\mathcal{P}^+$ with a novel or an existing category;
11:    $\mathcal{U} = \mathcal{U} - \mathcal{P}^+$;
12: **until** all images are categorized

---

6 The *PicMarker* and User Intervention

Figure 3 demonstrates the user interface of *PicMarker*. As Fig. 3 shows, it consists of two parts, the loading zone on the left, and the category zone on the right. The images of selected primary cluster will be displayed in the loading zone. In the category zone, user can create a novel category or select any existing categories, and rename any categories.

When there are some images to be classified, their similarities will be first measured. With the similarity matrix, images will be clustered and the primary cluster is selected automatically. Users are then asked to mark its members. They can either (a) mark positive images which correspond to the largest subset of a particular category, others will remain negative by default, or (b) mark negative images that do not belong to the major category, others will remain positive. In practice, the user always choose to perform fewest operations to create the largest group with all images belong to a same category. Then the user can select a novel or an existing category to label the group of positive images. The system can achieve the goals of image categorization and category discovery simultaneously. Besides, as the category of every image is explicitly confirmed by the user, the result is very consistent with the user intention.

7 Experiments

7.1 Dataset and Evaluation

The proposed method is evaluated on three challenging image data sets: UIUC-Sports event dataset [20], MIT Scene 15 Dataset [17], and Scene 8 [24]. The event dataset is a dataset for event classification. It contains 1,579 images of 8 widely

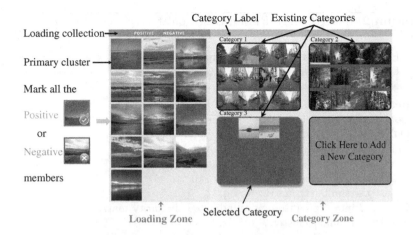

Fig. 3. A prototype system of PicMarker.

varied sport events. Scene 15 is a widely used dataset for scene categorization. It contains 4,485 gray-level images of 15 scenes. Scene 8 is the original 8 scenes of Scene 15, it contains totally 2,688 color images.

The effectiveness of the proposed iterative categorization method is measured by the user effort. The user effort is defined as the average number of operations required for labeling one image:

$$\mu = U/n, \tag{14}$$

where n is the number of all labeled images, and U is the total operation amount which includes two aspects: (1) the number of selected positive or negative member, and (2) one operation for labeling the positive group. When categorizing images one by one manually, the user effort is 1.

7.2   Evaluation of the Learned Similarity

In order to evaluate the quality of the learned similarity space, we follow Galleguillos et al. [13] to assess the average purity of all clusters generated from the similarity. The purity of a cluster B is defined as: $purity\,(B) = \max\limits_{l \in L} \dfrac{|\{x \in B \wedge l\,(x) = l\}|}{|B|}$, where L is the label set, and $l\,(x)$ is the groundtruth label of sample x.

In this experiment, we evaluate our method on the Scene 15 dataset. For each class, 50 images are picked out to learn the similarity metric. We compare the performance of the native metric (similarity based on the Euclidean distance), the learned metric by OASIS, the learned metric by the proposed OMSML, the learned metric by OMSML with constraint propagation (OMSML+CP), and the learned metric by offline method MLR [13]. For the native metric, OASIS and MLR, we uniformly concatenate the appearance features with the semantic features.

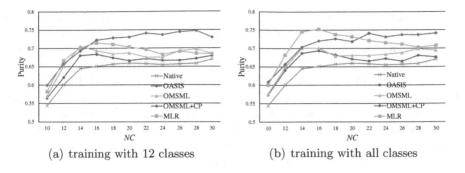

(a) training with 12 classes   (b) training with all classes

Fig. 4. The curve of the average cluster purity.

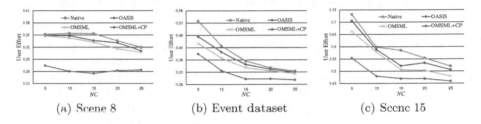

(a) Scene 8   (b) Event dataset   (c) Scene 15

Fig. 5. The user effort in different collections.

We vary the number of clusters from 10 to 30 (the step size is 2), and report the average purity. Besides, to evaluate whether the learned similarity can improve the clustering results when the evaluation set has some unknown categories, we also report the average purity with only 12 classes of images selected as training set. The resulting mean purity curves are reported in Fig. 4.

From Fig. 4, we observe that the curves achieved by learning based methods all lies above the native curve. The curve of OMSML lies above the curve of OASIS, which demonstrate the effectiveness of the proposed multi-view metric learning. And the curve of OMSML+CP lies close to the curve of MLR, and even above it in some cases, which implies that the adjusted online learned similarity has a comparable quality with the offline batch learned similarity.

7.3 The User Effort in Iterative Classification

In this section, we evaluate the efficiency of the proposed iterative categorization method by assessing the use effort according to Eq. 14 in all three dataset. We regard the groundtruth as users' labels to simulate the operations of real users. To evaluate the performance of the proposed method, we choose the native metric (similarity based on the Euclidean distance), OASIS, OMSML, and OMSML+CP for comparison. The setting of image features is the same as that in Sect. 7.2.

Before the categorization, the metric matrixes and combinational weights are initialized. For each dataset, we vary the number of clusters from 5 to 25 (the step size is 5) in iterations. Figure 5 demonstrates the results of the user effort in three datasets. It can be seen that, the learned metrics could reduce users' burden using the proposed framework to categorize images, because the user efforts of learned metrics are lower than those of the native metrics. The user effort using OMSML is slightly lower that using OASIS, which indicates that uniform concatenation is not optimal to combine different kinds of features, and the proposed multi-view metric learning has obvious effect to combine them. Of all the curves in three datasets, the user effort of OMSML+CP is the lowest. It implies that the proposed method can reduce user's burden for categorizing images obviously in an iterative framework.

7.4   The Adaption to Incremental and Large-Scale Collection

In practice, there may be some large scale collections or ever-increasing collections to be categorized. It is impractical to measure similarity between images in these collections. As online learning strategy is used in our method, we can split these collections into blocks and regard them as incremental data which arrives block by block, and categorize the blocks one by one with online learned similarity metrics. Accordingly, we evaluate that whether the proposed method can improve the efficiency when categorizing incremental blocks. The evaluation is done by assessing the user effort in successive blocks.

In our experiment, we partition the Scene 15 dataset into 6 blocks. The first block is regarded as the initial collection, and other blocks are incremental collections which will arrive one by one. Before the categorization of the initial collection, the similarity metrics will be initialized with identity matrixes and equal combinational weights. After one block categorized, the learned similarity metrics will be kept for the categorization of the next block. We record the user effort using the native metric, OMSML, and OMSML+CP with varying the number of cluster (NC) from 10 to 20 (the step size is 5).

Figure 6 shows the result. We can see that the user efforts using native metric in successive blocks are quite stable, which is understandable. But the user efforts using proposed methods are getting lower and lower. It implies that the learned metrics are effective for incremental collections. That is to say, using the proposed method, users can categorize large-scale and incremental collections with growing efficiencies.

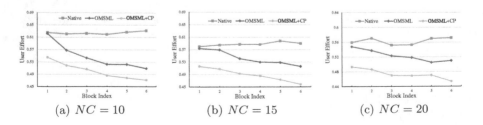

(a) $NC = 10$ (b) $NC = 15$ (c) $NC = 20$

Fig. 6. User effort for incremental collections.

7.5   User Study

Idealized experiments are not enough to prove that a newly proposed method is actually useful for real users in practice. Therefore, we perform a user study to evaluate the effectiveness of our method in real-world categorization tasks. We invite 12 people to use our *PicMarker* system to categorize images. Half of them use the proposed method, while the other half use the native similarity metric. They start to categorize images directly after our brief description about the usage. The image collections contains 538 images selected from the Scene 8 dataset. The user effort and time cost are recorded. In addition, we also try to categorize those images one by one manually without using the prototype system, and record the time cost.

Table 1 shows the average user effort and time cost. It can be seen that proposed framework have a nice efficiency even adopts the native similarity metric since it only needs 0.301 operations to label an image and 643 s to categorize all those 538 images on average, which is significantly less than 1 operation and about 1,200 s needed to categorize them one by one manually. And the proposed method outperform the native metric obviously in both user effort and time cost. Besides, we found that the classification results were not so consistent between participants, and the final category number differs ranging from 5 to 9. It demonstrates that proposed method can fit the needs of diverse needs, and users can classify image collections flexibly and freely using our system.

Table 1. The statistical result of user study.

	One by one	Native	Ours
User effort	1	0.301	0.264
Time (s)	∼1,200	643	561

8   Conclusion

In this paper, we propose a data-driven image categorization method based on iterative clustering which can get personalized result. The features of our method lies in three aspects. Firstly, users can categorize images group by group effectively with very little burden in an iterative manner. Secondly, users can categorize scalable and large-scale image collections with growing efficiency in an incremental manner. Thirdly, users can categorize images flexibly and freely as they wish. Experimental results shows that the proposed method improves both the effectiveness and efficiency of image categorization.

Acknowledgment. This work is supported by National High Technology Research and Development Program of China (No. 2007AA01Z334), National Natural Science Foundation of China (No. 61321491, 61272219), Innovation Fund of State Key Laboratory for Novel Software Technology (No. ZZKT2013A12, ZZKT2016A11), Program for New Century Excellent Talents in University of China (NCET-04-04605).

References

1. Akata, Z., Perronnin, F., Harchaoui, Z., Schmid, C.: Good practice in large-scale learning for image classification. IEEE Trans. Pattern Anal. Mach. Intell. **36**, 507–520 (2014)
2. Ballan, L., Bertini, M., Uricchio, T., Bimbo, A.D.: Data-driven approaches for social image and video tagging. Multimedia Tools Appl. **74**, 1443–1468 (2015)
3. Bergamo, A., Torresani, L., Fitzgibbon, A.: PICODES: learning a compact code for novel-category recognition. In: NIPS, pp. 2088–2096 (2011)
4. Bianco, S., Ciocca, G., Cusano, C.: CURL: image classification using co-training and unsupervised representation learning. Comput. Vis. Image Underst. **145**, 15–29 (2016)
5. Chechik, G., Sharma, V., Shalit, U., Bengio, S.: Large scale online learning of image similarity through ranking. J. Mach. Learn. Res. **11**, 11–14 (2009)
6. Cheng, M.M., Zhang, Z., Lin, W.Y., Torr, P.: BING: binarized normed gradients for objectness estimation at 300fps. In: CVPR, pp. 3286–3293. IEEE (2014)
7. Ciocca, G., Cusano, C., Santini, S., Schettini, R.: On the use of supervised features for unsupervised image categorization: an evaluation. Comput. Vis. Image Underst. **122**, 155–171 (2014)
8. Deng, J., Berg, A.C., Li, K., Fei-Fei, L.: What does classifying more than 10,000 image categories tell us? In: Daniilidis, K., Maragos, P., Paragios, N. (eds.) ECCV 2010. LNCS, vol. 6315, pp. 71–84. Springer, Heidelberg (2010). doi:10.1007/978-3-642-15555-0_6
9. Elhoseiny, M., Saleh, B., Elgammal, A.: Write a classifier: zero-shot learning using purely textual descriptions. In: ICCV, pp. 2584–2591. IEEE (2013)
10. Feng, Z., Jin, R., Jain, A.: Large-scale image annotation by efficient and robust kernel metric learning. In: ICCV, pp. 1609–1616 (2013)
11. Freund, Y., Schapire, R.E.: A decision-theoretic generalization of on-line learning and an application to boosting. J. Comput. Syst. Sci. **55**, 119–139 (1997)
12. Csurka, G., Dance, C.R., Fan, L., Jutta Willamowski, C.B.: Visual categorization with bags of keypoints. In: ECCV Workshop, pp. 1–22 (2004)
13. Galleguillos, C., McFee, B., Lanckriet, G.R.G.: Iterative category discovery via multiple kernel metric learning. Int. J. Comput. Vision **108**, 115–132 (2014)
14. Jin, R., Hoi, S.C.H., Yang, T.: Online multiple kernel learning: algorithms and mistake bounds. In: Hutter, M., Stephan, F., Vovk, V., Zeugmann, T. (eds.) ALT 2010. LNCS (LNAI), vol. 6331, pp. 390–404. Springer, Heidelberg (2010). doi:10.1007/978-3-642-16108-7_31
15. Krizhevsky, A., Sutskever, I., Hinton, G.E.: ImageNet classification with deep convolutional neural networks. In: NIPS, pp. 1–9 (2012)
16. Lampert, C.H., Nickisch, H., Harmeling, S.: Attribute-based classification for zero-shot visual object categorization. IEEE Trans. Pattern Anal. Mach. Intell. **36**, 453–465 (2014)
17. Lazebnik, S., Schmid, C., Ponce, J.: Beyond bags of features: spatial pyramid matching for recognizing natural scene categories. In: CVPR, pp. 2169–2178. IEEE (2006)
18. Lee, Y.J., Grauman, K.: Learning the easy things first: self-paced visual category discovery. In: CVPR, pp. 1721–1728. IEEE (2011)
19. Lee, Y.J., Grauman, K.: Object-graphs for context-aware visual category discovery. IEEE Trans. Pattern Anal. Mach. Intell. **34**, 346–358 (2012)

20. Li, L.J., Fei-Fei, L.: What, where and who? Classifying events by scene and object recognition. In: ICCV, pp. 1–8. IEEE (2007)
21. Lu, Z., Ip, H.H.S.: Constrained spectral clustering via exhaustive and efficient constraint propagation. In: Daniilidis, K., Maragos, P., Paragios, N. (eds.) ECCV 2010. LNCS, vol. 6316, pp. 1–14. Springer, Heidelberg (2010). doi:10.1007/978-3-642-15567-3_1
22. Mensink, T., Verbeek, J., Perronnin, F., Csurka, G.: Distance-based image classification: generalizing to new classes at near zero cost. IEEE Trans. Pattern Anal. Mach. Intell. **35**, 1–14 (2013)
23. Naphade, M., Smith, J., Tesic, J., Chang, S.-F., Hsu, W., Kennedy, L., Hauptmann, A., Curtis, J.: Large-scale concept ontology for multimedia. IEEE Multimedia **13**, 86–91 (2006)
24. Oliva, A., Hospital, W., Ave, L., Torralba, A.: Modeling the shape of the scene: a holistic representation of the spatial envelope. Int. J. Comput. Vision **42**, 145–175 (2001)
25. Park, S.H., Yun, I.D., Lee, S.U.: Data-driven interactive 3D medical image segmentation based on structured patch model. In: Gee, J.C., Joshi, S., Pohl, K.M., Wells, W.M., Zöllei, L. (eds.) IPMI 2013. LNCS, vol. 7917, pp. 196–207. Springer, Heidelberg (2013). doi:10.1007/978-3-642-38868-2_17
26. Ristin, M., Guillaumin, M., Gall, J., Gool, L.V.: Incremental learning of NCM forests for large-scale image classification. In: CVPR, pp. 3654–3661. IEEE (2014)
27. Rohrbach, M., Stark, M., Szarvas, G., Gurevych, I., Schiele, B.: What helps where - and why? Semantic relatedness for knowledge transfer. In: CVPR, pp. 910–917. IEEE (2010)
28. Royer, A., Lampert, C.H.: Classifier adaptation at prediction time. In: CVPR, pp. 1401–1409. IEEE (2015)
29. Sánchez, J., Perronnin, F., Mensink, T., Verbeek, J.: Image classification with the fisher vector: theory and practice. Int. J. Comput. Vision **105**, 222–245 (2013)
30. Li, X., Guo, Y., Schuurmans, D.: Semi-supervised zero-shot classification with label representation learning. In: ICCV, pp. 4211–4219. IEEE (2015)
31. Lu, Z., Ip, H.: Combining context, consistency, and diversity cues for interactive image categorization. IEEE Trans. Multimedia **12**, 194–203 (2010)
32. Su, Y., Jurie, F.: Learning compact visual attributes for large-scale image classification. In: Fusiello, A., Murino, V., Cucchiara, R. (eds.) ECCV 2012. LNCS, vol. 7585, pp. 51–60. Springer, Heidelberg (2012). doi:10.1007/978-3-642-33885-4_6
33. Szegedy, C., Liu, W., Jia, Y., Sermanet, P., Reed, S., Anguelov, D., Erhan, D., Vanhoucke, V., Rabinovich, A.: Going deeper with convolutions. In: CVPR, pp. 1–9. IEEE (2015)
34. Torresani, L., Szummer, M., Fitzgibbon, A.: Efficient object category recognition using classemes. In: Daniilidis, K., Maragos, P., Paragios, N. (eds.) ECCV 2010. LNCS, vol. 6311, pp. 776–789. Springer, Heidelberg (2010). doi:10.1007/978-3-642-15549-9_56
35. Wang, M., Lai, Y.K., Liang, Y., Martin, R.R., Hu, S.M.: BiggerPicture: data-driven image extrapolation using graph matching. ACM Trans. Graph. **33**, 1–13 (2014)
36. Xu, X., Shimada, A., Nagahara, H., Taniguchi, R.: Learning multi-task local metrics for image annotation. Multimedia Tools Appl. **75**, 2203–2231 (2014)
37. Ye, Z., Liu, P., Tang, X., Zhao, W.: May the torcher light our way: a negative-accelerated active learning framework for image classification. In: ICIP, pp. 1658–1662. IEEE (2015)

Image Alignment

Adaptive Direct RGB-D Registration and Mapping for Large Motions

Renato Martins[1,2(✉)], Eduardo Fernandez-Moral[1], and Patrick Rives[1]

[1] Inria, Université Côte d'Azur, Sophia Antipolis, France
[2] MINES ParisTech, PSL Research University, Sophia Antipolis, France
renato-jose.martins@inria.fr

Abstract. Dense direct RGB-D registration methods are widely used in tasks ranging from localization and tracking to 3D scene reconstruction. This work addresses a peculiar aspect which drastically limits the applicability of direct registration, namely the weakness of the convergence domain. First, we propose an activation function based on the conditioning of the RGB and ICP point-to-plane error terms. This function strengthens the geometric error influence in the first coarse iterations, while the intensity data term dominates in the finer increments. The information gathered from the geometric and photometric cost functions is not only considered for improving the system observability, but for exploiting the different convergence properties and convexity of each data term. Next, we develop a set of strategies as a flexible regularization and a pixel saliency selection to further improve the quality and robustness of this approach.

The methodology is formulated for a generic warping model and results are given using perspective and spherical sensor models. Finally, our method is validated in different RGB-D spherical datasets, including both indoor and outdoor real sequences and using the KITTI VO/SLAM benchmark dataset. We show that the different proposed techniques (weighted activation function, regularization, saliency pixel selection), lead to faster convergence and larger convergence domains, which are the main limitations to the use of direct methods.

1 Introduction

Feature based registration methods have bigger convergence domains (if feature matching is successful) but are locally less precise and more sensitive to outliers than direct dense methods [1,2]. Feature-based methods (e.g. [3–6]) rely on an intermediary estimation process based on thresholding [7,8] before requiring matching between frames to recover camera motion. This feature extraction and matching process is often badly conditioned, noisy and not robust, and therefore it must rely on higher level robust estimation techniques and on filtering.

Direct approaches (*image-based*), however, do not rely on this feature extraction and matching process. The camera motion is directly estimated by minimising a non-linear intensity error between images, via a parametric warping function. In this way, the matching and the motion estimation are performed simultaneously at each step of the optimisation. Classically direct approaches have

© Springer International Publishing AG 2017
S.-H. Lai et al. (Eds.): ACCV 2016, Part IV, LNCS 10114, pp. 191–206, 2017.
DOI: 10.1007/978-3-319-54190-7_12

focused on region-of-interest tracking whether they are modelled by affine [9], planar [10–12], or multiple-plane tracking [13,14]. In [15] direct approaches were generalized to use the full-image densely and track 6 DOF pose using stereo cameras whilst mapping the environment through dense stereo matching.

In general, registration is performed only between close frames (small displacements), since dense registration tasks are particularly sensible to the local convexity of the cost error function. The error function convergence depends on a number of parameters including: the inherent noise in the photometric and geometric sensor measurements, the resolution (sampling at different scales), the scene configuration (symmetry) and the scene stationarity (illumination changes or moving objects). Even though a mathematical condition for the convergence cannot be established, some effort was done in estimating convergence envelops for teaching and repeating techniques [16,17].

We are interested in applying direct registration in larger displacements which is useful for re-location tasks, where the current trajectory may not be "close enough" to the trajectory where the model was learned and/or because the conditions of observation have changed: lighting, occlusions and dynamic objects. This problem also occurs in large scale scenes due to storage capability and complexity of configurations. In these cases RGB registration techniques have their performance challenged, since convergence is likely to happen only for small displacements (for instance see Fig. 1). This work addresses a contribution in this direction by considering the information gathered from ICP point-to-plane [18] and photometric error direct cost functions [19], not only for improving ranking conditioning as in [20,21], but for taking into consideration the convexity of both terms to achieve a larger convergence domain and smaller number iterations.

The main contribution of this work is an adaptive RGB-D error cost function that has a larger convergence domain and a faster convergence in both simulated and real data. This formulation employs the relative condition number metric to update the weighting of the RGB and depth costs. We show that this significantly improves the convergence stability and the speed of convergence. A set of strategies are also presented to further increase the robustness of the system. First, we discuss a regularization of the geometry in planar patches that reduces the spurious noise (specially at non-textured regions) and that generates a confidence index for each pixel; Second, we present a coherent pixel selection from saliency that ensures good observation properties of each DOF for both RGB and ICP registration tasks.

The remainder is organized as follows. First, we review recent related works in Sect. 1.1. Next, we introduce the basic classical method of dense registration and our adapted formulation in Sects. 2.1 and 2.2. Section 2.3 describes further improvements in accuracy by a regularization of the depth information and the extension of a saliency concept for computational efficiency. Lastly, we present experimental results in Sect. 3 for indoor (simulated and real) and outdoor contexts, and to conclude the paper in Sect. 4.

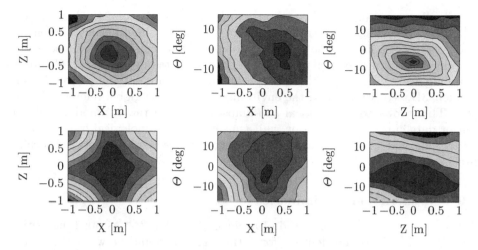

Fig. 1. Intensity RGB level curves (first row) and ICP ploint-to-plane (second row) for a typical corridor frame at the Sponza Atrium model. The costs are evaluated in the simplified case of 3DOF (one rotation and two translations) and the corresponding level curves are from the surfaces $C(\mathbf{x})$ with $\mathbf{x} = [x\ 0\ z\ \mathbf{0}_{(1\times3)}]^T$ (left column), $\mathbf{x} = [x\ 0\ 0\ \theta\ 0\ 0]^T$ (middle) and $\mathbf{x} = [0\ 0\ z\ \theta\ 0\ 0]^T$ (right column). The ICP point-to-plane cost is flatter near the solution. (Color figure online)

1.1 Main Related Works

Large image displacement is an active area of research in the optical flow community [1,2,22]. Variational optimization methods are typically applied to constrain the flow estimation (each pixel has two degrees of freedom) in a dense framework. In addition, [1] jointly consider features (i.e. taking advantage of the invariance and stability of SIFT and SURF to scene changes) in cases of large motions. These approaches are not suitable in our context since we aim to keep the direct estimation concept (no matching stage).

This work is mainly related to direct RGB-D motion estimation techniques, being close to [20,23]. An important issue raised in these previous works is the scaling of the geometric and photometric cost terms for ensuring nice convergence properties. The interesting work of [23] applies a smooth steep function to weight the influence of the RGB and ICP terms during optimization. Although sharing a similar framework and initial conclusions (which were founded independently from their work), we propose an additional equivalent activation function based on the conditioning of the error terms. This formulation is more stable and capable of dealing with cross-peak instabilities. The work of [20] adopts λ_D to scale the photometric error (in pixels) to the geometric error (in meters) by taking $\lambda_D = median(\mathcal{I}^*)/median(\mathcal{D}^*)$. This metric ensures better ranking conditions (e.g. in cases of non-textured regions) with similar convergence rate, but fails to handle basic cases of bimodal pixel/depth intensities and the convergence properties of both costs.

Finally, our regularization method, which is an extension of [24, 25], is directly related to [26] which perform a region growing using simultaneously intensity and geometric contours. The regularization is also particularly useful in compact mapping techniques (even if not being explicitly treated in this work). Compact mapping deals with the problem of representing the world without performing an explicit 3D reconstruction of the environment in a single global reference frame [27]. This allows to create local sub-maps or to store raw (unmodified) local sensor data in the representation whilst maintaining a topological framework at large-scale that is accurate enough to ensure the connectivity between locally precise frames.

1.2 Notation and Preliminaries

A frame $\mathcal{F} = \{\mathcal{I}, \mathcal{D}\}$ is composed of an image $\mathcal{I} \in [0, 1]^{m \times n}$ as pixel intensities and $\mathcal{D} \in \mathbb{R}_+^{m \times n}$ as the depth information. The mapping between the image pixel coordinates $\mathbf{p} \in \mathbb{P}^2$ and depth to 3D cartesian coordinates is given by the sensor projection $g : \mathbb{P}^2 \times \mathbb{R}_+ \mapsto \mathbb{R}^3$. The sensor projection model of interest is the perspective and spherical model (the images are projected in the unit sphere $\mathbb{S}^2$). Point coordinates correspondences between frames are given by the warping function $w : \mathbb{P}^2 \times \mathbb{R}_+ \times \mathbb{SE}(3) \mapsto \mathbb{P}^2$, under observability conditions at different viewpoints. Denoting $\mathbf{K}$ the intrinsic sensor model and $\mathbf{q}_S \in \mathbb{S}^2$ being the unitary vector, the corresponding warping functions are given by:

$$
\begin{aligned}
&\bullet \text{ Perspective: } w(\mathbf{p}, \mathcal{D}(\mathbf{p}), \mathbf{T}) = \frac{\mathcal{D}(\mathbf{p})\mathbf{K}\mathbf{R}\mathbf{K}^{-1}\mathbf{p}+\mathbf{K}\mathbf{t}}{[\mathcal{D}(\mathbf{p})\mathbf{K}\mathbf{R}\mathbf{K}^{-1}\mathbf{p}+\mathbf{K}\mathbf{t}]_3} \\
&\bullet \text{ Spherical: } w(\mathbf{p}, \mathcal{D}(\mathbf{p}), \mathbf{T}) = \mathbf{q}_S^{-1}\left(\frac{\mathcal{D}(\mathbf{p})\mathbf{R}\mathbf{q}_S(\mathbf{p})+\mathbf{t}}{||\mathcal{D}(\mathbf{p})\mathbf{R}\mathbf{q}_S(\mathbf{p})+\mathbf{t}||}\right)
\end{aligned}
\tag{1}
$$

where $\mathbf{q}_S^{-1}(\bullet)$ is the inverse unit sphere mapping to cartesian coordinates and the operator $[\bullet]_i$ selects the ith coordinate value. The pose $\mathbf{T}(\mathbf{x}) \in \mathbb{SE}(3)$ linking two frames (reference and target frame) is defined by the exponential map with six degrees of freedom (DOF) $\mathbf{x} \in \mathbb{R}^6$ (please see Appendix A for details). For notation convenience, in the rest of the paper $w(\mathbf{p}, \mathcal{D}(\mathbf{p}), \mathbf{T}) := w(\mathbf{p}, \mathbf{T})$. The normal vector of the surface s in the depth map $\mathcal{D}$, $s : \mathbb{R}^3 \mapsto \mathbb{R}; r - \mathcal{D}(\mathbf{p}) = 0$ is given by the gradient $\mathbf{n} = \nabla s(\mathbf{q})$, orthogonal to its tangent plane $\mathcal{P}(\mathbf{n}, d)$: $\mathbf{n}^T\mathbf{q} + d$ with $d = -\mathbf{n}^T\mathbf{q}_0, \forall \mathbf{q}_0 \in \mathcal{P}$.

2 Proposed Approach

Our adaptive RGB-D registration approach is based on classical direct VO [19] and ICP point-to-plane [18] strategies. In fact, the intensity and depth data error terms display different convergence properties for small and large motions. We aim to explore these complementary aspects, in terms of convergence, by using a modified cost function, where the geometric term prevails in the first coarse iterations, while the intensity data term dominates in the finer increments.

Next, we present additional aspects to improve the quality and robustness of this approach. They are particularly pertinent when performing localization to previously acquired frames (e.g. when locating a target frame to a local submap). At first, the frame depth estimates are refined by taking into account the geometric and photometric continuity of the scene (using superpixels). This is done by segmenting the scene in planar patches, which improves the depth accuracy considerably whilst allowing better normal surface estimation (specially in noisy measurements from stereo). The advantages are twofold: (i) the regularization improves and reduces the spurious noise, specially at non-textured regions; (ii) the generation of a confidence index that can be used further for pixel selection in an extended saliency concept.

2.1 Hybrid RGB-D Cost Function

The pose $\hat{T}T(x)$ between a reference and a target frame is performed iteratively from a linearised convex cost function of the following photometric and geometric errors

$$e_I(\mathbf{p}, \mathbf{x}) = \mathcal{I}(w(\mathbf{p}, \hat{T}T(\mathbf{x}))) - \mathcal{I}^*(\mathbf{p}) \tag{2}$$

$$e_D(\mathbf{p}, \mathbf{x}) = \lambda_D (\hat{R}R(\mathbf{x})\mathbf{n}^*(\mathbf{p}))^T (g(w(\mathbf{p}, \hat{T}T(\mathbf{x}))) - \hat{T}T(\mathbf{x})g^*(\mathbf{p})) \tag{3}$$

where $\hat{T}$ is the initial pose guess; $\mathbf{n}^*$ the normal surface vector calculated at the reference frame; $g(\bullet)$ is the inverse projection model; and λ_D is a tuning parameter for scaling the error terms. The intensity only (RGB) registration method is equivalent to consider $\lambda_D = 0$. Equation (2) is a classical optical flow constraint equation (OFCE) term (within the hypothesis of Lambertian surfaces) and (3) is equivalent to a flow point-to-plane ICP, both assuming predominant static surfaces. To ensure these assumptions, robust M-estimators (denoted as $\rho(\bullet)$) are applied for mitigating outliers influence [28]. This allows to reduce the effects of self occlusions, moving objects, illumination and interpolations errors in the direct estimation.

The classic RGB-D registration consists of using jointly (2) and (3) as

$$C(\mathbf{x}) = \sum_{\mathbf{p}} \rho_I(e_I(\mathbf{p}, \mathbf{x})) + \sum_{\mathbf{p}} \rho_D(e_D(\mathbf{p}, \mathbf{x})) \tag{4}$$

Choosing a large λ_D ($\lambda_D >> 1$) in (4) is equivalent to the direct ICP method, while $\lambda_D \approx 0$ corresponds to a classical dense VO. To increase the convergence rate, the optimization procedure is done considering multi-resolution Gaussian pyramidal images [19]. The optimization begins in the smallest resolution (pyramid at level n) to the bigger resolution (level 1). The corresponding Jacobians and framework is resumed in the Appendix A.

2.2 Adaptive Formulation

As stated previously in Sect. 2.1, a main concern with direct methods is about their convergence, since only local properties are settled from Eqs. (2), (3) and (4). We observed in both simulated and real sequences that the intensity and geometric terms have distinct convergence properties. While the convexity analysis of the cost terms cannot be established in general, the intensity RGB term has slower convergence (flatter) than the ICP point-to-plane cost, but its locally more precise when near the solution. This agrees with the findings of [23] in face tracking tasks. For illustration, we present typical shapes of the RGB and ICP cost terms (4) for 3DOF (two translations and one rotation) in Fig. 1 of a frame in the Sponza Atrium model dataset. The geometric error component (second row) is more discriminant than the intensity cost (first row) when further from the solution. Conversely, (as can be noticed in Fig. 1) ICP is less discriminant in the vicinity of the solution, meaning that the ICP point-to-plane is flatter than the RGB term for small interframe displacements. Besides, due to the scene symmetry along the Z axis (corridor-canyon like environment), the convergence rate might be slow if the task is restricted to this DOF (see Fig. 1 bottom right level plot).

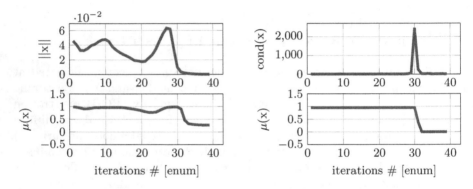

Fig. 2. Activation adaptive function $\mu(\mathbf{x})$ while performing registration in the KITTI outdoor dataset. The left column corresponds to (5) and the right to (6) with the tuning parameters of Table 1. The conditioning criteria (6) is easily detectable. Note that the norm of the pose increments using (5) are not monotonically decreasing (bottom left).

Based on these observations, we propose a modified cost function where the geometric term prevails in the first coarse iterations, while the intensity data term dominates in the finer increments (in the neighbourhood of the solution). A natural candidate activation function $\mu(\mathbf{x})$ in this context is the smoothed step

$$\mu(\mathbf{x}) = k_1 / \left(1 + \exp(-k_2(\|\mathbf{x}\| - c))\right) \tag{5}$$

depending on the size of the pose increments $\mathbf{x}$ along the minimization of (4). This is similar to the solution adopted in [23]. Note that selecting a high value to k_2 and c near the numeric iteration limit is equivalent to perform a sequential independent ICP and intensity RGB tasks (in cascade). Therefore, the activation (5) is particularly sensitive to the tuning parameters k_2 and c, which can induce oscillations (cross-peak) by transforming the original cost (4) in a non convex function. To address these issues, a second strategy is to analyse the costs relative behaviour along the optimization steps. The idea is that the relative conditioning number detects when the algorithm is in the vicinity of the solution (i.e. where the ICP cost is less discriminant). An equivalent adaptive function is then:

$$\mu(\mathbf{x}) = k_1 \mathbf{1}(cond_{\mathbf{x}}(C_I(\mathbf{x}))/cond_{\mathbf{x}}(C_D(\mathbf{x}))) \tag{6}$$

where the indicator function $\mathbf{1}(\bullet)$ is zero when $cond_{\mathbf{x}}(C_I(\mathbf{x}))/cond_{\mathbf{x}}(C_D(\mathbf{x})) >> 1$ and one otherwise and

$$cond_{\mathbf{x}}(C(\mathbf{x})) = \left| \frac{C(\mathbf{x}_0 \circ \mathbf{x}) - C(\mathbf{x}_0)}{C(\mathbf{x}_0)} \right| / \frac{||\mathbf{x}||}{||\mathbf{x}_0||} \tag{7}$$

being an estimate of the relative condition number of the RGB (C_I) and ICP (C_D) cost functions, with $\mathbf{x}_0 = se3^{-1}(\log(\hat{\mathbf{T}}))$ and $\circ$ is the additive Lie algebra action. We show in Fig. 2 typical curves using the KITTI VO/SLAM dataset [29] for both adaptive metrics – Eq. (5) is presented in the first column and (6) in the second. The parameters of each activation are detailed in Table 1. This activation proved to detect correctly the sensitivity of the costs, whilst being of easy tuning. The respective hybrid modified cost function is designed as a joint adaptive RGB-ICP cost

$$\tilde{C}(\mathbf{x}) = (1 - \mu(\mathbf{x})) \sum_{\mathbf{P}} \rho_I(e_I(\mathbf{p}, \mathbf{x})) + \mu(\mathbf{x}) \sum_{\mathbf{p}} \rho_D(e_D(\mathbf{p}, \mathbf{x})) \tag{8}$$

Where the respective Jacobians are linear combinations of the Jacobians from the original formulation: $\tilde{\mathbf{J}} = [\sqrt{(1 - \mu(\mathbf{x}))}\mathbf{J}^I \quad \sqrt{\mu(\mathbf{x})}\mathbf{J}^D]^T$. Finally, we adopted the Huber robust function for ρ_I, ρ_D for ensuring convexity properties when further from the solution [28]. To avoid outliers influence, the robust function is switched to Tukey when in the vicinity of the minimum (i.e. when the conditioning in (6) is large). The respective Jacobians and details about the optimization are given in the Appendix A. As it will be shown later, this formulation proved to have significant advantages in extensive tests performed in simulated and in real indoor and outdoor sequences.

2.3   Regularization in Planar Patches and Pixel Selection

Before estimating inter-frame poses, we can perform a regularization based on the assumption that the scene contains piecewise smooth surfaces. The frame depth map is then represented as a set of planes of variable size, where non-planar surfaces are approximated by a set of small planar patches. This assumption is applicable for any environment: structured and non-structured as long as the planar approximation error is smaller than the measurement error. This can be easily achieved by selecting the suitable parameters according to the sensor noise model (regardless the type of scene).

The plane segmentation is performed by region growing, starting from a set of seeds distributed around the image. The conditions are: (i) in order to

Fig. 3. Spherical intensity image (left), point cloud (middle) and segmented patches (right) of rendered indoor scene using the superpixels constraints. For display purposes, plane colors are the average color of each patch. The segmentation preserves edges as the calibration target and whilst reducing the noise in frontier stitched regions. (Color figure online)

add a pixel i into a group of neighbouring pixels representing a surface s, we verify if their normal vector have the similar direction; and (ii) that the 3D point corresponding to the pixel $\mathbf{p}$ lies approximately on the plane defined by s (orthogonal projection error) (Fig.3).

A last stage is carried out to merge the contiguous planar patches that lie approximately onto the same 3D plane. For that we exploit the color continuity of the images, where changes in color or texture often delimit different objects and different planar surfaces. This is used to guide the planar segmentation of the scene and to reduce the smoothing/aliasing effect of objects. The color image is segmented in superpixels with similar photometric properties (color and texture) using the single linear iterative clustering (SLIC) algorithm [30], which encodes nice properties as strong adherence to boundaries and compactness. Afterwards, the mean patch $(d_s, \mathbf{n}_s)$ related to each superpixel region is extracted considering all the patches that are mostly englobed by that superpixel. The assumption about the planar continuity of the scene will help to reduce the uncertainty since new information is provided. Thus, the uncertainty of a depth measurement (pixel in the depth image) belonging to a plane is modelled as:

$$\Sigma_{\mathcal{D}}(\mathbf{p}) = \frac{r_\sigma(\mathbf{p})}{|\mathbf{n}_s^T \mathbf{q}_S(\mathbf{p})|} \tag{9}$$

where r_σ is the ratio between the smallest and the bigger eigenvalue of the covariance of the patch (the smallest eigenvalue from the singular value decomposition of the covariance matrix of the 3D points).

Observability and Information Selection. In this section we derive the concepts for exploiting the certainty index gathered from the regularization along with the pixel selection. Since only a subset of the image information is useful for pose computation, e.g. non textured regions has no influence in RGB registration cost term – no photometric gradient. The problem is then to select points that best constraints the cost functions and discard the ones who does not This formulation is equivalent to ensure that the Fisher Information Matrix (FIM)

$\mathbf{J}^T\mathbf{J}$ is well conditioned. This is performed by simply analysing the magnitude of the analytical Jacobians columns (see Appendix A for details). This is the underlying idea presented in the works of [18] for the ICP cost and in [31] for the RGB cost (2). Hence, the update row ranking considering the geometric confidence is done under the following modified Jacobian

$$\bar{\mathbf{J}} = \frac{1}{\Sigma_D}\left[(\|\mathbf{J_1}^I\| + \|\mathbf{J_1}^D\|) \;\; \|\mathbf{J_2}^I\| + \|\mathbf{J_2}^D\|) \;\; \|\mathbf{J_3}^I\| + \|\mathbf{J_3}^D\|) \atop \|\mathbf{J_4}^I\| + \|\mathbf{J_4}^D\|) \;\; \|\mathbf{J_5}^I\| + \|\mathbf{J_5}^D\|) \;\; \|\mathbf{J_6}^I\| + \|\mathbf{J_6}^D\|) \right] \quad (10)$$

with Σ_D the plane uncertainty index (9). The advantages are twofold: (i) to exclude the points that does not contributes directly to the estimation (the valid information can be masked from the spurious noise in the redundant not useful information); (ii) computation efficiency whilst keeping the precision.

3   Experiments and Results

We evaluate the technique in sequences of indoor and outdoor images using perspective and spherical RGB-D sensors. We consider the average of the rotation relative error (RRE), translation relative error (TRE) and number of iterations to convergence as quantitative metrics. In cases of lack of ground truth (e.g. in the real indoor spherical sequence), a qualitative analysis is done for the intensity and depth errors. Unless specified, the term Adaptive RGB D corresponds to the cost (8) using (6).

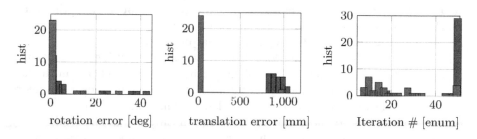

Fig. 4. Rotation, translation errors and number of iterations for a fixed resolution for the simulated testbed dataset with gap of 10 frames. RGB-D dense registration errors in rotation (left in degrees), translation (center in millimetres) and number of iterations to converge (right figure). The registration considering the classical RGB-D is presented in blue, while the adaptive formulation using the conditioning (6) is in red. The precision and convergence rate are substantially improved when exploiting the activation factor (in red). (Color figure online)

Implementation Aspects. The iterative pose estimation algorithm (see Appendix A) is said to have converged (either to a global or to a local minimum) when the norm of pose increments $\mathbf{x}$ are bellow a fixed threshold in

Table 1. Parameters in the activation functions.

	Parameters	Typical range
Meth. 1 [20]	$\lambda_D = med(\mathcal{I})/med(\mathcal{D})$, $\mu = 0.5$	$\lambda_D \in [5, 50]$
Adapt. 1 (5)	$\lambda_D = 1$, $k_1 = 1 - 10^{-5}$, $k_2 = 100$, $c = 0.001$	$\mu \in [0, k_1]$
Adapt. 2 (6)	$\lambda_D = 1$, $k_1 = 1 - 10^{-5}$	$\mu \in \{0, k_1\}$

successive iterations (10^{-5} for the rotation and 10^{-3} for translation). The parameters employed (λ_D and activation function (5)) are described in Table 1. We used the same parameters in all the next experiments.

- **Spherical Simulated Sequence:**
  At first, we evaluate our approach in controlled conditions using 500 RGB-D spherical synthesized images from the Sponza Atrium model. We start in a fixed resolution for evidencing the differences between the classic RGB-D and the adaptive formulation. The maximum number of iterations is 50. To emulate different motion speeds, only a sub-set of the frames is picked up (gaps of 5, 10 and 15 frames) – Fig. 4 shows the pose errors and the number of iterations for a gap of 10 spheres. The registration results are synthesized in Table 2. The distances between frames are in average of 0.15 [m] and of 4 degrees in rotation. The convergence was achieved even in cases considering translations and rotations of around 2.5[m] and 60[deg]. The convergence failed in less than 10% of trials in the furthest experiment (gap of 15). These cases happened when the reference scene was almost completely occluded in the target scene (e.g. corridor 90 degrees turns) and are expected to happen since the direct method's hypothesis of overlapping is not fulfilled.
- **Spherical Indoor and Outdoor Real Sequences:**
  The spherical indoor and outdoor frames are acquired using two designed spherical RGB-D sensors [31,32]. The indoor images were captured in the hall and offices of the Inria building using a set of eight Asus sensors. A more qualitative analysis is done due to the lack of ground truth. With a separation of five frames, the method did not converge in only 9% of the trials. They correspond mostly to the cases where the maximum number of iterations was reached – in this case 150 iterations as in Fig. 5. The adaptive solution had a better performance with a much smaller number of iterations. Note that we apply the same experiment to the RGB only cost function (i.e. Eq. (4) with $\lambda_D = 0$), but it either reached a local minima or did not converge for most of the frames (black trajectory in Fig. 5 left). The same experiment was performed with the regularized+saliency criteria selecting the most informative points. Although the convergence success rate remained invariant, when compared to the adaptive one, it gives more stability and efficiency to the optimization since a reduced number of iterations are performed (see right Fig. 5). Finally, we depict two experiments in the presence of large motions and dynamics objects (see Fig. 6). The data was acquired in the two different regions of a building with many occlusions, large rotations, and with dynamic

objects. The same conclusions were obtained from the outdoor data, acquired in an urban/semi-urban area using a spherical stereo system [31]. A qualitative view of the respective intensity and geometric errors during a registration task can be seen in Fig. 7.

– **KITTI Outdoor Perspective Sequence**
We also provide results for the perspective outdoor sequence of the KITTI Visual Odometry/SLAM benchmark. It is a challenging dataset since the scene is mainly semi-structured (roads in an urban area) and with a travel speed up to 60 km/h. We observed that in the outdoor scenarios the overlapping regions are much sparser because only the road plane is the persistent

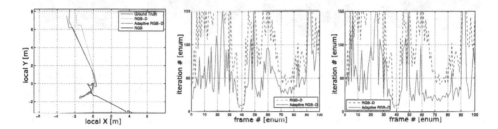

Fig. 5. Trajectories (left) using RGB, RGB-D and adaptive RGB-D over a sequence of 500 frames. A gap of five frames were used to compute each task. The ground truth was obtained using the RGB with step of one frame. The number of iterations to convergence with the adaptive RGB-D (red curves in the center) is significantly smaller than when considering the classic method (in blue). The number of iterations is reduced when taking into account the regularization+saliency stage. (Color figure online)

Fig. 6. Intensity and geometry errors between indoor frames with a gap of fifteen frames for two different scenes. Each row is composed of two pairs of errors: the classic RGB-D and the adaptive formulation in the form $(e_I(x), e_D(x))$.

Fig. 7. Intensity and depth errors between outdoor frames with a gap of fifteen frames. The first two columns are of a classic RGB-D and the last two columns correspond to the adaptive approach. Each row is composed of the final errors: intensity $(\mathbf{e_I(x)}, \mathbf{e_D(x)})$. The adaptive RGB-D (last columns) has smaller geometric and photometric errors (bigger errors are encoded with lighter colors). The regularization + saliency had particularly improved the resulting pose computation for the outdoor case (second row).

overlapping surface. The respective error metrics are displayed in Table 3 for a fixed resolution.

Lastly, we combined the adaptive formulation with a multi-resolution Gaussian pyramid of four levels (the higher the level, the smaller the image resolution is) to assets the efficiency of the approach in this context (see Table 4). The maximum number of iterations was of 50 at each pyramid level. To account the different computational cost of one iteration between the levels, we define the total number of iterations as $\sum_{i=1}^{4} l_i (2^{4-i})^2$ (with l_i the number of iterations at level i). The adaptive formulation is still more efficient and precise, although the discrepancy between the methods is reduced.

Table 2. Quantitative simulated spherical indoor sequence in a fixed resolution: average RRE[deg]/RTE[mm]/Iterations.

	Gap = 5	Gap = 10	Gap = 15
Meth. 1 [20]	3.67/423/47.3	7.80/1104/48.4	11.7/1520/48
Adapt. 1 (5)	0.68/96.4/31.2	1.11/466/32.9	2.17/833/34.8
Adapt. 2 (6)	**0.03/88.6/31**	**0.04/182/26.5**	**0.05/523/20.7**

Table 3. Quantitative KITTI outdoor sequence in fixed resolution: average RRE[deg]/RTE[mm]/Iterations.

	Gap = 1	Gap = 2	Gap = 3
Meth. 1 [20]	0.51/219/45.6	1.83/1071/49	2.75/1846/50
Adapt. 1 (5)	0.27/120/36.5	1.12/557/45	2.34/1101/**46.7**
Adapt. 2 (6)	**0.08/35.1/33.5**	**0.42/192/41.7**	**1.79/825**/47

Table 4. Quantitative KITTI outdoor sequence results using multi-resolution (pyramid of four levels): average RRE[deg]/RTE[mm]/Iterations.

	Gap = 1	Gap = 2	Gap = 3
Meth. 1 [20]	0.08/23.1/**447**	0.78/268/980	3.68/1059/1872
Adapt. 1 (5)	**0.06/16.5/704**	**0.19/81.4/856**	**0.83/251/1078**
Adapt. 2 (6)	**0.06/16.4**/1102	0.37/**47.5**/1269	1.05/**238**/1473

4   Conclusions

In this paper, we have presented an efficient RGB-D registration approach in the context of large inter-frame displacements. The technique exploits adaptively the photometric and geometric error terms based on their convergence characteristics. Additional aspects as a two step regularization and an extended pixel saliency selection improved the quality and robustness of this approach. Despite its simplicity, this technique was capable to deal with large rotations, occlusions and moving objects in real indoor and outdoor scenarios.

Future directions includes: (i) the formal characterization of the convergence domain for different symmetries and noise statistics for both intensity and geometry data terms; and (ii) finding convex (quasi-convex) dual formulations for adding more stable dense/semi-dense features as planes, edgelets and image moments in both intensity and geometric terms.

Acknowledgements. The authors thank Josh Picard and Paolo Salaris for the discussions/proof reading of the manuscript, and the reviewers for their thoughtful comments. This work is funded by CNPq of Brazil under contract number 216026/2013-0.

Appendix A: Error Jacobians and Optimization

The pose $\mathbf{T}(\mathbf{x}) \in \mathbb{SE}(3)$ is parametrized as function of angular and linear velocities $\mathbf{x} = (\boldsymbol{v}\delta t, \boldsymbol{\omega}\delta t) \in \mathbb{R}^6$ and the optimization will be related to this twist parametrization. The pose is related to the twist velocities by the exponential mapping $\mathbf{T}(\mathbf{x}) = \exp(se3(\mathbf{x}))$, with

$$se3(\mathbf{x}) = \begin{bmatrix} \mathbf{S}(\boldsymbol{\omega})\delta t & \boldsymbol{v}\delta t \\ \mathbf{0}_{(1\times3)} & 0 \end{bmatrix} \in \mathfrak{se}(3) \tag{11}$$

which is the Lie algebra of $\mathbb{SE}(3)$ at the identity element, $\mathbf{S}(\mathbf{z})$ represents the skew symmetric matrix associated to vector $\mathbf{z}$ and $\delta t = 1$.

The respective Jacobians will be derived following this parametrization. We ask the reader to see [19] for details about the photometric Jacobian $\mathbf{J}^I$. Next, for the geometric point-to-plane direct Jacobian $\mathbf{J}^D \in \mathbb{R}^{1 \times 6}$, we denote the 3D point error $\zeta(\mathbf{x})$:

$$
\begin{aligned}
\zeta(\mathbf{x}) &= -\hat{\mathbf{T}}\mathbf{T}(\mathbf{x}) \begin{bmatrix} g^*(\mathbf{p})) \\ 1 \end{bmatrix} + g(w(\mathbf{p}, \hat{\mathbf{T}}\mathbf{T}(\mathbf{x}))) \\
&= -\hat{\mathbf{R}}\mathbf{R}(\mathbf{x})g^*(\mathbf{p}) - \hat{\mathbf{R}}\mathbf{t}(\mathbf{x}) - \hat{\mathbf{t}} + g(w(\mathbf{p}, \hat{\mathbf{T}}\mathbf{T}(\mathbf{x})))
\end{aligned}
\tag{12}
$$

From Eqs. (3), (12) and the product rule:

$$
\mathbf{J}^D(\mathbf{0}) = \lambda_D \mathbf{n}^{*T} \left(\frac{\partial(\mathbf{R}(\mathbf{x})^T \hat{\mathbf{R}}^T \zeta(\mathbf{z}))}{\partial \mathbf{x}} \Bigg|_{\mathbf{z}=\mathbf{x}} + \mathbf{R}(\mathbf{x})^T \hat{\mathbf{R}}^T \frac{\partial(\zeta(\mathbf{x}))}{\partial \mathbf{x}} \right) \Bigg|_{\mathbf{x}=\mathbf{0}}
\tag{13}
$$

For clarity, the first term in Eq. (13) is $\mathbf{J_{d1}}$ and we decompose the second term in two Jacobians $\mathbf{J_{d2}}$ and $\mathbf{J_{d3}}$, such as $\mathbf{J}^D(\mathbf{0}) = \lambda \mathbf{n}^{*T}(\mathbf{J_{d1}}(\mathbf{0}) + \mathbf{J_{d2}}(\mathbf{0}) + \mathbf{J_{d3}}(\mathbf{0}))$. From $\frac{\partial(\mathbf{R}(\mathbf{x})\zeta)}{\partial \mathbf{x}} = \frac{\partial(\mathbf{R}(\mathbf{x})\zeta)}{\partial \mathbf{R}(\mathbf{x})} \frac{\partial \mathbf{R}(\mathbf{x})}{\partial \mathbf{x}}$ the first term is

$$
\mathbf{J_{d1}}(\mathbf{0}) = \begin{bmatrix} \mathbf{0}_{3 \times 3} & \mathbf{S}(\hat{\mathbf{R}}^T \zeta(\mathbf{0})) \end{bmatrix}
\tag{14}
$$

The second term is decomposed in two Jacobians

$$
\mathbf{J_{d2}}(\mathbf{0}) = \begin{bmatrix} -\mathbf{I}_{3 \times 3} & \mathbf{S}(g^*(\mathbf{p})) \end{bmatrix}
\tag{15}
$$

And finally the last Jacobian is the one corresponding to $\frac{\partial(g(w(\mathbf{p}, \hat{\mathbf{T}}\mathbf{T}(\mathbf{x}))))}{\partial \mathbf{x}}$. This derivative can be seen as an extended version of the image photometric gradient $\mathbf{J}^I$, for each component of $g(w(\mathbf{p}, \hat{\mathbf{T}}\mathbf{T}(\mathbf{x})))$. Then

$$
\mathbf{J_{d3}}(\mathbf{0}) = \begin{bmatrix} \mathbf{J_g} \big|_{[g(\mathbf{p}_w)]_1}^T & \mathbf{J_g} \big|_{[g(\mathbf{p}_w)]_2}^T & \mathbf{J_g} \big|_{[g(\mathbf{p}_w)]_3}^T \end{bmatrix}^T \mathbf{J_w} \mathbf{J_T}
\tag{16}
$$

And $\mathbf{J_g} \big|_{[g(\mathbf{p}_w)]_i}$ is the image gradient (as in the photometric term) of an image produced with the ith-coordinate of $g(w(\mathbf{p}, \hat{\mathbf{T}}\mathbf{T}(\mathbf{0})))$. Note that this Jacobian is small for points belonging to planar surfaces. Therefore, $\mathbf{J_{d3}}$ is neglected since only a fraction of the scene is on geometric discontinuities and since these points have higher sensitivity to depth error estimates and self-occlusions effects. Finally, we use the ESM formulation [19] for defining the optimization step for the RGB cost, while a Gauss-Newton step is employed for the geometric Jacobian. The reader is asked to see [33] for more details on the different optimization available techniques.

References

1. Brox, T., Malik, J.: Large displacement optical flow: descriptor matching in variational motion estimation. IEEE PAMI **33**, 500–513 (2011)
2. Braux-Zin, J., Dupont, R., Bartoli, A.: A general dense image matching framework combining direct and feature-based costs. In: IEEE ICCV (2013)
3. Howard, A.: Real-time stereo visual odometry for autonomous ground vehicles. In: IEEE IROS (2008)
4. Davison, A., Murray, D.: Simultaneous localization and map-building using active vision. IEEE TPAMI **24**, 865–880 (2002)
5. Nistér, D., Naroditsky, O., Bergen, J.: Visual odometry. In: IEEE CVPR (2004)
6. Kitt, B., Geiger, A., Lategahn, H.: Visual odometry based on stereo image sequences with ransac-based outlier rejection scheme. In: IEEE IV (2010)
7. Harris, C., Stephens, M.: A combined corner and edge detector. In: 4th Alvey Vision Conference (1988)
8. Lowe, D.: Distinctive image features from scale-invariant keypoints. IJCV **60**, 91–110 (2004)
9. Hager, G., Belhumeur, P.: Efficient region tracking with parametric models of geometry and illumination. IEEE TPAMI **20**, 1025–1039 (1998)
10. Lucas, B.D., Kanade, T.: An iterative image registration technique with an application to stereo vision. In: IJCAI (1981)
11. Irani, M., Anandan, P.: Robust multi-sensor image alignment. In: ICCV (1998)
12. Baker, S., Matthews, I.: Equivalence and efficiency of image alignment algorithms. In: IEEE CVPR (2001)
13. Mei, C., Benhimane, S., Malis, E., Rives, P.: Constrained multiple planar template tracking for central catadioptric cameras. In: BMVC (2006)
14. Caron, G., Marchand, E., Mouaddib, E.: Tracking planes in omnidirectional stereovision. In: IEEE ICRA (2011)
15. Comport, A., Malis, E., Rives, P.: Accurate quadrifocal tracking for robust 3d visual odometry. In: IEEE ICRA (2007)
16. Churchill, W., Tong, C., Gurau, C., Posner, I., Newman, P.: Know your limits: Embedding localiser performance models in teach and repeat maps. In: IEEE ICRA (2015)
17. Furgale, P., Barfoot, T.: Visual teach and repeat for long-range rover autonomy. JFR **27**, 534–560 (2010)
18. Gelfand, N., Ikemoto, L., Rusinkiewicz, S., Levoy, M.: Geometrically stable sampling for the icp algorithm. In: 3DIM (2003)
19. Comport, A., Malis, E., Rives, P.: Real-time quadrifocal visual odometry. IJRR **29**, 245–266 (2010)
20. Tykkala, T., Audras, C., Comport, A.: Direct iterative closest point for real-time visual odometry. In: ICCV Workshops (2011)
21. Kerl, C., Sturm, J., Cremers, D.: Dense visual SLAM for RGB-D cameras. In: IEEE IROS (2013)
22. Timofte, R., Gool, L.V.: Sparse flow: sparse matching for small to large displacement optical flow. In: IEEE WCACV (2015)
23. Morency, L., Darrell, T.: Stereo tracking using icp and normal flow constraint. In: ICPR (2002)
24. Martins, R., Fernandez-Moral, E., Rives, P.: Dense accurate urban mapping from spherical RGB-D images. In: IEEE IROS (2015)

25. Gokhool, T., Martins, R., Rives, P., Despre, N.: A compact spherical RGBD keyframe-based representation. In: IEEE ICRA (2015)
26. Weikersdorfer, D., Gossow, D., Beetz, M.: Depth-adaptative superpixels. In: IEEE ICP (2013)
27. Fernandez-Moral, E., Mayol-Cuevas, W., Arevalo, V., Gonzalez-Jimenez, J.: Fast place recognition with plane-based maps. In: IEEE ICRA (2013)
28. Zhang, Z.: Parameter estimation techniques: A tutorial with application to conic fitting. Technical report 2676, Inria (1995)
29. Geiger, A., Lenz, P., Urtasun, R.: Are we ready for autonomous driving? the kitti vision benchmark suite. In: IEEE CVPR (2012)
30. Achanta, R., Shaji, A., Smith, K., Lucchi, A., Fua, P., Susstrunk, S.: SLIC superpixels compared to state-of-the-art superpixels methods. IEEE Trans. PAMI **34**, 2274–2282 (2012)
31. Meilland, M., Comport, A., Rives, P.: Dense omnidirectional RGB-D mapping of large-scale outdoor environments for real-time localization and autonomous navigation. JFR **32**, 474–503 (2015)
32. Fernandez-Moral, E., Gonzalez-Jimenez, J., Rives, P., Arevalo, V.: Extrinsic calibration of a set of range cameras in 5 seconds without pattern. In: IEEE IROS (2014)
33. Barker, S., Matthews, I.: Lucas-kanade 20 years on: a unifying framework. IJCV **56**, 221–255 (2006)

Deep Discrete Flow

Fatma Güney[1(✉)] and Andreas Geiger[1,2]

[1] Autonomous Vision Group, MPI for Intelligent Systems, Tübingen, Germany
`fatma.guney@tue.mpg.de`
[2] Computer Vision and Geometry Group, ETH Zürich, Zürich, Switzerland

Abstract. Motivated by the success of deep learning techniques in matching problems, we present a method for learning context-aware features for solving optical flow using discrete optimization. Towards this goal, we present an efficient way of training a context network with a large receptive field size on top of a local network using dilated convolutions on patches. We perform feature matching by comparing each pixel in the reference image to every pixel in the target image, utilizing fast GPU matrix multiplication. The matching cost volume from the network's output forms the data term for discrete MAP inference in a pairwise Markov random field. We provide an extensive empirical investigation of network architectures and model parameters. At the time of submission, our method ranks second on the challenging MPI Sintel test set.

1 Introduction

Despite large progress, optical flow is still an unsolved problem in computer vision. Challenges provided by autonomous driving applications [1–3] or current benchmarks like KITTI [4,5] and MPI Sintel [6] include large motions, appearance changes, as well as uniform image regions. While the predominant paradigm for estimating optical flow is based on continuous optimization [7–9] with coarse-to-fine warping [10], recent approaches leverage discrete optimization strategies [11–14] in order to overcome local minima and to gain robustness.

While these approaches have shown promising results, their performance still falls considerably behind the state-of-the-art in stereo matching [15,16]. While 2D flow estimation is an inherently more difficult problem than 1D matching along the epipolar line, most existing works on discrete optical flow optimization exploit hand-crafted features for calculating the matching costs. In contrast, the most successful approaches in stereo matching exploit a combination of learned local feature representations and global discrete optimization [16–18].

In this paper, we investigate the utility of feature learning for discrete optical flow, see Fig. 1 for an illustration. In particular, we modify the "DiscreteFlow" framework of Menze et al. [11] by replacing their hand-crafted descriptors with

Electronic supplementary material The online version of this chapter (doi:10. 1007/978-3-319-54190-7_13) contains supplementary material, which is available to authorized users.

© Springer International Publishing AG 2017
S.-H. Lai et al. (Eds.): ACCV 2016, Part IV, LNCS 10114, pp. 207–224, 2017.
DOI: 10.1007/978-3-319-54190-7_13

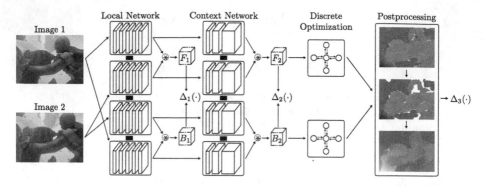

Fig. 1. Deep Discrete Flow. The input images are processed in forward order (top stream) and backward order (bottom stream) using local and context Siamese convolutional neural networks, yielding per-pixel descriptors. We then match points on a regular grid in the reference image to every pixel in the other image, yielding a large tensor of forward matching costs (F_1/F_2) and backward matching costs (B_1/B_2). Matching costs are smoothed using discrete MAP inference in a pairwise Markov random field. Finally, a forward-backward consistency check removes outliers and sub-pixel accuracy is attained using the EpicFlow interpolator [20]. We train the model in a piece-wise fashion via the loss functions.

learned feature representations. In particular, we investigate two types of networks: a local network with a small receptive field consisting of 3×3 convolutions followed by non-linearities as well as a subsequent context network that aggregates information over larger image regions using dilated convolutions [19]. As naïve *patch-based* training with dilated convolutions is computationally very expensive, we propose an efficient implementation based on regular strided convolutions. For efficient learning of the whole pipeline, we specify auxiliary loss functions akin to [16] and train the model in a piece-wise fashion.

We provide a detailed empirical analysis of the impact of each of the components of the pipeline. More specifically, we compare a large number of different local and context architectures with respect to each other and to traditional hand-crafted features. Further, we compare the results of the best-performing systems after discrete optimization and sub-pixel interpolation, and qualitatively visualize the results with their corresponding error images at every stage.

2 Related Work

In this section, we survey the most important related works. We first provide an overview of related optical flow approaches with a particular focus on recent discrete and mixed discrete/continuous approaches that attain state-of-the-art performance on current benchmarks. In the second part, we review current feature learning approaches for correspondence estimation.

Optical Flow: The classical formulation for estimating optical flow [7,9] involves solving a continuous variational optimization problem. To cope with displacements larger than a few pixels, coarse-to-fine estimation heuristics are commonly employed [10, 21–27]. Unfortunately, coarse-to-fine schemes often lead to blurred object boundaries on current benchmarks [4, 6, 28] due to their susceptibility to local minima in the energy function.

Thus, discrete formulations have recently gained popularity. One line of work incorporates pre-estimated sparse feature correspondences into the optimization process [20, 29–32]. To allow for more robust estimates, a second line of work directly formulates optical flow estimation as a discrete optimization problem [33], e.g., in terms of MAP inference in a Markov random field under appropriate flow priors. These approaches can be further categorized into epipolar constrained methods [34, 35], methods which estimate the most likely flow field based on a small set of dense flow field proposals [13, 36–38] and methods that estimate flow directly at the pixel level [11, 12, 14, 39, 40].

More specifically, Menze et al. [11] establish a sparse set of 500 flow proposals by matching Daisy descriptors [41] using fast approximate nearest neighbor techniques. Exploiting the truncation property of their pairwise potentials, they efficiently approximate the MAP solution using belief propagation. Chen et al. [12] extend the efficient min-convolution algorithm to 2D flow fields and optimize a discretized version of the classical variational objective using normalized cross-correlation as data term. For sub-pixel accuracy and to extrapolate into occluded regions, both approaches exploit an additional extrapolation and variational post-processing step [20].

While all aforementioned methods focus on the optimization of an energy function based on hand-crafted local feature descriptors, in this work, we investigate the benefits of learning feature representations for discrete optical flow estimation. In particular, we leverage the framework of [11] and replace their features using non-local pixel representations trained for predicting optical flow.

Feature Learning for Correspondence Estimation: Motivated by the success of deep learning in image classification and object recognition [42], a number of papers have tackled the problem of correspondence estimation by learning deep convolutional representations.

Recently, Fischer, Mayer et al. [43, 44] have demonstrated dense end-to-end flow prediction using a deep convolutional neural network which takes as input two images and directly outputs a flow map. While impressive performance has been demonstrated, the method does not attain state-of-the-art performance on current leaderboards. One difficulty is the model's high capacity and the associated large amount of data required to train it.

An alternative approach, which we follow in this paper, is to learn per-pixel feature representations using Siamese networks which can be fed into a winner-takes-all selection scheme or - as in our case - into a discrete optimization algorithm. While the learned representations tend to be more local, they are also less prone to overfitting. Importantly, even small datasets such as KITTI [4] or MPI Sintel [6] provide millions of training points as every pixel provides a training example. This is in contrast to dense approaches [43] where hundreds of thousands images with ground truth flow maps are required for obtaining reliable representations.

A number of approaches [45,46] aim for descriptor learning for sparse feature matching. Due to the relatively small number of interest points per image, metric learning networks can be exploited for this task. However, sparse feature matching approaches do not benefit from spatial smoothness priors which we incorporate into dense correspondence estimation via discrete optimization.

For the problem of binocular stereo matching, Zbontar et al. [16], Chen et al. [17] and Luo et al. [18] have demonstrated state-of-the-art performance by combining deep feature representations with discrete optimization. In similar spirit, Zagoruyko and Komodakis [47] learn Siamese matching networks for wide-baseline stereo matching. Motivated by this success, here we leverage feature learning and discrete optimization to tackle the more challenging problem of unconstrained 2D flow estimation.

Very recently, Bai et al. [48] have extended the approach of [18] to segment-wise epipolar flow where motion stereo has been estimated separately for each independently moving vehicle in the KITTI dataset [4]. In contrast, in this paper we neither assume rigidly moving objects nor the availability of highly accurate semantic instances. Thus, our method is also applicable to more general scenes as occuring, e.g., in the MPI Sintel optical flow challenge [6].

In [49], Siamese networks for optical flow computation have been combined with winner-takes-all matching and smoothing of the resulting correspondence field. While they use patch-wise max pooling operations to increase the size of the receptive field, we exploit computationally efficient dilated convolutions for this purpose. Furthermore, we investigate the usefulness of spatial priors and present a detailed empirical analysis of network architectures and settings.

3   Deep Discrete Flow

Menze et al. [11] formulate optical flow estimation as discrete MAP inference in a Markov random field with pairwise smoothness priors, followed by sub-pixel interpolation [20]. We follow their framework, but replace their hand-crafted Daisy features [41] with learned local and non-local representations to investigate the effect of feature learning on this framework as illustrated in Fig. 1. In Sect. 3.1, we first describe our local and context network architecture and provide details about training and inference. For completeness, we briefly review the discrete optimization framework [11] in Sect. 3.2.

3.1   Feature Learning Using Dilated Convolutions

The classical approach to establish correspondences between two images is to search for the most similar patch in the target image, given a particular patch in the reference image, assuming that corresponding regions appear more similar than non-corresponding regions. Popular similarity measures for optical flow include brightness and gradient constancy [24], normalized cross-correlation [12], SIFT [30], Daisy [11] and hierarchical histograms of oriented gradients [32].

Following recent trends in computer vision [16, 18, 45–49], we use deep convolutional neural networks in order to learn better representations tailored for the task. In particular, we use Siamese architectures to process a pair of patches and produce a matching score as an indication of their level of similarity. In addition to traditional local 3×3 convolutional layers, we integrate context information by adopting dilated convolutions [18], which have recently demonstrated great performance in semantic segmentation. Compared to increasing the receptive field size using max-pooling operations, dilated convolutions have the advantage of not decreasing the image resolution, thus allowing for efficient dense inference with reuse of computation. In addition, patch-based dilated convolution networks can be efficiently trained as we demonstrate in this section.

For efficiency and due to the difficulty of training CNN-CRF models jointly, we train our model in a piece-wise fashion using auxiliary loss functions. That is, as illustrated in Fig. 1, we first train the local architecture using Δ_1, followed by the context architecture using Δ_2, and finally the CRF as well as hyperparameters of the post-processing stage using Δ_3. We also tried joint training on top of the pre-trained local and context networks, but observed no significant improvements. This agrees with the observations reported in [19].

Network Architecture: We use a Siamese network architecture composed of two shared-weight branches, one for the reference patch and one for the target patch. As we are also interested in calculating the backward flow, we have an additional backward Siamese network which shares weights with the forward network as illustrated in Fig. 1. Each of the branches consists of several building blocks where each block is defined as convolution, Batch Normalization, and ReLU for non-linearity except the last one which contains only a convolutional layer. The unit-length normalized output of the last layer is used as a feature vector of the patch. The similarity s between image pixels is calculated as the dot product between the respective feature vectors. As opposed to current trends in feature learning for stereo matching [16], we do not exploit fully connected layers for score computation as the large set of potential correspondences renders this computationally intractable (i.e., one network evaluation for each pixel pair).

Local Network: Our local network leverages 3×3 convolution kernels. The hyper-parameters of the network are the number of layers and the number of feature maps in each layer as specified in our experimental evaluation. We call this network local, because the size of each feature's receptive field is relatively small (i.e., $2n + 1$ where n denotes the number of blocks).

Context Network: Deeper architectures with more convolutional layers increase the receptive field size, possibly leading to improved performance. However, complex high capacity models are also hard to train and require a lot of data. Our context network increases the size of the receptive field with only modest increase in complexity by exploiting dilated convolutions [19]. In contrast to normal convolutions, dilated convolutions read the input feature maps at locations with a spatial stride larger than one. Thus, they take more contextual information into account while not increasing the number of parameters with respect to regular (i.e., 1-dilated) convolutions. In contrast to pooling operations, spatial information is not lost.

Training: We consecutively train the local and the context network using the same auxiliary losses $\Delta_1 = \Delta_2$. As loss function, we leverage the hinge loss function which is defined for a positive-negative pair to penalize when the similarity score of the positive is not greater than the similarity score of the negative at least by margin: $\Delta_1(s_-, s_+) = \Delta_2(s_-, s_+) = \max(0, m + s_- - s_+)$. Here s_- denotes the the score of a wrong correspondence, s_+ denotes the score of a correct correspondence and m is the margin. We extract positive and negative patch pairs around points with valid ground-truth. Each positive is defined by the ground-truth flow with a perturbation of up to 1 pixel for robustness of the resulting feature representation. Unfortunately, the candidate set for the negative is the whole target image except the ground-truth matching point and thus intractable. Following [16], we sample negatives in a small circular region around each positive, keeping a minimum distance from the ground truth location. In particular, we use a threshold of 3 pixels for the minimum distance and a threshold of 6 pixels for the maximum distance to the ground truth flow. This ensures that the training set is composed of patches which are non-trivial to separate.

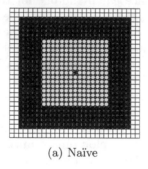 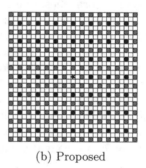

(a) Naïve   (b) Proposed

Fig. 2. Dilated Convolution Implementations. This figure shows the dilated convolution centers on a patch for the context network 2 and 12 with dilation factors 2, 4 and 8 as specified in Table 3. The center of the patch is marked with a red star and each color corresponds to a convolution center for a specific dilation factor, red for 4 dilations (shown in green), green for 2 dilations (shown in blue) and yellow for both. In other words, red dots show the convolution centers (outputs) for the 4-dilated convolutions which read their input values at the green points. Note that yellow points are only visible in (a) as red and green dots do not overlap in (b) due to the sparsity exploited by our approach. (Color figure online)

As illustrated in Fig. 2, a naïve implementation of dilated convolutions for training with patches would result in unnecessary computations. As we only need to forward/backward propagate information to/from the center of the patch, we can back-trace the source locations through the dilation hierarchy. Thus, we can

Sub-sampling with $dilations[1]$
for $i = 1$ to $\#dilations$ **do**
   if $i == \#dilations$ **then**
      $stride = 1$
   else
      $stride = \frac{dilations[i+1]}{dilations[i]}$
Convolution with $stride$
if $i < \#dilations$ **then**
   Batch Normalization, ReLU

for $i = 1$ to $\#dilations$ **do**
   DilatedConvolution with $dilations[i]$
   if $i < \#dilations$ **then**
      Batch Normalization, ReLU

Fig. 3. Fast Patch-based Training of Dilated Convolutional Networks. Left: A naïve implementation requires dilated convolution operations which are computationally less efficient than highly optimized **cudnn** convolutions without dilations. **Right:** The behavior of dilated convolutions can be replicated with regular convolutions by first sub-sampling the feature map and then applying 1-dilated convolutions with stride. Here *dilations* is denoting an array that specifies the dilation factor of the dilated convolution in each convolutional layer.

implement the dilated convolution operation with sub-sampling and strides as a regular convolution as shown in Fig. 2.

Furthermore, we are able to exploit the fact that dilated convolutions on patches can be expressed as regular convolutions with strides as illustrated by our pseudo-code in Fig. 3. Our experiments show that this reduces computation time as state-of-the-art implementations of regular convolutions (using cudnn) are significantly faster compared to dilated ones. This makes training with patch-based dilated convolutional networks much faster. At test time, we reuse the computations by dense convolutions over the image domain in traditional manner.

Inference: Differently from training with image patches, at test time the outputs of both branches of the network are computed for each point only once in a single forward pass of the full image, thereby reusing computations. The score computation between multiple reference points and every point on the target image can be performed efficiently as a single big matrix multiplication on the GPU. The first matrix is constructed by stacking reference feature descriptors as rows and the second matrix is built by stacking the target feature descriptors as columns. This waives the need for approximate search strategies as required in CPU-only model [11]. In our implementation, we handle the large GPU memory requirements by dividing the first matrix into individual chunks, balancing memory usage and computation time. We are able to further cut down inference time, as the post-processing stage which we use requires only every fourth pixel to be matched.

3.2 Discrete Optimization and Postprocessing

We follow the Discrete Flow approach [11] to aggregate information while respecting uncertainty in the matching. More specifically, we select the 300 best feature match hypotheses for each pixel on a regular 4-spaced grid, subject to mild non-maximum-suppression constraints (threshold 2 pixels), as input to a 4-connected MRF with pairwise smoothness constraints [11]. We find an approximate MAP solution using max-product belief propagation and the efficient robust pairwise potentials of [11].

As some pixels are occluded (i.e., not matchable) and due to the occurrence of outliers, we post-process our results using a forward-backward consistency check after discretely optimizing the forward and the backward flow. We further remove unlikely small segments from the solution using connected-component analysis. The resulting semi-dense flow map is fed into Epic Flow [20] for further refinement to sub-pixel accuracy. We optimize the parameters of the MRF and the post-processing stage using block coordinate descent with a $0/1$ outlier loss $\Delta_3 = [\|\mathbf{f} - \hat{\mathbf{f}}\|_2 > 3\,\mathrm{Px}]$, averaged over all unoccluded pixels. Here, $\mathbf{f}$ is the ground truth flow vector, $\hat{\mathbf{f}}$ denotes its prediction and $[\cdot]$ is the Iverson bracket.

4   Experimental Results

We evaluate the performance of different local and context architectures, as well as the whole Deep Discrete Flow pipeline on MPI Sintel [6], KITTI 2012 [4] and KITTI 2015 [5]. Towards this goal, we trained separate networks for Sintel and KITTI, but merged the training sets of KITTI 2012 and KITTI 2015. For our internal evaluations, we follow the KITTI and MPI Sintel protocols: we split the training set into a training and a validation set using every fifth image for validation and the remaining images for training. While the images in MPI Sintel are temporally correlated, we found that Siamese networks without fully-connected layers do not suffer from over-fitting (i.e., the training and the validation errors behave similarly).

Note that the MPI Sintel and KITTI datasets leverage different evaluation metrics, average endpoint error (EPE) and outlier ratio, respectively. We follow each dataset's criteria for the final results, but report 3 pixel outlier ratios in all non-occluded regions for comparing the raw output of different network architectures, since the primary goal of our learned patch representations is to reduce the number of outliers.

Before passing them to the network, we normalize each image to zero mean and unit variance. Following common wisdom [50], we set the kernel size to 3 and use stride 1 convolutions unless otherwise specified. We start the training with standard uniform initialization in Torch and monitor the average outlier ratio in non-occluded regions on a subset of the validation set to stop training. We use stochastic gradient descent with momentum 0.9 for optimization, a batch size of 128, a hinge loss margin of 0.2 and a learning rate of 0.002 without any decay. We observe no over-fitting for neither our local nor our context networks. A detailed run-time analysis is provided in the supplementary material.

Table 1. Local Architectures. RFS denotes the receptive field size in pixels.

Arch.	Layers	Feature maps									RFS
1	5	64	64	64	64	64					11
2	7	64	64	64	64	64	64	64			15
3	7	64	64	64	128	128	128	64			15
4	9	64	64	64	64	64	64	64	64	64	19
5	9	32	32	32	64	64	64	128	128	128	19

Table 2. Comparison of Local Architectures. Figure (a)+(b) show the performance of different local architectures using winner-takes-all on the validation sets of MPI Sintel and KITTI, respectively. As baseline, Fig. (c) shows the performance of matching Daisy features on both datasets using the parameter setting of [11] in the first column and our re-optimized parameters in the second column. All numbers are percentages of non-occluded bad pixels as defined by the KITTI evaluation protocol.

Arch.	Out-Noc
1	24.61 %
2	20.54 %
3	20.69 %
4	19.34 %
5	18.31 %

(a) **MPI Sintel**

Arch.	Out-Noc
1	34.60 %
2	29.71 %
3	29.89 %
4	30.37 %
5	27.36 %

(b) **KITTI**

	DF [11]	Optimized
MPI Sintel	29.97 %	19.16 %
KITTI	34.29 %	22.59 %

(c) **Daisy**

4.1 Baseline for Feature Matching

We leverage the winner-takes-all solution of Daisy [41] feature matching as pursued in Discrete Flow [11] as local baseline for our learned feature representations. For a fair comparison, we optimized the hyper-parameters of the Daisy feature descriptor [41] on a subset of the MPI Sintel training set using block coordinate descent to minimize the ratio of outliers. The results are shown in Table 2c. More details and the resulting combinations are provided in the supplementary material. Note that the optimized Daisy descriptor has 264 dimensions and a receptive field of approximately 40×40 pixels while Discrete flow [11] uses Daisy descriptors of length 68 with a receptive field of approximately 20×20 pixels. All numbers correspond to the WTA solution calculated over the full target image using exact matching.

4.2 Comparison of Local Architectures

We first compare five different local network architectures, including some of the architectures proposed in the literature for feature matching. Our starting point is the simple 5-layer architecture of [16] (architecture 1). We create additional architectures by changing the number of layers and feature maps in each layer as shown in Table 1. Architecture 5 corresponds to the recently proposed

Table 3. Context Architectures. This table shows different context architectures and their receptive field sizes (RFS). We list the architectures that share the same set of dilations (and consequently RFS) in one row. Architectures in the same row differ solely by the number of feature maps in each layer. Receptive field sizes are added (+) to the RFS size of the respective local architecture.

Arch.	Feature maps	Arch.	Feature maps					Dilations						RFS
1	all 64	11	64	128	256	512		2	4	8	16			+60
2		12	64	128	256			2	4	8				+28
3		13	64	128				2	4					+12
4		14	64	128	256			4	8	16				+56
5		15	64	128				8	16					+48
6		16	128	128				4	4					+16
7		17	64	64	128	128		2	2	4	4			+24
8		18	64	64	128	128	256 256	2	2	4	4	8	8	+28
9		19	128	128	256	256		8	8	16	16			+48

Table 4. Comparison of Context Architectures. This table shows the performance of different context architectures on top of local architecture 1 on the validation sets of MPI Sintel and KITTI. "Out-Noc" is defined as in Table 2.

Arch.	Out-Noc	Arch.	Out-Noc	Arch.	Out-Noc	Arch.	Out-Noc
1	16.32 %	11	11.92 %	1	30.16 %	11	24.28 %
2	14.51 %	12	12.19 %	2	25.82 %	12	20.28 %
3	15.65 %	13	14.32 %	3	24.67 %	13	21.39 %
4	15.27 %	14	12.53 %	4	29.40 %	14	25.29 %
5	15.66 %	15	13.34 %	5	28.54 %	15	24.89 %
6	15.10 %	16	13.59 %	6	25.11 %	16	21.13 %
7	15.68 %	17	13.69 %	7	26.78 %	17	22.93 %
8	15.50 %	18	13.01 %	8	31.63 %	18	24.68 %
9	20.04 %	19	14.55 %	9	40.12 %	19	34.43 %

| (a) **MPI Sintel** | (b) **KITTI** |

9 layer network for stereo and flow in [18,48]. Adding more layers changes network's receptive field size and has a clear effect on the performance as shown in Table 2. However, compared to our local architectures, the Daisy descriptor is fairly competitive. We attribute this to its larger receptive field size. In the next section, we explore context architectures to increase the receptive field size of our learned representations.

4.3   Comparison of Context Architectures

Towards this goal, we fix the local architecture to the simplest one (architecture 1), and train different context architectures on top of this network. In a later section, we also show the performance of the best context architecture trained on top of the best local architecture. We create two types of context architectures, one by fixing the number of feature maps to 64, and one by changing the number of feature maps in each layer as summarized in Table 3.

Again, we compare the winner-takes-all performance in Table 4. We first note that the outlier ratio is significantly lower than the outlier ratio of the local architectures and the Daisy baseline shown in Table 2. Secondly, architectures which double the number of feature maps consistently outperform their respective constant counterparts. Finally, the 3-layer context architectures 2 and 12 are the best performing models in our set.

4.4   Evaluation of Model Components

Table 5 compares (from left to right) the results of winner-takes-all (WTA), discrete optimization and the full pipeline including post-processing and Epic Flow interpolation with respect to each other. For this experiment, we selected the simplest local architecture 1 as well as the top performing local architecture 5, both with and without context. For comparison, we also show the results of Discrete Flow with Daisy features as baseline ("DF [11]").

We first note that for WTA (first column), the context architectures improve the outlier ratio significantly with respect to local architectures for both datasets. However, this improvement is less visible after spatial smoothing (second and third column). We conclude that the gain of leveraging a larger receptive field can be partially compensated by using a spatial smoothing stage.

Table 5. Comparison of Model Components. This table shows our results after winner-takes-all feature matching, after discrete optimization and the results of the full pipeline including post-processing and sub-pixel interpolation. We report end-point-errors (EPE) and outliers ratios (Out) both in non-occluded (Noc) and in all image regions (Occ) on the respective validation sets.

Local	Context	Winner-takes-All				Discrete Optimization				Full Pipeline			
		Noc		Occ		Noc		Occ		Noc		Occ	
		Out	EPE	Out	EPE	Out	EPE	Out	EPE	Out	EPE	Out	EPE
DF [11]		26.36 %	27.92	29.85 %	33.93	10.45 %	7.44	14.82 %	13.05	8.20 %	2.77	11.28 %	4.61
1	-	24.67 %	49.86	28.45 %	62.05	12.06 %	10.27	16.55 %	17.60	7.27 %	2.78	10.14 %	4.41
1	12	12.24 %	25.70	16.76 %	39.69	8.95 %	8.73	13.44 %	16.03	6.93 %	2.61	9.92 %	4.18
5	-	18.36 %	42.51	22.64 %	56.30	10.75 %	10.76	15.29 %	18.61	7.28 %	2.83	10.12 %	4.37
5	12	12.13 %	27.94	16.66 %	42.42	8.75 %	9.12	13.26 %	16.70	7.07 %	2.73	10.02 %	4.29

(a) **MPI Sintel**

Local	Context	Winner-takes-All				Discrete Optimization				Full Pipeline			
		Noc		Occ		Noc		Occ		Noc		Occ	
		Out	EPE	Out	EPE	Out	EPE	Out	EPE	Out	EPE	Out	EPE
DF [11]		33.01 %	30.21	40.99 %	49.16	10.84 %	3.73	21.81 %	22.38	8.55 %	1.76	18.43 %	4.49
1	-	34.38 %	69.45	42.00 %	99.55	13.38 %	8.75	24.00 %	29.92	8.35 %	2.14	16.73 %	4.44
1	12	20.00 %	41.67	29.33 %	77.21	13.26 %	9.90	23.54 %	31.00	9.75 %	2.57	18.27 %	5.35
5	-	27.18 %	58.58	35.69 %	92.92	13.07 %	9.63	23.63 %	31.70	8.74 %	2.38	17.12 %	4.77
5	12	22.09 %	55.46	31.10 %	92.85	14.01 %	12.78	24.16 %	34.72	10.62 %	2.99	19.10 %	5.95

(b) **KITTI**

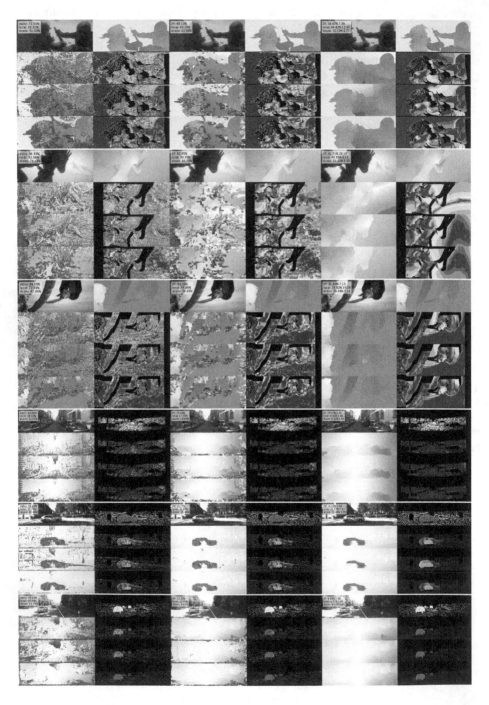

Fig. 4. Qualitative Results. See Sect. 4.6 for details.

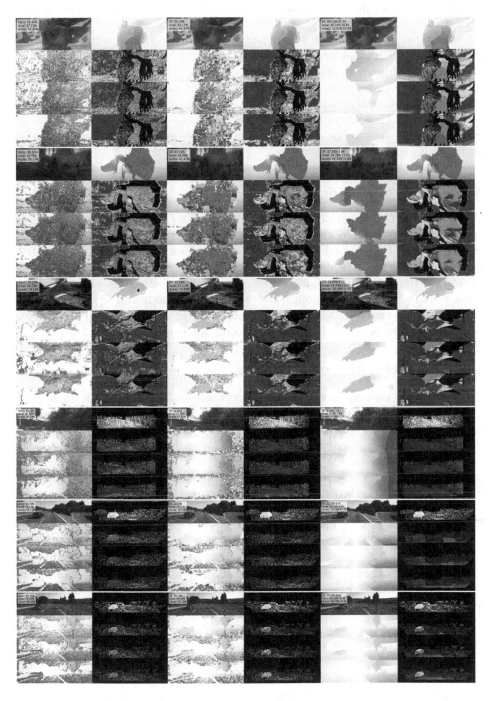

Fig. 5. Qualitiative Results. See Sect. 4.6 for details.

On the KITTI dataset, the improvements are less pronounced than on the MPI Sintel dataset. Here, our smallest local architecture (second row) outperforms Discrete Flow [11] slightly. Interestingly, the context architectures improve performance when considering the winner-takes-all (WTA) solution, but perform on par or even lead to degradation after spatial smoothing (second and third column). Our investigations revealed that the reason for this is the scale changes which are prominently present on KITTI (but less so on MPI Sintel) and which the networks have difficulty to cope with. We thus conclude that progress in invariant deep representations (in particular scale invariance) is necessary to address this issue.

4.5   Results on Test Set

We submitted our results to the MPI Sintel and KITTI 2012 and 2015 evaluation servers. We picked the best row for each dataset according to the results in Table 5, i.e., local model 1 in combination with context model 12 for the MPI Sintel and local model 1 alone for both KITTI datasets. In accordance with our results on the training/validation split, we obtain good results on MPI Sintel (best performing method amongst the published methods and second best performing method overall) while we are slightly better than Discrete Flow [11] on KITTI 2012 and KITTI 2015. We refer to the benchmark websites for details[1].

4.6   Qualitative Results

Figure 4 shows visualizations of the different stages of our approach for several selected images from both MPI Sintel (top) and KITTI (bottom). Some failure cases are shown in Fig. 5. Each sub-figure shows from top-to-bottom: the input image and the ground-truth flow, Discrete Flow with Daisy features, our local architecture 1, our architectures 1 + 12. For each sub-figure, the first double column shows the WTA result on the grid, the second the results of discrete optimization and the last double column shows the final result.

As evidenced from these results, the proposed feature learning approach handles object boundaries more precisely and in general leads to lower errors for all inliers. However, these advantages diminish after discrete optimization and in particular Epic Flow interpolation as non-matched regions are responsible for the largest portion of the remaining errors. From Fig. 5, it is clearly visible that the learned representations suffer from strong scale changes which need to be addressed to further improve performance.

5   Conclusion and Future Work

We presented an efficient way of learning features for optical flow in a discrete framework by showing that dilated convolutions can be implemented efficiently

---

[1] http://sintel.is.tue.mpg.de/, http://www.cvlibs.net/datasets/kitti/.

for patch-based training. Learning features with context networks improves feature matching performance with respect to local architectures and manually engineered features for both the MPI Sintel and KITTI datasets. Although our experiments demonstrated that learning features with context is crucial for reducing outliers in the WTA solution of the network, large gains mostly diminish in the later stages of our pipeline. We found that large changes in scale pose problems to current feature learning approaches, prompting for the development of inherently scale invariant deep features. Finally, we remark that our current model's performance is hampered by piece-wise training. We therefore plan to investigate end-to-end training by back-propagating errors through all stages of our pipeline.

References

1. Zhang, H., Geiger, A., Urtasun, R.: Understanding high-level semantics by modeling traffic patterns. In: Proceedings of the IEEE International Conference on Computer Vision (ICCV) (2013)
2. Schönbein, M., Geiger, A.: Omnidirectional 3d reconstruction in augmented manhattan worlds. In: Proceedings of the IEEE International Conference on Intelligent Robots and Systems (IROS) (2014)
3. Geiger, A., Lauer, M., Wojek, C., Stiller, C., Urtasun, R.: 3D traffic scene understanding from movable platforms. IEEE Trans. Pattern Anal. Mach. Intell. (PAMI) **36**, 1012–1025 (2014)
4. Geiger, A., Lenz, P., Urtasun, R.: Are we ready for autonomous driving? the KITTI vision benchmark suite. In: Proceedings of the IEEE Conference on Computer Vision and Pattern Recognition (CVPR) (2012)
5. Menze, M., Geiger, A.: Object scene flow for autonomous vehicles. In: Proceedings of the IEEE Conference on Computer Vision and Pattern Recognition (CVPR) (2015)
6. Butler, D.J., Wulff, J., Stanley, G.B., Black, M.J.: A naturalistic open source movie for optical flow evaluation. In: Fitzgibbon, A., Lazebnik, S., Perona, P., Sato, Y., Schmid, C. (eds.) ECCV 2012. LNCS, vol. 7577, pp. 611–625. Springer, Heidelberg (2012). doi:10.1007/978-3-642-33783-3_44
7. Horn, B.K.P., Schunck, B.G.: Determining optical flow. Artif. Intell. (AI) **17**, 185–203 (1981)
8. Lucas, B.D., Kanade, T.: An iterative image registration technique with an application to stereo vision. In: Proceedings of the International Joint Conference on Artificial Intelligence (IJCAI) (1981)
9. Black, M.J., Anandan, P.: A framework for the robust estimation of optical flow. In: Proceedings of the IEEE International Conference on Computer Vision (ICCV) (1993)
10. Brox, T., Bruhn, A., Papenberg, N., Weickert, J.: High accuracy optical flow estimation based on a theory for warping. In: Pajdla, T., Matas, J. (eds.) ECCV 2004. LNCS, vol. 3024, pp. 25–36. Springer, Heidelberg (2004). doi:10.1007/978-3-540-24673-2_3
11. Menze, M., Heipke, C., Geiger, A.: Discrete optimization for optical flow. In: Gall, J., Gehler, P., Leibe, B. (eds.) GCPR 2015. LNCS, vol. 9358, pp. 16–28. Springer, Cham (2015). doi:10.1007/978-3-319-24947-6_2

12. Chen, Q., Koltun, V.: Full flow: optical flow estimation by global optimization over regular grids. In: Proceedings of the IEEE Conference on Computer Vision and Pattern Recognition (CVPR) (2016)
13. Wulff, J., Black, M.J.: Efficient sparse-to-dense optical flow estimation using a learned basis and layers. In: Proceedings of the IEEE Conference on Computer Vision and Pattern Recognition (CVPR) (2015)
14. Horáček, M., Besse, F., Kautz, J., Fitzgibbon, A., Rother, C.: Highly overparameterized optical flow using patchmatch belief propagation. In: Fleet, D., Pajdla, T., Schiele, B., Tuytelaars, T. (eds.) ECCV 2014. LNCS, vol. 8691, pp. 220–234. Springer, Cham (2014). doi:10.1007/978-3-319-10578-9_15
15. Güney, F., Geiger, A.: Displets: resolving stereo ambiguities using object knowledge. In: Proceedings of the IEEE Conference on Computer Vision and Pattern Recognition (CVPR) (2015)
16. Žbontar, J., LeCun, Y.: Stereo matching by training a convolutional neural network to compare image patches. J. Mach. Learn. Res. (JMLR) **17**, 1–32 (2016)
17. Chen, Z., Sun, X., Wang, L., Yu, Y., Huang, C.: A deep visual correspondence embedding model for stereo matching costs. In: Proceedings of the IEEE International Conference on Computer Vision (ICCV) (2015)
18. Luo, W., Schwing, A., Urtasun, R.: Efficient deep learning for stereo matching. In: Proceedings of the IEEE Conference on Computer Vision and Pattern Recognition (CVPR) (2016)
19. Yu, F., Koltun, V.: Multi-scale context aggregation by dilated convolutions. In: Proceedings of the International Conference on Learning Representations (ICLR) (2016)
20. Revaud, J., Weinzaepfel, P., Harchaoui, Z., Schmid, C.: EpicFlow: edge-preserving interpolation of correspondences for optical flow. In: Proceedings of the IEEE Conference on Computer Vision and Pattern Recognition (CVPR) (2015)
21. Bruhn, A., Weickert, J., Schnörr, C.: Lucas/Kanade meets Horn/Schunck: combining local and global optic flow methods. Int. J. Comput. Vis. (IJCV) **61**, 211–231 (2005)
22. Demetz, O., Stoll, M., Volz, S., Weickert, J., Bruhn, A.: Learning brightness transfer functions for the joint recovery of illumination changes and optical flow. In: Fleet, D., Pajdla, T., Schiele, B., Tuytelaars, T. (eds.) ECCV 2014. LNCS, vol. 8689, pp. 455–471. Springer, Cham (2014). doi:10.1007/978-3-319-10590-1_30
23. Ranftl, R., Bredies, K., Pock, T.: Non-local total generalized variation for optical flow estimation. In: Fleet, D., Pajdla, T., Schiele, B., Tuytelaars, T. (eds.) ECCV 2014. LNCS, vol. 8689, pp. 439–454. Springer, Cham (2014). doi:10.1007/978-3-319-10590-1_29
24. Sun, D., Roth, S., Black, M.J.: A quantitative analysis of current practices in optical flow estimation and the principles behind them. Int. J. Comput. Vis. (IJCV) **106**, 115–137 (2013)
25. Werlberger, M., Trobin, W., Pock, T., Wedel, A., Cremers, D., Bischof, H.: Anisotropic Huber-L1 optical flow. In: Proceedings of the British Machine Vision Conference (BMVC) (2009)
26. Zach, C., Pock, T., Bischof, H.: A duality based approach for realtime TV-L1 optical flow. In: Pattern Recognition Letters, pp. 214–223. Springer, Heidelberg (2007)
27. Zimmer, H., Bruhn, A., Weickert, J.: Optic flow in harmony. Int. J. Comput. Vis. (IJCV) **93**, 368–388 (2011)

28. Baker, S., Scharstein, D., Lewis, J., Roth, S., Black, M., Szeliski, R.: A database and evaluation methodology for optical flow. Int. J. Comput. Vis. (IJCV) **92**, 1–31 (2011)
29. Braux-Zin, J., Dupont, R., Bartoli, A.: A general dense image matching framework combining direct and feature-based costs. In: Proceedings of the IEEE International Conference on Computer Vision (ICCV) (2013)
30. Brox, T., Malik, J.: Large displacement optical flow: descriptor matching in variational motion estimation. IEEE Trans. Pattern Anal. Mach. Intell. (PAMI) **33**, 500–513 (2011)
31. Timofte, R., Gool, L.V.: Sparse flow: sparse matching for small to large displacement optical flow. In: Proceedings of the IEEE Winter Conference on Applications of Computer Vision (WACV) (2015)
32. Weinzaepfel, P., Revaud, J., Harchaoui, Z., Schmid, C.: DeepFlow: large displacement optical flow with deep matching. In: Proceedings of the IEEE International Conference on Computer Vision (ICCV) (2013)
33. Steinbrücker, F., Pock, T., Cremers, D.: Large displacement optical flow computation without warping. In: Proceedings of the IEEE International Conference on Computer Vision (ICCV), pp. 1609–1614 (2009)
34. Yamaguchi, K., McAllester, D., Urtasun, R.: Robust monocular epipolar flow estimation. In: Proceedings of the IEEE Conference on Computer Vision and Pattern Recognition (CVPR) (2013)
35. Yamaguchi, K., McAllester, D., Urtasun, R.: Efficient joint segmentation, occlusion labeling, stereo and flow estimation. In: Fleet, D., Pajdla, T., Schiele, B., Tuytelaars, T. (eds.) ECCV 2014. LNCS, vol. 8693, pp. 756–771. Springer, Cham (2014). doi:10.1007/978-3-319-10602-1_49
36. Lempitsky, V.S., Roth, S., Rother, C.: Fusionflow: discrete-continuous optimization for optical flow estimation. In: Proceedings of the IEEE Conference on Computer Vision and Pattern Recognition (CVPR) (2008)
37. Chen, Z., Jin, H., Lin, Z., Cohen, S., Wu, Y.: Large displacement optical flow from nearest neighbor fields. In: Proceedings of the IEEE Conference on Computer Vision and Pattern Recognition (CVPR) (2013)
38. Yang, J., Li, H.: Dense, accurate optical flow estimation with piecewise parametric model. In: Proceedings of the IEEE Conference on Computer Vision and Pattern Recognition (CVPR) (2015)
39. Mozerov, M.: Constrained optical flow estimation as a matching problem. IEEE Trans. Image Process. (TIP) **22**, 2044–2055 (2013)
40. Besse, F., Rother, C., Fitzgibbon, A., Kautz, J.: PMBP: patchmatch belief propagation for correspondence field estimation. Int. J. Comput. Vis. (IJCV) **110**, 2–13 (2014)
41. Tola, E., Lepetit, V., Fua, P.: Daisy: an efficient dense descriptor applied to wide baseline stereo. IEEE Trans. Pattern Anal. Mach. Intell. (PAMI) **32**, 815–830 (2010)
42. Krizhevsky, A., Sutskever, I., Hinton, G.E.: Imagenet classification with deep convolutional neural networks. In: Advances in Neural Information Processing Systems (NIPS) (2012)
43. Fischer, P., Dosovitskiy, A., Ilg, E., Häusser, P., Hazirbas, C., Smagt, V.G.P., Cremers, D., Brox, T.: FlowNet: learning optical flow with convolutional networks. arXiv.org:1504.06852 (2015)
44. Mayer, N., Ilg, E., Haeusser, P., Fischer, P., Cremers, D., Dosovitskiy, A., Brox, T.: A large dataset to train convolutional networks for disparity, optical flow, and scene flow estimation. In: CVPR (2016)

45. Han, X., Leung, T., Jia, Y., Sukthankar, R., Berg, A.C.: Matchnet: unifying feature and metric learning for patch-based matching. In: Proceedings of the IEEE Conference on Computer Vision and Pattern Recognition (CVPR) (2015)
46. Simo-Serra, E., Trulls, E., Ferraz, L., Kokkinos, I., Fua, P., Moreno-Noguer, F.: Discriminative learning of deep convolutional feature point descriptors. In: Proceedings of the IEEE International Conference on Computer Vision (ICCV) (2015)
47. Zagoruyko, S., Komodakis, N.: Learning to compare image patches via convolutional neural networks. In: Proceedings of the IEEE Conference on Computer Vision and Pattern Recognition (CVPR) (2015)
48. Bai, M., Luo, W., Kundu, K., Urtasun, R.: Deep semantic matching for optical flow. arXiv.org:1604.01827 (2016)
49. Gadot, D., Wolf, L.: Patchbatch: a batch augmented loss for optical flow. In: Proceedings of the IEEE Conference on Computer Vision and Pattern Recognition (CVPR) (2016)
50. Simonyan, K., Zisserman, A.: Very deep convolutional networks for large-scale image recognition. In: Proceedings of the International Conference on Learning Representations (ICLR) (2015)

Dense Motion Estimation for Smoke

Da Chen[1]([✉]), Wenbin Li[2], and Peter Hall[1]

[1] Department of Computer Science, University of Bath, Bath, UK
d.chen@bath.ac.uk
[2] Department of Computer Science, University College London, London, UK

Abstract. Motion estimation for highly dynamic phenomena such as smoke is an open challenge for Computer Vision. Traditional dense motion estimation algorithms have difficulties with non-rigid and large motions, both of which are frequently observed in smoke motion. We propose an algorithm for dense motion estimation of smoke. Our algorithm is robust, fast, and has better performance over different types of smoke compared to other dense motion estimation algorithms, including state of the art and neural network approaches. The key to our contribution is to use skeletal flow, without explicit point matching, to provide a sparse flow. This sparse flow is upgraded to a dense flow. In this paper we describe our algorithm in greater detail, and provide experimental evidence to support our claims.

1 Introduction

Accurate dense motion estimation is one of the fundamental problems in Computer Vision. The problem is especially marked for highly dynamic phenomena, such as smoke. Yet a solution for fluids is of value to a diverse set of areas of study. In Computer Graphics there are applications in both post-production [1–3] and model acquisition [4]; In Atmospheric research, there are applications for storm identification and forecast [5], forecast and tracking the evolution of convective systems [6] and rain cloud tracking [7].

Our contribution is a method which gives simple, fast, robust, and accurate motion estimation of smoke. Our approach differs from most others because it do not rely on the constant brightness assumption, nor does it use feature matching of any kind. Smoke is partially transparent, it diffuses and aggregates as it moves. These characteristics breach the assumptions that typically underpin dense motion algorithms. This may explain why they under-perform in the case of smoke when compared to other object classes.

The key assumption in our motion estimation method is that the structure of the smoke density distribution does not change much from frame to frame. This allows us to estimate motion smoke in a global sense (described below) and thereby avoid the error-prone procedure of matching points in video frames based on local appearance, which is in constant flux.

S.-H. Lai et al. (Eds.): ACCV 2016, Part IV, LNCS 10114, pp. 225–239, 2017.
DOI: 10.1007/978-3-319-54190-7_14

We now describe our approach briefly, considering a pair of video frames. Our algorithm estimates a dense flow using a sparse flow as a starting point. Our key assumption (global smoke density changes little from frame to frame) is responsible for the sparse estimation, which begins by assuming that pixel intensity and smoke density are directly proportional. In each frame, we construct a skeleton that characterises the smoke density. In this paper, we use density ridges as the basis for skeleton construction. We then estimate the motion of each skeletal point on skeleton one using all points on skeleton two. This gives a sparse flow that is upgraded to a dense flow by interpolation [8,9] and refined using an existing energy minimisation approach [10].

It is also worth noting that we do not use graph matching to map skeletons: graph matching is NP hard and it is difficult to account for any changes in structure that do occur. Local appearance information (*i.e.* similarity measures with features or patches as input) plays no role at all in our motion estimator: we have already observed appearance changes that may confuse standard motion estimation methods. Instead we independently map each skeletal point, $\mathbf{x}$, in skeleton one, S_1, in frame one to its expected location, $E[\mathbf{y}|\mathbf{x}] = \sum_{\mathbf{y} \in S_2} \mathbf{y} p(\mathbf{y}|\mathbf{x})$, when all points $\mathbf{y}$ in skeleton two, S_2, are considered.

We describe our algorithm in detail in Sect. 3. Experiments against both classical and a range of state of the art alternatives (including latest CNN based methods) show our method outperforms all of them, see Sect. 4. We wrap up the paper in Sect. 5 by pointing to both limitations and future applications of the work.

2   Related Work

Optical flow pioneered by Horn and Schunck [11] is probably the best known and widely used general method for motion estimation. They assumed brightness constancy and local smoothness, leading to a regularised minimisation problem over the whole image. Although this method is generally applicable, it faces many challenges such as large motion displacement, motion discontinuities, *etc.* [12]. Advanced methods [13–15] were proposed to solve the specific challenge in optical flow. With the deep learning trend, CNN based methods [16] are also proposed to deal with these challenges.

Unfortunately, none of these methods can properly deal with all of the many challenges that exist simultaneously in highly dynamic natural phenomena where the objects have no definitive features (*e.g.* smoke, water) or many similar features (*e.g.* leaves on trees). Brightness constancy is routinely violated in such cases, and if local features do exist they can move rapidly compared to their size. Furthermore, regularisors typically assume smoothness in the flow field; again an assumption that is often violated by natural phenomena. In Sect. 4, we evaluate these methods with smoke cases.

Faced with these problems, the literature has seen a steady stream of papers that proposed solutions dedicated to the motion estimation of fluids. In this paper, we cannot mention all of them. Here we will discuss the most relevant and state of the art techniques. Mémin and Pérez provide an early example [17]; they construct a dense flow by fitting parametric models to *critical points* (vortices of different classes) and interpolating between them.

In recent years, fluid motion estimation methods have been proposed that combine appearance based model with optical flow framework. These methods [18–21] add constraints to prefer the fluid like motion in the energy minimization process. Corpetti *et al.* [19], for example, apply a divergence-curl-type smoothness to replace the original smoothness term in optical flow framework. Auroux *et al.* [18] propose to use a group of appearance feature based terms as candidates to replace the smoothness term. These methods, however, are all limited by the basic assumptions of optical flow.

In addition to appearance based fluid motion estimation methods, physical based fluid motion estimation methods are also proposed. Most of these methods [4,22–24] are based on the Navier-Stoke(NS) equation to describe the fluid motion. For example, Doshi *et al.* [24] proposed a process of optical flow based on the NS equation. However, this method only optimises for the smoothness of the flow field and ignores the accuracy of the estimation without an appropriate evaluation. Anumolu *et al.* [22] provided a framework for smoke simulation with strict physical regularities considering 'vorticity confinement' [25]. Nevertheless, this framework only works for the interaction boundary between rigid object and smoke. Li *et al.* [23] proposed a brightness constancy constraint(BCC) based framework that combined the NS equations with the 3D potential flow. Although it could be extended to other physical models, this framework is not compatible with rotational flow and is limited by the physical model they used.

Apart from the methods based on NS equations, many novel methods based on other physical properties have been proposed. For example, Sakaino [26] proposed a motion estimation method based on the physical properties of waves. Those properties are components of a wave model which is generated by a combination of sinusoidal functions. Some recently proposed methods such as [27,28] are based on refractive properties, while [29] is based on light path Snell's law. These methods, though, are limited to laboratory environments and require specific equipment configurations.

We provide an algorithm that makes general assumptions: No strong physical model is applied, instead, we assume that a skeletal structure can be used to characterise broad appearance and skeletal matching can be used to characterise broad motion. This general idea has been used by [7] for rain cloud tracking with radar image; we differ in the way in which both construct and match skeletons. Noted that the proposed method is much simpler compare to [7].

3   Method

Our motion estimation method for smoke assumes that: (1) the density of smoke, ρ, at time t is related to density at time $t + dt$ by

$$\rho_{t+dt}(\mathbf{y}) = \int_{\mathbf{x} \in \Re^3} \phi(\mathbf{y}, \mathbf{x}) \rho_t(\mathbf{x}) d\mathbf{x} \qquad (1)$$

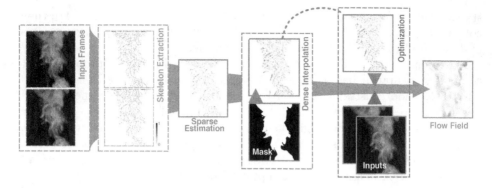

Fig. 1. System overview. In our system pipeline, we first extract the smoke skeletons and estimate a sparse vector field in between (Sect. 3.1). This sparse information is then interpolated to a dense motion field (Sect. 3.2) which is further refined within a typical optical flow framework (Sect. 3.3).

in which ϕ is a mass transfer function ($\rho_t(\mathbf{x})d\mathbf{x}$ being the mass) that includes systematic motion and diffusion; (2) that fluid motion is smooth if looked at globally, but possibly turbulent when looked at locally; and (3) the observed density of smoke and pixel intensity are in direct proportion to one another. These assumptions are justified by observing smoke and other fluid in motion: we see a two dimensional projection of a three dimensional phenomena in which local motion details are superimposed on a global motion.

Ideally, we would like to solve for the transfer function $\phi(\mathbf{y}, \mathbf{x})$, over all $\Re^3$, given a pair of input frames; but this is not possible. Instead we estimate the conditional density $p(\mathbf{y}|\mathbf{x})$, for a sparse set of points in $\Re^2$ that are chosen to characterise both the global and local motion of the flow, and then upgrade to a dense flow. In fact, our motion estimation algorithm has three main parts as shown in Fig. 1: (1) we produce a sparse estimation; (2) we interpolate the sparse estimation into a dense estimation; (3) we refine the dense estimation. Each part is now described in turn.

3.1   Sparse Estimation

A novel sparse estimator is the key to our method and is the technical contribution of this paper. It is here that we approximate the transfer function with a conditional probability, for a sparse point set, in a manner that is sensitive to both global and local motion. Our sparse estimator has three substeps: (1) segment the smoke from the background; (2) fit a skeleton to each frame; (3) estimate sparse frame-to-frame motion using adjacent skeletons.

Algorithms exist that segment smoke and other natural phenomena from general backgrounds [30–33]. These methods provide various ways to detect areas with dynamic texture based on spatiotemporal filters [30], optical flow motion field [31]. But to focus on motion estimation alone, we use video acquired in laboratory conditions in which smoke appears to be light on a black background. This makes segmenting smoke simple: we threshold at some low value ε to make a mask $M(\mathbf{x}) = f(\mathbf{x}) > \varepsilon$. For pixel values in $[0, 255]$ we typically set $\varepsilon = 1$.

The next step is skeleton fitting. The purpose of the skeleton is to characterise the density distribution of smoke particles so that global information can be used when obtaining a sparse estimate for motion. The skeleton we use traces ridge lines of smoke density. It is a compound skeleton that is built from several individual skeletons, each characterising density in a different scale.

An individual skeleton comprises the ridge points of density (intensity) image that has been Gaussian blurred at scale σ. Let $g_\sigma = G_\sigma \odot f$ be the original image f convolved with blur kernel G_σ. We locate ridges at pixels in g_σ that are locally maximal in either the 'x' or 'y' direction, but we do not depend on this particular skeleton, we could also use $e.g.$ a watershed skeleton [34]. The objective is to obtain a binary image $h_\sigma(\mathbf{x})$, which is a characteristic skeleton at scale σ. The skeleton is masked, using M, to ensure it lies only within the segmented smoke. The small scale skeletons capture local details, whereas the large scale skeletons capture global structure.

In principle it would be possible to use skeletons at different scales to obtain a motion estimate using a coarse-to-fine algorithm of some kind. We avoid that complication, instead preferring to build a compound skeleton by aggregation:

$$h(\mathbf{x}) = \frac{1}{N} \sum_{i=1}^{N} h_{\sigma_i}(\mathbf{x}) \tag{2}$$

The result of this is a multi-valued skeleton that tends to emphasise stable structures within the smoke density: higher values of the skeleton indicate locations that are more stable over scale. Figure 2 illustrates the whole skeleton building process.

With a multi-valued skeleton in hand, we can continue to estimate sparse motion between skeletons in adjacent frames. Trying to match points 1-1 between skeletons is inappropriate: in principle because of the diffusion of smoke and in

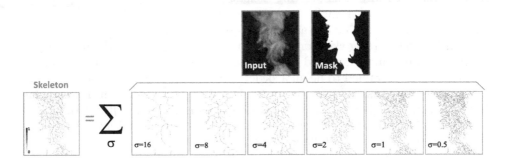

Fig. 2. Making a multi-valued smoke skeleton. Given an input smoke frame and a segmented mask, the skeleton can be aggregated by the sub-skeletons with different scale σ.

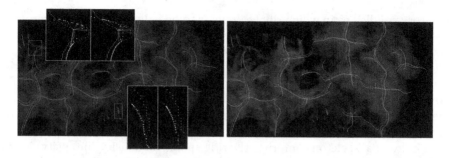

Fig. 3. Left: frame one with starting skeleton (red) and attracting skeleton (green) in second frame. Detail at top shows one point attracted to nearby points, expected motion is shown in yellow; right box shows the same, but for an isotropic Gaussian attractor. The bottom detail shows that an anisotropic Gaussian (left) leads to a better coverage of the attracting skeleton (green crosses) by the expected point destinations (yellow points) compared to an isotropic Gaussian (left). Right of figure shows the start skeleton (red), the attracting skeleton (green) and the sparse motion estimate (yellow). Please zoom in to see the details. (Color figure online)

practice because the number of skeletal points is likely to change frame to frame. Iterative closest point [35] or similar technique to (for example) warp one skeleton into another requires the motion to be parametrisable: a strong assumption we wish to avoid. Using graph matching techniques is also not appropriate; graph matching is NP-hard, therefore pixels should not be nodes. The alternative is to link skeletal pixels into lines, but that is ad-hoc and complex.

Our solution is very simple. We specify the probability that a skeletal point $\mathbf{x}$ in frame at time t has moved to skeletal point $\mathbf{y}$ in frame at time $t + dt$. This is a conditional probability, here denoted $p(\mathbf{y}|\mathbf{x})$. Then we compute the expected location at frame $t + dt$ of the point $\mathbf{x}$. For all $\mathbf{x} \in S_1$

$$E[\mathbf{y}|\mathbf{x}] = \sum_{\mathbf{y} \in S_2} \mathbf{y} p(\mathbf{y}|\mathbf{x}). \tag{3}$$

We note that this way of associating skeletal points $\mathbf{x}$ with destination points $\mathbf{y}$ imposes no strong constraints. There may be points $\mathbf{x}$ for which $\max p(\mathbf{y}|\mathbf{x}) < \epsilon$; these points can be filtered out as mapping to nothing. Conversely, there may be no point on the starting skeleton that maps to a given point on the destination skeleton, equally, many $\mathbf{x}$ could map to the same $\mathbf{y}$. Indeed, an expected point $E[\mathbf{y}|\mathbf{x}]$ is not constrained to lie on a skeleton. For these reasons we cannot properly describe our sparse estimator as being a matching algorithm, instead the destination skeleton is used as an attractor to construct a sparse flow estimation.

The conditional $p(\mathbf{y}|\mathbf{x})$ is defined using spatial distance and intensity of the skeletal pixels, similar to a bilateral filter [36]

$$p(\mathbf{y}|\mathbf{x}) \propto \mathcal{N}(\mathbf{x}|\mathbf{y}, C_y)\mathcal{N}(h(\mathbf{x})|h(\mathbf{y}), \sigma_v). \tag{4}$$

in which $h(.)$ is the value at a point in a multi-valued skeletal image. The standard deviation σ_v is set to be related to the number of skeletal levels, typically

$\sigma_v = 2/(N-1)$, where N is the number of scales in skeleton searching. The covariance matrix C_y is a non-isotropic Gaussian:

$$C_y = ULU^T \tag{5}$$

$$U = [\hat{\mathbf{n}}, \ \hat{\mathbf{t}}] \tag{6}$$

$$L = \sigma \text{diag}([1, \eta]) \tag{7}$$

in which U is an orthonormal matrix that orients the Gaussian using the unit normal, $\hat{\mathbf{n}}$, and unit tangent $\hat{\mathbf{t}}$ to the skeleton S_2 at $\mathbf{y}$. The diagonal matrix L, squashes the distribution by a factor η in the direction of the tangent to the skeleton, so that it narrows over the skeleton line. Typically $\eta = 1/10$. The reason for using an anisotropic, oriented Gaussian is to keep the distribution $\sum_{\mathbf{y} \in S_2} \mathcal{N}(\mathbf{z}|\mathbf{y}, c_y)$ reasonably flat for all points $\mathbf{z} \in S_2$.

If a non-isotropic Gaussian is used, the cumulative distribution will tend to peak in the middle of skeletal lines, with the result that expected values $E[\mathbf{y}|\mathbf{x}]$ will tend to crowd there also. This is seen in Fig. 3 which also illustrates that points in skeleton two attract points in skeleton one; skeletons are not explicitly matched.

It could be argued that we should make more use of the information in the multi-valued skeleton. In particular we could use the "stability" value, $v(\mathbf{x})$ in the multi-valued skeleton, similar to a coarse-to-fine strategy but "stable to less-stable". However, the use of a bilateral filter tends to make sure points in the starting skeleton are attracted to points of similar stability in the end skeleton; and given we are obtaining high quality results, this complication has not yet been pursued.

3.2 Dense Interpolation

Given a sparse flow estimation, the problem now is to upgrade this to a dense flow. More exactly, we want to estimate a vector field $\mathbf{v}(\mathbf{x})$ for all points $\mathbf{x}$ in frame one using the partial estimate $S(\mathbf{x}) \cdot \mathbf{u}(\mathbf{x})$ in which $S(\mathbf{x})$ is now a binary mask that locates points in the multi-valued skeleton, and $\cdot$ denotes the Hadamard product (also called Schur product or entrywise product); and $\mathbf{u}(\mathbf{x}) = E[\mathbf{y}|\mathbf{x}] - \mathbf{x}$ for all $\mathbf{x} \in S$.

We interpolate flow using a method due to Garcia et al. [9], which is designed for natural phenomenon. We consider each element of the vector field independently. Letting $\mathbf{u} = [u_1, u_2]^T$ and $\mathbf{v} = [v_1, v_2]^T$ denote the fields, the interpolation process yields a new field v_k by the following energy minimization problem:

$$\operatorname*{argmin}_{v_k} \left\| M^{\frac{1}{2}} \cdot (v_k - u_k) \right\|^2 + \lambda \left\| \triangledown^2 v_k \right\|^2 \tag{8}$$

in which M is a mask over the smoke pixels to indicate the smoke area for sparse to dense interpolation, and λ is a Lagrange multiplier that controls smoothness. Following Strang [37], Garica [8] shows that this least squares problem can be equivalently expressed using the discrete cosine transform (DCT) and its inverse (IDCT)

$$\text{IDCT}(\Gamma \cdot \text{DCT}(M \cdot (u_k - v_k) + v_k)) \tag{9}$$

where Γ is a 2D "filtering tensor" defined as:

$$\Gamma_{i_1, i_2} = \left(1 + \lambda \left(\sum_{j=1}^{2} \left(2 - \cos\left(\frac{(i_j - 1)\pi}{n_j}\right)\right)\right)^2\right)^{-1} \tag{10}$$

where i_j locates the ith element along the jth dimension, and n_j is the width of that dimension. The algorithm proposed by Garica [8] fixes the smoothing parameter, λ.

We recognise that this method of interpolation ignores any correlation between vector components. Even so our results are sufficiently good that we have opted to leave such matters aside for now.

3.3   Optical Flow Energy and Optimization

The dense flow estimated so far yields good results as shown in Sect. 4. It shows improvement over all tested methods. Nevertheless, qualitative evaluation shows the field exhibits some noise. We now improve the motion estimation by smoothing it using variational optical flow energy, which requires we solve:

$$E(\mathbf{v}) = \int_{\Omega} \underbrace{\phi(\|f_1(\mathbf{x} + \mathbf{v}) - f_2(\mathbf{x})\|^2)}_{Brightness\ Constancy} + \alpha \underbrace{\phi(\|\nabla f_1(\mathbf{x} + \mathbf{v}) - \nabla f_2(\mathbf{x})\|^2)}_{Gradient\ Constancy} d\mathbf{x}$$

$$+ \gamma \int_{\Omega} \underbrace{\phi(\|\nabla v_1\|^2 + \|\nabla v_2\|^2)}_{Smoothness\ Constraint} d\mathbf{x} \tag{11}$$

where f_* denotes the input images and $\mathbf{v}$ represents the smoothed flow field in between; $\nabla = (\partial_{xx}, \partial_{yy})^T$ is a spatial gradient and $\phi(s)$ penalizes the flow gradient norm. The energy function was defined as a combination of a data term (brightness constancy and gradient constancy) [38] and a smoothness term [39].

In this case, given a full-size dense initial motion field, $\mathbf{v}(\mathbf{x})$, one level energy minimization is supposed to give better performance and precision comparing to the conventional coarse-to-fine scheme when it comes to the specific scenarios with boundary overlapping or thin objects [10]. Those difficulties often occur in smoke motion estimation.

To optimize our proposed energy, we follow the same process in [10] by initializing the solution with our dense motion field from previous step and apply the fixed point iterations [38] without the coarse-to-fine scheme. The optimal flow field is obtained by solving the final linear systems using 30 iterations of the successive over relaxation method. All the parameters here are applied as same as [10].

Table 1. Quantitative measure on interpolation error. We perform Interpolation Error (IE) metric on our smoke dataset (DLT, DMT, LLT and LMT) by comparing our method (Ours and noEF) to eight other baselines *i.e.* EpicFlow (EF), Horn & Schunck (HS), Black & Anandan (BA), Class+NL (NL), LDOF, MDP and FlowNet (F_C and F_S).

IE	Ours	noEF	EF	HS	BA	NL	LDOF	MDP	F_C	F_S
DLT	**0.093**	0.102	0.150	0.153	0.156	0.153	0.257	0.155	0.153	*0.148*
DMT	**0.116**	**0.116**	0.301	*0.174*	0.260	0.297	0.449	0.259	0.312	0.317
LLT	**0.033**	0.040	0.052	0.054	0.054	0.053	0.053	*0.048*	0.050	0.053
LMT	**0.072**	0.074	0.132	*0.102*	0.103	0.106	0.113	0.111	0.106	0.115

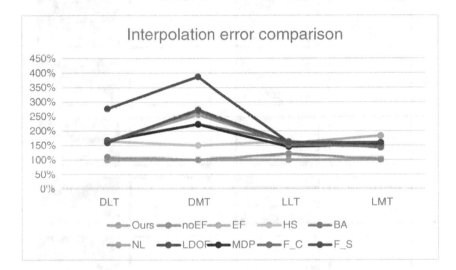

Fig. 4. Interpolation error comparison: interpolation error for 4 test cases where all baselines are highlighted using different colours. The plotting is visually connected for the same algorithm. Noted that these results are all normalized by the best result shown in Table 1

4   Evaluation

In this section, we evaluate our proposed method by comparing to eight baselines on four real-world sequences using the interpolation error (IE).

Given a ground truth benchmark [12,40], the motion estimation algorithms are commonly evaluated using quantitative metrics *i.e.* endpoint error (EE) and angular error (AE). However, it is difficult to obtain ground truth motion field from the real-world smoke objects due to their transparency and high dynamic properties. To solve this issue, Baker *et al.* [40] provide another quantitative measure *i.e.* interpolation error(IE), for evaluating optical flow methods with a lack of ground truth. They applies the flow field result to warp the first input

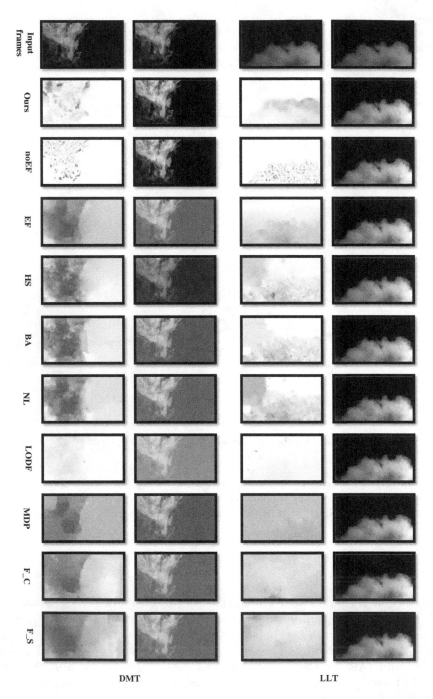

Fig. 5. Flow fields and warping results. The top row shows two sequences of input frames as DMT and LLT. Other figures are the results using all the methods in Table 1

frame to estimate the second frame; then measure the interpolation error by comparing the warping result with the real second frame.

Moreover, we consider the real-world sequence for all of our comparison because the synthetic smoke object is not able to fully reflect the dynamics of real-world motion behaviour. There is also no benchmark publicly available for providing real-world smoke data together with the ground truth motion. In this case, the interpolation error (IE) is supposed to be the only quantitative measure for such specific smoke scenario.

In the following measure, we compare two implementations (ours and noEF) of our method to both classic and the state-of-the-art motion estimation methods as follows. Horn and Schunck [11] (HS) is considered to be the first optical flow method in the field, while the Black and Anandan [14] (BA) and Classic+NL [15] provide extra robustness to motion discontinuity by using motion regularization and median filtering respectively. Furthermore, Brox et al. (LDOF) [13] address the large motion displacement issue using the additional feature matching, while Xu et al. (MDP) [39] is currently one of the state-of-the-art methods with a leading performance on the Middleburry benchmark [40]. EpicFlow [10] and FlowNet [16] further improve the performance by using the deep neural networks. Note that we do not consider the fluid motion estimators in context as the related source code is not publicly available to us.

To evaluate the performance of our algorithm, we picked four sequences that represent different real-world smoke i.e. dense but low texture smoke (DLT), dense but more texture smoke (DMT), light low texture smoke (LLT) and light but more texture smoke (LMT). The quantitative comparison is presented in Table 1 in which the first and second columns show our results with (Ours) and without (noEF) smoothing, respectively. In Fig. 4, we also visualize the IE results which are normalised by our best result. The proposed method outperforms all the tested approaches and significantly improves the motion precision in all different trials. On average, our result with smoothing is 149% better than the second best state of art result and 108% better than our result without smoothing. Figure 5 shows the flow field and warping results for DMT and LLT using all the baselines. Our methods yield visually more realistic interpolation result as well as the preservation on the smoke boundary.

Figure 6 highlights a visual comparison of the flow fields produced by three different processes. Compared with MDP (a state-of-the-art), both of our proposed methods preserve the fine smoke motion details. By applying the flow field to warp the former input frame, the estimation results for the second frame are shown in Fig. 7. Compared with the second input frame, all the methods generate visually plausible structures of smoke since the structures between two input frames are similar. However, in the aspect of smoke detail capture, our algorithm outperforms MDP.

A key observation is that DMT is the hardest case for all methods based on Fig. 4 and Table 1. We speculate this is because the DMT case contains lots of similar textures that confuse standard methods. Baselines such as EpicFlow(EF) [10] or BA [14], strongly rely on object edges. With complicated

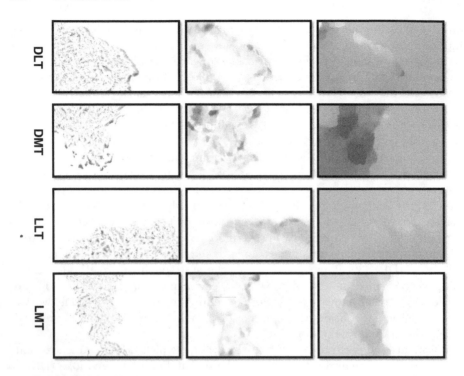

Fig. 6. Flow field comparison: Left: Proposed method without energy minimization. Middle: Proposed method with energy minimization. Right: MDP method [39]

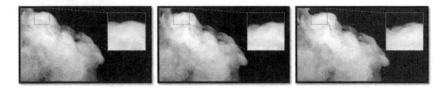

Fig. 7. Warping results comparison: Left: Ground truth frame Middle: Warping result using proposed method. Right: Warping result using MDP method [39]

and deformable textures, the edge constraint is easily violated. Our skeletal method does not rely on texture or appearance. In both quantitative results and qualitative results, we are clearly better compared to all the other methods.

5   Conclusion

We have described a smoke motion estimation algorithm for various smoke types. It makes no assumptions about brightness, local appearance, or physics; it does assume global appearance is similar between frames. It assumes global appearance can be characterised by a skeleton, and 'skeleton matching' is used to construct a sparse flow that initiates a dense flow algorithm. Results on various kinds of smoke show we outperform a gamut of both classical and state of the art dense motion estimation methods, including CNN based methods.

The most significant limitation of the method is that it is two dimensional. However, the underlying mathematics of our sparse flow algorithm is not complicated and should extend to three dimensions. A second important limit is that so far we have used only video captured under laboratory conditions; but work exists to segment smoke and other dynamic phenomena from real world scenes (*e.g.* [30]). Finally, we designed the motion estimation method for smoke and have not yet tested it on other highly dynamics objects such as fire.

Nonetheless, our motion estimation method is simpler and more robust than all algorithms we tested against, and it yields a model of smoke that is useful in Computer Graphics applications; an area we plan to explore.

Acknowledgement. The authors are supported by the EPSRC project OAK EP/K02339X/1. We thank the reviewers for constructive comments, and also thank H. Gong for his helpful suggestion and comments.

References

1. Gregson, J., Ihrke, I., Thuerey, N., Heidrich, W., et al.: From capture to simulation-connecting forward and inverse problems in fluids. ACM Trans. Graph. **33**, 1 (2014)
2. Okabe, M., Anjyo, K., Onai, R.: Extracting fluid from a video for efficient post-production. In: Proceedings of the Digital Production Symposium, pp. 53–58. ACM (2012)
3. Ghanem, B., Ahuja, N.: Extracting a fluid dynamic texture and the background from video. In: IEEE Conference on Computer Vision and Pattern Recognition, CVPR 2008, pp. 1–8. IEEE (2008)
4. Li, C., Pickup, D., Saunders, T., Cosker, D., Marshall, D., Hall, P., Willis, P.: Water surface modeling from a single viewpoint video. IEEE Trans. Vis. Comput. Graph. **19**, 1242–1251 (2013)
5. Lakshmanan, V., Rabin, R., DeBrunner, V.: Multiscale storm identification and forecast. Atmos. Res. **67**, 367–380 (2003)
6. Vila, D.A., Machado, L.A.T., Laurent, H., Velasco, I.: Forecast and tracking the evolution of cloud clusters (ForTraCC) using satellite infrared imagery: methodology and validation. Weather Forecast. **23**, 233–245 (2008)
7. Barbaresco, F., Monnier, B.: Rain clouds tracking with radar image processing based on morphological skeleton matching. In: 2001 International Conference on Image Processing, vol. 1, pp. 830–833, Proceedings. IEEE (2001)
8. Garcia, D.: Robust smoothing of gridded data in one and higher dimensions with missing values. Comput. Stat. Data Anal. **54**, 1167–1178 (2010)
9. Wang, G., Garcia, D., Liu, Y., De Jeu, R., Dolman, A.J.: A three-dimensional gap filling method for large geophysical datasets: application to global satellite soil moisture observations. Environ. Model. Softw. **30**, 139–142 (2012)
10. Revaud, J., Weinzaepfel, P., Harchaoui, Z., Schmid, C.: EpicFlow: edge-preserving interpolation of correspondences for optical flow. In: Proceedings of the IEEE Conference on Computer Vision and Pattern Recognition, pp. 1164–1172 (2015)

11. Horn, B.K., Schunck, B.G.: Determining optical flow. In: 1981 Technical Symposium East, International Society for Optics and Photonics, pp. 319–331 (1981)
12. Butler, D.J., Wulff, J., Stanley, G.B., Black, M.J.: A naturalistic open source movie for optical flow evaluation. In: Fitzgibbon, A., Lazebnik, S., Perona, P., Sato, Y., Schmid, C. (eds.) ECCV 2012. LNCS, vol. 7577, pp. 611–625. Springer, Heidelberg (2012). doi:10.1007/978-3-642-33783-3_44
13. Brox, T., Malik, J.: Large displacement optical flow: descriptor matching in variational motion estimation. IEEE Trans. Pattern Anal. Mach. Intell. **33**, 500–513 (2011)
14. Black, M.J., Anandan, P.: The robust estimation of multiple motions: parametric and piecewise-smooth flow fields. Comput. Vis. Image Underst. **63**, 75–104 (1996)
15. Sun, D., Roth, S., Black, M.J.: Secrets of optical flow estimation and their principles. In: 2010 IEEE Conference on Computer Vision and Pattern Recognition (CVPR), pp. 2432–2439. IEEE (2010)
16. Dosovitskiy, A., Fischer, P., Ilg, E., Hausser, P., Hazirbas, C., Golkov, V., van der Smagt, P., Cremers, D., Brox, T.: FlowNet: learning optical flow with convolutional networks. In: Proceedings of the IEEE International Conference on Computer Vision, pp. 2758–2766 (2015)
17. Mémin, E., Pérez, P.: Fluid motion recovery by coupling dense and parametric vector fields. In: The Proceedings of the Seventh IEEE International Conference on Computer Vision, vol. 1, pp. 620–625. IEEE (1999)
18. Auroux, D., Fehrenbach, J.: Identification of velocity fields for geophysical fluids from a sequence of images. Exp. Fluids **50**, 313–328 (2011)
19. Corpetti, T., Mémin, E., Pérez, P.: Estimating fluid optical flow. In: 15th International Conference on Pattern Recognition, vol. 3, pp. 1033–1036, Proceedings. IEEE (2000)
20. Corpetti, T., Mémin, É., Pérez, P.: Dense estimation of fluid flows. IEEE Trans. Pattern Anal. Mach. Intell. **24**, 365–380 (2002)
21. Corpetti, T., Heitz, D., Arroyo, G., Memin, E., Santa-Cruz, A.: Fluid experimental flow estimation based on an optical-flow scheme. Exp. Fluids **40**, 80–97 (2006)
22. Anumolu, L., Ahmed, I.: Simulation of smoke in OpenFOAM framework (2012)
23. Li, F., Xu, L., Guyenne, P., Yu, J.: Recovering fluid-type motions using navier-stokes potential flow. In: Computer Vision and Pattern Recognition (CVPR), 2010 IEEE Conference on, IEEE 2448–2455(2010)
24. Doshi, A., Bors, A.G.: Robust processing of optical flow of fluids. IEEE Trans. Image Process. **19**, 2332–2344 (2010)
25. Steinhoff, J., Underhill, D.: Modification of the Euler equations for vorticity confinement: application to the computation of interacting vortex rings. Phys. Fluids **6**, 2738–2744 (1994)
26. Sakaino, H.: Fluid motion estimation method based on physical properties of waves. In: IEEE Conference on Computer Vision and Pattern Recognition, CVPR 2008, pp. 1–8. IEEE (2008)
27. Ji, Y., Ye, J., Yu, J.: Reconstructing gas flows using light-path approximation. In: 2013 IEEE Conference on Computer Vision and Pattern Recognition (CVPR), pp. 2507–2514. IEEE (2013)
28. Xue, T., Rubinstein, M., Wadhwa, N., Levin, A., Durand, F., Freeman, W.T.: Refraction wiggles for measuring fluid depth and velocity from video. In: Fleet, D., Pajdla, T., Schiele, B., Tuytelaars, T. (eds.) ECCV 2014. LNCS, vol. 8691, pp. 767–782. Springer, Cham (2014). doi:10.1007/978-3-319-10578-9_50

29. Ding, Y., Li, F., Ji, Y., Yu, J.: Dynamic fluid surface acquisition using a camera array. In: 2011 IEEE International Conference on Computer Vision (ICCV), pp. 2478–2485. IEEE (2011)
30. Teney, D., Brown, M., Kit, D., Hall, P.: Learning similarity metrics for dynamic scene segmentation. In: Proceedings of the IEEE Conference on Computer Vision and Pattern Recognition, pp. 2084–2093 (2015)
31. Chen, J., Zhao, G., Salo, M., Rahtu, E., Pietikainen, M.: Automatic dynamic texture segmentation using local descriptors and optical flow. IEEE Trans. Image Process. **22**, 326–339 (2013)
32. Haindl, M., Mikeš, S.: Unsupervised dynamic textures segmentation. In: Wilson, R., Hancock, E., Bors, A., Smith, W. (eds.) CAIP 2013. LNCS, vol. 8047, pp. 433–440. Springer, Heidelberg (2013). doi:10.1007/978-3-642-40261-6_52
33. Chan, A.B., Vasconcelos, N.: Variational layered dynamic textures. In: IEEE Conference on Computer Vision and Pattern Recognition, pp. 1062–1069, CVPR 2009. IEEE (2009)
34. Strahler, A.N.: Quantitative analysis of watershed geomorphology. EOS Trans. Am. Geophys. Union **38**, 913–920 (1957)
35. Zhang, Z.: Iterative point matching for registration of free-form curves and surfaces. Int. J. Comput. Vis. **13**, 119–152 (1994)
36. Tomasi, C., Manduchi, R.: Bilateral filtering for gray and color images. In: Sixth International Conference on Computer Vision, pp. 839–846. IEEE (1998)
37. Strang, G.: The discrete cosine transform. SIAM Rev. **41**, 135–147 (1999)
38. Brox, T., Bruhn, A., Papenberg, N., Weickert, J.: High accuracy optical flow estimation based on a theory for warping. In: Pajdla, T., Matas, J. (eds.) ECCV 2004. LNCS, vol. 3024, pp. 25–36. Springer, Heidelberg (2004). doi:10.1007/978-3-540-24673-2_3
39. Xu, L., Jia, J., Matsushita, Y.: Motion detail preserving optical flow estimation. IEEE Trans. Pattern Anal. Mach. Intell. **34**, 1744–1757 (2012)
40. Baker, S., Scharstein, D., Lewis, J., Roth, S., Black, M.J., Szeliski, R.: A database and evaluation methodology for optical flow. Int. J. Comput. Vis. **92**, 1–31 (2011)

Data Association Based Multi-target Tracking Using a Joint Formulation

Jun Xiang[1,2], Jianhua Hou[2], Changxin Gao[1], and Nong Sang[1(✉)]

[1] National Key Laboratory of Science and Technology on Multispectral Information Processing, School of Automation, Huazhong University of Science and Technology, Wuhan, China
{junxiang,cgao,nsang}@hust.edu.cn
[2] Hubei Key Laboratory of Intelligent Wireless Communications, South-Central University for Nationalities, Wuhan, China
hou8781@126.com

Abstract. We revisit the classical conditional random filed based tracking-by-detection framework for multi-target tracking, in which function factors associating pairs of short tracklets in a long term are modeled to produce final tracks. Unlike most previous approaches which only focus on modeling feature difference for distinguishing pairs of targets, we propose to directly model the joint formulation of pairs of tracklets for association in the CRF framework. To this end, we use a Hough Forest (HF) based learning framework to effectively learn a discriminative codebook of features among tracklets by utilizing appearance and motion cues stored in the leaf nodes. Given the learned codebook, the joint formulation of tracklet pairs can be directly modeled in a nonparametric manner by defining a sharing and excluding matrix. Then all of the statistics required in CRF inference can be directly estimated. Extensive experiments have been conducted on several public datasets, and the performance is comparable to the state of the art.

1 Introduction

Multi-target tracking is a fundamental task in computer vision with a wide range of applications, including visual surveillance, human behavior analysis and video retrieval. Although many effective approaches have been proposed, robust tracking remains a challenging problem due to poor image quality, high motion complexity, large appearance variation, frequent occlusions, etc.

Recently, tracking-by-detection approaches have become increasingly popular thanks to the significant improvements in object detection techniques [1–4]. In those approaches (also called data association based method), detection responses generated by a pre-trained detector are associated to different unique IDs corresponding to different targets. Such association problem can usually be solved in a Maximum-A-Posterior (MAP) estimation framework or, equivalently, an energy minimization framework using a carefully designed cost function.

Affinity model design, as one of the fundamental problems of data association is critical for tracking. An effective affinity model should provide good generalization to be robust to intra-target variation, yet remaining discriminative for

© Springer International Publishing AG 2017
S.-H. Lai et al. (Eds.): ACCV 2016, Part IV, LNCS 10114, pp. 240–255, 2017.
DOI: 10.1007/978-3-319-54190-7_15

the extra-target variation. Many efforts have been made to model the feature difference to produce discriminative models. For example, within the CRF based tracking methods [5,6], potential functions are modeled by taking into account several feature difference terms, such as position distance, Bhattacharyya distance of appearance and motion. Similar strategies can be found in the cost-flow network based tracking framework which builds linking cost between detections depending on several feature difference.

As stated in [7], the task to model the feature difference is equivalent to first projecting all 2D points on a 1D line and then performing association or, equivalently classification in 1D. While such projection can capture the major discriminative information, it may reduce the "separability". Therefore, data association framework may be limited by discarding the discriminative information.

Another aspect of tracking approach is the problem of model parameter learning. A straightforward formulation is to directly model the feature difference distribution as a Gaussian [5,6,8]. Given labeled data, one supervised way to learning is to select the parameters such that the training data has the highest likelihood under the model. However, for tracking model with complex structure, exact maximum likelihood training is intractable.

In this paper we propose to model the joint information of tracklet pairs for the CRF based tracking problem under a Hough Forest learning framework. The CRF factors linking two tracklets is not depending on feature differences, but on joint feature configurations. There are at least three benefits of applying this framework to the multi-target tracking problem: (1) By modeling the joint formulation, more discriminative pair-wise models can be designed for differentiating targets without losing "separability". (2) The intractable CRF parameter learning process can be avoided because all of the joint feature distributions required in CRF inference can be directly estimated in a nonparametric manner. (3) Part based learning strategy in HF makes our tracking method robust to occlusions, since each part of target during CRF inference is contributing independently for association, which is still reliable when the target is partially occluded.

In summary, this paper makes the following contributions:

- A joint formulation of CRF framework in terms of factors between pair of tracklets is defined. Unlike traditional feature difference based modeling scheme, the joint model produce more discriminative separability in the original feature space.
- A novel Hough Forest learning scheme is designed to learn the joint model from training data by effectively combining appearance and motion information. The complex CRF parameter learning step is avoided because all of the joint statistics required in CRF inference can be estimated in a nonparametric manner by utilizing training sets collected in the leaf of forest. Since the training and matching against the forests codebook are very fast, such CRF learning and inference are very efficient.

The rest of the paper is organized as follow. Section 2 presents related work. Section 3 formulates the CRF framework. Potential functions of CRF are detailed in Sects. 4 and 5. Section 6 describes the optimization step. Finally, experimental results are presented in Sect. 7.

2   Related Work

Most current approaches for multi-target tracking are based on the tracking-by-detection strategy [5,6,8–10]. Given noisy observations, an important issue of these approaches is to design more accurate association models to handle occlusions, false alarms, missed detections, and to produce final trajectories for each target.

To reduce the computation, Huang et al. [8] developed a hierarchical association framework, in which low level reliable tracklets, i.e., a set of short yet confident tracks, are first generated. The resulting tracklets are then fed into a MAP association problem solved by the Hungarian algorithm at the middle level, and are further refined at a higher level to model scene exits and occluders. Following this hierarchical method, Kuo et al. [11] proposed an algorithm for online learning a discriminative appearance model to resolve ambiguities between the different targets. Other hierarchical methods are also considered in [12,13].

More recently, CRF based tracking methods [5,6,14,15] have been proposed to better distinguish difficult pairs of targets in crowded scenarios by considering long term dependency among the detections/tracklets. Under the CRF framework, the assumption that associations between detection/tracklet pairs are independent is relaxed, thus the dependencies among local associations can be exploited to improve the tracking performance. In [14], local associations, including motion dependency and occlusion dependency between two tracklets, are considered. The affinities and dependencies are represented by unary and pairwise energy terms respectively. Different from [5,14] focused on better distinction between difficult pairs of targets by considering discriminative features. The global descriptors for distinguishing different targets, and pairwise descriptors for differentiating difficult pairs, are incorporated into an online learned CRF framework. Instead of modeling the dependencies at tracklet level, Heili et al. [6] performed association at detection level. Specifically, the association of detection pairs is modeled not only based on a similarity measure but on a dissimilarity measure as well, and the model parameters are learned in an unsupervised data-driven fashion.

Another group of approach formulated multi-target tracking as a min-cost network optimization problem [16–18] where the optimal flow in a connected graph of detections encodes the selected track. While earlier min-cost network flow optimization methods have used linear programming, recently proposed solutions to the min-cost flow optimization include push-relabel methods [16], successive shortest paths [17,18], and dynamic programming [18].

As mentioned in Sect. 1, existing approaches try to distinguish pairs of target by modeling difference in the feature space that may reduce the separability.

And the model parameter learning process is often intractable for multi-target tracking in practice.

Differently, we alternatively propose to model the CRF based tracking method in a joint formulation. Specifically, our CRF formulation follows [6], in which pairwise factors linking pairs of tracklets and their hidden labels are modeled under two hypotheses. And a higher order potentials is defined in terms of label costs. The novel aspect of our framework is that the factor functions are modeled in a joint formulation. To this end, we introduce a Hough Forest learning method to obtain a discriminative codebook among tracklets. Then a sharing matrix indicating the degree of sharing between different tracklets, and an excluding matrix indicating the degree of excluding between different tracklets, can be used to model the factors in terms of joint feature distributions under two hypotheses. Another different trait of our method is that all of the statistics required in CRF inference can be directly estimated utilizing the tracklet-label configurations collected in the leaf of HF. Furthermore, part based HF framework could handle the occlusions because each part of target is exploited independently to evaluate the CRF inference, which is still reliable when the target is partially occluded.

It is noted that our work is motivated by the application of HF in [4,19]. Leaf nodes of HF collect the information about the locations and sizes of object bounding boxes in training images, and this information is used to predict a spatial distribution of bounding boxes in a test image. However, in this paper we instead use appearance, motion, and position information to predict the joint feature distributions of pairs of tracklet.

3   Problem Formulation

To reduce the labeling computation, the graphical model of our CRF is defined at the tracklet level. Given detection responses of a video scene, short but reliable tracks $\mathcal{T} = \{T_i\}$ are first generated by linking detection responses in consecutive frames using a rather conservative strategy [8]. Then we seek for the optimal label field $\{l_i\}_{i=1:N_T}$, where $\{l_i\}$ denotes the label of tracklet $\{T_i\}$, so that tracklets within the same track should be assigned the same label. We limit the number of tracklet pairs (T_i, T_j) by imposing constrain that the gap between any linkable pair of tracklets should satisfying:

$$\Omega = \{(i,j)|0 < t_j^s - t_i^e < T_{thr}\} \tag{1}$$

where T_{thr} is the maximal gap threshold between any linkable pairs ($T_{thr}=8$ in our implementation), t_j^s and t_i^e represent the beginning and ending of T_j and T_i respectively. The illustration of graphical model and factor graph is shown in Fig. 1, We model the posterior probability of the label field given all the observations as follows:

$$p(L|T) = \frac{1}{Z(T)}\Psi_{pair}(L,T)\Psi_L(L)$$
$$\propto \prod_{(i,j)\in\Omega} \varphi_{pair}(l_i, l_j, T_i, T_j) \cdot \Psi_L(L) \tag{2}$$

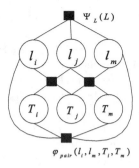

Fig. 1. Factor graph illustration of our conditional random field model.

Here $\Psi_L(L)$ acts as the regularizer penalizing complex solutions since we do not know in advance the number of targets N. $\varphi_{pair}(l_i, l_j, T_i, T_j)$ denotes the pairwise energy function which is constructed considering several terms jointly for pairs of targets in the feature space.

3.1   Factor Modeling

For each tracklet pair we introduce a factor term $\varphi_{pir}(\cdot)$, and a feature function $f(T_i, T_j)$ is defined to compute a joint measure between pair of tracklets in terms of appearance and motion. Then the corresponding CRF pairwise factor is defined under two different hypotheses to account the pairwise feature similarity as well as dissimilarity:

$$\varphi_{pir}(l_i, l_j, T_i, T_j) \overset{\Delta}{=} p(f(T_i, T_j)|H(l_i, l_j)) \tag{3}$$

And the two different hypotheses correspond to whether the labels of pair-wise tracklets are the same or not:

$$H(l_i, l_j) = \begin{cases} H_0 & if\ l_i \neq l_j \\ H_1 & if\ l_i = l_j \end{cases} \tag{4}$$

3.2   Equivalent Energy Minimization

Similar to [6], the maximization of the probability in Eq. 2 can be conducted by minimizing the following energy:

$$U(L) = (\sum_{(i,j)\in\Omega} \beta_{ij}\delta(l_i - l_j)) + \Lambda(L) \tag{5}$$

where $\delta(\cdot)$ denotes the Kronecker function ($\delta(a) = 1$ if $a = 0$, $\delta(a) = 0$ otherwise), and the Potts coefficient for each pairwise factor is defined as:

$$\beta_{ij} = \log[\frac{p(f(T_i, T_j)|H_0)}{p(f(T_i, T_j)|H_1)}] \tag{6}$$

Here the term $\Lambda(L) = -\log \Psi_L(L)$ is the negative logarithm of the label prior distribution.

The intuition behind Eq. 5 is that the penalty is taken only for associating tracklet pairs within the same track, i.e., $l_i = l_j$. To be clear, the Potts coefficient is defined by the log-likelihood ratio of the joint distribution of a pair under the two hypotheses, which can be seen as a cost for associating pair. That means such labeling process inclines towards to associate the pair of tracklets when the more negative coefficient will be (i,e. $p(f(T_i, T_j)|H_0) < p(f(T_i, T_j)|H_1)$), otherwise the more positive this coefficient is (i,e. $p(f(T_i, T_j)|H_0) > p(f(T_i, T_j)|H_1)$), the more likely the pair of tracklets should not be associated.

In the next section, the specific factor models will be defined and illustrated.

4   Potts Coefficient Factor Modeling: In a Joint Formulation

So far we have described the formulation of our model. In this Section, we specify more precisely the Potts coefficient factor modeling process. In practice, we employ the Hough Forest to capture intrinsic properties of trajectory by building a codebook among tracklets where codebook entries record tracklet-label posterior and their configurations' information. Then we use these evidences to estimate the two joint distributions in a nonparametric manner by defining a sharing and excluding matrix over the tracklet codebook.

4.1   Training the Multi-target Codebook

For training the multi-tracklet codebook, we modified the multi-class Hough Forests learning framework proposed by [4,19]. We choose features in terms of position, motion and appearance. For Hough Forests, each tree h is constructed based on the set of patches $\{S_i = (I_i, c_i, d_i, m_i, t_i, z_i)\}$ which are densely sampled from the detections in the set of reliable tracklets $\mathcal{T} = \{T_i\}$. Here, I_i, c_i , d_i, m_i and t_i respectively denotes the appearance, tracklet label, offset, motion and time index of the patch. And finally z_i is the centroid position of the detection where the patch is sampled from. Note that the m_i, t_i and z_i from the same detection have the same values. And patches sampled from detections within the same tracklet are assigned the same class label $c_i \in [1, |\mathcal{T}|]$. To deal with occlusion and false alarm, we sample the patches from background as negatives associated with class label $c_i = 0$. For a background patch, d_i, m_i, t_i and z_i are undefined.

The main task of HF is to cluster training samples $\{S_i\}$ into a codebook $\mathcal{L}$, and samples collected within the same leaf share certain properties. To this end, starting from the root with all samples, according to certain criteria, a binary test is assigned recursively to each node n that optimally splits samples S^n of that node into two sets S^n_{left} and S^n_{right} . The optimal binary tests are selected to minimize either offset or class uncertainties [4]. We extend the criteria by adding motion and position uncertainties which are formulated as:

$$U^m(S) = \sum_{j,c_j \neq 0}^{|S|} ||m_j - \bar{m}||_2^2 \tag{7}$$

$$U^z(S) = \sum_{j,c_j \neq 0}^{|S|} ||z_j' - \bar{z}'||_2^2 \tag{8}$$

where $\bar{m}$ is the mean vector of motion $\bar{m} = \frac{1}{|S|}\sum_j m_j$. z_j' is the prediction position of S_j at the mean time of samples, where $z_j' = z_j + m_j \cdot (\bar{t} - t_j)$, $\bar{t} = \frac{1}{|S|}\sum_j t_j$ and $\bar{z}'$ is the mean position of samples and $\bar{z}' = \frac{1}{|S|}\sum_j z_j'$. Equation 7 enforces that samples collected in the same node should be similar in motions, while Eq. 8 favors samples sharing their image locations. Optimizing the class, offset, motion and position uncertainties indeed prefers samples at each leaf node to satisfy both sharing properties.

4.2   Measuring Sharing and Excluding

By optimizing the class, offset, motion and position uncertainties, we obtain the discriminative mutil-tracklet codebook $\mathcal{L}$ trained over all tracklets. And in leaf nodes, HF records tracklet-label posterior of samples $p_L(c_i)$ (detailed later). Hence, we can obtain a sharing matrix $Q : (|\mathcal{T}| + 1) \times (|\mathcal{T}| + 1)$ ($|\mathcal{T}|$ number of tracklet class and one number of background class), which indicates the degree of sharing between different tracklets. An element $Q(c_i, c_j)$ of the sharing matrix can be calculated by

$$Q(c_i, c_j) = \frac{1}{\varsigma} \sum_{L=1}^{|L|} \sum_{g=1}^{|p_L(c_i)|} \sum_{e=1}^{|p_L(c_j)|} p_L(c_j) \cdot \theta(d_{c_i,g}^L, d_{c_j,e}^L) \cdot \vartheta(z_{c_i,g}^L, z_{c_j,e}^L) \tag{9}$$

where, $\varsigma = \sum_j^{|\mathcal{T}|+1} Q(c_i, c_j)$ is a normalization factor and $\theta(d_{c_i,g}^L, d_{c_j,e}^L)$ denotes the threshold of offset function, which is one if $\left\| d_{c_i,g}^L - d_{c_j,e}^L \right\|$ is smaller than a threshold (set to 10 pixels in our implementation). Similarly $\vartheta(z_{c_i,g}^L, z_{c_j,e}^L)$ equals to one if $\left\| z_{c_i,g}^L - z_{c_j,e}^L \right\|_2$ is smaller than a threshold (set to 35 pixels in experiments). Hence, two samples of different tracklet-labels are considered sharing, only when they are clustered into the same leaf and with similar offset and position.

We can also obtain an excluding matrix $Q' : (|\mathcal{T}| + 1) \times (|\mathcal{T}| + 1)$ which indicates the degree of excluding between different tracklets. An element $Q'(c_i, c_j)$ of the excluding matrix is calculated by

$$Q'(c_i, c_j) = \frac{1}{\varsigma'}(Q(c_i, c_i) - Q(c_i, c_j)) \tag{10}$$

where $\varsigma' = \sum\limits_{j}^{|\mathcal{T}|+1} Q'(c_i, c_j)$ is a normalization factor. The intuition behind $Q'(c_i, c_j)$ is that the greater occurrence difference of two tracklets falling in the same leaf nodes, the more repulsion between the two tracklets.

4.3   Estimation Potts Coefficient Factor

For every leaf node L of HF, we have obtained tracklet-label posterior of training samples $\{p_L(c_i)\}$ after HF learning. Specifically, we compute tracklet class label count $\phi_L = \{\eta(c_i) : c_i = 0, \cdots |\mathcal{T}|\}$, where $\eta(c_i)$ is the number of samples belonging to class c_i that reached L. Normalizing $\eta(c_i)$ over the total number of samples in L gives an estimate of the class posterior. As mentioned above, that during HF training step, a class label $c_i \in [1, |\mathcal{T}|]$ is assigned to each tracklet T_i, and we add background label as a especial case defined by $c_i = 0$. To estimate the joint distributions under two hypotheses, we first define two temporal relationship between tracklet classes: conflicting relationship means that no time overlapping exists for two tracklets ; and uncertain relationships means that time overlapping exists. And the temporal relationship between two classes c_i and c_j is denoted as (c_i, c_j).

In the following, we explain how to compute the Potts coefficient factor in our CRF inference.

Given two tracklets (T_i, T_j), we first sampled M patch pairs from them, denoted as $\{(x_i^m, x_j^m)\}$. Each pair is "dropped" down through H decision trees of HF. Suppose x_i^m and x_j^m reach leaf nodes L_i^h and L_j^h in tree $h = 1, \cdots \mathcal{H}$. Then, we use the tracklet class posterior and configurations stored in L_i^h and L_j^h, as well as the learned sharing matrix and excluding matrix, to compute the two joint distributions. Specifically, we directly proposed the joint distribution under H_0 of the tree h as:

$$
\begin{aligned}
p_h(x_i^m, x_j^m | H_0) = &\sum_{(c_u, c_v) = conf} p_{L_i^h}(c_u) \cdot p_{L_j^h}(c_v) \cdot Q'(c_u, c_v) \\
+ &\sum_{(c'_u, c'_v) = unc} p_{L_i^h}(c'_u) \cdot p_{L_i^h}(c'_v) \cdot Q'(c'_u, c'_v)
\end{aligned}
\tag{11}
$$

where c_u and c_v denote conflicting tracklet class labels respectively in L_i^h and L_j^h, i.e., $c_u \in \{C_{L_i^h}\}, c_v \in \{C_{L_j^h}\}$, with temporal relationship $(c_u, c_v) = $ "conf". Where $C_{L_i^h}$ and $C_{L_j^h}$ are the class label sets in L_i^h and L_j^h. Similarly, c'_u and c'_v denote uncertain class label pairs in L_i^h and L_j^h.

And the joint distribution under H_1 of the tree h is formulated as:

$$
\begin{aligned}
p_h(x_i^m, x_j^m | H_1) = &\sum_{c_u = c_v} p_{L_i^h}(c_u) \cdot p_{L_j^h}(c_v) \cdot Q(c_u, c_v) \\
+ &\sum_{(c'_u, c'_v) = unc} p_{L_i^h}(c'_u) \cdot p_{L_j^h}(c'_v) \cdot Q(c'_u, c'_v)
\end{aligned}
\tag{12}
$$

where c_u and c_v denote the same class labels respectively in L_i^h and L_j^h, and c'_u and c'_v denote uncertain class labels in L_i^h and L_j^h. The straightforward understanding of Eqs. 11 and 12 is that all available causes enumerated under different hypotheses are contributing to the corresponding distribution.

Finally, the Potts coefficient factor during CRF inference for T_i and T_j is approximated by averaging all of the sampled patch pairs within the two tracklets over the trees:

$$\beta_{ij} = \log\left[\frac{\sum\limits_{m=1}^{M}\sum\limits_{h=1}^{\mathcal{H}} P_h(x_i^m, x_j^m | H_0)}{\sum\limits_{m=1}^{M}\sum\limits_{h=1}^{\mathcal{H}} P_h(x_i^m, x_j^m | H_1)}\right] \tag{13}$$

5   Label Cost

For modeling the label cost energy function of Eq. 5, we employ the technique proposed by [6] to impose penalty of model complexity. This technique can avoid having too many labels and obtain coherent tracks from the scene viewpoint.

6   Optimization

It can be proved that the energy expression in Eq. 5 does not follow the submodularity principle and hence can not be solved using global graph cut optimization techniques. Instead, we directly introduce a heuristic algorithm to find a good labeling solution. We apply the sliding window based block Iterated Conditional Modes (ICM) optimization method for inference. For each window span t_w, let us define the set of new tracklets (initialized as new tracks) after time t as T_t' in which each pair of tracklets has the conflicting relationship ($T_i \in T_t', T_j \in T_t', T_i$ and T_j is time overlapping). And we also define current labeled tracklets before t as T_t. N^t denotes the set of active labels, that have already been assigned in T_t . For ICM inference, each of the current new tracklets can potentially take a label among the N^t active labels or take a new label. In this way tracks before t can be extended(assign labels to new tracklets) or stopped.

The inference procedure is similar to [6] and Fig. 2 illustrates how the t_w window is used for labeling. The dotted lines show the links that will be used to compute the assignment costs (including coefficient factors as well as label costs) for inference.

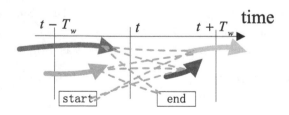

Fig. 2. Bolock ICM at time t

7   Experiments

We evaluate our approach on five challenging public datasets: TUD data [20], PETS [21], ETH [22], and Town Centre [23]. and MOTChallenge [24]. When comparing with CRF based methods [5,6] (high relation to our method) on date set TUD, PETS, ETH, and Town Centre, the evaluation software available at http://iris.usc.edu/people/yangbo/data/EvaluationTool.zip is used such as [5,6]. And for comparing with more recent methods, we submit our tracker to https://motchallenge.net and present the performance on 2DMOT2015. We show quantitative comparisons with the state-of-the-art methods, as well as visualized results of our approach.

7.1   Implementation Details

For fair comparison, when testing on TUD, ETH and PET, this paper directly uses the same detection results and ground truth as used in [5,25]. When testing on Town Centre, we use detection results and ground truth files provided by the official. A pre-processing step is taken to filter detections so as to remove those of unrealistic size. And final results are obtained by using a post step which uses tracking information to interpolate tracks for missed detections and removes short ones for false alarms.

All detections are normalized into 24×58 for HF training and testing. About 60% of detections within a tracklet are employed for training where 50 patches of size 24×58 are randomly selected for each detection. 10 decision trees are trained where 2000 binary tests are generated randomly for each node and a leaf node is formed when the tree depth 20 is reached, or less than 10 patches for only one training class are left. We use appearance features as the same as [4]. For motion feature of detections within a tracklet, we employ the point based Median Flow (MF) tracking method [26] to extract reliable point pairs and obtain sparse motion flow histograms. The motion flows are quantified into 18 orientation bins. We count the average angle of motion flows falling in the largest orientation bin, and the average motion displacement as the final motion vector of detections.

7.2   Effectiveness of the Sharing Matrix

In this section we explain the effectiveness of the sharing matrix Q. As mentioned above that sharing matrix indicates the degree of sharing between different tracklet classes. We directly obtain the association results over tracklets by clustering sharing matrices using the complete-linkage agglomerative clustering and the quantitative results are provide in Table 1. Intuitively, during the clustering experiments, a small threshold of linkage rate would possibly result in increasing of Frags, and a large threshold would be inefficient for identity distinction. We conducted several experiments and present the best performance by considering the trade-off between Frag and IDS. From the results we find that the sharing matrix indeed indicate the degree of similarity between targets.

Table 1. Effectiveness of the sharing matrix

Dataset	Recall	Prec	MT	ML	Frag	IDS
TUD	0.74	0.95	0.7	0	5	2
PET	0.84	0.94	0.75	0	12	3
ETH	0.71	0.95	0.45	0.18	44	11

Table 2. Comparison of results on TUD dataset

Method	Rec	Prec	MT	ML	MOTA	MOTP	IDS	Frag
Kuo [11]	78%	98.8%	0.6	0.0	#	#	0	3
Andriyenko [20]	#	#	0.6	0.1	0.67	0.66	7	4
Kuo [9]	81%	99.5%	0.6	0.1	#	#	1	0
Yang [25]	87%	96.7%	0.7	0.0	#	#	0	1
Andriyeko [27]	#	#	#	#	0.62	0.63	4	1
Zamir [13]	81%	96%	#	#	0.78	0.63	0	#
Heili [6]	81%	94%	0.7	0.0	0.9	0.84	0	1
Ours	83%	99%	0.8	0.0	0.95	0.92	0	1

Table 3. Comparison of results on PETS 2009

Method	Rec	Prec	MT	ML	MOTA	MOTP	IDS	Frag
Kuo [11]	90%	97.9%	0.74	0.0	#	#	1	28
Andriyenko [20]	#	#	0.83	0.0	0.81	0.76	15	21
Kuo [9]	89%	99.6%	0.79	0.0	#	#	1	23
Yang [25]	93%	95.3%	0.9	0.0	#	#	1	13
Andriyeko [28]	#	#	#	#	0.89	0.56	10	8
Breitenstein [29]	#	#	#	#	0.8	0.56	#	#
Zamir [13]	96%	94.0%	#	#	0.9	0.69	8	#
Heili [6]	87%	94.0%	0.75	0.0	0.89	0.66	0	3
Ours	82%	99.0%	0.63	0.0	0.80	0.58	0	7

Due to the simply clustering algorithm the matrix could not provide very good tracking results.

7.3   Comparison with the State of the Art

Tables 2, 3, 4 and 5 show the comparison with recent state of the art algorithms for the different datasets. Figure 3 shows some visual results of our tracker.

We first evaluate our approach on TUD-Stadtmitte sequence. Table 2 indicates that we outperform all approaches in terms of Recall, MT, MOTA and MOTP. We obtain 0 IDS as other four methods and the second best Frag.

Table 4. Comparison of tracking results on ETH

Method	Rec	Prec	MT	ML	Frag	IDS
Kuo [11]	52.0%	91.0%	25.8%	35.5%	20	25
Kuo [9]	76.8%	86.6%	58.4%	8.0%	23	11
Yang [25]	79.0%	90.4%	68.0%	7.2%	19	11
Ours	75%	92.0%	68.0%	7.2%	19	11

Table 5. Comparison results on towncenter

Method	Rec	Prec	MOTA	MOTP	Frag	IDS
Benfold [23]	0.77	0.88	0.65	0.80	#	#
Zamir [13]	#	#	0.76	0.72	#	#
Heili [6]	0.73	0.90	0.79	0.82	#	#
Ours	0.75	0.91	0.82	0.81	35	21

Table 3 shows the comparison results on PET dataset. PETS 09 S2.L1 is a video of 795 frames recorded at 7 fps, and presents a moderately crowded scene where 20 pedestrians are often crossing each others trajectories, creating inter-person occlusions. People are also often occluded by a street light in the middle of the scene, creating miss-detections. From the results, we can see that the proposed method has no significant improvement on PET dataset. This is because the detections of PET dataset are quite small relative to other dataset, and the power of part based HF learning framework may be limited.

We also evaluate our approach on the ETH data set captured by a pair of cameras on a moving stroller in busy street scenes. The stroller is mostly moving forward, but sometimes has panning motions, which makes the motion affinity between two tracklets less reliable. It is shown in Table 4 that our method is comparable to the state-of-the-art approaches.

Finally, we compare our tracking results on Town Centre. The results are summarized in Table 5. On the Town Centre sequence, the recall and precision of our detections are similar to detections provided by the authors of [23]. We outperform all method in terms of MOTA and obtain the second best MOTP (0.81 vs 0.82).

To further show more meaningful quantitative evaluation of the proposed method, we evaluate it on the recent 2DMOT2015 Benchmark. The number of test sequences in this benchmark is 11, in which 5 of them are taken by moving cameras and 6 of them are taken by static cameras. Table 6 shows our performance along with the three competitors which are belong to the class of batch processing techniques just like us. And the performance is comparable to the state of the art (excluding deep neural network based approaches), if not better.

Fig. 3. Qualitative results of our tracker on experimented sequences.

Table 6. Comparison results on 2DMOT2015

Method	MOTA	MOTP	GT	MT	ML	FP	FN	IDs
CEM [10]	19.3	70.7	721	8.5	46.5	14180	34591	813
SegTrack [30]	22.5	71.7	721	5.8	63.9	7890	39020	697
MotiCon [31]	23.1	70.9	721	4.7	52	10404	35844	1018
Ours	21.4	68.6	721	5.3	53.3	7102	40736	445

7.4   Running Time

For HF training, the most computational cost of our method comes from feature extractions for the appearance and motion model. While for CRF inference, the computation cost only depends on the number of tracklet (the graph construction), feature extractions (without motion), and ICM inference. In fact, forests are very efficient at runtime, since matching a sample against a tree is logarithmic in the number of leaves. We believe the running time of our algorithm is at least comparable to the state of the art. To be specific, our experiments were performed on a laptop with Intel 2.8 GHz CPU and 2 GB memory, and the code is implemented in C++. Given the detection responses, for the less crowded TUD, PETS, and ETH datasets, our current implementation runs at 2.0, 1.8, and 1.0 fps, respectively, not including the cost of HF training time; for Town Center the speed is 0.27 fps.

8   Conclusions

In this paper, we have presented a joint formulation based CRF framework for multi-target tracking, in which the pairwise factors of CRF are modeled by a discriminative codebook obtained by a Hough Forest learning framework. In this sense, we try to overcome the limitation of traditional feature difference based algorithm in separability between targets. Benefiting from the special expressions of Hough Forest, our approach also avoids the tedious parameter learning step, and all of the statistics required in CRF inference can be directly estimated with a nonparametric manner. Experimental results support the effectiveness and validity of the method. In the feature to promote the tracking performance, we will plan to integrate more explicit occlusion reasoning and missed detections recovering model to the energy function. Then more discriminative features with more physical constraints should be employed and more effective model design techniques and optimization strategy should be considered.

Acknowledgement. This work was supported by the National Natural Science Foundation of China under Grant 61271328, 61671484 and 61401170.

References

1. Viola, P., Jones, M.: Rapid object detection using a boosted cascade of simple features. In: Proceedings of the 2001 IEEE Computer Society Conference on Computer Vision and Pattern Recognition, CVPR 2001, vol. 1, pp. I–511. IEEE (2001)
2. Dalal, N., Triggs, B.: Histograms of oriented gradients for human detection. In: IEEE Computer Society Conference on Computer Vision and Pattern Recognition, CVPR 2005, vol. 1, pp. 886–893. IEEE (2005)
3. Felzenszwalb, P.F., Girshick, R.B., McAllester, D., Ramanan, D.: Object detection with discriminatively trained part-based models. IEEE Trans. Pattern Anal. Mach. Intell. **32**, 1627–1645 (2010)
4. Gall, J., Lempitsky, V.: Class-specific hough forests for object detection. In: Proceedings of the IEEE Conference Computer Vision and Pattern Recognition (2009)
5. Yang, B., Nevatia, R.: Multi-target tracking by online learning a CRF model of appearance and motion patterns. Int. J. Comput. Vis. **107**, 203–217 (2014)
6. Heili, A., López-Méndez, A., Odobez, J.M.: Exploiting long-term connectivity and visual motion in CRF-based multi-person tracking. IEEE Trans. Image Process. **23**, 3040–3056 (2014)
7. Chen, D., Cao, X., Wang, L., Wen, F., Sun, J.: Bayesian face revisited: a joint formulation. In: Fitzgibbon, A., Lazebnik, S., Perona, P., Sato, Y., Schmid, C. (eds.) ECCV 2012. LNCS, vol. 7574, pp. 566–579. Springer, Heidelberg (2012). doi:10.1007/978-3-642-33712-3_41
8. Huang, C., Wu, B., Nevatia, R.: Robust object tracking by hierarchical association of detection responses. In: Forsyth, D., Torr, P., Zisserman, A. (eds.) ECCV 2008. LNCS, vol. 5303, pp. 788–801. Springer, Heidelberg (2008). doi:10.1007/978-3-540-88688-4_58
9. Kuo, C.H., Nevatia, R.: How does person identity recognition help multi-person tracking?. In: 2011 IEEE Conference on Computer Vision and Pattern Recognition (CVPR), pp. 1217–1224. IEEE (2011)

10. Milan, A., Roth, S., Schindler, K.: Continuous energy minimization for multitarget tracking. IEEE Trans. Pattern Anal. Mach. Intell. **36**, 58–72 (2014)
11. Kuo, C.H., Huang, C., Nevatia, R.: Multi-target tracking by on-line learned discriminative appearance models. In: 2010 IEEE Conference on Computer Vision and Pattern Recognition (CVPR), pp. 685–692. IEEE (2010)
12. Bak, S., Chau, D.P., Badie, J., Corvee, E., Bremond, F., Thonnat, M.: Multi-target tracking by discriminative analysis on Riemannian manifold. In: 2012 19th IEEE International Conference on Image Processing (ICIP), pp. 1605–1608. IEEE (2012)
13. Zamir, A.R., Dehghan, A., Shah, M.: GMCP-tracker: global multi-object tracking using generalized minimum clique graphs. In: Fitzgibbon, A., Lazebnik, S., Perona, P., Sato, Y., Schmid, C. (eds.) Computer Vision – ECCV 2012. LNCS, vol. 7573, pp. 343–356. Springer, Heidelberg (2012)
14. Yang, B., Huang, C., Nevatia, R.: Learning affinities and dependencies for multi-target tracking using a CRF model. In: 2011 IEEE Conference on Computer Vision and Pattern Recognition (CVPR), pp. 1233–1240. IEEE (2011)
15. Heili, A., Chen, C., Odobez, J.: Detection-based multi-human tracking using a CRF model. In: 2011 IEEE International Conference on Computer Vision Workshops (ICCV Workshops), pp. 1673–1680. IEEE (2011)
16. Zhang, L., Li, Y., Nevatia, R.: Global data association for multi-object tracking using network flows. In: IEEE Conference on Computer Vision and Pattern Recognition, CVPR 2008, pp. 1–8. IEEE (2008)
17. Berclaz, J., Fleuret, F., Turetken, E., Fua, P.: Multiple object tracking using k-shortest paths optimization. IEEE Trans. Pattern Anal. Mach. Intell. **33**, 1806–1819 (2011)
18. Pirsiavash, H., Ramanan, D., Fowlkes, C.C.: Globally-optimal greedy algorithms for tracking a variable number of objects. In: 2011 IEEE Conference on Computer Vision and Pattern Recognition (CVPR), IEEE (2011)
19. Nima Razavi, J.G., Gool, L.V.: Scalable multi-class object detection. In: 2011 IEEE Conference on Computer Vision and Pattern Recognition (CVPR), pp. 1505–1512. IEEE (2011)
20. Andriyenko, A., Schindler, K.: Multi-target tracking by continuous energy minimization. In: 2011 IEEE Conference on Computer Vision and Pattern Recognition (CVPR), pp. 1265–1272. IEEE (2011)
21. Pets 2009 dataset. (http://www.cvg.rdg.ac.uk/PETS2009)
22. Ess, A., Leibe, B., Schindler, K., Van Gool, L.: Robust multiperson tracking from a mobile platform. IEEE Trans. Pattern Anal. Mach. Intell. **31**, 1831–1846 (2009)
23. Benfold, B., Reid, I.: Stable multi-target tracking in real-time surveillance video. In: 2011 IEEE Conference on Computer Vision and Pattern Recognition (CVPR), pp. 3457–3464. IEEE (2011)
24. Leal-Taixe, L., Milan, A., Reid, I., Roth, S., Schindler, K.: Motchallenge 2015: Towards a benchmark for multi-target tracking
25. Yang, B., Nevatia, R.: An online learned CRF model for multi-target tracking. In: 2012 IEEE Conference on Computer Vision and Pattern Recognition (CVPR), pp. 2034–2041. IEEE (2012)
26. Kalal, Z., Mikolajczyk, K., Matas, J.: Forward-backward error: automatic detection of tracking failures. In: 2010 20th International Conference on Pattern Recognition (ICPR), pp. 2756–2759. IEEE (2010)
27. Li, Y., Huang, C., Nevatia, R.: Learning to associate: hybridboosted multi-target tracker for crowded scene. In: IEEE Conference on Computer Vision and Pattern Recognition, CVPR 2009, pp. 2953–2960. IEEE (2009)

28. Andriyenko, A., Schindler, K., Roth, S.: Discrete-continuous optimization for multi-target tracking. In: 2012 IEEE Conference on Computer Vision and Pattern Recognition (CVPR), pp. 1926–1933. IEEE (2012)
29. Breitenstein, M.D., Reichlin, F., Leibe, B., Koller-Meier, E., Van Gool, L.: Online multiperson tracking-by-detection from a single, uncalibrated camera. IEEE Trans. Pattern Anal. Mach. Intell. **33**, 1820–1833 (2011)
30. Milan, A., Leal-Taixe, L., Schindler, K., Reid, I.: Joint tracking and segmentation of multiple targets. In: 2015 IEEE Conference on Computer Vision and Pattern Recognition (CVPR), pp. 5397–5406. IEEE (2015)
31. Leal Taixe, L., Fenzi, M., Kuznetsova, A., Rosenhahn, B., Savarese, S.: Learning an image-based motion context for multiple people tracking. In: 2014 IEEE Conference on Computer Vision and Pattern Recognition (CVPR), pp. 3542–3549. IEEE (2014)

Combining Texture and Shape Cues for Object Recognition with Minimal Supervision

Xingchao Peng[✉] and Kate Saenko

Computer Science Department, Boston University, Boston, USA
{xpeng,saenko}@bu.edu

Abstract. We present a novel approach to object classification and detection which requires minimal supervision and which combines visual texture cues and shape information learned from freely available unlabeled web search results. The explosion of visual data on the web can potentially make visual examples of almost any object easily accessible via web search. Previous unsupervised methods have utilized either large scale sources of texture cues from the web, or shape information from data such as crowdsourced CAD models. We propose a two-stream deep learning framework that combines these cues, with one stream learning visual texture cues from image search data, and the other stream learning rich shape information from 3D CAD models. To perform classification or detection for a novel image, the predictions of the two streams are combined using a late fusion scheme. We present experiments and visualizations for both tasks on the standard benchmark PASCAL VOC 2007 to demonstrate that texture and shape provide complementary information in our model. Our method outperforms previous web image based models, 3D CAD model based approaches, and weakly supervised models.

1   Introduction

Object classification and detection are fundamental tasks in computer vision. Previous mainstream object detectors based on hand-designed features (HOG [6]) and classifiers like linear discriminant analysis (LDA) [37] or Support Vector Machines [2,11] required training of limited parameters, and were thus sustainable with small datasets. More recent Deep Convolutional Neural Network (DCNN)-based object detectors [14,26] utilize more powerful DCNN features and yield a significant performance boost, both for image classification [16,18,19,34] and object detection [14,26]. Nevertheless, deep convolutional neural networks need lots of labeled images to train their millions of parameters. Collecting these images and annotating the objects is cumbersome and expensive. Even the most popular detection datasets provide a limited number of labeled categories, e.g., 20 classes in PASCAL VOC [9] and 200 in ImageNet [7]. Hence, a question arises: is it possible to avoid the frustrating collection and annotation process and still train an effective object classifier or detector?

To achieve this goal, researchers have proposed several recent models for object model training. [17] introduces a transfer learning method that gains an

© Springer International Publishing AG 2017
S.-H. Lai et al. (Eds.): ACCV 2016, Part IV, LNCS 10114, pp. 256–272, 2017.
DOI: 10.1007/978-3-319-54190-7_16

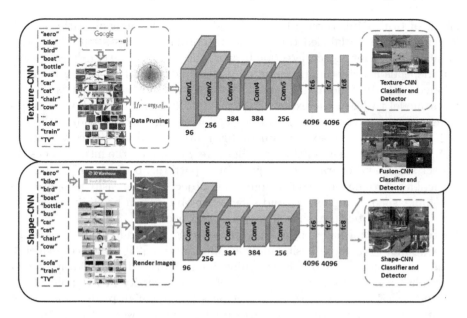

Fig. 1. Two-Stream Texture and Shape Model: We propose a framework to combine texture cues and shape information for object recognition with minimal supervision. The Texture-CNN stream is trained on images collected from on-line image search engines and the Shape-CNN stream is separately trained on images generated from domain-specific 3D CAD models. To combine the two streams, an average of the two last layers' activations is computed to create the Fusion-CNN object classifier and detector. In the test phrase, the model will forward an image patch through the two networks simultaneously and compute average fusion of the activations from the last layers. (Best viewed in color.) (Color figure online)

object detector by transferring learned object knowledge from a classifier. [28, 29] propose to train an object localization and detection model with image-level labels. However, these methods still count on per-image manual supervision. In contrast, the methods in the previous literature that assume no per-image labeling data can be categorized into two groups: (1) methods that utilize on-line search results or an existing unlabeled dataset [3, 8, 40]; (2) methods that render domain-specific synthetic images [23, 32]. For example, [3, 8] propose to learn a visual representation and object detectors from on-line images. [40] leverages a concept learner to discover visual knowledge from weakly labeled images (weak labels can be in the form of keywords or short description). On the other hand, [23, 32] proposed to generate synthetic training images from 3D CAD models. Such synthetic data is shown to be useful for augmenting small amounts of real labeled data to improve object detection, as bounding boxes can be obtained "for free" and objects can be rendered in arbitrarily many viewpoints.

While these approaches can work effectively in some cases, there are still many challenges that need to be addressed:

- **Lack of bounding boxes:** Unsupervised machine learning algorithms tackle learning where no labeled training data is provided. This setting leads to a great challenge for object detection because the performance of a detection system depends heavily on differentiated positive and negative training examples labeled with tight bounding boxes. Without such annotations, it is difficult for a model to learn the extent and shape of objects, i.e. which parts of the image correspond to the target object and which are background.
- **Missing shape or texture cues:** Prior literature uses either web search [3,8] or synthetic images [23,31,32], but rarely combines the intrinsic cues like object shape and characteristic appearance patterns, or "texture", which are critical for recognition systems. Some rigid objects can be easily recognized by its shape, such as aeroplanes and sofa; other objects can be easily recognized by their unique texture, such as leopards and bees.
- **Domain Shift:** Images from different sources have different statistics for background, texture, intensity and even illumination [24], which consequently results in domain shift problems. Unlike images taken in the wild, most photographs returned by image search engines lack diversity in viewpoint, background and shape, for image search engines follow a high-precision low-recall regime. In addition, synthetic images used in current work [23] are far from photorealism in terms of intensity contrast, background, and object texture.

In this paper, we address these shortcomings by proposing a two-stream DCNN architecture (See Fig. 1) that decomposes the input into texture and shape feature streams. The texture stream learns realistic texture cues from images downloaded from the Internet. Web images (which contain noise from backgrounds and unrelated results) are collected by searching for the names of categories in image search engines, i.e. Google Image Search. We prune the noise data and use the cleaned data to train the texture-based stream for classification or detection. The shape stream is trained exclusively on shape information from 2D images rendered from CAD models annotated with category labels. Note that images from web search also contain shape information, but due to the lack of bounding box annotations, it is not accurate enough for localization models. Synthetic images can also be generated from 3D CAD models with added texture mapping, but the result is non-photorealistic and lacks real-world variety. Therefore, synthetic images rendered by 3D CAD models can be viewed as primarily shape-oriented and the web-search images can be viewed as primarily texture-oriented. The outputs of the two streams are combined through averaging the two top layers' activations. Our method requires no tedious manual bounding box annotation of object instances and no per-image category labeling and can generate training data for almost any novel category. Table 1 shows a comparison of the amount of supervision with other methods. The only supervision in our work comes from labeling the CAD models as positive examples vs. "outliers" while downloading them, and choosing proper textures for each category while generating the synthetic images.

We evaluate our model for object classification and detection on the standard PASCAL VOC 2007 dataset, and show that the texture and shape cues can

Table 1. Comparison of supervision between our method and others. The supervision in our work only comes from labeling the CAD models and their texture when generating the synthetic images. The second column indicates whether in-domain data (VOC *train+val* set) is used in the methods. The amount of supervision used in each method increases from top to bottom.

Type	VOC training data used	Supervision
CAD supervision, Ours, [23]	NO	Labeling 3D CAD models and their texture
Webly supervision [3,8]	NO	Semantic labeling
Selected supervision [40]	NO	Per-image labeling With Text
Weak supervision [28,29]	YES	Per-image labeling
Full supervision [14]	YES	Per-instance labeling

reciprocally compensate for each other during recognition. In addition, our detector outperforms existing webly supervised detectors [8], the approach based on synthetic images only [23], and the weakly supervised learning approach of [28], despite using less manual supervision.

To summarize, this work contributes to the computer vision community in the following three aspects:

- we propose and implement a recognition framework that decomposes images into their shape and texture cues;
- we show that combining these cues improves classification and detection performance while using minimal supervision;
- we present a unified schema for learning from both web images and synthetic CAD images.

2   Related Work

Webly Supervised Learning. The explosion of visual data on the Internet provides important sources of data for vision research. However, cleaning and annotating these data is costly and inefficient. Researchers have striven to design methods that learn visual representations and semantic concepts directly from the unlabeled data. Because the detection task requires stronger supervision than classification, most previous research work [1,12,20,25] involving web images only tackles the object classification task. Some recent work [4,8] aims at discovering common sense knowledge or capturing intra-concept variance. In the work of [3,8], webly supervised object detectors are trained from image search results. We follow a similar approach as in [3] to train our texture model from web search data, but also add shape information using a CAD-based CNN.

Utilization of CAD Models. CAD models had been used by researchers since the early stages of computer vision. Recent work involving 3D CAD models focuses on pose prediction [21,30,31,33]. Other recent work applied CAD models to 2D object detection [23,32] by rendering synthetic 2D images from 3D CAD models and using them to augment the training data. The main drawback of these methods is that the rendered images are low-quality and lack real texture, which significantly hurts their performance. In contrast, we propose a two-stream architecture that adds texture information to the CAD-based shape channel. [23] explored several ways to simulate real images, by adding real-image background and textures onto the 3D models, but this requires additional human supervision to select appropriate background and texture images for each category. In this work, we propose more effective ways to simulate real data with less supervision.

Two-stream Learning. The basic aim of two-stream learning is to model two-factor variations. [35] proposed a bilinear model to separate "style" and "content" of an image. In [22], a two-stream architecture was proposed for fine-grained visual recognition, and the classifier is expressed as a product of two low-rank matrices. [13,27] utilized two-stream architectures to model the temporal interactions and aspect of features. We propose a CNN-based two-stream architecture that learns intrinsic properties of objects from disparate data sources, with one stream learning to extract texture cues from real images and the other stream learning to extract shape information from CAD models.

3   Approach

Our ultimate goal is to learn a good object classifier and object detector from the massive amount of visual data available via web search and from synthetic data. As illustrated in Fig. 1, we introduce a two-stream learning architecture to extract the texture cues and shape information simultaneously. Each stream consists of three parts: the data acquisition component, the DCNN model and the object classifier or detector. The intuition is to utilize texture-oriented images from the web to train the texture stream and correspondingly, use shape-oriented images rendered from 3D CAD models to train the shape stream.

3.1   DCNN-based Two-Stream Model

The history of two-stream learning can be traced back to over a decade ago when a "bilinear model" was proposed by [35] to separate the "style" and "content" of an image. More recent use [13,22,27] of two-stream learning is based on a similar philosophy: employ different modalities to model different intrinsic visual properties, e.g. spatial features and temporal interactions. Inspired by this idea, we propose a two-stream learning architecture, with one stream modeling real image texture cues and the other modeling 3D shape information. We demonstrate that the texture and shape cues can reciprocally compensate for each other's errors through late fusion.

For fair comparison with other baselines, within each stream, we use the eight-layer "AlexNet" architecture proposed by [19]. It contains five convolutional layers, followed by two fully connected layers ($fc6$, $fc7$). After $fc7$, another fully connected layer ($fc8$) is applied to calculate the final class predictions.

The network adopts "dropout" regularization to avoid overfitting and non-saturating neurons (*ReLU* layers) to increases the nonlinear properties of the decision function and to speed up the training process. The network is trained by stochastic gradient descent and takes raw RGB image patches of size 227×227.

The last layer (*fc8*) in each stream is represented by a softmax decision function. To combine the learned texture cues and shape information, we fuse the streams to render the final prediction as follows:

$$P(I = j | \mathbf{x}) = \frac{e^{\mathbf{x}^T \mathbf{w}_j^t}}{2 \sum_{i=1}^{N} e^{\mathbf{x}^T \mathbf{w}_i^t}} + \frac{e^{\mathbf{x}^T \mathbf{w}_j^s}}{2 \sum_{i=1}^{N} e^{\mathbf{x}^T \mathbf{w}_i^s}} \tag{1}$$

where $P(I = j | \mathbf{x})$ denotes the probability that image I belongs to category j given feature vector $\mathbf{x}$ (*fc7* feature in this case); $\mathbf{x}^T$, N, $\mathbf{w}_i^t$, $\mathbf{w}_i^s$ are the transpose of $\mathbf{x}$, the number of total categories, weight vector for category i in Texture CNN, weight vector for category i in Shape CNN, respectively.

The final probability $P(I = j | \mathbf{x})$ is used as the score for Two-Stream classifier and detector.

3.2 Texture CNN Stream

Previous work [23,39] has shown that discriminative texture information is crucial for object classification and object detection systems. The challenge is how to obtain large scale accurate texture data with the least effort and how to prune the noisy images from unrelated search results. Previous approaches [3,4,8] have tried various search engines to form the texture bank, while other research work [10,38] attempt to clean the data.

Noise Data Pruning. We assume the distribution of features of the higher CNN layers follows a multivariate normal distribution, thus we can fit the data from each class to the domain-specific Gaussian distribution as follows:

$$f_{\mathbf{x}}(x_1, x_2...x_k) = \frac{1}{\sqrt{(2\pi)^k |\sum|}} * exp(-\frac{1}{2}(\mathbf{x} - u)^T \sum{}^{-1}(\mathbf{x} - u)) \tag{2}$$

where $\mathbf{x}$ is an k-dimensional feature vector and $\sum$ is the covariance matrix.

To remove outliers, for each category j ($j \in [1, N]$, N is the category number), we start from the downloaded image set S_j with noise data and an empty set T_j. For each image i, we perform outlier removal by

$$T_j = \begin{cases} T_j \cup S_j(i), & P(S(i) = j | u^j, \sum{}^j) \geqslant \varepsilon^j \\ T_j, & P(S(i) = j | u^j, \sum{}^j) < \varepsilon^j \end{cases} \tag{3}$$

where $P(S(i) = j | u, \sum)$, ε^j, u^j, $\sum{}^j$ are the probability that image i belongs to category j, the pruning threshold for category j, the mean and covariance matrix of domain specific Gaussian distribution, respectively. We then use $\{T_1, T_2, ..., T_N\}$ to train the Texture CNN.

3.3   Shape CNN Model

Crowdsourced 3D models are easily accessible online and can be used to render unlimited images with different backgrounds, textures, and poses [23,31]. The widely used 3D Warehouse[1], Stanford Shapenet[2] provide numerous 3D CAD models for research use. Previous work has shown the great potential of synthetic images rendered from 3D CAD models for object detection [23] and pose estimation [31].

The flexibility and rigidity give 3D CAD models the unique merit for monitoring image properties such as background, texture and object pose, with constant shape information. However, the drawbacks of synthetic data are also obvious:

1. **Lack of realism:** Sythetic images generally lack realistic intensity information which explicitly reflect fundamental visual cues such as texture, illumination, background.
2. **Statistic Mismatch:** The statistics (eg. edge gradient) of synthetic image are different from realistic images. Thus the discriminatory information preserved by the DCNN trained on synthetic images may lose its effect on realistic images.

Simulating Real-Image Statistics:
One flaw of synthetic images is that the instance is inconsistent with the background. Thus, even rendered with very sophisticated parameters (pose variation, illumination, etc.), the statistics mismatch in intensity level still remains. Figure 2 illustrates the difference of edge gradients between synthetic images rendered with white backgrounds and real images.

After analysing the synthetic data, we find the objects in the synthetic images tend to have higher contrast edges compared to real images taken in the wild. Adding more realistic backgrounds [23] is a good way to decrease the contrast, but may obscure the object if the background is not chosen carefully. Instead, in this work, for each image I, we process I by:

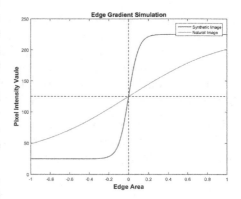

Fig. 2. Illustration of Edge Gradient. Synthetic images rendered with white backgrounds tend have higher contrast edges around the outline of the object than natural images taken in the wild. Note that this figure is an illustration, not representing real pixel intensities.

$$I' = \psi_{\mathbf{G}}(I) + \xi_G \qquad (4)$$

---

[1] https://3dwarehouse.sketchup.com.
[2] http://shapenet.cs.stanford.edu/.

where $\psi_{\mathbf{G}}(.)$ is a smoothing function based on a Gaussian filter and ξ_G is a Gaussian noise generator. $\psi_{\mathbf{G}}(.)$ is used to mitigate the sharp edge contrast and ξ_G to increase the intensity variations.

4   Experiments

In this section, we describe our experimental settings in detail. We start from downloading texture images via web search and rendering shape data from 3D CAD models. We evaluate our Two-Stream CNN classifier and detector on the standard benchmark PASCAL VOC 2007 [9] dataset.

4.1   Data Acquisition

Texture Data. As illustrated in Fig. 1, we leverage a text-based image search engine (Google image search engine in our experiments) to collect the image data. Most of the images returned by Google contain a single object centered in the picture. This is good news for an algorithm attempting to learn the main features of a certain category. However, the drawback is the returned data is noisy and highly biased. For example, the top results returned by searching "aeroplane" may contain many toy aeroplane and paper plane images. To make matters worse, some returned images contain no "aeroplane", but objects from other categories.

We use the name of each object category as the query for Google image search engine to collect the training images. After removing unreadable images, there are about 900 images for each category and 18212 images in total.

With millions of parameters, the CNN model easily overfits to small dataset. Therefore, data augmentation is valuable. Since most of the images are object-centered, we crop 40 patches by randomly locating the top-left corner (x_1, y_1) and bottom-right corner (x_2, y_2) by the following constraint:

$$
\begin{cases}
x_1 \in [\frac{W}{20}, \frac{3W}{20}], & y_1 \in [\frac{H}{20}, \frac{3H}{20}] \\
x_2 \in [\frac{17W}{20}, \frac{19W}{20}], & y_2 \in [\frac{17H}{20}, \frac{19H}{20}]
\end{cases}
\tag{5}
$$

(W, H) are the width and height of original image. The constraint ensures that 49%–81% of the center area of the image is reserved. This image subsampling process leaves us about 0.7 million images to train the Texture CNN.

We further utilize the approach illustrated in Sect. 3.2 to remove outliers from the downloaded data. For each category j ($j \in [1, N]$, N is the class number), we denote all the image patches after image subsampling as S_j. We adopt a DCNN architecture, known as "AlexNet" to extract $fc7$ feature for each patch $i \in S_j$ to form the $fc7$ feature set F_j. We fit the F_j to a multivariate normal distribution $\mathcal{N}(u, \sum)$ and compute the probability of each image path i. Through the fitting process we can find domain-specific variables u^j and $\sum^j$. The threshold ε^j in Eq. 3 is set so that the probabilities of 80% of patches from S_j are larger than it.

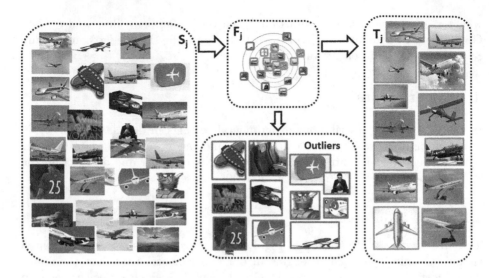

Fig. 3. Illustration Of Noisy Data Pruning. We fit the downloaded data to a multivariate normal distribution and remove the outliers if their probability is less than a learned threshold. (Best viewed in color) (Color figure online)

Figure 3 shows some samples which have been pruned out from the keyword search for "aeroplane".

Shape Data. 3D CAD models of thousands of categories are available online. We utilize the 3D CAD models provided by [23] to generate our training images. These 3D CAD models were downloaded from 3D Warehouse by querying the name of the target categories. However, these models contain many "outliers" (eg. tire CAD models while searching car CAD models). To solve this, we manually selected the positive examples and delete the "outliers". To be consistent with our target dataset, we only adopt 547 3D CAD models for the 20 categories in PASCAL 2007, ranging from "aeroplane" to "tv-monitor".

AutoDesk 3ds MAX[3] is adopted to generate the synthetic images, with the entire generation process completed automatically by a 3ds Max Script. The rendering process is almost the same as [23] (we refer the reader to this work for more details), except that our approach generates possible poses by exhaustively selecting (X, Y, Z) rotation parameters, where X, Y and Z are the degree that the 3D CAD model needs to rotate around X-axis, Y-axis, Z-axis. As shown in Fig. 4, for each 3D CAD model, we first align the model to front view and set rotation parameters (X, Y, Z). The 3D CAD model is then rotated by the chosen parameters. In our experiments, we only increment one variable from (X, Y, Z) by 2 degrees at one time, constrained by $X, Y \in [-10, 10]$ and $Z \in [70, 110] \cup [250, 290]$ to cover possible intra-category variations. In total, we generate 833,140 synthetic training images to train our Shape CNN model.

---

[3] http://www.autodesk.com/store/products/3ds-max.

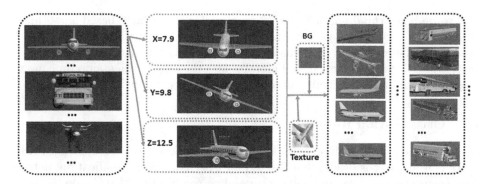

Fig. 4. Synthetic Data Rendering Process. For each 3D CAD model, we first align the model to the front view and set rotation parameters (X, Y, Z). The 3D CAD model is then rotated by the chosen parameters. To mitigate the intensity contrast around object edges, we add a background and texture to the final synthetic image.

To mitigate the intensity contrast around object edges, we add the mean image of ImageNet [7] as the background to the final synthetic images. In our ablation study, we try either texture-mapping objects with real image textures as in [23], or using uniform gray (UG) texture.

As addressed in Sect. 3.3, one shortcoming of synthetic data is the statistic mismatch, especially for the intensity contrast around the object edges. Figure 5 shows the difference in edge gradient between synthetic images with white background and images taken in the wild. Recent work on DCNN visualization [39] has shown that parameters in lower layers are more sensitive to edges. Thus matching the statistics of synthetic images to those of real images is a good

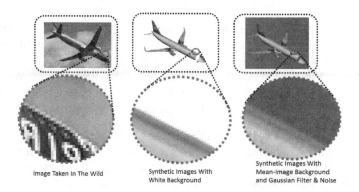

Image Taken In The Wild   Synthetic Images With White Background   Synthetic Images With Mean-Image Background and Gaussian Filter & Noise

Fig. 5. Illustration of different edge gradients. The edge gradients of synthetic images with white background differ drastically from images taken in the wild. In the third image, we replace the white background with a mean image computed from ImageNet [7] images and apply $\psi_G(.)$ to smooth the edge contrast. We further add some Gaussian noise ξ_G to the image to increase background variation.

way to decrease the domain shift. We apply a Gaussian filter based smoothing function $\psi_G(.)$ for every synthetic image, and then add Gaussian noise ξ_G to the smoothed images. In our experiment, we use $\mathcal{N}(0, 1)$ for $\psi_G(.)$ and $\mathcal{N}(0, 0.01)$ for ξ_G. The third image in Fig. 5 illustrates how the edge gradients become more similar to real image gradients after $\psi_G(.)$ and ξ_G are applied.

4.2 Classification Results

We evaluate our approach on standard benchmark PASCAL VOC 2007 [9]. PASCAL VOC dataset is originally collected for five challenges: classification, detection, segmentation, action classification and person layout. PASCAL VOC 2007 has 20 categories ranging from people, animals, plant to man-made rigid objects, and contains 5011 training/validation images and 4952 testing images. In our experiments, we only use testing images for evaluation.

Among the 4952 testing images, 14976 objects are annotated with tight bounding boxes. In our classification experiments, we crop 14976 patches (one patch for one object) with the help of these bounding boxes to generate test set.

The DCNN in each stream is initialized with the parameters pre-trained on ImageNet [7]. The last output layer is changed from a 1000-way classifier to a 20-way classifier and is randomly initialized with $\mathcal{N}(0, 0.01)$. The two DCNNs are trained with same settings, with the base learning rate to be 0.001, momentum to be 0.9 and weight decay to be 0.0005. Two dropout layers are adopted after $fc6, fc7$ with the dropout ratio to be 0.5.

The results in Table 2 show when adding texture cues to Shape-CNN with Eq. 1, the performance rises from 28.3%, 29.9%, 31.7% to 38.1%, 38.7%, 39.3%, respectively. On the other side, adding shape cues to Texture-CNN can boost the performance from 36.7% to 39.3%, with $\psi_G(.)$ and ξ_G applied. The results also demonstrate that simulating the real statistics can benefit the classification results.

To better analyze how texture cues and shape information compensate for each other, we plot the confusion matrix (with X-axis representing ground truth labels and Y-axis representing predictions) for Texture-CNN, Shape-CNN, Fusion-CNN in Fig. 6 (Networks with "-G" or "-GN" have similar results). For the convenience of comparison, we re-rank the order of categories and plot per-category accuracy in Fig. 6. There are some interesting findings:

- The top-right confusion matrix (for Shape-CNN) in Fig. 6 shows that CNN trained on synthetic images mistakes most of the "train" images for "bus", and mistakes "horse" and "sheep" images for "cow", which is not very surprising because they share similar shape visual information.
- From two confusion matrices on the top, we can see that Texture-CNN trained on web images tends to mistake other images for "plant" and "TV", while Shape-CNN is keen on categories like "cow", "bus", "bottle" etc.
- The last sub-figure in Fig. 6 is per-category accuracy. We re-rank the order of categories for inspection convenience. Shape-CNN (green line) tends to perform well for the categories presented on the left and Texture-CNN (blue line)

Table 2. Classification Results. Prefix "S-", "T-" and "F-" denote "Shape", "Texture", "Fusion", respectively. Suffix "-G", "-N", "-UG" indicate synthetic data is smoothed with $\psi_G(.)$, is colored with Gaussian noise ξ_G, and is generated with uniform gray texture. The results show Fusion-CNN model outperforms Shape-CNN model and Texture-CNN model and simulating real statistics benefit object classification system with minimal supervision. Note the last column is not mean accuracy, but accuracy over all test set.

Method	aer	bik	bir	boa	bot	bus	car	cat	chr	cow	tbl	dog	hrs	mbk	prs	plt	shp	sof	trn	tv	All
T-CNN	**61**	**76**	**36**	39	12	44	**53**	68	12	52	**16**	**46**	**58**	56	19	**86**	**63**	22	**76**	89	36.7
S-CNN	37	24	18	50	**60**	**82**	36	54	16	78	2	10	3	56	19	31	6	26	7	77	28.3
S-CNN-G	31	20	21	58	44	80	44	42	20	73	3	9	3	66	21	42	8	34	6	80	29.9
S-CNN-GN	32	20	20	51	53	80	47	47	**25**	68	2	12	5	**69**	21	53	8	**37**	3	85	31.7
F-CNN	52	66	30	50	42	74	45	**70**	16	**84**	13	33	41	59	24	71	53	27	50	89	38.1
F-CNN-G	56	65	31	**59**	27	76	48	63	18	77	14	35	46	64	**25**	71	55	30	51	89	38.7
F-CNN-GN	56	65	31	48	32	74	51	65	21	68	13	35	47	67	**25**	77	54	31	46	90	**39.3**
S-CNN-UG	30	25	16	51	20	80	40	11	13	49	1	5	16	76	27	64	1	24	35	46	27.2
S-CNN-G-UG	25	16	15	43	20	86	43	8	15	57	1	3	5	81	21	53	1	30	14	38	27.5
S-CNN-GN-UG	35	23	13	55	33	82	37	9	22	34	0	4	3	78	29	56	0	29	19	39	29.4
F-CNN-UG	56	68	30	52	16	70	50	61	14	57	12	38	49	70	28	83	55	23	68	81	39.3
F-CNN-G-UG	50	59	29	48	16	79	50	62	15	61	11	35	46	78	26	80	54	26	56	81	38.1
F-CNN-GN-UG	62	65	28	55	20	74	49	61	19	50	13	38	48	75	32	81	56	26	62	84	41.4

is more likely to get a high performance for the categories presented on the right. Taking a closer look at the categories, we find that Shape-CNN will work well for shape-oriented or rigid categories, e.g. bus, bottle, motorbike, boat, etc. Inversely, Texture-CNN is more likely to obtain higher performance on texture-oriented categories, such as cat, sheep, plant, horse, bird etc. The performance of Fusion-CNN is mostly between Shape-CNN and Texture-CNN and never gets a very poor result, which is why it can work better than CNNs based on single cues. We also tried performing a max fusion over the two streams, but the performance improvement is not comparable with average fusion.

We also try removing the texture on the object from the synthetic image by replacing it with uniform gray (UG) pixels. As shown in Table 2, this achieves similar results, indicating that synthetically adding real texture to CAD models may not be important for this classification task.

4.3 Object Detection Results

In our detection experiments, we find "S-CNN" ("F-CNN"), "S-CNN-G" ("F-CNN-G") and "S-CNN-GN" ("F-CNN-GN") get comparable results, thus we only report result for "S-CNN" ("F-CNN") in this section.

For detection, we followed the standard evaluation schema provided by [9]: a prediction bounding box P is considered to be a valid detection if and only if the area of overlap IoU exceeds 0.5. The IoU is denoted with the following formula:

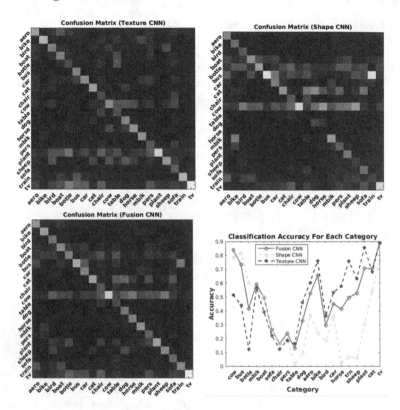

Fig. 6. Confusion matrix and classification results. The confusion matrix has been normalized by the number of total images per category. From up to bottom, left to right, the four figures are: confusion matrix of texture CNN, confusion matrix of shape CNN, confusion matrix of fusion CNN, classification accuracy for each category. (Best viewed in color!) (Color figure online)

$IoU = \frac{area(B_p \cap B_{gt})}{area(B_p \cup B_{gt})}$, where $B_p \cap B_{gt}$ denotes the intersection of the predicted bounding box and the ground truth bounding box and $B_p \cup B_{gt}$ their union.

Region Proposal. An excellent region proposal method contributes to the performance of both supervised and unsupervised learning. In our experiments, we adopt EdgeBox [41] to generate region proposals. EdgeBox is a efficient region proposal algorithm which generates bounding box proposals from edge maps obtained by contour detector. The bounding boxes are scored by the number of enclosed contours inside the boxes.

Like in the classification task, for each region proposal, we pass it to Shape-CNN and Texture-CNN simultaneously, and fuse the last layers' activations. Similar to [3], we randomly crop patches from YFCC [36] as the negative samples. Further, we follow the schema of R-CNN [14] to compute mAP. We compare our method to following baselines.

- **VCNN(ICCV'15)** [23] In this work, the authors propose to render domain-specific synthetic images from 3D CAD models and train an R-CNN [14] based object detector. Some results may involve minor supervision, e.g. selecting background and texture. We compare to their W-RR model (white background, real texture) where the amount of supervision is almost the same as in this work.

- **LEVAN(CVPR'14)** [8] LEVAN uses items in Google N-grams as queris to collect training images from Internet. They propose a fully-automated approach to organize the visual knowledge about a concept and further apply their model to detection task on PASCAL VOC 2007.
- **Webly Supervised Object Detection(ICCV'15)** [3] The webly supervised learning approach collects images from Google and Flickr by searching for the name of a certain category and utilizes Examplar-LDA [15] and agglomerative clustering [5] to generate the potential "ground truth" bounding box. For fair comparison, we only compared to their results of images downloaded from Google.
- **Concept Learner(CVPR'15)** [40] Concept learner is designed to discover thousands of visual concepts automatically from webly labeled images. It first trains a concept learner on the SBU dataset and selects the learned concept detectors to compute the average precision.

Results listed in Table 3 demonstrate that combining real-image texture information with Shape-CNN will boost the mAP from 15.0 to 19.7, a relative 31.3% increase! Inversely, adding shape information to Texture-CNN boost the mAP from 18.1 to 19.7, which shows texture cues and shape information can compensate for each other in detection task. Despite the minimal amount of required

Table 3. Illustration Of Detection Results. Methods are categorized by their supervision type. "Webly", "Selected", "Weakly", "Full" represent webly supervision, selected supervision, weak supervision, full supervision, respectively. The definitions of these supervisions are the same as in [40]. The supervision in our work only comes from labeling the CAD models and choosing proper texture for the CAD models when generating synthetic images. The supervision used in [23] is almost the same as in our approach, except that they also labeled pose for the CAD models. The results demonstrate that our Fusion-CNN model outperforms methods based on single visual cues and other methods with similar or higher required supervision effort.

Method	aer	bik	bir	boa	bot	bus	car	cat	chr	cow	tbl	dog	hrs	mbk	prs	plt	shp	sof	trn	tv	mAP
T-CNN	22	20	19	18	9	35	28	21	9	13	4	16	29	31	6	11	15	11	25	19	18.1
S-CNN	20	18	18	15	9	29	23	4	9	16	0	13	19	26	13	9	14	12	4	27	15.0
F-CNN	29	23	19	22	9	41	29	17	9	21	1	20	23	33	9	9	17	13	16	27	19.7
S-CNN-UG	22	17	13	12	9	23	26	2	2	13	0	6	11	29	5	9	2	10	1	17	11.4
F-CNN-UG	25	18	15	18	10	38	30	14	3	18	1	17	20	31	8	10	12	12	11	19	17
VCNN [23]	36	23	17	15	12	25	35	21	11	16	0.1	16	16	29	13	9	4	10	0.6	29	17
Levan, Webly [8]	14	36	13	10	9	35	36	8	10	18	7	13	31	28	6	2	19	10	24	16	17.2
Chen's Webly [3]	35	39	18	15	8	31	39	20	16	13	15	4	21	34	9	17	15	23	28	19	20.9
Zhou's, Selected [40]	30	34	17	13	6	44	27	23	7	16	10	21	25	36	8	9	22	17	31	18	20.5
Siva's, Weakly [28]	13	44	3	3	0	31	44	7	0	9	10	2	29	38	5	0	0	4	34	0	13.9
Song's, Weakly [29]	8	42	20	9	10	36	39	34	1	21	10	28	29	39	9	19	21	17	36	7	22.7
RCNN, Full [14]	58	58	39	32	24	51	59	51	20	51	41	46	52	56	43	23	48	35	51	57	44.7

supervision, our Fusion-CNN also obtains higher performance than a purely 3D CAD model based method like [23], a webly supervised approach like [8] and a weakly supervised method where in-domain training data from PASCAL VOC 2007 is available [8]. The results show that Fusion-CNN outperforms DCNN based on single visual cues and other methods where similar or higher levels of supervision are adopted.

As an ablation study, we perform the same experiments on synthetic images generated without texture ("S-CNN-UG", "F-CNN-UG" in Table 3). The results reveal that, unlike classification, adding some texture into the synthetic images helps to boost performance for the detection task.

5   Conclusion

In this work, we proposed and implemented a novel minimally-supervised learning framework that decomposes images into their shape and texture and further demonstrated that texture cues and shape information can reciprocally compensate for each other. Furthermore, a unified learning schema, including pruning noise web data and simulating statistics of real images is introduced, both for web image based learning and synthetic image based learning. Finally, our classification and detection experiments on VOC 2007 show that our Fusion-CNN with minimal supervision outperforms DCNNs based on single cues (only shape, only texture) and previous methods that require similar or more supervision effort. We believe our model is valuable for scaling recognition to many visual object categories and can be generalized to other generic tasks such as pose detection, robotic grasping and object manipulation.

Acknowledgement. This research was supported by NSF award IIS-1451244 and a generous donation from the NVIDIA corporation.

References

1. Bergamo, A., Torresani, L.: Exploiting weakly-labeled web images to improve object classification: a domain adaptation approach. In: Advances in Neural Information Processing Systems, pp. 181–189 (2010)
2. Boser, B.E., Guyon, I.M., Vapnik, V.N.: A training algorithm for optimal margin classifiers. In: Proceedings of the Fifth Annual Workshop on Computational Learning Theory, pp. 144–152 (1992)
3. Chen, X., Gupta, A.: Webly supervised learning of convolutional networks. In: Proceedings of the IEEE International Conference on Computer Vision, pp. 1431–1439 (2015)
4. Chen, X., Shrivastava, A., Gupta, A.: Neil: extracting visual knowledge from web data. In: Proceedings of the IEEE International Conference on Computer Vision, pp. 1409–1416 (2013)
5. Chen, X., Shrivastava, A., Gupta, A.: Enriching visual knowledge bases via object discovery and segmentation. In: Proceedings of the IEEE Conference on Computer Vision and Pattern Recognition, pp. 2027–2034 (2014)

6. Dalal, N., Triggs, B.: Histograms of oriented gradients for human detection. In: Proceedings of IEEE Conference Computer Vision and Pattern Recognition (2005)
7. Deng, J., Dong, W., Socher, R., Li, L.-J., Li, K., Fei-Fei, L.: Imagenet: a large-scale hierarchical image database. In: IEEE Conference on Computer Vision and Pattern Recognition, pp. 248–255 (2009)
8. Divvala, S., Farhadi, A., Guestrin, C.: Learning everything about anything: webly-supervised visual concept learning. In: Proceedings of the IEEE Conference on Computer Vision and Pattern Recognition, pp. 3270–3277 (2014)
9. Everingham, M., Van Gool, L., Williams, C.K.I., Winn, J., Zisserman, A.: The pascal visual object classes (VOC) challenge. Int. J. Comput. Vision **88**, 303–338 (2010)
10. Fan, J., Shen, Y., Zhou, N., Gao, Y.: Harvesting large-scale weakly-tagged image databases from the web. In: CVPR, pp. 802–809 (2010)
11. Felzenszwalb, P.F., Girshick, R.B., McAllester, D., Ramanan, D.: Object detection with discriminatively trained part based models. IEEE Trans. Pattern Anal. Mach. Intell. **32**, 1627–1645 (2010)
12. Fergus, R., Fei-Fei, L., Perona, P., Zisserman, A.: Learning object categories from internet image searches. Proc. IEEE **98**, 1453–1466 (2010)
13. Fragkiadaki, K., Arbelaez, P., Felsen, P., Malik, J.: Learning to segment moving objects in videos. In: 2015 IEEE Conference on Computer Vision and Pattern Recognition (CVPR), pp. 4083–4090 (2015)
14. Girshick, R., Donahue, J., Darrell, T., Malik, J.: Rich feature hierarchies for accurate object detection and semantic segmentation. arXiv preprint arXiv:1311.2524 (2013)
15. Hariharan, B., Malik, J., Ramanan, D.: Discriminative decorrelation for clustering and classification. In: Fitzgibbon, A., Lazebnik, S., Perona, P., Sato, Y., Schmid, C. (eds.) ECCV 2012. LNCS, vol. 7575, pp. 459–472. Springer, Heidelberg (2012). doi:10.1007/978-3-642-33765-9_33
16. He, K., Zhang, X., Ren, S., Sun, J.: Deep residual learning for image recognition. arXiv preprint arXiv:1512.03385 (2015)
17. Hoffman, J., Guadarrama, S., Tzeng, E., Donahue, J., Girshick, R.B., Darrell, T., Saenko, K.: LSDA: Large Scale Detection Through Adaptation. CoRR abs/1407.5035 (2014)
18. Simonyan, K., Zisserman, A.: Very Deep Convolutional Networks for Large-Scale Image Recognition. CoRR abs/1409.1556 (2014)
19. Krizhevsky, A., Sutskever, I., Hinton, G.E.: Imagenet classification with deep convolutional neural networks. In: Advances in Neural Information Processing Systems, pp. 1097–1105 (2012)
20. Li, L.-J., Fei-Fei, L.: Optimol: automatic online picture collection via incremental model learning. Int. J. Comput. Vis. **88**, 147–168 (2010)
21. Liebelt, J., Schmid, C.: Multi-view object class detection with a 3D geometric model. In: 2010 IEEE Conference on Computer Vision and Pattern Recognition (CVPR), pp. 1688–1695 (2010)
22. Lin, T.-Y., RoyChowdhury, A., Maji, S.: Bilinear CNN models for fine-grained visual recognition. In: Proceedings of the IEEE International Conference on Computer Vision, pp. 1449–1457 (2015)
23. Peng, X., Sun, B., Ali, K., Saenko, K.: Learning deep object detectors from 3D models. In: Proceedings of the IEEE International Conference on Computer Vision, pp. 1278–1286 (2015)

24. Saenko, K., Kulis, B., Fritz, M., Darrell, T.: Adapting visual category models to new domains. In: Daniilidis, K., Maragos, P., Paragios, N. (eds.) ECCV 2010. LNCS, vol. 6314, pp. 213–226. Springer, Heidelberg (2010). doi:10.1007/978-3-642-15561-1_16
25. Schroff, F., Criminisi, A., Zisserman, A.: Harvesting image databases from the web. IEEE Trans. Pattern Anal. Mach. Intell. **33**, 754–766 (2011)
26. Ren, S., He, K., Girshick, R., Sun, J.: Faster R-CNN: towards real-time object detection with region proposal networks. In: Advances in Neural Information Processing Systems (NIPS 2015) (2015)
27. Simonyan, K., Zisserman, A.: Two-stream convolutional networks for action recognition in videos. In: Advances in Neural Information Processing Systems, pp. 568–576 (2014)
28. Siva, P., Xiang, T.: Weakly supervised object detector learning with model drift detection. In: IEEE International Conference on Computer Vision (ICCV 2011), pp. 343–350 (2011)
29. Song, H.O., Girshick, R., Jegelka, S., Mairal, J., Harchaoui, Z., Darrell, T.: On learning to localize objects with minimal supervision. arXiv preprint arXiv:1403.1024 (2014)
30. Stark, M., Goesele, M., Schiele, B.: Back to the future: learning shape models from 3D CAD data. In: BMVC, vol. 2 (2010)
31. Su, H., Qi, C.R., Li, Y., Guibas, L.J.: Render for CNN: viewpoint estimation in images using CNNs trained with rendered 3D model views. In: Proceedings of the IEEE International Conference on Computer Vision, pp. 2686–2694 (2015)
32. Sun, B., Saenko, K.: From virtual to reality: fast adaptation of virtual object detectors to real domains. In: BMVC (2014)
33. Sun, M., Su, H., Savarese, S., Fei-Fei, L.: A multi-view probabilistic model for 3D object classes. In: IEEE Conference on Computer Vision and Pattern Recognition, CVPR 2009, pp. 1247–1254 (2009)
34. Szegedy, C., Liu, W., Jia, Y., Sermanet, P., Reed, S., Anguelov, D., Erhan, D., Vanhoucke, V., Rabinovich, A.: Going deeper with convolutions. arXiv preprint arXiv:1409.4842 (2014)
35. Tenenbaum, J.B., Freeman, W.T.: Separating style and content with bilinear models. Neural Comput. **12**(6), 1247–1283 (2000)
36. Thomee, B., Shamma, D.A., Friedland, G., Elizalde, B., Ni, K., Poland, D., Borth, D., Li, L.-J.: The new data and new challenges in multimedia research. arXiv preprint arXiv:1503.01817 (2015)
37. Welling, M.: Fisher Linear Discriminant Analysis. Department of Computer Science, University of Toronto, vol. 3 (2005)
38. Xia, Y., Cao, X., Wen, F., Sun, J.: Well begun is half done: generating high-quality seeds for automatic image dataset construction from web. In: Fleet, D., Pajdla, T., Schiele, B., Tuytelaars, T. (eds.) ECCV 2014. LNCS, vol. 8692, pp. 387–400. Springer, Cham (2014). doi:10.1007/978-3-319-10593-2_26
39. Zeiler, M.D., Fergus, R.: Visualizing and understanding convolutional networks. In: Fleet, D., Pajdla, T., Schiele, B., Tuytelaars, T. (eds.) ECCV 2014. LNCS, vol. 8689, pp. 818–833. Springer, Cham (2014). doi:10.1007/978-3-319-10590-1_53
40. Zhou, B., Jagadeesh, V., Piramuthu, R.: Conceptlearner: discovering visual concepts from weakly labeled image collections. In: Proceedings of the IEEE Conference on Computer Vision and Pattern Recognition, pp. 1492–1500 (2015)
41. Zitnick, C.L., Dollár, P.: Edge boxes: locating object proposals from edges. In: Fleet, D., Pajdla, T., Schiele, B., Tuytelaars, T. (eds.) ECCV 2014. LNCS, vol. 8693, pp. 391–405. Springer, Cham (2014). doi:10.1007/978-3-319-10602-1_26

Video Temporal Alignment for Object Viewpoint

Anestis Papazoglou[1]([✉]), Luca Del Pero[1,2], and Vittorio Ferrari[1]

[1] University of Edinburgh, Edinburgh, Scotland
a.papazoglou@sms.ed.ac.uk, vittorio.ferrari@ed.ac.uk
[2] Blippar, London, UK
luca.delpero@blippar.com

Abstract. We address the problem of temporally aligning semantically similar videos, for example two videos of cars on different tracks. We present an alignment method that establishes frame-to-frame correspondences such that the two cars are seen from a similar viewpoint (e.g. facing right), while also being temporally smooth and visually pleasing. Unlike previous works, we do not assume that the videos show the same scripted sequence of events. We compare against three alternative methods, including the popular DTW algorithm, on a new dataset of realistic videos collected from the internet. We perform a comprehensive evaluation using a novel protocol that includes both quantitative measures and a user study on visual pleasingness.

1 Introduction

Temporal alignment of videos is often a key step in several popular tasks, such as video morphing [1], video mosaicking and stitching [2], video compositing [3], video summarisation [4], action recognition and video retrieval [5] and High Dynamic Range (HDR) video [6]. Much previous work on temporal alignment focuses on videos of the same scene recorded from multiple cameras [7–14]. Instead, we want to align videos that are only weakly related: we simply require that their main object belongs to the same semantic class. For example, two videos of different cars driving along different tracks, and backgrounds.

Our alignment method establishes frame-to-frame correspondences such that the two cars are seen from a similar viewpoint (e.g. facing right) while also enforcing temporal smoothness, i.e. we preserve the temporal order of the frames in the original videos as much as possible (Fig. 1). Our key intuition is that the object viewpoint is a good indicator of whether two individual frames showing different cars are aligned correctly. Temporal smoothness promotes consistency at a larger temporal scale (i.e. an entire left turn, Fig. 1), which is more robust to noise in individual frames, and also makes the alignments more visually pleasing.

A few previous works [15–17] have tackled aligning semantically similar videos. However, they typically assume that the videos show a scripted sequence of events (e.g. drinking motion [17], hand waving [15]), possibly out of phase (*i.e.* the events occur at different, varying speeds). Under this assumption, finding an optimal alignment can be solved using Dynamic Time Warping (DTW) [18]

© Springer International Publishing AG 2017
S.-H. Lai et al. (Eds.): ACCV 2016, Part IV, LNCS 10114, pp. 273–288, 2017.
DOI: 10.1007/978-3-319-54190-7_17

Fig. 1. Viewpoint-driven temporal alignment. The goal of this task is to align the two videos so that both of them show the object from the same viewpoint frame-by-frame as shown above. This example alignment was produced by our method.

(as in [15,17]). However, this assumption is unrealistic for most real-world videos, where events may occur in a different order, or some occuring in only one of the videos.

Here, we present a method that is able to cope with such challenging videos. Our assumption is that we can decompose videos into contiguous temporal segments, and put them into correspondence so that each pair of corresponding segments (rather than the entire videos) show the same sequence of events (Fig. 3). The main contribution of our approach is to solve the temporal segmentation and the correspondence problems jointly. For this, we use a principled probabilistic model defined over the space of all possible temporal segmentations and correspondences (Sect. 3). A likelihood function promotes putting in correspondence segments showing similar viewpoints, while other components favour temporal consistency and smoothness. Inference in our model is a computationally intractable combinatorial problem. Therefore, we present a Markov Chain Monte Carlo (MCMC) sampling [19] procedure to search its complex parameter space efficiently (Sect. 4).

We test our method on a set of 22 videos of cars racing in rally competitions collected from the internet, where we have manually annotated the viewpoint in each frame for evaluation (Sect. 6.1). These videos are challenging, showing fast motion, complex backgrounds and different car models. We automatically split them into different shots using [20], but they are otherwise untrimmed and unedited. This is different from videos used in previous work [15–17], which are trimmed so that they show the exact same sequence of events in their entirety. We release this dataset at http://calvin.inf.ed.ac.uk/datasets/videoalignment.

In our videos, events are often in a different order and occur a different number of times. Hence, determining their optimal alignment can be ambiguous, i.e. we cannot define a unique ground-truth alignment as in [17]. For instance, if a certain viewpoint appears only once in a video and multiple times in the other, there are multiple valid ways of aligning them (e.g. Fig. 2). To address this, we perform a comprehensive evaluation that takes into account several different factors: viewpoint similarity, temporal consistency and visual pleasingness.

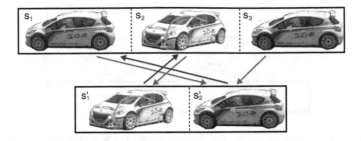

Fig. 2. Example alignment between two videos. The first video shows the same events as the second (going straight, turning left) but in a different sequence. These videos cannot be aligned by just stretching and shrinking the time domain of the videos (as DTW does), but our method can cope with it.

We evaluate these factors quantitatively on our dataset using a new carefully designed evaluation protocol, as well as with a substantial user study on visual pleasingness (in contrast to previous works that are mostly evaluated qualitatively on a few videos, *e.g.* [15,16]). Our results show that our method is superior to three alternative alignment methods (Sect. 6.2), including the popular DTW [18].

2 Related Work

Previous works on temporal alignment can be categorised based on their assumptions about the input videos.

Videos of the same scene from different views. Most previous works, e.g. [7–12] focus on joint spatio-temporal alignment of videos of the same dynamic scene, recorded by two uncalibrated cameras placed at different viewpoints (typically stationary). [21] also attempts to spatio-temporally align videos of a single dynamic scene, but they jointly process videos from multiple cameras instead of just two. The work of [22] also assumes a single dynamic scene recorded by multiple cameras, but focuses on temporal alignment only.

Videos of the same scene at different times. A few works [13,14,23] focus on spatio-temporal alignment of videos of the same scene, but taken at a different time. To compensate for the lack of temporal overlap between the input videos, these works assume the cameras follows roughly the same trajectory.

Videos of semantically similar scenes. Our work falls in the category of temporal alignment of videos that do not show the same scene, but rather semantically similar content (e.g. drinking motion [17], hand waving [15]). Typically, the videos depict people performing some scripted sequence of actions, such as drinking or waving [15–17], and the goal is to put in correspondence frames showing the same body pose. These approaches typically align short videos showing the exact same sequence of events (possibly at different speed) and cannot handle challenging videos showing events in a different order like we do.

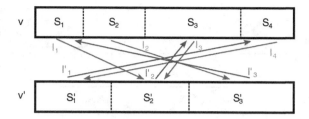

Fig. 3. One possible configuration (Sect. 3) of temporal segmentations S, S' and correspondences l between two videos v, v'. Each video is partitioned into a series of contiguous temporal segments: $S = \{s_1, s_2, s_3, s_4\}$ and $S' = \{s'_1, s'_2, s'_3\}$. Each segment has a correspondence in the other video, denoted by l (arrows). Note that the correspondences are not necessarily mutual (e.g. $s_2 \rightarrow s'_3$, but $s'_3 \rightarrow s_1$).

Frame descriptors for video retrieval and classification. Some works [24–26] design good frame descriptors for video retrieval and classification. These methods do not consider aligning videos like we do (but [25] considers the somewhat related problem of recovering the temporal order of the jumbled frames of a single video). While the frame descriptor we use focuses on viewpoint similarity (Sect. 5), our video alignment formulation is general and can use any other frame descriptor. In the experiments (Sect. 6.3) we compare our viewpoint descriptor to the descriptor from [25], which achieves state-of-the-art on several retrieval tasks by encoding temporal context.

3   Temporal Alignment Model

Our goal is to align two videos where different events may appear in a different order. Figure 2 shows a simple example, featuring two types of events: going straight (s_1, s_3, s'_2) and turning left (s_2, s'_1). Ideally, we would like to match s'_1 to s_2 and s'_2 to either s_1 or s_3 (both would be valid). Note that the problem is not symmetric: when aligning the second video to the first, we would like to align s_1 to s'_2, s_2 to s'_1, and s_3 to s'_2 again. Aligning this example requires shuffling the temporal order of the videos, and re-using some of its segments.

An additional challenge is that the temporal segmentation of the videos into different events is also not known in advance. Our method solves the temporal segmentation and the segment correspondence problems jointly, using a single probabilistic model over the two tasks, which we now define formally.

Let v and v' be the two videos we want to align. $S = \{s_1, ..., s_N\}$ is the set of *contiguous* temporal segments composing v. The temporal segmentation S' of v' is defined analogously. The correspondence l_i indicates which segment from v' is matched to the i-th segment in v (l'_j is defined analogously; note that $l_i = j \nRightarrow l'_j = i$, Fig. 3). The model parameters Θ include the temporal segments of both videos $\mathcal{S} = \{S, S'\}$ and the set $\mathcal{L}$ containing all correspondences (Fig. 3). Note how both the segmentations S, S' and the correspondence $\mathcal{L}$ have a variable number of elements, as the number of segments in each video is not predefined. It is another parameter to be searched over during inference.

We define the posterior distribution over the parameters to be

$$p(\mathcal{L}, \mathcal{S}|D) = p(\mathcal{L}|\mathcal{S}, D) \cdot p(\mathcal{S}|D) \tag{1}$$

where D are appearance descriptors extracted for all frames in the videos. Since we want to align the videos so that they show the same viewpoint, we use state-of-the-art CNN descriptors [27] which we specifically fine-tuned to classify different viewpoints (Sect. 5). The two factors in the posterior compete to allow our model to find alignments which put similar viewpoints in correspondence, while also being temporally smooth. The *correspondence likelihood* $p(\mathcal{L}|\mathcal{S}, D)$ promotes putting into correspondence temporal segments (across videos) that are consistently similar in appearance through time. The *temporal segmentation likelihood* $p(\mathcal{S}|D)$ promotes having few temporal segments. Having too many segments can cause the alignment to look jerky due to the frequent segment switches over time, which is not visually pleasing. Furthermore, it promotes that each temporal segment is homogeneous in appearance (within a video). A homogeneous segment is likely to contain a single viewpoint, which makes it a good unit for matching across videos. We now discuss each factor in more detail.

Correspondence likelihood. We define the correspondence likelihood to be

$$p(\mathcal{L}|\mathcal{S}, D) = \prod_i p(l_i|\mathcal{S}, D) \cdot \prod_j p(l'_j|\mathcal{S}, D) \tag{2}$$

where each $p(l_i = k|\mathcal{S}, D)$ evaluates the likelihood of $l_i = k$ according to the appearance similarity of s_i and s'_k (these factors are conditionally independent). We define the probability of one correspondence l_i to be

$$p(l_i = s'_j|\mathcal{S}, D) \propto \exp\left(-\alpha_M \frac{||s_i||}{||v||} d(s_i, s'_j)\right) \tag{3}$$

where α_M is a scalar weight, $||s_i||$ is the length of segment s_i (*i.e.*, the number of frames in it), $||v||$ is the length of video v, and $d(s_i, s'_j)$ denotes the appearance distance between the segments s_i and s'_j.

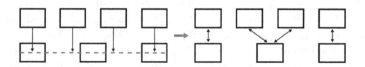

Fig. 4. Our appearance distance d (4) measures the similarity in appearance between two segments of potentially different length (Sect. 3). We first put the segment frames in one-to-one correspondence. For this, we project the longest segment onto the shorter one (top), and put each frame in the longest segment in correspondence with the frame closest to the projection (bottom). d is the distance in appearance averaged over all corresponding frames (4).

We designed d so that it can evaluate whether the appearance of the segments is consistently similar through time. Since the segments can have different length (*i.e.*, different speed), we first put their frames in one-to-one correspondence, denoted by $(f \to f')$ (see Fig. 4). We can now compute

$$d(s_i, s'_j) = \frac{\sum_{f \to f'} \text{a}(f, f')}{\max(||s_i||, ||s'_j||)} \tag{4}$$

where $\text{a}(f, f')$ denotes the appearance distance between frames f and f' (Sect. 5). Note that DTW [18] is a reasonable alternative segment distance, as it also measures similarity through time. However, we found that d produces comparable results to DTW, while being computationally more efficient.

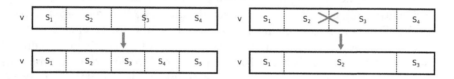

Fig. 5. (Left) Split move (Sect. 4). In this example, temporal segment s_3 is split in half, creating two new segments. **(Right)** Merge move (Sect. 4). In this example, temporal segments s_2 and s_3 are merged into a single segment.

Temporal segmentation likelihood. The temporal segmentation likelihood $p(\mathcal{S}|D)$ promotes having a small number of segments that are homogeneous in terms of appearance.

$$p(\mathcal{S}|D) \propto \exp\left(-\alpha_T \sum_i \frac{||s_i||}{||v||} \Delta_i\right) \exp\left(-\alpha_P(||S||^2 + ||S'||^2)\right) \tag{5}$$

where α_T, α_P are scalar weights, Δ_i is the appearance distance a averaged over all pairs of frames within s_i (Sect. 5), and $||S||$ and $||S'||$ are the number of segments in v and v', respectively. Note that we can compute Δ_i in constant time by using summed area tables [28]. The ratio $\frac{||s_i||}{||v||}$ ensures that the contribution of each temporal segment is proportional to its length.

The first factor in Eq. 5 promotes segments that are homogeneous in terms of appearance. The second factor acts as a prior, promoting having a small number of segments. These two factors and the correspondence likelihood compete in order to strike a balance on the optimal number of segments. On one hand, having many short segments results in a high $p(\mathcal{S}|D)$, which is trivially maximised when each frame forms its own segment (which is maximally homogeneous in appearance). In this limit case, $p(\mathcal{L}|\mathcal{S}, D)$ reduces to a nearest-neighbour matching between individual frames in the two videos, which results in a low average appearance distance, but is also sensitive to noisy appearance descriptors. On the other hand, having a few long segments brings temporal smoothness and produces a more visually pleasing alignment. However, having corresponding segments that show the same sequence of viewpoints is more unlikely.

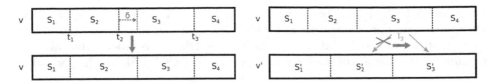

Fig. 6. (Left) Perturbation move (Sect. 4). In this example, we propose moving the delimiter t_2 between temporal segments s_2 and s_3 by an offset δ (with the constraint that $t_1 < (t_2 + \delta) < t_3$). **(Right)** Correspondence move (Sect. 4). In this example, the correspondence l_3 for temporal segment s_3 (green arrow) changes from s_2' to s_3' (Color figure online).

4   Inference

Maximising the posterior (Eq. 1) with respect to Θ is a hard combinatorial problem, since we jointly optimise over the number of segments N, N', the position of their delimiters S, S', as well as the set of correspondences $\mathcal{L}$. Furthermore, the posterior (Eq. 1) is a complex distribution which we cannot evaluate analytically. Thus, we use Markov Chain Monte Carlo (MCMC) sampling [19] to search the parameter space.

Following the standard formulation, at each iteration we propose a new sample Θ' from the current sample Θ using a proposal distribution $q(\Theta'|\Theta)$. Θ' is then accepted with probability

$$A(\Theta', \Theta) = \min\left(1, \frac{p(\Theta'|D) \cdot q(\Theta'|\Theta)}{p(\Theta|D) \cdot q(\Theta|\Theta')}\right) \tag{6}$$

If Θ' is accepted, it becomes the current sample, otherwise we keep Θ. Our proposal distribution uses four different kind of moves, each sampling over a subset of Θ. For each move, we change a single model parameter while keeping all other parameters fixed. Finally, we select the sample with the highest posterior (Eq. 1) as our output.

Perturbation move. We define a perturbation as changing the position of one of the current delimiters t by an offset δ (Fig. 6 left). We construct Θ' from the current sample with $f(\Theta, t, \delta)$, which replaces t with $t + \delta$. We choose (t, δ) from the space of all possible perturbations (t', δ') conditioned on the current positions of the delimiters in Θ. For this, we sample from

$$q_P(t, \delta|\Theta) = \frac{p(f(\Theta, t, \delta)|D)}{\sum_{(t', \delta')} p(f(\Theta, t', \delta')|D)} \tag{7}$$

Merge and split move. The merge move proposes merging a pair of subsequent segments into a single one (Fig. 5 right). We select which segments to merge from all possible merges given the delimiters in the current sample, using a proposal constructed analogously to (7). The complementary split move proposes splitting a segment in half, yielding two segments (Fig. 5 left). Both merge and split moves change the number of segments in a video. We note that merge/split moves are quite a standard tool in MCMC methods for solving association problems, for example in the domain of tracking multiple objects (*e.g.* MCMCDA [29] or [30]).

Correspondence move. This move chooses a segment s_i in a video and proposes to change its matching segment in the other video (*i.e.*, it changes l_i, Fig. 6 right). We choose s_i and the new value for l_i from all possible alternatives given the current segmentation $\mathcal{S}$. Again, we use a proposal constructed analogously to (7).

The way we constructed the proposals above increases the acceptance ratio of the moves, which improves mixing. For example, choosing (t, δ) in the perturbation move from a uniform distribution would result in a low acceptance ratio (which significantly improves using q_P). Note that, while our proposals need to compute a large number of posteriors during each move, they can still do it efficiently since most of the terms are shared between these computations, and need to be computed only once.

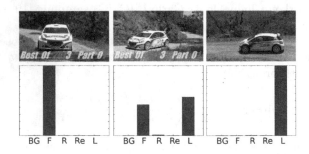

Fig. 7. Three example outputs of the softmax layer from cars in 90, 150 *and* 180 degree viewpoints respectively. The labels correspond to background (BG), front 90 degrees (F), right 0 degrees (R), rear 270 degrees (Re) and left 180 degrees (L).

Initialisation. Starting from a Θ sample with a reasonably good posterior reduces the amount of time wasted in regions of low probability at the beginning of the sampling process (compared to random initialisation). We begin by individually decomposing each video into homogeneous temporal segments, without considering any correspondences. This is achieved by optimising the temporal segmentation likelihood $p(\mathcal{S}|D)$ using just perturbation, merge and split moves. Since there is no correspondence likelihood involved, $p(\mathcal{S})p(\mathcal{S}|D)$ can be optimised independently for each video efficiently.

Having the initial temporal segmentation $\mathcal{S}$, we then find the optimal correspondence between these segments. This corresponds to optimising the correspondence likelihood $p(\mathcal{L}|\mathcal{S}, D)$. Since the correspondences are conditionally independent under our model, we can find the exact optimal set of correspondences with a nearest neighbour approach.

5   Appearance Descriptors

We now discuss the appearance distance $a(f, f')$ that we use to compute the distance between frames as part of our likelihood (Sect. 3). We designed it to capture the difference in viewpoint between two frames.

Appearance distance. Modern appearance classifiers based on Convolutional Neural Networks (CNN) are state-of-the-art for whole image classification [31] and object detection [27]. However, they are optimised to differentiate between objects of different classes, and they actually strive for invariance to viewpoint differences. Therefore they are not ideal for our problem. Instead, we train a CNN classifier, based on the AlexNet architecture [31], to distinguish among viewpoints of the car class (which we use in our experiments). We start from a CNN pre-trained for image classification on the ILSVRC 2012 [31,32] and fine-tune it to classify 4 car viewpoints (front, left, right, rear) and the background. As training data we use the PASCAL VOC 2012 [33] dataset which has car images with viewpoint annotations. In order to focus on the appearance of the cars and not on the background, we crop the cars from the image using the provided bounding-box annotations.

After training, we apply the CNN viewpoint classifier on a video frame and use the output of the softmax layer as our frame descriptor (i.e. a 5D vector, summing to 1). The intuition is that if the viewpoint of the input frame matches one of the training viewpoints from PASCAL VOC closely, the softmax output vector will be peaked on one of the 4 viewpoints. Instead, if the viewpoint of the frame lies in-between the training viewpoints, the output probability mass should be spread between two viewpoints (Fig. 7). We then define the distance $a(f, f')$ between two frames as the histogram intersection of their frame descriptors.

Fig. 8. A visualisation of the segmentation pipeline. We segment the cars in the videos using video foreground segmentation [34] (Sect. 5). Here we show: object proposals from selective search [35] (top left), the pixel-wise probability map produced by the car detector [36] (top right), the resulting segmentation (bottom left), and the bounding box of the segmentation is used to extract the viewpoint descriptor (bottom right).

Object localisation in the videos. To compute the viewpoint descriptor on a video frame, we first need to localise the car up to a bounding-box. This focuses the descriptor on the car and matches the kind of data the CNN was trained on.

We start from the video segmentation technique of [34], which can handle unconstrained video and reliably segments objects even under rapid motion and against cluttered backgrounds. This method uses a spatio-temporal Markov Random Field with unary potentials derived from motion, and a pairwise potential enforcing spatial and temporal smoothness. Here, we add a unary potential derived from a car detector trained on PASCAL VOC 2012 dataset [33]. While [34] is class agnostic and segments arbitrary foreground objects, the car detector injects domain-specific knowledge to anchor the segmentation on cars.

More precisely, for each video frame we extract object proposals using Selective Search [35] and score them with an R-CNN [27] car detector pre-trained on PASCAL VOC. Next, we score each pixel by the sum of the scores of all the proposals containing it. This results in a pixel-wise 'heatmap' that we use as the additional unary potential (Fig. 8).

We evaluated the method on our dataset using the CorLoc [37] performance measure as in [34]. Adding the car detector performs significantly better than [34], which is class agnostic with (88.5% accuracy versus 73.2%) respectively. Furthermore, the class detector [27] alone achieves just 69.8%, which shows that using video object segmentation can significantly improve the accuracy over using just a detector.

6   Experimental Evaluation

In this section we first introduce the data used for evaluation (Sect. 6.1). Second, we present the methods we compare against (Sect. 6.2). Next, we present our evaluation protocol (Sect. 6.3) and finally discuss our results (Sect. 6.4).

6.1   Data

We assembled a novel dataset of 22 video sequences depicting cars on racing sequences collected from YouTube, each 5–30 s long. These videos are challenging, showing different cars in different races, with fast motion, fast moving camera and cluttered backgrounds.

Our data contains viewpoint annotations for each frame. For this, we use an annotation protocol that reduces the amount of manual effort as follows. We first define a set of 16 canonical viewpoints, spaced by 22.5 degrees (starting from full frontal). We manually annotate all the frames showing one of them. Then, we automatically annotated the rest of the frames by linearly interpolating the manual annotations.

We evaluate our model on a pair of videos v, v' only if at least 50% of frames in v show a viewpoint that also appears in v' (up to a difference of 10 degrees). This leads to 251 pairs of videos out of the 484 possible pairs. We plan to release this dataset along with the ground truth annotations upon acceptance.

6.2   Alternative Methods

We compare our model against three alternatives: a nearest neighbour model, an MRF model promoting temporal smoothness, and Dynamic Time Warping (DTW). Note that as input, all methods use the same appearance distance $a(f, f')$ used in our model.

Nearest neighbour. The nearest neighbour model matches each frame in v to its closest neighbour in $v\prime$ according to the appearance distance a (Sect. 5). This simple model has no notion of temporal smoothness and it allows us to verify what we can achieve using appearance alone.

This approach corresponds to the following model:

$$\mathcal{L}^* = \operatorname*{argmin}_{\mathcal{L}} \sum_f U(l_f) \tag{8}$$

where U is a unary cost of matching frame f in v to a frame $f\prime$ in $v\prime$. Note that in this case, l denotes the correspondence between individual frames instead of entire temporal segments. We set U between $f, f\prime$ to be equal to their appeance distance $a(f, f\prime)$.

MRF model. As a second method, we augment the nearest neighbour model by adding temporal smoothness between subsequent frames. For this we use a Markov Random Field (MRF) with pairwise terms that promote that consecutive frames in v are in correspondence with frames in v' that are also close in time (Fig. 9). More precisely, we solve the following optimisation problem:

$$\mathcal{L}^* = \operatorname*{argmin}_{\mathcal{L}} \sum_f U(l_f) + \alpha_W \sum_f W(l_f, l_{f+1}) \tag{9}$$

where α_W is a weight term, and U is the same unary term used in the nearest neighbour model. We define the pairwise potential W to be:

$$W(l_f, l_{f+1}) = (|l_f - l_{f+1}| - 1)^2 \tag{10}$$

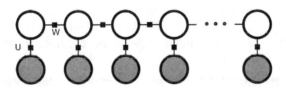

Fig. 9. The MRF model corresponding to Eq. 9. Each hidden node (in white) corresponds to a single frame in video v. The observable states (in grey) correspond to the frames in video $v\prime$.

Dynamic Time Warping. Dynamic Time Warping (DTW) [18] is a popular sequence alignment algorithm. Assuming that the two sequences show the same event transition with the only variable being the speed of each, DTW can compute an optimal alignment between them. However, this assumption does not necessarily hold true for realistic video sequences. As input, we used the appearance distance a to compute the distance between individual frames.

Table 1. Comparative user study. The table shows the comparative results between the different methods: nearest neighbour (NN), the MRF model, DTW and our model. The value in a cell shows the percentage of videos for which the participants preferred the method of the corresponding row over that of the column. For example, the participants preferred our method over the nearest neighbour approach for 90.9% of the videos.

Method	NN	MRF	DTW	Ours
NN	-	12.7	34.8	9.1
MRF	87.3	-	47.2	15.6
DTW	64.2	52.8	-	23.4
Ours	**90.9**	**84.4**	**76.6**	-

6.3   Evaluation Protocol

Comparative user study. We performed a substantial user study to verify that our results are indeed visually more appealing to humans compared to the alternative methods. We performed a "blind taste test" in which participants are presented with the same pair of sequences aligned by two different methods, and asked which alignment they think is better, *i.e.*, it is consistent in terms of viewpoint and also looks realistic. In our setup the participant is shown an original video and how it was aligned to a second video by two different alignment methods. The original video is displayed in the centre of the screen, the two alignments are on each side, being played simultaneously (we randomly choose on which side we put them). The participant then has to decide which one they think is better. Note that we never reveal to the user which method was used to produce the alignments we display.

We use this protocol to compare all of the alignment methods in pairs (*e.g.* our full method vs DTW, DTW vs MRF, etc.). We ensure that each pair of methods is shown to at least 3 different participants for each pair of videos and we aggregate the results (Table 1).

Quantitative criteria. We identify several properties that correspond to what humans perceive as attractive alignments. First, frames in correspondence should be displaying the same viewpoint. Second, the viewpoint transitions should be temporally consistent. Third, long sequences of correct correspondences are preferable. Based on this observation, we propose two measures to evaluate an alignment quantitatively:

Percentage of correct correspondences: This measures the percentage of frames that are in correct correspondence, *i.e.* difference in viewpoint is 22.5 degrees or less (which is equal to the spacing we use to manually annotate keypoints, Sect. 6.1), and the difference in the viewpoint derivative is 5 degrees/frame or less. Intuitively, the viewpoint difference ensures that aligned frames show the same viewpoint, while the derivative difference ensures that the viewpoint transition is smooth.

Longest correct sequence: This measures the length of the longest sequence of correct correspondences in each video, normalised by the length of the video. Intuitively, alignments that are correct for large periods of time are visually preferable to alignments with alternating correct and erroneous correspondences.

Quantitative criteria vs user study. We analyse how well these two evaluation criteria capture what humans perceive as a good temporal alignment, by verifying how accurately they can predict the results of the user study. We do this as follows. Given two methods, we predict that the user will choose the alignment found by the method that scores higher according to the evaluation criteria. We then report the *prediction accuracy* of each criterion, *i.e.* the number of times the prediction made using that criterion is correct, averaged over all possible pairs of methods and videos (Table 3).

Fig. 10. Qualitative results. Each pair of rows (a–d) shows an original video (top) and a second video aligned to it by our method (bottom).

6.4   Results and Discussion

Comparative user study. Table 1 shows the results of the user study. The value in a cell shows the percentage of videos for which the participants preferred

Table 2. Quantitative results. Comparison of the different video alignment methods: nearest neighbour (NN), MRF model, DTW and our model. The first two columns show the percentage of correct correspondences when we use our appearance descriptor (Sect. 5) and when we use the descriptor from [25]. The next two columns show performance on longest correct sequence (Sect. 6.3).

	Correct correspondence %		Longest correct sequence	
	Our descriptor	Descriptor [25]	Our descriptor	Descriptor [25]
NN	29.8	16.8	9.4	8.7
MRF	46.0	18.9	27.5	12.7
DTW	40.8	15.8	26.4	11.7
Our alignment model	**47.8**	20.0	**30.3**	13.6

Table 3. Evaluation of criteria. Each value corresponds to the accuracy of a quantitative criterion when trying to predict the results of our user study (Sect. 6.3).

	Human agreement
Correct correspondence %	70.7
Longest correct sequence	77.7

the method of the corresponding row over that of the column. Our method substantially outperforms all three alternative methods (last row).

The nearest neighbour model produces very jittery alignments, as it does not enforce any temporal smoothness. As a consequence, the participants do not find the output visually pleasing. Thanks to pairwise temporal smoothness, the MRF model partly alleviates this problem. However, the smoothness is promoted only at a local level (between consecutive frames). Hence, the MRF is unable to capture smooth transitions of viewpoints on a larger time scale. Instead, our model enforces smoothness at the level of temporal segments, leading to large, piece-wise smooth alignments.

Interestingly, the participants clearly prefer DTW over the nearest neighbour model, but results are comparable with respect to the MRF model. As mentioned before, DTW makes the strong assumption that both videos show the exact same sequence of events, possibly occurring at varying speeds. When this assumption holds, DTW can produce an optimal temporal alignment, and the participants prefer it over MRF. However, in the scenario where this assumption does not hold, the participants consistently prefer MRF. They however clearly prefer our method over both DTW and MRF, as our model can handle both scenarios thanks to the temporal segmentation.

Quantitative criteria. Table 2 shows the performance of each method according to our two quantitative criteria (Sect. 6.3) using our descriptor (Sect. 5) and the state-of-the-art descriptor [25]. Our alignment method outperforms all of the alternatives for both criteria and both descriptors. Moreover, our descriptor outperforms [25] on both criteria, probably because object viewpoint is a powerful cue for our task. In contrast to the user study results, DTW performs significantly worse than the MRF model with respect to percentage of correct correspondence. This indicates that humans tend to prefer smoother alignments, even if the aligned frames exhibit a larger difference in viewpoint. While the quantitative performance difference between the MRF model and ours is rather mild, users prefer our method over the MRF on 84% of the videos, showing that our alignments are much more visually pleasing. Figure 10 shows some qualitative results produced by our method.

Quantitative criteria vs user study. Table 3 shows an analysis of how well our evaluation criteria can predict what humans perceive as visually pleasing alignments. As can be seen from the results, both criteria show a strong correlation to what the participants prefer, in particular the longest correct sequence.

References

1. Liao, J., Lima, R.S., Nehab, D., Hoppe, H., Sander, P.V.: Semi-automated video morphing. In: Eurographics Symposium on Rendering (2014)
2. Agarwala, A., Zheng, K.C., Pal, C., Agrawala, M., Cohen, M., Curless, B., Szeliski, R.: Panoramic video textures. In: SIGGRAPH (2005)
3. Ruegg, J., Wang, O., Smolic, A., Gross, M.: Ducttake: spatiotemporal video compositing. Comput. Graph. Forum (Proc. Eurograph.) **32**, 51–61 (2013)
4. Ngo, C., Ma, Y., Zhang, H.: Video summarization and scene detection by graph modeling. IEEE Trans. Circ. Syst. Video Technol. **15**, 296–305 (2005)
5. Jiang, Y., Ngo, C., Yang, J.: Towards optimal bag-of-features for object categorization and semantic video retrieval. In: International Conference on Image and Video Retrieval (2007)
6. Kang, S.B., Uyttendaele, M., Winder, S., Szeliski, R.: High dynamic range video. ACM Trans. Graph. **26**(3), 760–768 (2007)
7. Caspi, Y., Irani, M.: A step towards sequence-to-sequence alignment. In: CVPR (2000)
8. Caspi, Y., Irani, M.: Spatio-temporal alignment of sequences. IEEE Trans. PAMI **24**, 1409–1424 (2002)
9. Caspi, Y., Irani, M.: Alignment of non-overlapping sequences. In: ECCV (2001)
10. Caspi, Y., Simakov, D., Irani, M.: Feature-based sequence-to-sequence matching. IJCV **68**, 53–64 (2006)
11. Wolf., L., Zomet, A.: Wide baseline matching between unsynchronized video sequences. IJCV (2006)
12. Tuytelaars, T., van Gool, L.: Synchronizing video sequences. In: CVPR (2004)
13. Evangelidis, G.D., Bauckhage, C.: Efficient subframe video alignment using short descriptors. IEEE Trans. PAMI (2013)
14. Wang, O., Schroers, C., Zimmer, H., Gross, M., Sorkine-Hornung, A.: Videosnapping: Interactive synchronization of multiple videos. ACM Trans. Graph. (2014)
15. Rao, C., Gritai, A., Shah, M.: View-invariant alignment and matching of video sequences. In: ICCV (2003)
16. Ukrainitz, Y., Irani, M.: Aligning sequences and actions by maximizing space-time correlations. In: Leonardis, A., Bischof, H., Pinz, A. (eds.) ECCV 2006. LNCS, vol. 3953, pp. 538–550. Springer, Heidelberg (2006). doi:10.1007/11744078_42

17. Dexter, E., Perez, P., Laptev, I.: Multi-view synchronization of human actions and dynamic scenes. In: BMVC (2009)
18. Sakoe, H., Chiba, S.: Object segmentation by alignment of poselet activations to image contours. IEEE Trans. Acoust. Speech Signal Proc. (1978)
19. Neal, R.M.: Probabilistic inference using markov chain Monte Carlo methods. Technical Report CRG-TR-93-1, University of Toronto (1993)
20. Kim, W.H., Kim, J.N.: An adaptive shot change detection algorithm using an average of absolute difference histogram within extension sliding window. In: ISCE (2009)
21. Padua, F.L.C., Carceroni, R.L.: Linear sequence-to-sequence alignment. IEEE Trans. PAMI (2009)
22. Douze, M., Revaud, J., Verbeek, J., Jegou, H., Schmid, C.: Circulant temporal encoding for video retrieval and temporal alignment. IJCV (2016)
23. Diego, F., Serrat, J., Lpez, A.M.: Joint spatio-temporal alignment of sequences. IEEE Trans. Multimedia (2013)
24. Simonyan, K., Zisserman, A.: Very deep convolutional networks for large-scale image recognition. arXiv abs/1409.1556 (2014)
25. Ramanathan, V., Tang, K., Mori, G., Fei-Fei, L.: Learning temporal embeddings for complex video analysis. In: ICCV (2015)
26. Zha, S., Luisier, F., Andrews, W., Srivastava, N., Salakhutdinov, R.: Exploiting image-trained cnn architectures for unconstrained video classification. In: BMVC (2015)
27. Girshick, R., Donahue, J., Darrell, T., Malik, J.: Rich feature hierarchies for accurate object detection and semantic segmentation. In: CVPR (2014)
28. Crow, F.C.: Summed-area tables for texture mapping. In: SIGGRAPH (1984)
29. Oh, S., Russell, S.J., Sastry, S.: Markov chain monte carlo data association for multi-target tracking. IEEE Trans. Autom. Control **54**, 481–497 (2009)
30. Brau, E., J., G., Simek, K., Del Pero, L., Dawson, C.R., Barnard, K.: Bayesian 3D tracking from monocular video. In: ICCV (2013)
31. Krizhevsky, A., Sutskever, I., Hinton, G.E.: Imagenet classification with deep convolutional neural networks. In: NIPS (2012)
32. Russakovsky, O., Deng, J., Su, H., Krause, J., Satheesh, S., Ma, S., Huang, Z., Karpathy, A., Khosla, A., Bernstein, M., Berg, A., Fei-Fei, L.: ImageNet large scale visual recognition challenge. IJCV (2015)
33. Everingham, M., Van Gool, L., Williams, C.K.I., Winn, J., Zisserman, A.: The PASCAL Visual Object Classes Challenge 2012 (VOC2012) Results (2012). http://www.pascal-network.org/challenges/VOC/voc2012/workshop/index.html
34. Papazoglou, A., Ferrari, V.: Fast object segmentation in unconstrained video. In: ICCV (2013)
35. Uijlings, J.R.R., van de Sande, K.E.A., Gevers, T., Smeulders, A.W.M.: Selective search for object recognition. IJCV (2013)
36. Girshick, R.: Fast R-CNN. In: ICCV (2015)
37. Prest, A., Leistner, C., Civera, J., Schmid, C., Ferrari, V.: Learning object class detectors from weakly annotated video. In: CVPR (2012)

3D Computer Vision

Recovering Pose and 3D Deformable Shape from Multi-instance Image Ensembles

Antonio Agudo$^{(\boxtimes)}$ and Francesc Moreno-Noguer

Institut de Robòtica i Informàtica Industrial (CSIC-UPC), Barcelona, Spain
aagudo@iri.upc.edu

Abstract. In recent years, there has been a growing interest on tackling the Non-Rigid Structure from Motion problem (NRSfM), where the shape of a deformable object and the pose of a moving camera are simultaneously estimated from a monocular video sequence. Existing solutions are limited to single objects and continuous, smoothly changing sequences. In this paper we extend NRSfM to a multi-instance domain, in which the images do not need to have temporal consistency, allowing for instance, to jointly reconstruct the face of multiple persons from an unordered list of images. For this purpose, we present a new formulation of the problem based on a dual low-rank shape representation, that simultaneously captures the between- and within-individual deformations. The parameters of this model are learned using a variant of the probabilistic linear discriminant analysis that requires consecutive batches of expectation and maximization steps. The resulting approach estimates 3D deformable shape and pose of multiple instances from only 2D point observations on a collection images, without requiring pre-trained 3D data, and is shown to be robust to noisy measurements and missing points. We provide quantitative and qualitative evaluation on both synthetic and real data, and show consistent benefits compared to current state of the art.

1 Introduction

The joint estimation of 3D shape and camera pose from a collection of images either acquired from different viewpoints or by a single moving camera is one of the most active areas in computer vision. In the last two decades, many works have addressed this problem under the assumption of a rigid scene [1–4] (see Fig. 1-Left). More recently, a number of approaches have been proposed to tackle the non-rigid case, and estimating a deforming 3D shape together with the camera pose from solely 2D observations [5–9]. This is the so-called Non-Rigid Structure from Motion (NRSfM) problem, which is inherently ambiguous

---

Electronic supplementary material The online version of this chapter (doi:10.1007/978-3-319-54190-7_18) contains supplementary material, which is available to authorized users.

© Springer International Publishing AG 2017
S.-H. Lai et al. (Eds.): ACCV 2016, Part IV, LNCS 10114, pp. 291–307, 2017.
DOI: 10.1007/978-3-319-54190-7_18

Rigid SfM   Non-Rigid SfM   Manifold Non-Rigid SfM

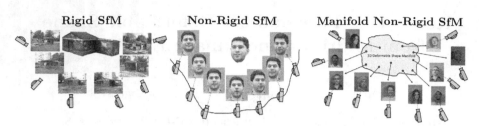

Fig. 1. Schematic comparison of our approach (denoted as Manifold Non-Rigid SfM) against standard Rigid SfM and Non-Rigid SfM. Left: In Rigid SfM pipelines, a collection of images, either acquired from different cameras –even with a different calibration– and viewpoints or from a single moving camera, is used to recover pose and 3D shape of a rigid scene. **Center:** Non-Rigid SfM normally handles objects that deform between consecutive frames, but the collection of images normally needs to be temporally consistent and the object remain the same. **Right:** Our Manifold Non-Rigid SfM is applicable to image collections that do not retain temporal consistency, and most importantly, can estimate the shape of different deforming instances in the same family.

and requires introducing several priors. The most standard assumption holds on that the observed objects do not arbitrarily change their shapes, and that deformations can be ruled by low-rank models. For instance, low-rank shape models span the shape using a linear combination of rigid and global basis, weighted by time-varying coefficients [10–12]. This has led to a number of solutions for sparse [7,13,14], dense [15], and sequential [16,17] reconstruction. A common characteristic of all these NRSfM algorithms is that the input images belong to one single object viewed from consecutive and smoothly changing viewpoints (see Fig. 1-Center), which has been exploited to introduce further constraints about temporal smoothness on the shape deformations [5,11,18], on the point trajectories [19,20] and on the camera motion [21]. Temporal smoothness priors have proven to be a powerful constraint on sequential NRSfM [16,22,23], giving consistent and accurate solutions.

In this paper, we depart from the assumptions of previous NRSfM approaches, by proposing a solution that does not require temporal consistency of the input images, and most importantly, that it can be applied to simultaneously recover the 3D deformable shape and pose of *different instances* of the same class of objects. In essence, we bring the standard scenario of the rigid structure from motion depicted in Fig. 1-Left to a non-rigid domain, as shown in Fig. 1-Right, in which we learn pose and a 3D non-rigid shape manifold. To do so, we expand the NRSfM formulation using a dual low-rank shape model that independently represents the deformations between- and within-object instances. Each of these components and their corresponding set of weights is learned by means of a variant of the Probabilistic Linear Discriminant Analysis (PLDA) [24] and iterating between partial expectation and maximization steps. The resulting approach estimates all this from the sole input of 2D input tracks on a collection

of images without requiring pre-trained 3D data, it is robust to noisy observations and it can handle missing tracks due to occlusions or lack of visibility.
We demonstrate the effectiveness on both synthetic and real image collections,
showing the advantages of the proposed approach with respect to state-of-the-art
techniques. We believe our model opens up the NRSfM topic to a series of new
problems in which the single instance and smooth camera viewpoint changes are
no longer a requirement.

2   Related Work

Reconstructing the shape of a non-rigid object while recovering the camera
motion from only 2D point trajectories is known to be a severely under-
constrained problem that requires prior information in order to be solved. The
prior most widely used in NRSfM consists in constraining the shape to lie on
a global low-rank shape subspace, that can be computed over a set of training
data [25], applying modal [26,27] or spectral [28] analysis over a rest shape, or
estimating it on-the-fly [10,12,29]. Most approaches build upon the well-known
closed-form factorization technique used for rigid reconstruction [30], enforcing
camera orthonormality constraints. This is also done in [5,11,18,31] by incorporating temporal and spatial smoothness constraints on top of the low-rank ones.
More recently, temporal smoothness is enforced by means of differentials over
the 3D shape matrix by directly minimizing its rank [13,15], or by means of a
union of temporal low rank shape subspaces [14].

Alternatively, pre-defined trajectory basis have been introduced to constrain the trajectory of every object point, turning the original trilinear problem to a bilinear one [19]. In [32], trajectory priors were used in terms of 3D
point differentials. Subsequent works have combined shape and trajectory constraints [6,33,34]. More recently, both low-rank shape and trajectory subspaces
have been linked to a force subspace, giving them a physical interpretation [9].
In any event, while achieving remarkable results, all previous approaches aim at
modeling one single object, observed from smoothly changing viewpoints. The
approach we propose here, gets rid of both these limitations.

We would also like to mention that our approach can be somewhat related
to methods that model low-dimensional shape manifolds using, e.g., Gaussian
Mixtures [35] or Gaussian Processes [36,37]. However, note that all these techniques assume again smoothly changing video sequences, and the 3D shape to
be aligned with the camera. Additionally, none of these approaches tackles the
problem of besides retrieving 3D shape, estimating the camera pose, as we do.

Contributions. In short, we propose a novel NRSfM solution which brings
together a number of characteristics not found in other methods: (1) It recovers
3D non-rigid shape and camera motion from image collections that do not exhibit
temporal continuity, i.e., our approach does not require monocular videos as
input; (2) It jointly encodes between- and within-object deformations; (3) It can
simultaneously model several instances of a same family; and (4) the number

of instances does not need to be known in advance. Our method is robust to artifacts such as noise or discontinuities due to missing tracks, and yields accurate reconstructions.

3   3D Deformable Shape Manifold Model

This section describes the proposed low-rank shape model, and specifically focuses on highlighting the main differences with respect to previous similar formulations.

Let $\mathbf{s}^k = [(\mathbf{x}_1^k)^\top, \ldots, (\mathbf{x}_N^k)^\top]^\top$ be the $3N$-dimensional representation of the shape at the k-th frame of a collection of K images, with $\mathbf{x}_i^k = [x_i^k, y_i^k, z_i^k]^\top$ denoting the 3D coordinates of the i-th point. Traditional low-rank shape methods [5,11,22] approximate the shape $\mathbf{s}^k$ by a linear combination of a shape at rest $\mathbf{s}_0$ and Q rigid shapes $\mathbf{F} \in 3N \times Q$, weighted by time-varying coefficients $\boldsymbol{\gamma}^k = [\gamma_1^k, \ldots, \gamma_Q^k]^\top$:

$$\mathbf{s}^k = \mathbf{s}_0 + \mathbf{F}\boldsymbol{\gamma}^k. \tag{1}$$

In all theses approaches the collection of images is assumed to be temporally ordered (i.e., the superscript ^{k} conveys time information), such that the deformation between two consecutive frames k and $k+1$ changes smoothly. Additionally, it is assumed that the shapes $\mathbf{s}^k$ for $k = \{1, \ldots, K\}$ belong to the same object.

The proposed new formulation does not impose both these constraints: we let the K images of the collection to be acquired from different viewpoints that do not follow a smooth path, and the images may belong to C different instances of the same class of object (e.g., faces of different individuals). For doing so, we draw inspiration on the LDA [38,39], and propose a model that besides the term $\mathbf{F}\boldsymbol{\gamma}^k$ in Eq. (1) representing the deformations undergone by one single object (we call them *within-object* deformations), it incorporates a term that approximates the deformation among different objects of the same class (we call them *between-object* deformations). More specifically, an object instance $c \in \{1, \ldots, C\}$ at image frame k can be approximated as:

$$\mathbf{s}^{k,c} = \mathbf{s}_0 + \mathbf{B}\boldsymbol{\psi}^c + \mathbf{F}\boldsymbol{\gamma}^k, \tag{2}$$

where $\mathbf{B}$ is a $3N \times B$ matrix containing the B shape basis vectors of the between-individual subspace and $\boldsymbol{\psi}^c = [\psi_1^c, \ldots, \psi_B^c]^\top$ are their corresponding class-varying weight coefficients.

Equation (2) can be interpreted as follows: the term $\mathbf{s}_0 + \mathbf{B}\boldsymbol{\psi}^c$ allows encoding the 3D shape manifold, but not the object particularities of the k-th frame, and thus, we do not index it with the superscript ^{k}. On the other hand, the term $\mathbf{F}\boldsymbol{\gamma}^k$ is intended to encode the within-individual deformations, which are specific for each frame k. Additionally, note that the formulation considers a vector of coefficients $\boldsymbol{\psi}^c$ specific per each object c. This assumes that the partition of the K images into C object classes is known in advance. If this is not possible, or simply if the number of classes is unknown, we set $C = 1$.

Eventually, in the following section, we may compactly represent Eq. (2) as:

$$\mathbf{s}^{k,c} = \mathbf{s}_0 + \mathbf{M}_s \mathbf{m}^{k,c}, \tag{3}$$

where we define a $3N \times (B+Q)$ matrix $\mathbf{M}_s \equiv [\mathbf{B}, \mathbf{F}]$ and a vector of coefficients $\mathbf{m}^{k,c} \equiv [\boldsymbol{\psi}^{c\top}, \boldsymbol{\gamma}^{k\top}]^\top$.

4 Learning 3D Deformable Manifold, Shape and Motion

We now describe our approach to simultaneously learn the 3D deformable shape manifold $\mathbf{B}$, the shape basis $\mathbf{F}$ to code the time-varying deformations, their corresponding coefficients $\{\boldsymbol{\psi}, \boldsymbol{\gamma}\}$ and the camera motion from a collection of images.

4.1 Problem Formulation

Let us consider the $3N$-dimensional shape $\mathbf{s}^{k,c}$ of Eq. (2) is observed by an orthographic camera. The projection onto the image plane of the 3D points in frame k can be written as a $2N$ vector $\mathbf{w}^{k,c}$:

$$\mathbf{w}^{k,c} = \mathbf{G}^k \mathbf{s}^{k,c} + \mathbf{h}^k + \mathbf{n}^k, \tag{4}$$

where $\mathbf{G}^k = \mathbf{I}_N \otimes \mathbf{R}^k$ is the $2N \times 3N$ camera motion matrix, $\mathbf{I}_N$ is the N-dimensional identity matrix, $\mathbf{R}^k$ are the first two rows of a full rotation matrix, and $\otimes$ denotes the Kronecker product. Similarly, $\mathbf{h}^k = \mathbf{1}_N \otimes \mathbf{t}^k$ is a $2N$ vector resulting from concatenating N bidimensional translation vectors $\mathbf{t}^k$, and $\mathbf{1}_N$ is a N-vector of ones. Finally, $\mathbf{n}^k$ is a $2N$ dimensional vector of Gaussian noise that accounts for the unexplained data variation.

We can then define our problem as that of estimating for $k = \{1, \ldots, K\}$, the shape $\mathbf{s}^{k,c}$ and camera pose parameters $\{\mathbf{R}^k, \mathbf{t}^k\}$, given the 2D point observations $\mathbf{w}^{k,c}$ corrupted by noise $\mathbf{n}^k$ and the list the classes $c = \{1, \ldots, C\}$.

In order to make the problem tractable, we constrain the shape $\mathbf{s}^{k,c}$ to lie on the dual low-rank shape subspace defined by the manifold $\{\mathbf{B}, \mathbf{s}_0\}$ and by the within-object subspace $\mathbf{F}$. We therefore inject Eq. (2) into Eq. (4) to rewrite our observation model as:

$$\mathbf{w}^{k,c} = \mathbf{G}^k(\mathbf{s}_0 + \mathbf{B}\boldsymbol{\psi}^c + \mathbf{F}\boldsymbol{\gamma}^k) + \mathbf{h}^k + \mathbf{n}^k. \tag{5}$$

4.2 Probabilistic Formulation of the Problem

To simultaneously learn the 3D deformable shape manifold, the instance-specific shape and the camera pose from 2D point correspondences as described in Eq. (5), we propose an algorithm similar to the Probabilistic LDA approach used to represent shape distributions [24, 40, 41]. However, these previous formulations are intended to retrieve mappings that do not change the dimensionality of the input data. In our case, we aim at estimating a mapping that brings the

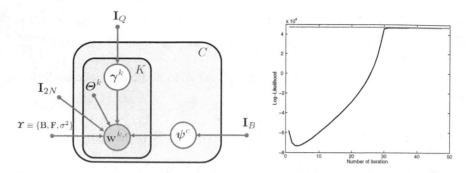

Fig. 2. Left: Graphical representation of our probabilistic NRSFM formulation with a dual low-rank shape model. Given the 2D observations $\mathbf{w}^{k,c}$ of K shapes belonging to C different object instances, the proposed approach learns, for each image frame, the pose parameters Θ^k and a shape model. The shape is represented by two low-rank matrices $\mathbf{B}$ and $\mathbf{F}$ approximating the deformation between- and within-objects, with their corresponding weights ϕ^c and γ^k, respectively. These latent variables and the 2D observations are assumed to be normally distributed with covariances $\mathbf{I}_B$, $\mathbf{I}_Q$ and $\sigma^2 \mathbf{I}_{2N}$, respectively, also learned from data. **Right:** Evolution of the log-likelihood function in Eq. (8) as a number of iterations, for a specific problem with 200 images and 157 points.

2D observations to 3D interpretations, i.e., we solve an inverse problem. This will need from a substantially different methodology.

In order to proceed we assume the between- and within-object coefficient vectors to be normally distributed, i.e., $\psi^c \sim \mathcal{N}(\mathbf{0}; \mathbf{I}_B)$ and $\gamma^k \sim \mathcal{N}(\mathbf{0}; \mathbf{I}_Q)$, respectively. Assuming these probabilistic priors, both vectors become latent variables that can be marginalized out and never need to be explicitly computed. We can then propagate the previous distributions to the deforming shapes on Eq. (2), yielding:

$$\mathbf{s}^{k,c} \sim \mathcal{N}\left(\mathbf{s}^{k,c}|\mathbf{s}_0; \mathbf{B}\mathbf{B}^\top + \mathbf{F}\mathbf{F}^\top\right). \tag{6}$$

Let us also consider a Gaussian prior distribution with variance σ^2 to model the noise over the shape observations, such that $\mathbf{n}^k \sim \mathcal{N}(\mathbf{0}; \sigma^2 \mathbf{I}_{2N})$. Any remaining variation on the observations that is not explained by the shape parameters is described as noise. Since both latent variables follow a Gaussian prior distribution, the distribution of the observed variables $\mathbf{w}^{k,c}$ on Eq. (5) is also Gaussian:

$$\mathbf{w}^{k,c} \sim \mathcal{N}\left(\mathbf{w}^{k,c}|\mathbf{G}^k\mathbf{s}_0 + \mathbf{h}^k; \mathbf{G}^k\mathbf{B}\mathbf{B}^\top\mathbf{G}^{k^\top} + \mathbf{G}^k\mathbf{F}\mathbf{F}^\top\mathbf{G}^{k^\top} + \sigma^2\mathbf{I}_{2N}\right). \tag{7}$$

In order to learn the parameters of this distribution we use an Expectation-Maximization (EM) algorithm, as done in other NRSfM approaches [5,7,9]. However, this approach will need a bit more of machinery, as previous methods did only consider one single latent variable. Here we are estimating two latent variables (γ^k and ψ^c) which will require to re-define the algorithm by including multiple consecutive E-steps.

4.3   Expectation Maximization

We next describe the details of the EM algorithm to learn the PLDA-inspired model from 2D point correspondences. Let us denote by $\boldsymbol{\Theta}^k \equiv \{\mathbf{R}^k, \mathbf{t}^k\}$ the set of pose parameters that need to be estimated per frame, and by $\boldsymbol{\Upsilon} \equiv \{\mathbf{B}, \mathbf{F}, \sigma^2\}$ the set of parameters that are common for all frames of the collection. Regarding the latent space, let $\boldsymbol{\Psi} = [\boldsymbol{\psi}^1, \ldots, \boldsymbol{\psi}^C]$ and $\boldsymbol{\Gamma} = [\boldsymbol{\gamma}^1, \ldots, \boldsymbol{\gamma}^K]$ be the between- and within-object latent variables, respectively.

Our problem consists in estimating the parameters $\boldsymbol{\Theta} = \{\boldsymbol{\Theta}^1, \ldots, \boldsymbol{\Theta}^K\}$ and $\boldsymbol{\Upsilon}$, given the set of 2D positions of all points $\mathbf{w} = \{\mathbf{w}^{1,\mathcal{C}(1)}, \ldots, \mathbf{w}^{K,\mathcal{C}(K)}\}$, where $\mathcal{C}(k) = c$ is a function that returns the instance label c associated to the k-th frame. Recall that we do not assume temporal coherence between two consecutive observations $\mathbf{w}^{k,\mathcal{C}(k)}$ and $\mathbf{w}^{k+1,\mathcal{C}(k+1)}$. See Fig. 2-Left for a representation of the problem as a graphical model.

The corresponding data likelihood we seek to maximize is therefore given by:

$$p(\mathbf{w}|\boldsymbol{\Theta}, \boldsymbol{\Upsilon}) \sim \prod_{k=1}^K p\left(\mathbf{w}^{k,\mathcal{C}(k)}|\mathbf{R}^k, \mathbf{t}^k, \mathbf{B}, \mathbf{F}, \sigma^2\right). \tag{8}$$

In order to maximize this equation, the EM algorithm we propose iteratively alternates between two steps: the E-step to obtain the distribution over latent coordinates and the M-step to update the model parameters. However, since our model contains two types of latent variables and several model parameters, we use partial E- and M- steps, as we next explain.

E-Steps: To estimate the posterior distribution over the latent variables $\boldsymbol{\psi}^c$ and $\boldsymbol{\gamma}^k$ given the current model parameters and observations, we propose executing two consecutive E-steps. Assuming independent and identically distributed random samples, and applying the Bayes' rule and the Woodbury's matrix identity [42], the distribution over $\boldsymbol{\Psi}$ can be shown to be:

$$p(\boldsymbol{\Psi}|\mathbf{w}, \boldsymbol{\Theta}, \boldsymbol{\Upsilon}) = \prod_{c=1}^C p(\boldsymbol{\psi}^c|\mathbf{w}, \boldsymbol{\Theta}, \boldsymbol{\Upsilon}) \sim \prod_{c=1}^C \mathcal{N}\left(\boldsymbol{\mu}_\psi^c; \boldsymbol{\Sigma}_\psi^c\right),$$

with:

$$\boldsymbol{\mu}_\psi^c = \boldsymbol{\Lambda}_\psi^c \sum_{k=1}^K \left(\mathbf{w}^{k,c} - \mathbf{G}^k \mathbf{s}_0 - \mathbf{h}^k\right) \mathbb{I}(k),$$

$$\boldsymbol{\Sigma}_\psi^c = \mathbf{I}_B - \boldsymbol{\Lambda}_\psi^c \left(\sum_{k=1}^K \mathbf{G}^k \mathbb{I}(k)\right) \mathbf{B},$$

$$\boldsymbol{\Lambda}_\psi^c = \mathbf{B}^\top \left(\sum_{k=1}^K \mathbf{G}^k \mathbb{I}(k)\right)^\top \left(\sigma^2 \mathbf{I}_{2N} + \left(\sum_{k=1}^K \mathbf{G}^k \mathbb{I}(k)\right)\left(\mathbf{F}\mathbf{F}^\top + \mathbf{B}\mathbf{B}^\top\right)\left(\sum_{k=1}^K \mathbf{G}^k \mathbb{I}(k)\right)^\top\right)^{-1},$$

where $\mathbb{I}(k) = 1$ if $\mathcal{C}(k) == c$, and zero otherwise. Note that this indicator function enforces computing $\boldsymbol{\mu}_\psi^c$, $\boldsymbol{\Sigma}_\psi^c$ and $\boldsymbol{\Lambda}_\psi^c$ using only the image frames k belonging to the object class c.

Once the distribution over $\boldsymbol{\Psi}$ is known, the distribution over $\boldsymbol{\Gamma}$ can be estimated by:

$$p\left(\boldsymbol{\Gamma}|\mathbf{w},\boldsymbol{\Theta},\boldsymbol{\Upsilon},\boldsymbol{\Psi}\right) = \prod_{k=1}^{K} p\left(\boldsymbol{\gamma}^{k}|\mathbf{w},\boldsymbol{\Theta},\boldsymbol{\Upsilon},\boldsymbol{\Psi}\right) \sim \prod_{k=1}^{K} \mathcal{N}\left(\boldsymbol{\mu}_{\gamma}^{k};\boldsymbol{\Sigma}_{\gamma}^{k}\right),$$

with:

$$\boldsymbol{\mu}_{\gamma}^{k} = \boldsymbol{\Lambda}_{\gamma}^{k}\left(\mathbf{w}^{k,\mathcal{C}(k)} - \mathbf{G}^{k}\mathbf{s}_{0} - \mathbf{G}^{k}\mathbf{B}\boldsymbol{\mu}_{\psi}^{c} - \mathbf{h}^{k}\right),$$

$$\boldsymbol{\Sigma}_{\gamma}^{k} = \mathbf{I}_{Q} - \boldsymbol{\Lambda}_{\gamma}^{k}\mathbf{G}^{k}\mathbf{F},$$

$$\boldsymbol{\Lambda}_{\gamma}^{k} = \sigma^{-2}\mathbf{F}^{\top}\mathbf{G}^{k^{\top}}\left(\mathbf{I}_{2N} - \sigma^{-2}\mathbf{G}^{k}\mathbf{F}\left(\mathbf{I}_{Q} + \sigma^{-2}\mathbf{F}^{\top}\mathbf{G}^{k^{\top}}\mathbf{G}^{k}\mathbf{F}\right)^{-1}\mathbf{F}^{\top}\mathbf{G}^{k^{\top}}\right).$$

M-Steps: We then replace the latent variables by their expected values and update the model parameters by optimizing the following negative log-likelihood function $\mathcal{A}(\boldsymbol{\Theta},\mathbf{w})$ with respect to the parameters $\boldsymbol{\Theta}^{k}$, for $k = \{1,\ldots,K\}$ and $\boldsymbol{\Upsilon}$:

$$\mathcal{A}(\boldsymbol{\Theta},\mathbf{w}) = \mathbb{E}\left[-\sum_{k=1}^{K}\log p(\mathbf{w}^{k,c}|\boldsymbol{\Theta}^{k},\boldsymbol{\Upsilon})\right]$$

$$= \frac{1}{2\sigma^{2}}\sum_{k=1}^{K}\mathbb{E}\left[\|\mathbf{w}^{k,c} - \mathbf{G}^{k}(\mathbf{s}_{0} + \mathbf{M}_{s}\mathbf{m}^{k,c}) - \mathbf{h}^{k}\|_{2}^{2}\right] + NK\log(2\pi\sigma^{2}).$$

Since this function cannot be minimized in closed form for all parameters, we perform partial M-steps over each of the model parameters. For doing so, we first consider the compact model on Eq. (3) and define the following expectations:

$$\boldsymbol{\mu}_{\mathrm{m}}^{k,c} = \mathbb{E}[\mathbf{m}^{k,c}] = \left[\boldsymbol{\mu}_{\psi}^{c}{}^{\top}\ \ \boldsymbol{\mu}_{\gamma}^{k}{}^{\top}\right]^{\top},\quad \hat{\boldsymbol{\mu}}_{\mathrm{m}}^{k,c} = [1\ \ \boldsymbol{\mu}_{\mathrm{m}}^{k,c}{}^{\top}]^{\top},$$

$$\boldsymbol{\phi}_{\mathrm{mm}}^{k,c} = \mathbb{E}[\mathbf{m}^{k,c}\mathbf{m}^{k,c}{}^{\top}] = \begin{bmatrix}\boldsymbol{\phi}_{\psi\psi}^{c} & \boldsymbol{\mu}_{\psi}^{c}\boldsymbol{\mu}_{\gamma}^{k}{}^{\top} \\ \boldsymbol{\mu}_{\gamma}^{k}\boldsymbol{\mu}_{\psi}^{c}{}^{\top} & \boldsymbol{\phi}_{\gamma\gamma}^{k}\end{bmatrix},\quad \hat{\boldsymbol{\phi}}_{\mathrm{mm}}^{k,c} = \begin{bmatrix}1 & \boldsymbol{\mu}_{\mathrm{m}}^{k,c}{}^{\top} \\ \boldsymbol{\mu}_{\mathrm{m}}^{k,c} & \boldsymbol{\phi}_{\mathrm{mm}}^{k,c}\end{bmatrix},$$

where $\boldsymbol{\phi}_{\psi\psi}^{c} = \mathbb{E}[\boldsymbol{\psi}^{c}\boldsymbol{\psi}^{c}{}^{\top}] = \boldsymbol{\Sigma}_{\psi}^{c} + \boldsymbol{\mu}_{\psi}^{c}\boldsymbol{\mu}_{\psi}^{c}{}^{\top}$ and $\boldsymbol{\phi}_{\gamma\gamma}^{k} = \mathbb{E}[\boldsymbol{\gamma}^{k}(\boldsymbol{\gamma}^{k})^{\top}] = \boldsymbol{\Sigma}_{\gamma}^{k} + \boldsymbol{\mu}_{\gamma}^{k}(\boldsymbol{\mu}_{\gamma}^{k})^{\top}$.

To update each of the individual model parameters, we set $\partial\mathcal{A}/\partial\boldsymbol{\Theta} = 0$ for each parameter on $\boldsymbol{\Theta}$. The update rules can be shown to be:

$$\mathrm{vec}(\mathbf{M}_{s}) \leftarrow \left(\sum_{k=1}^{K}\left(\boldsymbol{\phi}_{\mathrm{mm}}^{k,c}{}^{\top}\otimes\left(\mathbf{G}^{k}{}^{\top}\mathbf{G}^{k}\right)\right)\right)^{-1}\mathrm{vec}\left(\sum_{k=1}^{K}\mathbf{G}^{k}{}^{\top}\left(\mathbf{w}^{k,c} - \mathbf{G}^{k}\mathbf{s}_{0} - \mathbf{h}^{k}\right)\boldsymbol{\mu}_{\mathrm{m}}^{k,c}{}^{\top}\right),$$

$$\mathbf{R}^{k} \leftarrow \underset{\mathrm{s.t.}\mathbf{R}^{k}\mathbf{R}^{k^{\top}}=\mathbf{I}_{2}}{\arg\min}\ \|\mathbf{R}^{k}\sum_{i=1}^{N}\left(\mathbf{M}_{s}\hat{\boldsymbol{\phi}}_{\mathrm{mm}}^{k,c}\mathbf{M}_{s}^{\top}\right)_{i} - \sum_{i=1}^{N}(\mathbf{w}_{i}^{k,c} - \mathbf{t}^{k})\left(\mathbf{M}_{s}\hat{\boldsymbol{\mu}}_{\mathrm{m}}^{k,c}\right)_{i}^{\top}\|_{\mathcal{F}}^{2},$$

$$\mathbf{t}^k \leftarrow \frac{1}{N} \sum_{i=1}^{N} \left(\mathbf{w}_i^{k,c} - \mathbf{R}^k (\mathbf{s}_{0,i} + (\mathbf{M}_s \boldsymbol{\mu}_{\mathbf{m}}^{k,c})_i) \right),$$

$$\sigma^2 \leftarrow \frac{1}{2NK} \sum_{k=1}^{K} \left(\| \mathbf{w}^{k,c} - \mathbf{G}^k \mathbf{s}_0 - \mathbf{h}^k \|^2 - 2 \left(\mathbf{w}^{k,c} - \mathbf{G}^k \mathbf{s}_0 - \mathbf{h}^k \right)^\top \times \right.$$

$$\left. \times \mathbf{G}^k \mathbf{M}_s \boldsymbol{\mu}_{\mathbf{m}}^{k,c} + \mathrm{tr} \left(\mathbf{M}_s^\top \mathbf{G}^{k\top} \mathbf{G}^k \mathbf{M}_s \boldsymbol{\phi}_{\mathbf{mm}}^{k,c} \right) \right),$$

where $\mathbf{w}^{k,c} = [(\mathbf{w}_1^{k,c})^\top, \ldots, (\mathbf{w}_N^{k,c})^\top]^\top$ is a $2N$-dimensional vector, $\mathbf{w}_i^{k,c}$ are 2D coordinates of the i-th point in frame k, and $(\mathbf{M}_s \boldsymbol{\mu}_{\mathbf{m}}^{k,c})_i$ is the i-th 3D point of the $3N$ vector $\mathbf{M}_s \boldsymbol{\mu}_{\mathbf{m}}^{k,c}$. To solve the optimization for $\mathbf{R}^k$ we use a non-linear minimization routine.

The overall process is quite efficient, and requires, in average, a few tens of iterations to converge for collections of a few hundreds of images. Figure 2-Right plots the evolution of the log-likelihood of Eq. (8) for a collection of 200 images with 157 points each. In this case, the algorithm converged in 30 iterations, taking 323 seconds on a laptop with an Intel Core i7 processor at 2.4 GHz.

We initialize motion parameters by rigid factorization (similar to shape-based NRSfM approaches [5,12,15]) and the dual low-rank model by means of a coarse-to-fine approach, where each basis that is added explains as much of the deformable motion variance as possible. Additionally, since we are estimating global models we can handle occlusions, and missing observations can be easily inferred from the observed data. We will demonstrate this robustness in the results section.

5   Experiments

We next report quantitative and qualitative results of our method on face reconstruction. These results can be best viewed in the supplemental video[1]. For the quantitative results, we report the mean 3D reconstruction error [%] as defined in [12,15].

5.1   Synthetic Images

To quantitatively validate our method, we first consider a synthetic sequence with 3D ground truth. From the real and dense mocap data of [43], we render a sequence of 200 frames and 157 points per frame (denoted in the following as *Face* sequence), in which one person performs several face movements and gestures. We randomly shuffle the frames ordering in order to build a collection of images without temporal coherence. For this specific experiment we do not test the multi-object instance scenario. This is not a problem for our model though, despite it considers the between- and within-object subspaces. We could have forced the between-object subspace to be zero, but decided not doing so, and show that our approach can generalize from one to several objects.

---

[1] Videos can be found on website: http://www.iri.upc.edu/people/aagudo.

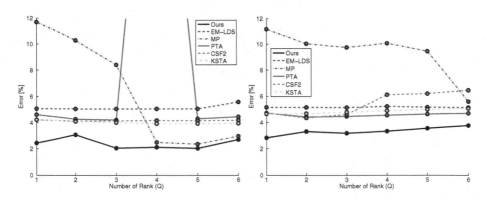

Fig. 3. Mean 3D reconstruction error in the *Face* sequence, as a function of basis rank Q for state-of-the-art methods: EM-LDS [5], MP [12], PTA [19], CSF2 [20], KSTA [6] and EM-PND [7]; and our approach. For the EM-PND [7], the 3D reconstruction error was 21.0% and 21.2% for the noise-less and noisy cases, respectively. For our approach, the between-object rank was set to a fixed value of $B = 3$. **Left:** Noise-less 2D measurements. **Right:** Noisy 2D measurements. Best viewed in color.

We use this sequence to compare the proposed method against several low-rank shape, trajectory and shape-trajectory methods in state of the art. In particular, we consider: EM-LDS [5], the metric projections MP [12], SPM [13] and EM-PND [7] for shape space; the point trajectory approach (PTA) [19] for trajectory space; the column space fitting (CSF2) [20] and the kernel shape trajectory approach (KSTA) [6] for shape-trajectory methods. The parameters of these methods were set as suggested in the original papers. In our case, we only have to set the rank B and Q of the between- and within-object subspaces, respectively.

We conduced experiments with and without noise in the observations. For the noisy case, we corrupted the measurements using additive zero-mean Gaussian noise with standard deviation $\sigma_{noise} = 0.02\kappa$, where κ denotes the maximum distance of a 2D point observation to the mean position of all observations. In Fig. 3, we plot the mean 3D reconstruction error as a function of the within-object basis rank Q, for our method and the other seven methods aforementioned. In our formulation the between-object basis rank was not accurately tuned and it was set to a constant value of $B = 3$. Observe that our approach consistently outperforms the rest of competing approaches for both the noise and noiseless experiments. Since EM-PND [7] does not need to set the basis rank, we did not include this method in the graph, and just report its reconstruction error, which is of 21.0%, far above from the rest of methods. Regarding SPM [13], the approach is not applicable to larger ranks as the number of linear-matrix-inequality constraints is not sufficient to solve this case, obtaining an error of 10.69% for $Q = 1$. It is worth noting that our results for $Q = 1$ are remarkably better than other approaches for $Q \geq 4$, that is our equivalent rank if we consider the 3 vectors of B. In Fig. 4 we represent the significance of the reconstruction error

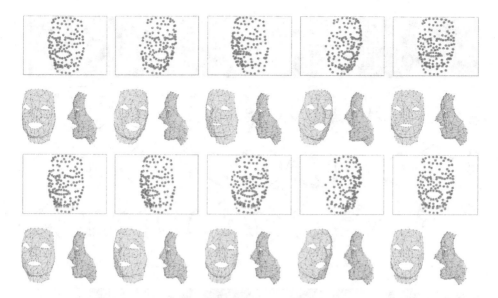

Fig. 4. Synthetic results on the *Face* sequence. We show the input images at the top, and at the bottom a frontal and side views of the reconstructed shapes. For all cases, we display the results with $Q = 3$ and $B = 3$. Best viewed in color.

values, and show some qualitative results, including the 2D input data and our reconstructed 3D shape.

5.2   Real Images

For the real data we consider two experiments with human faces of two or more individuals.

In the first scenario we process an American Sign Language (ASL) database, that consists in a collection of 229 images belonging to 2 subjects (male and female) with 77 feature points per frame. One of the challenges of this dataset is that some of the frames are heavily occluded by one or two hands, or by the own rotation of the face. The dataset is built from two sequences previously used to test NRSfM algorithms: the ASL1, consisting of 115 frames with a 17.4% of missing data [33], and the ASL2 sequence, consisting of 114 frames with a 11.5% of missing data [20]. Before processing the collection of images, we shuffle the frames to break the temporal continuity. In Fig. 5 we show two views per image of the shape estimated by our method, CSF2 [20] and KSTA [6]. For our approach we set $C = 1$. By doing this we ensure a fair comparison with the other two approaches, as we do not exploit the fact that the identity of each individual is known. Additionally we set $B = 2$ and $Q = 3$. From Fig. 5 we can observe that our model yields a qualitatively correct estimation of the shape. However, while CSF2 [20] provides very good results when processing the two sequences ASL1 and ASL2 independently, it is prone to fail when merging their

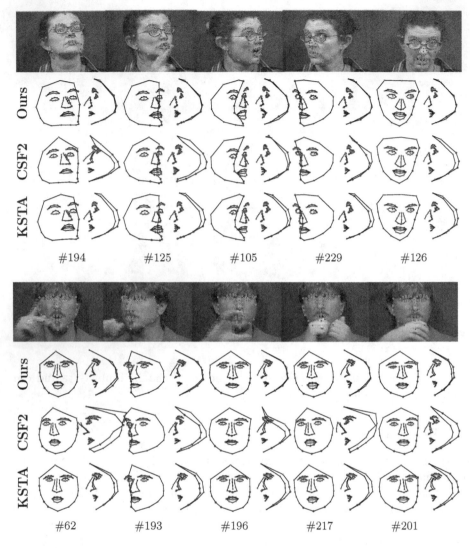

Fig. 5. ASL database. In each row we show the same information. **Top:** 2D tracking data and reconstructed 3D shape reprojected into several images with green circles and red squares, respectively. Blue squares correspond to missing points. **Bottom:** Camera frame and side-views of the reconstructed 3D shape: our solution, CSF2 [20] and KSTA [6], respectively. We also represent the number of image k in the input data, showing as different objects are intercalated. Best viewed in color.

data and shuffling the frames, using exactly the same rank of the subspace. Note the completely wrong estimation of the nose in some of the frames. This is relieved by KSTA [6], although by providing a quasi-rigid solution with almost no deformation adaption. Observe, for instance, that the lips remain always closed (see for example frames #105 and #62 of the upper and lower sequence in Fig. 5). Although only qualitatively, our approach correctly retrieves these deformations.

In order to bring some quantitative results to this analysis, we have built a pseudo-ground truth of this database (real 3D ground truth is not available) by independently processing ASL1 and ASL2 using the NRSfM approach proposed in [9] –the method that seems to provide best performance for these sequences– and compared its 3D reconstructions with the ones obtained when simultaneously processing the faces of the two persons. A summary of these results are provided in Table 1. In this case, we obtain the following errors: CSF2 [20] (14.93%), KSTA [6] (3.62%), EM-PFS [9] (8.37%), our approach (2.66%). If we specifically focus on the lips reconstruction, which is highly deformable, as expected the differences become more clear: CSF2 [20] (12.74%), KSTA [6] (11.08%), EM-PFS [9] (7.79%), our approach (4.53%). It is worth to point that the performance of EM-PFS [9] degrades when jointly processing images of ASL1 and ASL2. We presume the intrinsic physical model considered by this approach is sensitive to the differences between the two individuals. Additionally, we can exploit the list the classes c in our formulation. In this case, we obtain more accurate solutions: 2.50% considering all object shape and 4.35% for lips reconstruction.

Table 1. Quantitative comparison on ASL database. Comparison of our approach against CSF2 [20], KSTA [6] and EM-PFS [9] for the full face shape and the corresponding lips area in terms of 3D error [%]. In all cases we show the minimum error with the number of rank Q in the subspace (in brackets).

Data	Method				
	CSF2 [20]	KSTA [6]	EM-PFS [9]	Ours ($C=1$)	Ours ($C=2$)
Face	14.93(6)	3.62(6)	8.37(5)	**2.66(3)**	**2.50(3)**
Lips	12.74(6)	11.08(6)	7.79(5)	**4.53(3)**	**4.35(3)**

In the final experiment we evaluate our approach on a subset of the MUCT face database [44]. We gather an heterogeneous collection of 302 images belonging including very different face morphologies, poses and expressions. We ensure the input images contain similar numbers of males and females, and a cross section of ages and races. Once the images are chosen, we obtain the 2D observations using an off-the-shelf 2D active appearance model [45]. Again, in order to highlight the generality of our approach, we do not take advantage of the fact of knowing the object label of each input image, and we set $C = 1$. Figure 6 shows the qualitative results of our approach. We can observe how our method provides results that seem very realistic and correlate with the appearance of the images. Note, for instance that some of the faces have an overall thin shape (see woman in the first column) while other convey a quite round shape (see woman in the second column).

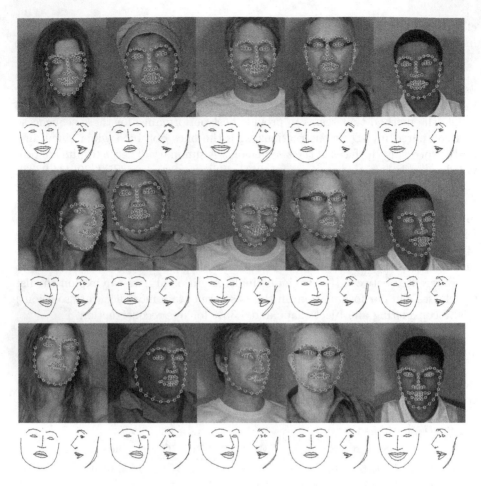

Fig. 6. MUCT database. In each row we show the same information. **Top:** 2D tracking data and reconstructed 3D shape reprojected into several images with green circles and red squares, respectively. **Bottom:** Camera frame and side-views of the reconstructed 3D shape. Best viewed in color.

6   Conclusion

In this paper we have proposed a new formulation of the NRSfM problem that allows dealing with collections of images with no temporal coherence, and including several instances of the same class of object. In order to make this possible we have proposed a dual low-rank space that separately models the deformations within each specific object and the deformations between individuals. These low-rank subspaces are learned using a variant of the probabilistic linear discriminant analysis, and an EM strategy that consecutively executes several expectation and maximization steps. We validate the approach in synthetic image collections –with one single object instance– and show improved results compared to state-of-the-art. Real results on datasets including faces of two or more persons, depict an even larger gap with previous NRSfM approaches. In the future we aim at extending our approach to different types of datasets, not only human faces. For instance, we believe we can readily exploit our formulation to learn deformable shape manifolds of human full body motions. Other fields, such as computer graphics animation could also benefit from this approach and transfer motion styles between different characters.

Acknowledgments. This work has been partially supported by the Spanish Ministry of Science and Innovation under project RobInstruct TIN2014-58178-R; by the ERA-net CHISTERA projects VISEN PCIN-2013-047 and I-DRESS PCIN-2015-147. The authors also thank Gerard Canal for fruitful discussions.

References

1. Hartley, R., Zisserman, A.: Multiple View Geometry in Computer Vision. Cambridge University Press, Cambridge (2000)
2. Triggs, B., McLauchlan, P.F., Hartley, R.I., Fitzgibbon, A.W.: Bundle adjustment - a modern synthesis. Vis. Algorithms Theory Pract. **1883**, 298–372 (2000)
3. Agarwal, S., Snavely, N., Simon, I., Seitz, S.M., Szeliski, R.: Building Rome in a day. In: ICCV (2009)
4. Lim, J., Frahm, J., Pollefeys, M.: Online environment mapping. In: CVPR (2011)
5. Torresani, L., Hertzmann, A., Bregler, C.: Nonrigid structure-from-motion: estimating shape and motion with hierarchical priors. TPAMI **30**, 878–892 (2008)
6. Gotardo, P.F.U., Martinez, A.M.: Kernel non-rigid structure from motion. In: ICCV (2011)
7. Lee, M., Cho, J., Choi, C.H., Oh, S.: Procrustean normal distribution for non-rigid structure from motion. In: CVPR (2013)
8. Chhatkuli, A., Pizarro, D., Bartoli, A.: Non-rigid shape-from-motion for isometric surfaces using infinitesimal planarity. In: BMVC (2014)
9. Agudo, A., Moreno-Noguer, F.: Learning shape, motion and elastic models in force space. In: ICCV (2015)
10. Bregler, C., Hertzmann, A., Biermann, H.: Recovering non-rigid 3D shape from image streams. In: CVPR (2000)
11. Bartoli, A., Gay-Bellile, V., Castellani, U., Peyras, J., Olsen, S., Sayd, P.: Coarse-to-fine low-rank structure-from-motion. In: CVPR (2008)
12. Paladini, M., Del Bue, A., Stosic, M., Dodig, M., Xavier, J., Agapito, L.: Factorization for non-rigid and articulated structure using metric projections. In: CVPR (2009)
13. Dai, Y., Li, H., He, M.: A simple prior-free method for non-rigid structure from motion factorization. In: CVPR (2012)
14. Zhu, Y., Huang, D., De La Torre, F., Lucey, S.: Complex non-rigid motion 3D reconstruction by union of subspaces. In: CVPR (2014)
15. Garg, R., Roussos, A., Agapito, L.: Dense variational reconstruction of non-rigid surfaces from monocular video. In: CVPR (2013)
16. Paladini, M., Bartoli, A., Agapito, L.: Sequential non rigid structure from motion with the 3D implicit low rank shape model. In: ECCV (2010)
17. Agudo, A., Montiel, J.M.M., Agapito, L., Calvo, B.: Online dense non-rigid 3D shape and camera motion recovery. In: BMVC (2014)

18. Lee, M., Choi, C.H., Oh, S.: A procrustean Markov process for non-rigid structure recovery. In: CVPR (2014)
19. Akhter, I., Sheikh, Y., Khan, S., Kanade, T.: Non-rigid structure from motion in trajectory space. In: NIPS (2008)
20. Gotardo, P.F.U., Martinez, A.M.: Non-rigid structure from motion with complementary rank-3 spaces. In: CVPR (2011)
21. Agudo, A., Moreno-Noguer, F., Calvo, B., Montiel, J.M.M.: Sequential non-rigid structure from motion using physical priors. TPAMI **38**, 979–994 (2016)
22. Agudo, A., Agapito, L., Calvo, B., Montiel, J.M.M.: Good vibrations: a modal analysis approach for sequential non-rigid structure from motion. In: CVPR (2014)
23. Agudo, A., Moreno-Noguer, F.: Simultaneous pose and non-rigid shape with particle dynamics. In: CVPR (2015)
24. Li, P., Fu, Y., Mohammed, U., Elder, J.H., Prince, S.J.D.: Probabilistic models for inference about identity. TPAMI **34**, 144–157 (2012)
25. Blanz, V., Vetter, T.: A morphable model for the synthesis of 3D faces. In: SIGGRAPH (1999)
26. Agudo, A., Montiel, J.M.M., Agapito, L., Calvo, B.: Modal space: a physics-based model for sequential estimation of time-varying shape from monocular video. JMIV **57**(1), 75–98 (2016)
27. Barbic, J., James, D.: Real-time subspace integration for St. Venant-Kirchhoff deformable models. TOG **24**, 982–990 (2005)
28. Agudo, A., Montiel, J.M.M., Calvo, B., Moreno-Noguer, F.: Mode-shape interpretation: re-thinking modal space for recovering deformable shapes. In: WACV (2016)
29. Xiao, J., Chai, J., Kanade, T.: A closed-form solution to non-rigid shape and motion. IJCV **67**, 233–246 (2006)
30. Tomasi, C., Kanade, T.: Shape and motion from image streams under orthography: a factorization approach. IJCV **9**, 137–154 (1992)
31. Del Bue, A., Llado, X., Agapito, L.: Non-rigid metric shape and motion recovery from uncalibrated images using priors. In: CVPR (2006)
32. Valmadre, J., Lucey, S.: General trajectory prior for non-rigid reconstruction. In: CVPR (2012)
33. Gotardo, P.F.U., Martinez, A.M.: Computing smooth time-trajectories for camera and deformable shape in structure from motion with occlusion. TPAMI **33**, 2051–2065 (2011)
34. Simon, T., Valmadre, J., Matthews, I., Sheikh, Y.: Separable spatiotemporal priors for convex reconstruction of time-varying 3D point clouds. In: Fleet, D., Pajdla, T., Schiele, B., Tuytelaars, T. (eds.) ECCV 2014. LNCS, vol. 8691, pp. 204–219. Springer, Cham (2014). doi:10.1007/978-3-319-10578-9_14
35. Sigal, L., Bhatia, S., Roth, S., Black, M.J., Isard, M.: Tracking loose-limbed people. In: CVPR (2004)
36. Wang, J.M., Fleet, D.J., Hertzmann, A.: Gaussian process dynamical models. In: NIPS (2005)
37. Urtasun, R., Fleet, D., Fua, P.: 3D people tracking with Gaussian process dynamical models. In: CVPR (2006)
38. Fisher, R.A.: The statistical utilization of multiple measurements. Ann. Eugenics **8**, 376–386 (1938)
39. Rao, C.R.: The utilization of multiple measurements in problems of biological classification. J. R. Stat. Soc. B **10**, 159–203 (1948)
40. Prince, S.J.D., Elder, J.H.: Probabilistic linear discriminant analysis for inferences about identity. In: ICCV (2007)

41. Ioffe, S.: Probabilistic linear discriminant analysis. In: Leonardis, A., Bischof, H., Pinz, A. (eds.) ECCV 2006. LNCS, vol. 3954, pp. 531–542. Springer, Heidelberg (2006). doi:10.1007/11744085_41
42. Woodbury, M.A.: Inverting modified matrices. Statistical Research Group, Memorandum Report 42 (1950)
43. Akhter, I., Simon, T., Khan, S., Matthews, I., Sheikh, Y.: Bilinear spatiotemporal basis models. TOG **31**, 17:1–17:12 (2012)
44. Milborrow, S., Morkel, J., Nicolls, F.: The MUCT landmarked face database. Pattern Recognition Association of South Africa (2010)
45. Cootes, T.F., Edwards, G.J., Taylor, C.J.: Active appearance models. In: Burkhardt, H., Neumann, B. (eds.) ECCV 1998. LNCS, vol. 1407, pp. 484–498. Springer, Heidelberg (1998). doi:10.1007/BFb0054760

Robust Multi-Model Fitting Using Density and Preference Analysis

Lokender Tiwari[1](✉), Saket Anand[1], and Sushil Mittal[2]

[1] IIIT-Delhi, New Delhi, India
lokendert@iiitd.ac.in
[2] Scibler Corporation, Santa Clara, USA

Abstract. Robust multi-model fitting problems are often solved using consensus based or preference based methods, each of which captures largely independent information from the data. However, most existing techniques still adhere to either of these approaches. In this paper, we bring these two paradigms together and present a novel robust method for discovering multiple structures from noisy, outlier corrupted data. Our method adopts a random sampling based hypothesis generation and works on the premise that inliers are densely packed around the structure, while the outliers are sparsely spread out. We leverage consensus maximization by defining the residual density, which is a simple and efficient measure of density in the 1-D residual space. We locate the inlier-outlier boundary by using preference based point correlations together with the disparity in residual density of inliers and outliers. Finally, we employ a simple strategy that uses preference based hypothesis correlation and residual density to identify one hypothesis representing each structure and their corresponding inliers. The strength of the proposed approach is evaluated empirically by comparing with state-of-the-art techniques over synthetic data and the AdelaideRMF dataset.

1   Introduction

Many computer vision applications require estimation of parameters of a mathematical model from a given set of observations. These observations (or features) are typically obtained through a process agnostic to the model being fit and therefore may contain gross outliers, i.e., points that do not belong to any structure. In order to fit the correct model and identify the inlier points, it is imperative for the estimator to be robust to gross outliers and have a high breakdown point. In cases when observations arise from multiple structures, inliers of one structure appear as outliers to the other structures. Such points have come to be called as pseudo-outliers. With multiple structures in the data, the fraction of outliers (both gross and pseudo) can quickly go in excess of 90%. This makes the requirement of a higher breakdown point more stringent.

Traditional robust regression techniques like least median of squares prove to be lacking due to low breakdown points. To achieve higher levels of robustness, random sampling based methods have become popular since RANdom SAmpling

© Springer International Publishing AG 2017
S.-H. Lai et al. (Eds.): ACCV 2016, Part IV, LNCS 10114, pp. 308–323, 2017.
DOI: 10.1007/978-3-319-54190-7_19

Consensus (RANSAC) algorithm [4]. RANSAC and its many variants rely on consensus set maximization for selecting a model from a set of model hypotheses generated by random sampling. In practice, RANSAC and its variants have been very successful at several geometric fitting problems in computer vision [7,14,18]. However, its performance crucially depends on a user-specified inlier threshold measuring the *scale* of the inlier noise.

Apart from RANSAC like *consensus* based methods, another class of robust mutli-model fitting algorithms that have emerged use *preference analysis* [1,3,8, 9,12,13]. These methods rely on rank-ordering the model hypotheses to generate a preference list for each data point. These preference lists define a feature space where an appropriate distance measure like Jaccard distance [13] or Tanimoto distance [8] could be used to cluster the data points. Preference analysis has also been applied successfully to guided sampling for hypothesis generation [3,15], where preference lists were used to compute similarity between points. In order to recover all the structures in the data, these methods also require a user-defined parameter that is commensurate to the scale of inlier noise [8,13].

User defined thresholds are not amenable to real-world scenarios where the inlier noise scale may vary over time or may not be known a priori. Moreover, in case of multiple structures, each structure may have different inlier noise distribution, necessitating a data-driven strategy for scale estimation. Inliers yield a small residual value whereas outliers (or pseudo-outliers) can have arbitrarily large residuals. Geometrically, we can say that inliers are densely packed around the regression surface, while outliers are spread sparsely in the ambient space. The density of points around the regression surface is a key factor in distinguishing between inliers and outliers.

Figure 1 uses an illustrative example to explain the different steps involved in our proposed technique. Figure 1a shows an example ('sene') from the

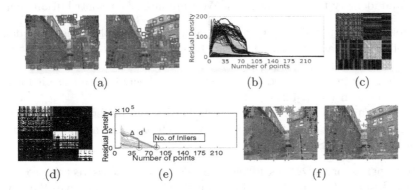

Fig. 1. An illustrative example - Density Preference Analysis (DPA), (a) Data, (b) Residual density profile with darker curves indicating selected candidate hypotheses, (c) Density based point correlation, (d) Strong inlier candidates of each hypothesis arrange in the order of structure membership (row wise) and point membership (column wise), (e) Inlier scale (fraction) estimation, (f) Model selection and final segmentation.

AdelaideRMF [16] planar segmentation dataset. The task is to fit two homographies corresponding to the two planes in the scene and identify the inliers in both cases. We first generate model hypotheses, either using a random sampling or a guided sampling based approach. For each hypothesis, the residual density profile (Sect. 3.1) is computed. Figure 1b shows the residual density plotted against the points ordered in increasing order of residuals. The grey curves show the residual density profiles for *all* model hypotheses, while the dark curves show those of selected candidate hypotheses (Sect. 4.1). We compute the density based point preferences (Sect. 3.3), which is used to obtain the point correlation matrix as shown in Fig. 1c. For better visualization, the points are ordered by group membership, with the bottom two blocks representing inliers corresponding to the two structures. The point correlation matrix and the selected hypotheses are used to identify strong inlier candidates for each hypothesis as shown in Fig. 1d. The rows correspond to the selected candidate hypotheses, while the columns represent the points. Again, for better visualization, both rows and columns have been reordered using ground truth group membership. For each row, the white pixel indicates a strong inlier candidate. We can see that several rows have strong inliers that comprise of points mostly from the same group. We use these points and the density drop rate to estimate the scale for each model hypothesis (Sect. 4.2). Figure 1e shows the number of inliers identified using the scale estimates. In Fig. 1f, we show the final segmentation obtained by applying our model selection algorithm (Sect. 4.3).

Consensus based methods capture the quality of each model hypothesis based on the consensus set maximization criterion. On the other hand, preference based methods naturally allow measuring the similarity (distance) between a pair of points or a pair of models. By leveraging both these strategies, our technique is able to automatically detect the number of structures in the data, estimate their scale of inlier noise for each structure, perform the point-model assignment of inliers, and identify the gross outliers. We summarize our contributions below:

- We define the residual density profile (Sect. 3.1), a simple measure of density that effectively captures the disparity between density of inliers and outliers for a given model.
- We define the density based point preferences (Sect. 3.3) and use it for inlier scale estimation (Sect. 4.2).
- We develop a simple algorithm that uses density and preference based hypothesis correlations for selecting one model for each structure and identifying their associated inliers (Sect. 4.3).

The paper is organized as follows. In Sect. 2 we discuss recent progress in robust multi-model fitting to put our contributions in perspective. We develop the preliminaries in Sect. 3 and describe our proposed method in Sect. 4. We show experimental results and comparisons on synthetic and real data in Sect. 5 and conclude with a discussion of future directions in Sect. 6.

2   Related Work

A large amount of work has been done in robust model fitting over the last few decades. In this section, we only discuss the work in context of the proposed method, mainly focusing on consensus based and preference based approaches for discovering *multiple* structures. We also discuss a few recent techniques that address multi-model fitting, albeit they don't strictly fall in these categories.

Consensus Based Methods: Consensus based methods like RANSAC [4] and its variants work on the premise that valid model hypotheses generate small residuals for inlier points, i.e., inlier points form dense clusters close to zero in the residual space. In [5], the RANSAC method was extended to extract multiple homographies by using a sequential *fit and remove* approach, where the algorithm detects one structure using RANSAC, removes the associated inliers and repeats the process until no more structures are found. This sequential approach has limitations because it can remove inliers erroneously (in case of incorrect scale estimates, or overlapping structures or both), leading to difficulties in recovery of other structures in the following repetitions. The multiRANSAC [18] method avoids the sequential approach by maintaining inlier sets of all structures in parallel. However, it requires the user to input the number of structure along with their respective inlier scales.

Torr et al. [14] proposed MLESAC and MSAC that use a maximum likelihood and M estimator based criterion respectively. The generalized projection based M-estimator (gpbM) [10] was designed to deal with *heteroscedastic*, i.e., point-dependent noise in the data, which is often encountered in geometric computer vision problems. It estimates the scale automatically and recovers the inlier structure by maximizing the heteroscedastic kernel density estimate in the residual space. Since a single hypothesis is selected based on the maximum density, gpbM is constrained to operate in a sequential *fit-and-remove* manner.

Preference Based Methods: Contrary to consensus based approaches, preference analysis reverse the roles of data points and model hypotheses. For each data point, residuals are computed with respect to a number of hypotheses. Given a data point, the data preference is defined as the set of hypotheses ordered by their residuals. Intuitively, the data point *prefers* a hypothesis if the corresponding residual is small.

In [19], data preferences are used to estimate the number of modes, followed by model selection and inlier recovery by analyzing histograms of residuals. Toldo and Fusiello [13] pointed out the difficulty of mode detection using residual histogram analysis in [19] and introduced a conceptual space, which represents each data point as a binary vector. A user-specified inlier threshold was used to generate the binary indicator vector identifying the preferred hypotheses. The Jaccard distance was then employed for agglomerative clustering in this conceptual space. In [1,2], the authors presented a data preferences based Mercer kernel, which following an outlier rejection step, was used for performing Kernel Principal Component Analysis (KPCA) based subspace clustering.

Wang et al. [17] introduced the Adaptive Kernel-Scale Weighted Hypotheses (AKSWH) algorithm by combining the Iterative K-th Order Scale Estimator (IKOSE) with a modified version of J-Linkage that clustered hypotheses instead of points. Following a post-processing hypothesis fusion step, the cluster representatives are chosen as the candidate models, which along with their IKOSE based scale estimates are used to identify the corresponding inliers.

More recently, T-linkage [8], a continuous version of J-linkage based on the Tanimoto distance was proposed. T-linkage also requires the user to adjust a tuning parameter approximating the scale of the inlier noise. Robust Preference Analysis (RPA) [9] used preference analysis along with low-rank approximation and nonnegative matrix factorization to effectively implement a robust form of spectral clustering. The affinity matrix for RPA is constructed by employing a Cauchy weighting function on the data preference lists. The Random Cluster Model Simulated Annealing (RCMSA)[11] formulated the multi-model fitting problem in a simulated annealing framework. It used data preferences by constructing a weighted graph, which was used to iteratively generate stronger model hypotheses using larger than minimal subsets.

In addition to robust multi-model fitting approaches, preference based methods have also been applied to guided sampling for hypothesis generation, which is an important component of any robust multi-model fitting method. Multi-GS [3], ITKSF [15] and DHF [15,16] are a few techniques that work well in practice.

3   Notation and Preliminaries

Consider a set of n data points $\mathcal{X} = \{\mathbf{x}_1, ..., \mathbf{x}_n\}$ in $\mathbb{R}^d$. Let the set of structures present in the data be denoted by $\mathcal{K} = \{(\boldsymbol{\theta}^{*1}, \mathcal{I}^{*1}), ..., (\boldsymbol{\theta}^{*k}, \mathcal{I}^{*k})\}$, where $\boldsymbol{\theta}^{*i}$ and $\mathcal{I}^{*i}$ denote the *true* model parameters and the index set of inlier points respectively of the ith structure. We define the *true* inlier fraction of each structure as $\eta^{*i} = \frac{|\mathcal{I}^{*i}|}{n}$, $i = 1, ..., k$ and $\eta^{*0} = 1 - \sum_{i=1}^{k} \eta^{*i}$ as the fraction of outliers.

As with most robust multi-model fitting methods, our method begins with a set of model hypotheses, generated by sampling minimal subsets of data. We denote this initial set of model hypotheses by $\vartheta_0 = \{\boldsymbol{\theta}^i | i = 1, ..., M_0\}$. In order to aid readability and comprehension, we will follow the convention where superscripts will always be indexed over the model hypotheses, while the subscripts will be indexed over the data points. Given a model hypothesis $\boldsymbol{\theta}^i$, the residual for point $\mathbf{x}_j$ is computed using the function $\phi(\boldsymbol{\theta}^i, \mathbf{x}_j) : \mathbb{R}^d \to \mathbb{R}_+$, where $\mathbb{R}_+$ denotes the nonnegative real half-line.

Our goal is to recover the correct number of structures k and label the data points in $\mathcal{X}$ as inliers of each of the structures by creating $\mathcal{L} = \{\ell_1, ..., \ell_n | \ell_i \in \{0, ..., k\}\}$, where the label 0 identifies the outliers. The inlier points can then be used for parameter estimation using an appropriate estimators such as least squares or M-estimators.

3.1   Residual Density Profile

For a given model hypothesis $\boldsymbol{\theta}^i$, we first compute the residual vector as

$$\mathbf{r}^i = [r^i_1 = \phi(\boldsymbol{\theta}^i, \mathbf{x}_1), ..., r^i_n = \phi(\boldsymbol{\theta}^i, \mathbf{x}_n)] \tag{1}$$

We find a permutation set $\mathbf{u}^i = [u^i_1, u^i_2 . . ., u^i_n]$ for model hypothesis $\boldsymbol{\theta}^i$ such that $r^i_{u^i_1} \leq r^i_{u^i_2} \leq . . ., \leq r^i_{u^i_n}$ and thus we get a *sorted residual* vector, which is then smoothed using an averaging filter of size $\lceil 0.025n \rceil$. The smoothed ordered residual vector $\boldsymbol{\rho}^i = [r^i_{u^i_1}, r^i_{u^i_2}, . . ., r^i_{u^i_n}]$ is used to define the residual density for $\boldsymbol{\theta}^i$ as

$$d^i_j = \frac{j}{\rho^i_j + \varepsilon} = \frac{j}{r^i_{u^i_j} + \varepsilon}, \qquad j = 1, ..., n \tag{2}$$

where ε is of the order of 10^{-4} and is used only to suppress very high densities. The corresponding residual density vector as $\mathbf{d}^i = [d^i_1, d^i_2, ..., d^i_n]$. This residual density estimate measures the number of points lying in a ball of radius ρ^i_j in the residual space of $\boldsymbol{\theta}^i$.

Using the residual density profile (2) has an advantage over residuals while dealing with outliers. The residual values of outliers are large, however, the magnitude varies with different model hypotheses. However, the residual density values for both gross outliers and pseudo-outliers are small as well as independent of the model hypothesis. This can be seen in the right part of Fig. 1b, where the variance of density values for points in the outlier region is very small compared to that of the inlier region. This property of the residual density profile helps us estimate the inlier scale accurately (Sect. 4.2).

3.2   Residual Based Hypothesis Preferences

Similar to [12,15], we use the residual based hypothesis preferences and define correlation between hypotheses. Given a hypothesis $\boldsymbol{\theta}^i$, its residual based preference list $\mathbf{u}^i$ is the rank ordering of points with smallest residual first. We compute the hypothesis correlation between two hypotheses $\boldsymbol{\theta}^i$ and $\boldsymbol{\theta}^j$ as

$$h(\boldsymbol{\theta}^i, \boldsymbol{\theta}^j) = \frac{1}{K} |u^i_{1:K} \cap u^j_{1:K}| \tag{3}$$

where $u^i_{1:K} = \{u^i_1, u^i_2, ..., u^i_K\}$ is the top-K residual based hypothesis preferences for $\boldsymbol{\theta}^i$. We set K to 10% of all the data points in our experiments.

Using (3), a pairwise hypothesis correlation matrix $\mathbf{H}$ can be computed with $\mathbf{H}_{ij} = h(\boldsymbol{\theta}^i, \boldsymbol{\theta}^j)$. However, random sampling based techniques usually lead to a large number of hypotheses. Since the complexity of computing $\mathbf{H}$ is quadratic in the number of hypotheses, it becomes impractical to compute the full $M \times M$ matrix $\mathbf{H}$. In Sect. 4.1 we suggest a way of selecting a small subset of promising hypotheses and mitigate this problem of complexity.

3.3   Density Based Point Preferences

In general, a point preference list for a given data point is a rank ordering of
the model hypotheses based on some criterion. In [3,15], a smallest residual first
criterion was used for ranking hypotheses. Chin et al. [3] made an important
empirical observation that inlier points belonging to the same structure have
highly correlated data preferences. As opposed to the traditional residual based
point preferences, we use *density* based point preferences, which in turn are used
to estimate correlation between data points.

We find a permutation $\mathbf{v}_j = [v_j^1, v_j^2, \ldots, v_j^{M_0}]$ for a data point $\mathbf{x}_j$ such that
$d_j^{v_j^1} \geq d_j^{v_j^2} \geq, \ldots, \geq d_j^{v_j^1}$. The permutation $\mathbf{v}_j$ induces a density based point preference
list for $\mathbf{x}_j$, with v_j^1 being its most preferred model hypothesis. We define the $n \times n$
point correlation matrix $\mathbf{P}$ by using the intersection kernel over the top-K density
based point preference lists

$$\mathbf{P}_{ij} = \frac{1}{K} |v_i^{1:K} \cap v_j^{1:K}| \tag{4}$$

where $v_i^{1:K} = \{v_i^1, v_i^2, \ldots, v_i^K\}$ is the top-K density based preferences for point $\mathbf{x}_i$
and K is a small number of top preferences, set to $\lceil 0.01 M_0 \rceil$ in all our experiments. As we use density based preferences, we need a much smaller K than
in case of residual based preferences. In Sect. 4.2, we use a variant of this point
correlation matrix $\mathbf{P}$ for identifying potential inliers of a given hypothesis.

4   Proposed Method: Density Preference Analysis

We will now develop our complete algorithm Density Preference Analysis[1] (DPA)
for recovering all inlier structures present in the data using the building blocks
described in the previous section.

4.1   Candidate Hypotheses Selection

We use the density based point preferences described in Sect. 3.3 to select promising hypotheses from ϑ_0. The residual density $\mathbf{d}_j^i$ given by (2) roughly measures
the likelihood of $\mathbf{x}_j$ being an inlier for the model hypothesis $\boldsymbol{\theta}^i$. If a data point
$\mathbf{x}_j$ has a hypothesis $\boldsymbol{\theta}^i$ in their top-5 density based preference list, we say the
point votes for $\boldsymbol{\theta}^i$. Since the points voting for $\boldsymbol{\theta}^i$ are likely to be inliers, a selected
hypothesis better represents the structure if more inliers vote for it. Recall that
the set $\mathbf{v}_j$ is a permutation of the indices of ϑ_0, i.e., $\{1, \ldots, M_0\}$. We create an
index set by taking the union of hypothesis indices in the top-5 preferences of
all points, i.e., $\bigcup_{j=1}^{n} v_j^{1:5}$. We then reject spurious hypotheses from this set by
eliminating the ones having fewer than two votes. We refer to this reduced set
of hypothesis indices as $\mathcal{I}_{\vartheta_s}$ and the corresponding set of selected hypotheses

---

[1] Matlab code available at https://www.iiitd.edu.in/~anands/files/code/dpa.zip.

as $\vartheta_s \subseteq \vartheta_0$ having cardinality $M_s \leq M_0$. The set ϑ_s contains hypotheses that appear in the top-5 preferences of at least two points and thus is expected to contain all the hypotheses that represent a structure well.

We emphasize that this step applies a conservative rule for rejecting poorly generated hypotheses and is primarily used for a computational advantage of processing fewer hypotheses in the following steps. The goal is to ensure retaining all the good hypotheses while rejecting many of the bad ones. We verified empirically that values between top-2 to top-10 preferences and 2–5 minimum votes do not affect the good hypothesis selection and thus the final accuracy significantly, and only affect the computational time for the next steps.

4.2   Inlier Scale Estimation

The residual density values drop significantly across the inlier-outlier boundary. We make use of this disparity in density to estimate the scale of inlier noise for a given hypothesis.

Hypothesis Refinement: We construct $\mathcal{J}^i$, $i \in \mathcal{I}_{\vartheta_s}$ as the potential inlier set for $\boldsymbol{\theta}^i$. $\mathcal{J}^i$ is initialized with all the points voting for $\boldsymbol{\theta}^i$. Note that $\mathcal{J}^i$ will always contain indices pointing to the original set $\mathcal{X}$ of data points. We normalize the point correlation matrix $\mathbf{P}$ (4), such that each columns has unit ℓ_1-norm. We compute the correlation score of all points with the points voting for $\boldsymbol{\theta}^i$ as

$$\bar{\mathbf{p}}^i = \prod_{j \in \mathcal{J}^i} \mathbf{r}_{\cdot j} \tag{5}$$

Points with a large $\bar{p}^i_j$ are highly correlated with the voting points in $\mathcal{J}^i$ and are strong contenders for inliers of $\boldsymbol{\theta}^i$. We define a threshold as $\tau^i = \min_{j \in \mathcal{J}^i} \bar{p}^i_j$ and impose the following criterion to include potential inliers in the set $\mathcal{J}^i$

$$\mathcal{J}^i = \{j \mid \bar{p}^i_j \geq \tau^i, \ j \in [n]\}, \quad \text{where}[n] = \{1, ..., n\}. \tag{6}$$

We create a similar set of potential outliers defined as the set of points that do not appear as strong inliers for any of the selected hypotheses

$$\mathcal{O} = [n] \setminus \left\{ \bigcup_{i=1}^{M_s} \mathcal{J}^i \right\} \tag{7}$$

The set $\mathcal{O}$ contains all the points that do not belong to any of the structures, and thus are gross outliers. However, since $\mathcal{J}^i$ only contain a subset of inliers, $\mathcal{O}$ may contain some inliers as well. We eliminate the hypotheses for which the largest residual of potential outliers is smaller than the largest residual of the strong inliers, i.e., reject $\boldsymbol{\theta}^i$ if $\max_{j \in \mathcal{O}} \rho^i_j \leq \max_{j \in \mathcal{J}^i} \rho^i_j$. This is a conservative step to reject poor hypotheses that have unreliable inliers.

We use the strong inlier points in $\mathcal{J}^i$ to obtain a refined estimate of $\boldsymbol{\theta}^i$ using least squares and recompute the corresponding residual density profile

(Sect. 3.1). This step has two positive effects: pure hypotheses that are representative of the inlier structure result in better residual density profiles with increased disparity between the densities of inliers and outliers. On the contrary, impure hypotheses generate density profiles with a lower density drop across the inlier-outlier boundary as will be seen in Sect. 4.3.

Density Drop Rate: Let $\mathcal{J}^i_{\uparrow 10} \subset \mathcal{J}^i$ contain 10% of points in $\mathcal{J}^i$ that have the largest residuals ρ^i_j. We expect these points to lie closer to the inlier-outlier boundary. Using these potential inlier points, we define the *density drop rate* as

$$\Delta d^i = \frac{\frac{1}{|\mathcal{J}^i_{\uparrow 10}|} \sum_{j \in \mathcal{J}^i_{\uparrow 10}} (d^i_{j_{max}} - d^i_j)}{\frac{1}{|\mathcal{J}^i_{\uparrow 10}|} \sum_{j \in \mathcal{J}^i_{\uparrow 10}} (\rho^i_j - \rho^i_{j_{max}})} \tag{8}$$

where $j_{max} = \arg\max_j d^i_j$ is the sorted residual index corresponding to maximum residual density. We use the density $d^i_{k_{max}}$, where $k_{max} = \arg\max_j \rho^i_j, j \in \mathcal{J}^i_{\uparrow 10}$ and the density drop rate Δd^i in (8) to linearly extrapolate and estimate the scale of inlier noise for $\boldsymbol{\theta}^i$ as

$$\widehat{\rho}^i = \frac{(d^i_{k_{max}} - d^i_{\mathcal{O}_{max}})}{\Delta d^i} + \rho^i_{k_{max}} \tag{9}$$

where $d^i_{\mathcal{O}_{max}} = max_{j \in \mathcal{O}} \, d^i_j$ is the maximum density of outlier points for $\boldsymbol{\theta}^i$. In the unlikely event when $\mathcal{O} = \{\emptyset\}$, we set $d^i_{\mathcal{O}_{max}} = 0$. The residual index $\widehat{j}^i = \arg\min_{j \in [n]} |\rho^i_j - \widehat{\rho}^i|$ can be used to compute the inlier fraction as $\widehat{\eta}^i = \widehat{j}^i/n$. The inlier scale estimation using linear extrapolation of the density works very well in practice as we observed in our experiments.

Using the scale estimates, we update the sets $\mathcal{J}^i$, $i \in \mathcal{I}_{\vartheta_s}$ to include all inliers

$$\mathcal{J}^i = \{j \mid r^i_j \leq \widehat{\rho}^i, \, j \in [n]\} \qquad \forall \, i \in \mathcal{I}_{\vartheta_s} \tag{10}$$

4.3   Model Selection and Labeling

We now have a tentative point-model hypothesis association in $\mathcal{J}^i, i \in \mathcal{I}_{\vartheta_s}$. From the candidate model hypotheses in $\mathcal{I}_{\vartheta_s}$, we need to identify *exactly* one model hypothesis for each structure. We further use the intuition that selected models should be able to explain *all* the inlier points. We describe a simple process (Algorithm 1) that uses the density drop rate and hypothesis preferences based correlation to identify a model corresponding to each structure, associate their inliers with the respective models and label the remaining points as gross outliers. While the process is greedy, we empirically show that it works well in practice.

We use the set of candidate models $\mathcal{I}_{\vartheta_s}$, the corresponding inlier sets $\mathcal{J}^i$, $i \in \mathcal{I}_{\vartheta_s}$ and the preference based hypothesis correlation matrix $\mathbf{H}$. We initialize the set of all inliers $\mathcal{J}$ and the set of final models $\mathcal{I}_{\vartheta}$ as empty sets. The output of the algorithm is a set of models, each corresponding to a structure in the data. When a candidate model hypothesis has an associated inlier set $\mathcal{J}^{i_t}$ with a sufficiently small overlap ($\leq \tau_o$) with $\mathcal{J}$, we refer to it as a novel hypothesis.

Algorithm 1. Model Selection

---

1 **Input:** variables - $\mathcal{I}_{\vartheta_s}$, $\mathcal{J}^i$ $(i \in \mathcal{I}_{\vartheta_s})$, $\mathbf{H}$
2   thresholds - τ_o, τ_h
3 **Output:** $\mathcal{I}_\vartheta$ - index set of selected models
4 **begin**
5   Initialize: $\mathcal{I}_\vartheta = \mathcal{J} = \{\emptyset\}$
6   **repeat**
7          $i_t \leftarrow \arg\max_{i \in \mathcal{I}_{\vartheta_s}} \Delta d^i$
8   **if** $|\mathcal{J} \cap \mathcal{J}^{i_t}| \leq \tau_o$ **then**
9              $\mathcal{I}_{\vartheta_t} = i_t \cup \{i \mid |\mathcal{J} \cap \mathcal{J}^i| \leq \tau_o, \mathbf{H}_{i_t i} \geq \tau_h, \forall i \in \mathcal{I}_{\vartheta_s}\}$
10             $\widehat{i_t} \leftarrow \arg\max_{i \in \mathcal{I}_{\vartheta_t}} \widehat{\eta}^i$
11             $\mathcal{I}_\vartheta = \mathcal{I}_\vartheta \cup \widehat{i_t}$
12             $\mathcal{J} = \mathcal{J} \cup \mathcal{J}^{i_t}$
13             $\mathcal{I}_{\vartheta_s} = \mathcal{I}_{\vartheta_s} \setminus \mathcal{I}_{\vartheta_t}$
14   **else**
15             $\mathcal{I}_{\vartheta_s} = \mathcal{I}_{\vartheta_s} \setminus \{i_t\}$
16   **until** $\mathcal{I}_{\vartheta_s} = \{\emptyset\}$
17 **return** $\mathcal{I}_\vartheta$

We use τ_o, which is typically set as a small percentage of points indicating the tolerable overlap between structures. The second threshold τ_h determines the smallest acceptable correlation between hypotheses. Both these thresholds are intuitive, easy to set and gracefully affect the performance accuracy. We identify a model hypothesis $\boldsymbol{\theta}^{i_t} \in \vartheta_s$ with the largest density drop rate (line 7). This aligns well with our intuition that a sharp density drop provides the strongest evidence of an inlier structure. Furthermore, if $\boldsymbol{\theta}^{i_t}$ is novel, we construct an index set $\mathcal{I}_{\vartheta_t}$ of model hypotheses that are novel *and* highly correlated with $\boldsymbol{\theta}^{i_t}$ (line 9), else we ignore it by eliminating the corresponding entry from $\mathcal{I}_{\vartheta_s}$ (line 15). Candidate models that are not novel are likely to arise from hypotheses that merge two distinct structures. From the subset $\mathcal{I}_{\vartheta_t}$, we select the model hypothesis with the largest fraction of inliers and use it as the representative for the corresponding structure (lines 10–11). We update the set of all inliers $\mathcal{J}$ and the set of candidate models $\mathcal{I}_{\vartheta_s}$ as in (lines 12–13). We repeat the process until all candidate models have been explored and $\mathcal{I}_{\vartheta_s}$ is empty and return the final set of models as $\mathcal{I}_\vartheta$. We post-process the set of models, $\mathcal{I}_\vartheta$, obtained from Algorithm 1 by ignoring the models that have a very small fraction of inliers $\frac{|\mathcal{J}^i|}{n} \leq 0.05$. Now the number of structures recovered is the cardinality of $\mathcal{I}_\vartheta$. Finally, due to the overlap tolerance τ_o, we have a small number of points that may be in multiple inlier sets. We reassign such points to models for which the residual density is the highest. This post-processing step results in a unique point-model assignment and all the unassigned points are labeled as outliers. Using the unique point-model assignment in $\mathcal{J}^i$, $i \in \mathcal{I}_\vartheta$, we can generate the label set $\mathcal{L}$. Our overall algorithm is summarized in Algorithm 2.

Algorithm 2. Density Preference Analysis(DPA)

---

1 **Input:** Data - $\mathcal{X}$, Model type -{*synthetic, homography, fundamental*}
2 **Output:** $\mathcal{L}$ - Labeled data points
3 **begin**
4     $\vartheta_0 \leftarrow$ generate model hypotheses using DHF [15].
5     $\mathbf{d}^i \leftarrow$ for each $\boldsymbol{\theta}^i$ compute residual density vector using (2).
6     $\mathbf{P} \leftarrow$ compute density based point correlation using (4).
7     $\vartheta_s \leftarrow$ select candidate hypotheses (Sect. 4.1).
8     $\mathcal{J}^i \leftarrow$ for each $\boldsymbol{\theta}^i \in \vartheta_s$ identify strong inliers using (6).
9     $\mathcal{O} \leftarrow$ identify potential outliers using (7).
10     $\widehat{\eta}^i \leftarrow$ for each $\boldsymbol{\theta}^i \in \vartheta_s$ estimate inlier scale and update $\mathcal{J}^i$ using (8-10)
11     $\{\mathcal{I}_\vartheta\} \leftarrow$ model selection (Algorithm 1).
12     $\mathcal{L} \leftarrow$ post-processing and point-model assignments (Sect. 4.3)

13 **return** $\mathcal{L}$

---

5   Experimental Results

In this section we evaluate our proposed approach, DPA, and compare with three recent robust multi-model fitting approaches RPA [9], T-Linkage [8] and RCMSA [11] using three different datasets. We used publicly available implementations provided by the respective authors and set the parameters as suggested in the corresponding publications.

We report the classification accuracy (CA) averaged over 10 runs on all our experiments. For our synthetic dataset, we use points along three concentric circles, with each structure corrupted with a different scale of noise. For the real experiments, we use the AdeleideRMF [16] data set, which consists of 19 image pairs each for homography based planar segmentation and fundamental matrix based motion segmentation. Each image pair is provided with SIFT [6] point matches corrupted with outliers and their ground truth labeling.

For DPA, we set the hypothesis correlation threshold τ_h to 0.6 for planar segmentation and 0.75 in case of motion segmentation and the synthetic experiments. We fixed the value of the overlap threshold $\tau_o = \lceil 0.025n \rceil$ for planar segmentation and concentric circles. It was set to $(\lceil 0.1n \rceil)$ for fundamental matrix based motion segmentation. A higher overlap threshold is justified as the epipolar constraint imposed by the fundamental matrix on each rigid object is a weaker constraint than the one from homography. We use DHF [15] for hypothesis generation for all experiments.

We emphasize that contrary to the competing methods, our method, DPA, does not require an estimate of the inlier noise scale. We follow [9] and provide the standard deviation of *all* inlier residuals to each of the competing methods (T-Linkage, RPA, and RCMSA). The parameters used are $\sigma_n = 0.0098$ for synthetic data, $\sigma_n = 0.013$ for planar segmentation and $\sigma_n = 0.005$ for motion segmentation. As we will show in the synthetic experiments, this dependency on the scale of noise is detrimental when different structures have significantly different noise. This may also be the reason of relatively poor performance of the competing methods on the planar segmentation experiments.

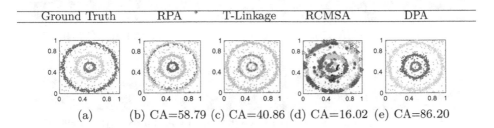

Ground Truth	RPA	T-Linkage	RCMSA	DPA

(a)　　　(b) CA=58.79 (c) CA=40.86 (d) CA=16.02 (e) CA=86.20

Fig. 2. Three Concentric Circles: (CA = Classification Accuracy(%)). Point membership is color coded. Outliers in green. (Color figure online)

5.1 Synthetic Example: Three Concentric Circles

In our synthetic experiment, we generate three concentric circles centered at $[0.5, 0.5]$ (radius - $0.08, 0.22, 0.45$), each having a different number of inlier points $(250, 450, 650)$ corrupted with Gaussian noise of different standard deviation $(0.01, 0.015, 0.018)$. We added uniformly distributed 300 gross outliers in the square region defined by $(0,0)$ and $(1,1)$. The data and the segmentation results of all the methods are shown in Fig. 2. This seemingly simple experiment resulted in poor performances by each of the methods RPA, T-Linkage and RCMSA, with the average CA as 58.79%, 40.86% and 16.02%. DPA managed to achieve an average CA of 86.20%. In the sample result shown in Fig. 2 both RPA and T-Linkage fail to find one circle, while oversegment the others. RCMSA results in a large oversegmentation and recovers 45 distinct structures, while RPA and T-Linkage recover two structures on average. DPA is able to recover all the three structures in each of the 10 runs. The poor performance of the competing methods is seemingly due to the variation in noise scale of different structures.

5.2 Planar Segmentation

We perform planar segmentation by fitting multiple homographies to the point matches. We use isotropic scaling of the point matches such that they have zero mean and have unit average distance from the origin. We use Sampson error as our residual function $\phi(\boldsymbol{\theta}, \mathbf{x})$. We report the quantitative results in Table 1 and see the average and median performance of DPA is superior than the other methods. We also show sample results on two examples in Fig. 3, where the other three competing methods perform poorly. The scale of noise in the sample sequences, *library* and *oldclassicwing*, are too high and too low respectively as compared to the input parameter σ_n [9]. The superior performance of DPA in these cases is largely due to its residual density based scale estimation.

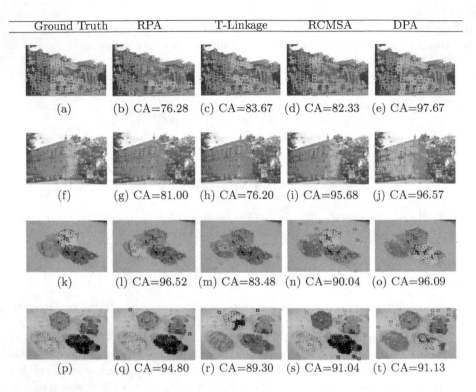

Ground Truth	RPA	T-Linkage	RCMSA	DPA

(a)   (b) CA=76.28   (c) CA=83.67   (d) CA=82.33   (e) CA=97.67

(f)   (g) CA=81.00   (h) CA=76.20   (i) CA=95.68   (j) CA=96.57

(k)   (l) CA=96.52   (m) CA=83.48   (n) CA=90.04   (o) CA=96.09

(p)   (q) CA=94.80   (r) CA=89.30   (s) CA=91.04   (t) CA=91.13

Fig. 3. [**Sample results**] **Planar Segmentation**: Row 1 - *library*, Row 2- *oldclassicswing*. **Motion Segmentation**: Row 3- *breadcubechips*,Row 4- *cubebreadtoychips*. (CA = Classification Accuracy (%)). Point membership is color coded. Outliers in green. (Color figure online)

5.3 Motion Segmentation

We perform motion segmentation on the AdelaideRMF images by fitting multiple fundamental matrices to the point matches. As preprocessing, we use isotropic scaling of the point matches such that they have zero mean and have unit average distance from the origin. We use the epipolar constraint to compute the residuals. We report the quantitative results in Table 2 and see the average and median performance of DPA is inferior only to RPA. We also show sample results on two examples in Fig. 3. We emphasize that despite not knowing the scale of inlier noise, DPA performs competitively resulting in average CA of excess of 90%.

6   Conclusion

We proposed a novel algorithm for the problem of robust multi-model fitting that leverages both consensus and preference analysis. To characterize the consensus based approach, we defined a residual density profile, which we further used for computing point preferences. We further leveraged the disparity between the density of potential inliers and outliers to estimate the scale of inlier noise

Table 1. Planar Segmentation. **Outlier%**- Percentage of outliers, **#S** - Number of true structures, Evaluation metric - Classification Accuracy (CA%) and corresponding running time (T) in seconds.

	Outlier%	#S	RPA	T	T-Link	T	RCMSA	T	DPA	T
barrsmith	68.87	2	62.90	201.67	57.93	62.71	84.81	2.88	**97.68**	35.79
bonhall	06.17	6	52.88	3712.26	60.41	1447.47	**81.67**	9.25	77.98	52.75
bonython	73.73	1	84.34	105.91	64.34	48.28	87.27	3.71	**96.57**	30.06
elderhalla	60.74	2	**99.07**	176.65	69.53	57.67	75.23	3.13	96.17	37.58
elderhallb	47.84	3	81.96	269.08	57.80	83.63	71.45	3.08	**85.90**	39.02
hartley	61.56	2	81.38	290.07	71.63	136.58	77.38	3.75	**96.91**	43.45
johnsona	20.91	4	**91.10**	535.49	57.80	179.43	83.00	3.94	87.08	63.94
johnsonb	12.01	7	66.84	1676.11	70.72	540.97	**79.41**	6.18	74.36	75.83
ladysymon	32.48	2	79.16	185.15	77.72	62.01	75.27	2.78	**90.46**	8.33
library	55.34	2	63.53	163.32	82.51	50.80	77.02	2.89	**95.21**	6.08
napiera	62.91	2	73.25	269.69	**81.26**	102.22	70.66	3.39	80.56	3.36
napierb	39.51	3	75.14	271.21	76.76	73.62	74.29	2.96	**83.63**	11.05
neem	36.51	3	78.51	247.19	53.03	63.91	71.87	3.23	**80.21**	7.79
nese	33.46	2	**99.21**	204.25	53.70	70.53	77.56	2.96	97.40	16.25
oldclassicswing	32.45	2	76.73	369.72	73.77	262.34	92.45	3.59	**96.33**	163.14
physics	45.28	1	**100.0**	40.12	68.49	26.54	54.53	15.85	98.40	11.84
sene	47.20	2	99.44	200.35	84.32	81.33	71.68	5.88	**99.76**	7.76
unihouse	16.55	5	88.00	9232.87	71.86	5890.28	**97.02**	16.37	93.17	85.83
unionhouse	76.50	1	76.14	226.33	77.29	116.83	90.06	5.70	98.34	16.22
mean			80.50	967.23	68.99	492.48	78.55	5.31	**90.90**	37.68
std			13.51	2174.70	9.92	1347.1	9.54	4.14	**8.10**	39.19
median			79.16	247.19	70.72	81.33	77.38	3.39	**95.21**	30.06

Table 2. Motion Segmentation. Notations are same as in Table 1

	Outlier%	#S	RPA	T	T-Link	T	RCMSA	T	DPA	T
biscuit	57.16	1	**98.36**	38.75	83.09	19.17	95.15	4.50	82.12	14.42
biscuitbook	47.51	2	96.42	49.42	**97.77**	20.37	92.52	5.21	97.24	123.66
biscuitbookbox	37.21	3	**95.83**	45.70	88.80	11.69	83.71	3.36	95.14	46.78
boardgame	42.48	1	**87.53**	46.13	83.73	13.53	78.46	3.79	83.69	29.52
book	21.48	1	**97.54**	16.27	82.57	6.31	94.01	12.62	90.16	94.86
breadcartoychips	35.20	4	**91.73**	45.31	80.51	9.75	78.82	3.24	91.56	15.53
breadcube	32.19	2	**95.95**	29.98	85.62	10.33	87.27	3.61	94.09	67.46
breadcubechips	35.22	3	**95.57**	36.91	82.00	9.66	83.17	3.38	94.61	24.64
breadtoy	37.41	2	**97.15**	44.39	96.81	14.61	78.37	3.77	90.59	15.45
breadtoycar	34.15	3	**92.17**	30.47	84.70	4.97	83.07	2.60	88.67	15.60
carchipscube	36.59	3	**94.30**	27.69	88.00	4.89	78.85	2.70	86.30	14.13
cube	69.49	1	**97.15**	33.65	46.29	15.86	87.98	5.79	96.89	9.81
cubebreadtoychips	28.03	4	**93.21**	67.63	80.18	19.12	81.62	3.83	87.28	43.75
cubechips	51.62	2	**96.48**	39.59	95.14	13.99	90.32	5.82	92.92	75.64
cubetoy	41.42	2	**96.31**	31.60	78.80	10.74	89.64	5.45	93.61	86.55
dinobooks	44.54	3	84.78	64.21	78.56	22.90	72.28	5.76	84.17	73.19
game	73.48	1	95.97	21.93	77.6	9.32	90.77	4.88	**97.47**	83.83
gamebiscuit	51.54	2	**96.95**	49.04	70.61	19.05	85.40	4.55	90.95	74.25
toycubecar	36.36	3	**91.70**	27.21	70.70	7.17	83.45	2.94	84.65	47.46
mean			94.47	39.25	81.65	12.81	84.99	4.62	90.64	50.34
std			**3.55**	13.21	11.36	5.44	6.16	2.20	4.86	34.03
median			**95.95**	38.75	82.57	11.69	83.71	3.83	90.95	46.78

for each model hypothesis. We devised a greedy scheme that uses preference based hypothesis similarity to identify a model hypothesis for representing each structure. Finally, we used a simple post-processing step to eliminate small, spurious structures and to generate a unique point to model assignment.

We showed empirical results using synthetic examples as well as the AdelaideRMF datasets. Without using any information about the scale of the noise, we showed competitive performance with other state-of-the-art multi-model fitting methods that primarily rely on preference analysis. Through our empirical analysis, we have shown that the residual density information judiciously combined with preference analysis can improve the performance in robust model fitting problems while reducing dependence on a priori knowledge about noise in data.

References

1. Chin, T.J., Wang, H., Suter, D.: The ordered residual kernel for robust motion subspace clustering. In: NIPS, pp. 333–341 (2009)
2. Chin, T.J., Wang, H., Suter, D.: Robust fitting of multiple structures: the statistical learning approach. In: ICCV, pp. 413–420 (2009)
3. Chin, T.J., Yu, J., Suter, D.: Accelerated hypothesis generation for multistructure data via preference analysis. IEEE TPAMI **34**, 625–638 (2012)
4. Fischler, M.A., Bolles, R.C.: Random sample consensus: a paradigm for model fitting with applications to image analysis and automated cartography. Commun. ACM **24**, 381–395 (1981)
5. Kanazawa, Y., Kawakami, H.: Detection of planar regions with uncalibrated stereo using distribution of feature points. In: BMVC, pp. 247–256 (2004)
6. Lowe, D.G.: Distinctive image features from scale-invariant keypoints. IJCV **60**, 91–110 (2004)
7. Lavva, I., Hameiri, E., Shimshoni, I.: Robust methods for geometric primitive recovery and estimation from range images. IEEE Trans. Syst. Man Cybern. Part B (Cybernetics) **38**, 826–845 (2008)
8. Magri, L., Fusiello, A.: T-Linkage: a continuous relaxation of j-linkage for multi-model fitting. In: CVPR, pp. 3954–3961 (2014)
9. Magri, L., Fusiello, A.: Robust multiple model fitting with preference analysis and low-rank approximation. BMVC **20**(1–20), 12 (2015)
10. Mittal, S., Anand, S., Meer, P.: Generalized projection-based M-estimator. IEEE TPAMI **34**, 2351–2364 (2012)
11. Pham, T.T., Chin, T.J., Yu, J., Suter, D.: The random cluster model for robust geometric fitting. IEEE TPAMI **36**, 1658–1671 (2014)
12. Raguram, R., Frahm, J.M.: RECON: scale-adaptive robust estimation via residual consensus. In: ICCV, pp. 1299–1306 (2011)
13. Toldo, R., Fusiello, A.: Robust multiple structures estimation with j-linkage. In: Forsyth, D., Torr, P., Zisserman, A. (eds.) ECCV 2008. LNCS, vol. 5302, pp. 537–547. Springer, Heidelberg (2008). doi:10.1007/978-3-540-88682-2_41
14. Torr, P.H.S., Zisserman, A.: MLESAC: a new robust estimator with application to estimating image geometry. CVIU **78**, 138–156 (2000)
15. Wong, H.S., Chin, T.J., Yu, J., Suter, D.: A simultaneous sample-and-filter strategy for robust multi structure model fitting. CVIU **117**, 1755–1769 (2013)
16. Wong, H.S., Chin, T.J., Yu, J., Suter, D.: Dynamic and hierarchical multi-structure geometric model fitting. In: ICCV, pp. 1044–1051 (2011)

17. Wang, H., Chin, T.J., Suter, D.: Simultaneously fitting and segmenting multiple-structure data with outliers. IEEE TPAMI **34**, 1177–1192 (2012)
18. Zuliani, M., Kenny, C.S., Manjunath, B.S.: The multiRANSAC algorithm and its application to detect planar homographies. In: IEEE ICIP, pp. 153–156 (2005)
19. Zhang, K., Kwok, J.T., Tang, M.: Accelarated convergence using dynamic mean shift. ECCV **2**, 257–268 (2006)

Photometric Bundle Adjustment
for Vision-Based SLAM

Hatem Alismail$^{(\boxtimes)}$, Brett Browning, and Simon Lucey

The Robotics Institute, Carnegie Mellon University, Pittsburgh, USA
{halismai,brettb,slucey}@cs.cmu.edu

Abstract. We propose a novel algorithm for the joint refinement of structure and motion parameters from image data directly without relying on fixed and known correspondences. In contrast to traditional bundle adjustment (BA) where the optimal parameters are determined by minimizing the reprojection error using tracked features, the proposed algorithm relies on maximizing the photometric consistency and estimates the correspondences implicitly. Since the proposed algorithm does not require correspondences, its application is not limited to corner-like structure; any pixel with nonvanishing gradient could be used in the estimation process. Furthermore, we demonstrate the feasibility of refining the motion and structure parameters simultaneously using the photometric error in unconstrained scenes and without requiring restrictive assumptions such as planarity. The proposed algorithm is evaluated on range of challenging outdoor datasets, and it is shown to improve upon the accuracy of the state-of-the-art VSLAM methods obtained using the minimization of the reprojection error using traditional BA as well as loop closure.

1 Introduction

Photometric, or image-based, minimization is a fundamental tool in a myriad of applications such as: optical flow [1], scene flow [2], and stereo [3,4]. Its use in vision-based 6DOF motion estimation has recently been explored demonstrating good results [5–8]. Minimizing the photometric error, however, has been limited to frame–frame estimation (visual odometry), or as a tool for depth refinement independent of the parameters of motion [9]. Consequently, in unstructured scenes, frame–frame minimization of the photometric error cannot reduce the accumulated drift. When loop closure and prior knowledge about the motion and structure are not available, one must resort to the Gold Standard: minimizing the reprojection error using bundle adjustment.

Bundle adjustment (BA) is the problem of jointly refining the parameters of motion and structure to improve a visual reconstruction [10]. Although BA is a versatile framework, it has become a synonym to minimizing the reprojection error across multiple views [11,12]. The advantages of minimizing the reprojection error are abundant and have been discussed at length in the literature [11,12]. In practice, however, there are sources of systematic errors

© Springer International Publishing AG 2017
S.-H. Lai et al. (Eds.): ACCV 2016, Part IV, LNCS 10114, pp. 324–341, 2017.
DOI: 10.1007/978-3-319-54190-7_20

in feature localization that are hard to detect and the value of modeling their uncertainty remains unclear [13,14]. For example, slight inaccuracies in calibration exaggerate errors [15], sensor noise and degraded frequency content of the image affect feature localization accuracy [16]. Even interpolation artifacts play a non-negligible role [17]. Although minimizing the reprojection is backed by sound theoretical properties [11], its use in practice must also take into account the challenges and nuances of precisely localizing keypoints [10].

In this work, we propose a novel method that further improves upon the accuracy of minimizing the reprojection error and even state-of-the-art loop closure [18]. The proposed algorithm brings back the image in the loop, and jointly refines the motion and structure parameters to maximize the photometric consistency across multiple views. In addition to improved accuracy, the algorithm does not require correspondences. In fact, correspondences are estimated automatically as a byproduct of the proposed formulation. The ability to perform BA without the need for precise correspondences is attractive because it can enable VSLAM applications where corner extraction is unreliable [19], as well as additional modeling capabilities that extend beyond geometric primitives [20,21].

1.1 Preliminaries and Notation

The Reprojection Error. Given an initial estimate of the scene structure $\{\boldsymbol{\xi}_j\}_{j=1}^N$, the viewing parameters per camera $\{\boldsymbol{\theta}_i\}_{i=1}^M$, and $\mathbf{x}_{ij}$ the projection of the jth point onto the ith camera, the reprojection error is given by

$$\epsilon_{ij}(\mathbf{x}_{ij}; \boldsymbol{\theta}_i, \boldsymbol{\xi}_j) = \|\mathbf{x}_{ij} - \pi\left(\mathbf{T}(\boldsymbol{\theta}_i), \mathbf{X}(\boldsymbol{\xi}_j)\right)\|, \tag{1}$$

where $\pi(\cdot, \cdot)$ is the image projection function. The function $\mathbf{T}(\cdot)$ maps the vectorial representation of motion to a rigid body transformation matrix. Similarly, $\mathbf{X}(\cdot)$ maps the parameterization of the point to coordinates in the scene.

In this work, we assume known camera calibration parameters as is often the case in VSLAM and parameterize the scene structure using the usual 3D Euclidean coordinates, where $\mathbf{X}(\boldsymbol{\xi}) := \boldsymbol{\xi}$, and

$$\boldsymbol{\xi}_j^\top = \left(x_j \; y_j \; z_j\right) \in \mathbb{R}^3. \tag{2}$$

The pose parameters are represented using twists [22], where the rigid body pose is obtained using the exponential map [23], *i.e.*:

$$\boldsymbol{\theta}_i^\top \in \mathbb{R}^6 \quad \text{and} \quad \mathbf{T}(\boldsymbol{\theta}) := \exp(\widehat{\boldsymbol{\theta}}) \in SE(3). \tag{3}$$

Our algorithm, similar to minimizing the reprojection error using BA, does not depend on the parameterization. Other representations for motion and structure have been studied in the literature and could be used as well [24–26].

Geometric Bundle Adjustment. Given an initialization of the scene points and motion parameters, we may obtain a refined estimate by minimizing the squared reprojection error in Eq. (1) across tracked features, *i.e.*:

$$\{\Delta\boldsymbol{\theta}_i^*, \Delta\boldsymbol{\xi}_j^*\} = \underset{\boldsymbol{\theta}_i, \boldsymbol{\xi}_j}{\text{argmin}} \sum_{i=1}^{M} \sum_{j=1}^{N} \frac{1}{2} \delta_{ij} \epsilon_{ij}^2(\mathbf{x}_{ij}, \Delta\boldsymbol{\theta}_i, \Delta\boldsymbol{\xi}_j), \tag{4}$$

where $\delta_{ij} = 1$ if the jth point is visible, or tracked, in the ith camera. We call this formulation *geometric* BA.

Minimizing the reprojection error in Eq. (4) is a large nonlinear optimization problem. Particular to BA is the sparsity pattern of its linearized form, which can be exploited beneficially for both large– and medium–scale problems [11].

1.2   The Photometric Error

The use of photometric information in Computer Vision has a long and rich history dating back to the seminal works of Lucas and Kanade [27] and Horn and Schunk [28]. The problem is usually formulated as a pairwise alignment of two images. One is the reference $\mathbf{I}_0$, while the other is the input $\mathbf{I}_1$. The goal is to estimate the parameters of motion $\boldsymbol{p}$ such that the sum of the squared intensity error is minimized

$$\boldsymbol{p}^* = \underset{\boldsymbol{p}}{\text{argmin}} \sum_{\mathbf{u} \in \Omega_0} \frac{1}{2} \|\mathbf{I}_0(\mathbf{u}) - \mathbf{I}_1(\mathbf{w}(\mathbf{u}; \boldsymbol{p}))\|^2, \tag{5}$$

where $\mathbf{u} \in \Omega_0$ denotes a subset of pixel coordinates in the reference image frame, and $\mathbf{w}(\cdot, \cdot)$ denotes the warping function [29]. Minimizing the photometric error has recently resurfaced as a robust solution to visual odometry (VO) [6,7,30]. Notwithstanding, minimizing the photometric error has not yet been explored for the *joint* optimization of the motion and structure parameters for VSLAM in unstructured scenes. The proposed approach fills in the gap by providing a photometric formulation for BA, which we call BA *without* correspondences.

2   Bundle Adjustment Without Correspondences

BA is not limited to minimizing the reprojection error [10]. We reformulate the problem as follows. First, we assume an initial estimate of the camera poses $\boldsymbol{\theta}_i$ as required by geometric BA. However, we do not require tracking information for the 3D points. Instead, for every scene point $\boldsymbol{\xi}_j$, we assign a *reference* frame denoted by $r(j)$. The reference frame is used to extract a fixed square patch denoted by $\boldsymbol{\phi}_j \in \mathbb{R}^D$ over a neighborhood denoted by $\mathcal{N}$. In addition, we compute an initial *visibility* list indicating the frames where the point may be in view. The visibility list for the jth point excludes the reference frame and is denoted by:

$$\mathbf{V}_j = \{k : k \neq r(j) \text{ and } \boldsymbol{\xi}_j \text{ is visible in frame } k\}, \text{ for } k \in [1, \ldots, M]. \tag{6}$$

Given this information and the input images $\{\mathbf{I}_i\}_{i=1}^M$, we seek to estimate an optimal update to the motion $\Delta\boldsymbol{\theta}_i{}^*$ and structure parameters $\Delta\boldsymbol{\xi}_j{}^*$ that satisfy

$$\{\Delta\boldsymbol{\theta}_i^*, \Delta\boldsymbol{\xi}_j^*\} = \operatorname*{argmin}_{\Delta\boldsymbol{\theta}_i, \Delta\boldsymbol{\xi}_j} \sum_{j=1}^N \sum_{k\in V(j)} \mathcal{E}(\boldsymbol{\phi}_j, \mathbf{I}_k; \Delta\boldsymbol{\theta}_k, \Delta\boldsymbol{\xi}_j), \text{ where} \tag{7}$$

$$\mathcal{E}(\boldsymbol{\phi}, \mathbf{I}'; \boldsymbol{\theta}, \boldsymbol{\xi}) = \sum_{\mathbf{u}\in\mathcal{N}} \frac{1}{2}\|\boldsymbol{\phi}(\mathbf{u}) - \mathbf{I}'(\pi(\boldsymbol{\theta}, \boldsymbol{\xi}) + \mathbf{u})\|^2. \tag{8}$$

The notation $\mathbf{I}'(\pi(\cdot, \cdot) + \mathbf{u})$ indicates sampling the image intensities in a neighborhood about the current projection of the point using an appropriate interpolation scheme (bilinear in this work). The objective is illustrated in Fig. 1.

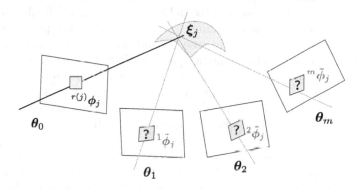

Fig. 1. Schematic of the proposed approach. We seek to optimize the parameters of motion $\boldsymbol{\theta}_i$ and structure $\boldsymbol{\xi}_j$ such that the photometric error with respect to a fixed patch at the reference frame is minimized; correspondences are estimated implicitly

Linearization and Sparsity. The optimization problem in Eq. (7) is nonlinear and its solution proceeds with standard techniques. Let $\boldsymbol{\theta}$ and $\boldsymbol{\xi}$ denote the current estimate of the camera and the scene point, and let the current projected pixel coordinate in the image plane be given by

$$\mathbf{u}' = \pi(\mathbf{T}(\boldsymbol{\theta}), \mathbf{X}(\boldsymbol{\xi})), \tag{9}$$

then taking the partial derivatives of the 1st-order expansion of the photometric error in Eq. (8) with respect to the motion and structure parameters we obtain:

$$\frac{\partial\mathcal{E}}{\partial\boldsymbol{\theta}} = \sum_{\mathbf{u}\in\mathcal{N}} \mathbf{J}^\top(\boldsymbol{\theta})\,|\boldsymbol{\phi}(\mathbf{u}) - \mathbf{I}'(\mathbf{u}' + \mathbf{u}) - \mathbf{J}(\boldsymbol{\theta})\Delta\boldsymbol{\theta}| \tag{10}$$

$$\frac{\partial\mathcal{E}}{\partial\boldsymbol{\xi}} = \sum_{\mathbf{u}\in\mathcal{N}} \mathbf{J}^\top(\boldsymbol{\xi})\,|\boldsymbol{\phi}(\mathbf{u}) - \mathbf{I}'(\mathbf{u}' + \mathbf{u}) - \mathbf{J}(\boldsymbol{\xi})\Delta\boldsymbol{\xi}|, \tag{11}$$

where $\mathbf{J}(\boldsymbol{\theta}) = \nabla \mathbf{I}(\mathbf{u}' + \mathbf{u})\frac{\partial \mathbf{u}'}{\partial \boldsymbol{\theta}}$, and $\mathbf{J}(\boldsymbol{\xi}) = \nabla \mathbf{I}(\mathbf{u}' + \mathbf{u})\frac{\partial \mathbf{u}'}{\partial \boldsymbol{\xi}}$. The partial derivatives of the projected pixel location with respect to the parameters are identical to those obtained when minimizing the reprojection error in Eq. (1), and $\nabla \mathbf{I} \in \mathbb{R}^{1 \times 2}$ denotes the image gradient. By equating the partial derivatives in Eqs. (10) and (11) to zero we arrive at the normal equations which can be solved efficiently using standard methods [31].

We note that the Jacobian involved in solving the photometric error has a higher dimensionality than its counterpart in geometric BA. This is because the dimensionality of intensity patches ($D \geq 3 \times 3$) is usually higher than the dimensionality of feature projections (typically 2 for a monocular reconstruction problem). Nonetheless, the Hessian remains *identical* to minimizing the reprojection error and the linear system remains sparse and is efficient to decompose. The sparsity pattern of the photometric BA problem is illustrated in Fig. 2.

Since the parameters of motion and structure are refined jointly, the location of the patch at the reference frame $\phi(\mathbf{u})$ in Eq. (8) will additionally depend on the pose parameters of the reference frame. Allowing the reference patch to "move" during the optimization adds inter-pose dependencies in the linear system and might cause the location of the reference patch to drift. For instance, the solution may be biased towards image regions with brighter absolute intensity values in an attempt to obtain the minimum energy in low-texture areas.

To address this problem, we fix the patch appearance at the reference frame by storing the patch values as soon as the reference frame is selected. This is equivalent to assuming a known patch appearance from an independent source. Under this assumption, the optimization problem now becomes: given a known and fixed patch appearance of a 3D point in the world, refine the parameters of the structure and motion such that photometric error between the fixed patch and its projection onto the other frames is minimized. This assumption has two advantages: (1) the Hessian sparsity pattern remains identical to the familiar form when minimizing the reprojection error using traditional BA, and (2) we can refine the three coordinates (or the full four projective coordinates [10]) of the scene points as opposed to only refining depth along a fixed ray in space.

In addition to improving the accuracy of VSLAM, the algorithm does not require extensive parameter tuning. This is possible by allowing the algorithm to determine the correct correspondences, hence eliminating the many steps required to ensure outlier-free correspondences with traditional BA. The current implementation of the proposed algorithms is controlled by the three parameters summarized in Table 1 and explained next.

Table 1. Configuration parameters for the proposed algorithm shown in Algorithm 1.

Parameter	Value
Patch radius	1 or 2
Non maxima suppression radius	1
Max distance to update V_j	2

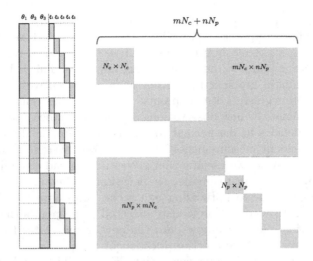

Fig. 2. Shown on the left is the form of the Jacobian for a photometric bundle adjustment problem consisting of 3 cameras, 4 points, and using a 9-dimensional descriptor, with $N_c = 6$ parameters per camera, and $N_p = 3$ parameters per point. The form of the normal equations is shown on the right. The illustration is not up to scale across the two figures.

Selecting Pixels. While it is possible to select pixel locations at every frame using a standard feature detector, such as Harris [32] or FAST [33], we opt to use a simpler and more efficient strategy based on the gradient magnitude of the image. This is performed by selecting pixels with a local maxima in a 3×3 neighborhood of the absolute gradient magnitude of the image. The rationale is that pixels with vanishing intensity gradients do not contribute to the linear system in Eqs. (10) and (11). Other strategies for pixel selection could used [34,35], but we found that the current scheme works well as it ensures an even distribution of coordinates across the field-of-view of the camera [36]. The proposed pixel selection strategy is also beneficial as it is not restricted to corner-like structure and allows us to use pixels from low-texture areas. We note that this pixel selection step selects pixels at integer locations; there is no need to compute accurate subpixel positions of the selected points at this stage.

In image-based (photometric) optimization there is always a distinguished reference frame providing fixed measurements [9,37,38]. Selecting a single reference in photometric VSLAM is unnecessary and may be inadvisable. It is unnecessary as the density of reconstruction is not our main goal. It is inadvisable because we need the scene points to serve as tie points [39] and to form a strong network of constraints [10]. Given the nature of camera motion in VSLAM, selecting points from every frame ensures the strong network of connections between the tie points. For instance, typical hand-held and ground robots motions are mostly forward with points leaving the field-of-view rapidly.

Nonetheless, selecting new scene points at every frame using the aforementioned non maxima suppression procedure has one caveat. If we always select pixels with strong gradients between consecutive frames, then we are likely to track previous scene points rather than finding new ones. This is because pixels with locally maximum gradient magnitude at the consecutive frame are most likely images of previously selected points. Treating projections of previously initialized scene points as new observations is problematic because it introduces unwanted dependencies in the normal equations and superficially increases the number of independent measurements in the linearized system of equations.

To address this issue, we assume that the structure and motion initial estimates are accurate enough to predict the location of the current scene points in the new frame. Prior to initializing new scene points, we use the provided pose initialization to warp all previously detected scene points that are active in the optimization sliding window onto the new frame. After that, we mark a 3×3 square area at the projection location of the previous scene points as an invalid location for initializing new points. Finally, The number of selected points per frame varies depending on the image resolution and image content. In our experiments, this number ranges between ≈ 4000–10000 points per image.

Determining Visibility. Ideally, we would like to assume that newly initialized scene points are visible in all frames and to rely on the algorithm to reliably determine if this is the case. However, automatically determining the visibility information along with structure and motion parameters is challenging, as many scene points quickly go out of view, or become occluded.

An efficient and reliable measure to detect occlusions and points that cannot be matched reliably is the normalized correlation. For all scene points that are close to the current frame i, we use the pose initialization $\mathbf{T}_i$ to extract a 5×5 intensity patch. The patch is obtained by projecting the scene points to the new frame and its visibility list is updated if the zero-mean normalized correlation score (ZNCC) is greater than 0.6. We allow ± 2 frames for a point to be considered close, $i.e. |i - r(j)| \leq 2$. This procedure is similar to determining visibility in multi-view stereo algorithms [4] and is best summarized in Algorithm 1.

Optimization Details. We use the Ceres optimization library [40] to optimize the objective in Eq. (7). We use the Levenberg-Marquardt algorithm [41,42] to minimize a Huber loss function instead of squared loss to improve robustness. Termination tolerances are set to 1×10^{-6}, and automatic differentiation facilities are used. The image gradients used in the linearized system in Eqs. (10) and (11) are computed using central-differences. Finally, we also make use of the Schur complement for a more efficient solution.

Since scene points do not remain in view for an extended period in most VSLAM datasets, the photometric refinement step is performed using a sliding window of five frames [43]. The motion parameters of the first frame in the sliding window is held constant to fixate the Gauge freedom [10]. The 3D parameters of the scene points in the first frame, however, are included in the optimization.

Algorithm 1. Summary of image processing in our algorithm

```
 1: procedure PROCESSFRAME(Iᵢ, Tᵢ)
 2:     Step 1: establish connections to the new frame
 3:     mask = all_valid(rows(I), cols(I))
 4:     for all scene points Xⱼ in sliding window do
 5:         if reference frame r(j) is too far from i then
 6:             continue
 7:         x := projection of Xⱼ onto image Iᵢ using pose Tᵢ
 8:         φ' := patch at x and φ := reference patch for Xⱼ
 9:         if zncc(φ, φ') > threshold then
10:             add frame i to visibility list Vⱼ
11:             mask(u) = invalid

12:     Step 2: add new scene points
13:     G := gradient magnitude of Iᵢ
14:     for all pixels u in Iᵢ do
15:         if u is a local maxima in G then
16:             if location u is valid in mask then
17:                 initialize a new point X with reference patch at I(u)
```

3 Experiments

In this section, we evaluate the performance of the proposed algorithm on two commonly used VSLAM benchmarks to facilitate comparisons with the state-of-the-art. The first is the KITTI benchmark [44], which contains imagery from an outdoor stereo camera mounted on a vehicle. The second is the Malaga dataset [45], which is particularly challenging for VSLAM because the baseline of the camera (12 cm) is small relative to the scene structure.

3.1 The KITTI Benchmark

Initializing with Geometric BA. Torr and Zisserman [12] convincingly argue that the estimation of structure and motion should proceed by feature extraction and matching to provide a good initialization for BA-based refinement techniques. Here, we use the output of ORB-SLAM [18], a recently proposed state-of-the-art VSLAM algorithm, to initialize our method. ORB-SLAM not only performs geometric BA, but also implements loop closure to reduce drift.

We only use the pose initialization from ORB-SLAM. We do not make use of the refined 3D points as they are available at selected keyframes only. This is because images in the KITTI benchmark are collected at 10 Hz, while the vehicle speed exceeds 80 km/h in some sections. Subsequently, the views are separated by a large baseline, which violates the small displacement assumption required for the validity of linearization in Eqs. (10) and (11).

Hence, to initialize 3D points we use the standard block matching stereo algorithm implemented in OpenCV. This is a winner-takes-all brute force search strategy based on the sum of absolute intensity differences (SAD). The algorithm is configured to search for 128 disparities using a 7×7 aggregation window.

The choice of initializing the algorithm with ORB-SLAM is intentional to assess the accuracy of the algorithm in comparison to the Gold Standard solution from traditional BA. We note, however, that a correspondence-free system is possible by initializing the pose parameters with a direct method [5], or possibly a low-quality GPS.

Performance of the algorithm is shown in Fig. 3 and not only does it outperform the accuracy of (bundle adjusted and loop-closed) ORB-SLAM, but it also outperforms other top performing algorithms, especially in the accuracy of estimating rotations. Compared algorithms include: ORB-SLAM [18], LSD-SLAM [5,30], VoBA [46], and MFI [47].

We note that sources of error in our algorithm are correlated with faster vehicle speeds. This is to be expected as the linearization of the photometric error holds only in a small neighborhood. This could be mitigated by implementing the algorithm in scale-space [48], or improving the initialization quality of the scene structure (either by better stereo, or better scene points obtained from a geometric BA refinement step). Interestingly, however, the rotation error is reduced at high speeds which can be explained by lack of large rotations. The same behavior can be observed with LSD-SLAM's performance as both methods rely on the photometric error, but our rate of error reduction is higher due to the joint refinement of pose and structure parameters.

Initializing with Frame–Frame VO. Surprisingly, and contrary to other image-based optimization schemes [15,50], our algorithm does not require an accurate initialization. Figure 5 demonstrates a significant improvement in accuracy when the algorithm is initialized using frame–frame VO estimates with unbounded drift. Here, we used a direct method to initialize the camera pose without using any feature correspondences [49].

Interestingly, however, when starting from a poor initialization our algorithm does not attain the same accuracy as when initialized using a better quality starting point as shown in Fig. 3. This leads us to conclude the algorithm is sensitive to the initialization conditions more so than traditional BA. Importantly, however, the algorithm is able to significantly improve upon a poor initialization.

Convergence Characteristics and Runtime. As shown in Fig. 6, most of the photometric error is eliminated during the first five iterations of the optimization. While this is by no means a metric of quality, it is reassuring as it indicates a well-behaved optimization procedure. The number of iterations and the cumulative runtime per sliding window of 5 frames is shown in Fig. 7. The median number of iterations is 34 with a standard deviation of ≈ 6. Statistics are computed on the KITTI dataset frames. The runtime is ≈ 2 s per sliding window (400 ms per frame) using a laptop with a dual core processor clocked at 2.8 GHz and 8 GB of RAM, which limits parallelism. We note that it is possible to improve the runtime of the proposed method significantly using the CPU, or the GPU. The bottleneck

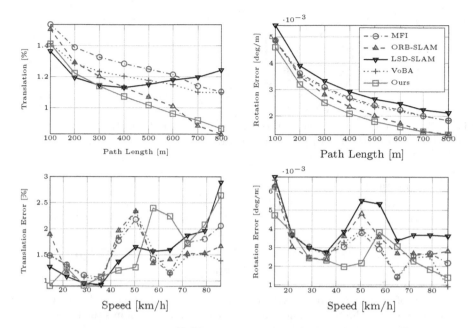

Fig. 3. Comparison to state-of-the-art algorithms on the KITTI benchmark. Our approach performs the best. Error in our approach correspond to segments of the data when the vehicle is driving at a high speed, which increases the magnitude of motion between frames and affects the linearization assumptions. No loop closure, or keyframing is performed using our algorithm. Improvement is shown qualitatively in Fig. 4

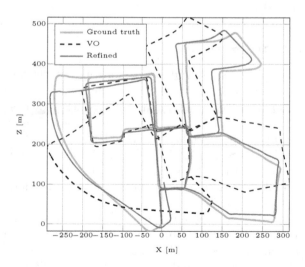

Fig. 4. Improvement starting from a poor initialization shown on the first sequence of the KITTI benchmark. Quantitative evaluation is shown in Fig. 3. We used a direct (correspondence-free) frame–frame VO method to initialize the pose parameters [49].

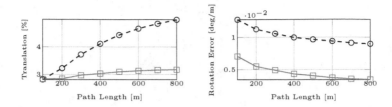

Fig. 5. Improvement in accuracy starting from a poor initialization using a frame–frame direct VO method with unbounded drift.

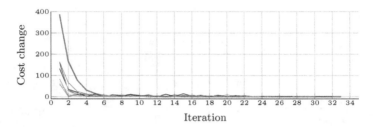

Fig. 6. Rate of error reduction at every iteration shown for the first 10 sliding windows, each with 5 frames. The thicker line shows the first bundle, which has the highest error. Most of the error is eliminated with the first 5 iterations.

of the proposed algorithm is image interpolation (which can be done efficiently with SIMD instructions) and the reliance on automatic differentiation (which limits any code optimization as the code must remain simple for automatic differentiation to work).

3.2   The Málaga Stereo Dataset

The Málaga dataset [47] is a particularly challenging dataset for VSLAM. The dataset features driving in urban areas using a small baseline stereo camera at resolution 800 × 600. The stereo baseline is 12 cm which provides little parallax for resolving distal observations. We use extracts 1, 3, and 6 in our evaluation.

Our experimental setup is similar to the KITTI dataset. However, we estimate the stereo using the SGM algorithm [51], as implemented in the OpenCV library. The stereo is used to estimate 16 disparities with a SAD block size of 5 × 5. We did not observe a significant difference in performance when using block matching instead of SGM.

The Malaga dataset provides GPS measurements, but they are not accurate enough for quantitative evaluation. The GPS path, however, is sufficient to qualitatively demonstrate precision. Results are shown in Fig. 8 in comparison with ORB-SLAM [18], which we used its pose output to initialize our algorithm. We note that in extract 3 of the Malaga dataset (shown on the left in Fig. 8), ORB-SLAM loses tracking during the turn and our algorithm continues *without* initialization.

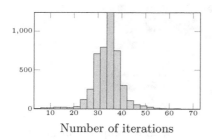
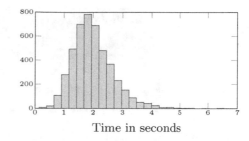

Number of iterations   Time in seconds

Fig. 7. Histogram of the number of iterations (on the left) and runtime (on the right). The median number of iterations is 34, with a standard deviation of 6.02. The median run time is 1.89, mean 1.98 and standard deviation of 0.69. The runtime is reported for sliding window of 5 frames on the KITTI benchmark.

Fig. 8. Our algorithm (magenta) compared with ORB-SLAM (dashed) against GPS (yellow) on extracts 3 and 6 of the Malaga dataset. For extract 3 ORB-SLAM loses tracking during the roundabout, where our algorithm continues without an initialization. Results for extract 6 are shown up to frame 3000 as ORB-SLAM loses tracking. The figure is best viewed in color. (Maps courtesy of Google Maps.) (Color figure online)

To assess the quality of pose estimates, we demonstrate results on a dense reconstruction procedure shown in Fig. 9. Using the estimated camera trajectory, we chain the first 6 m of the disparity estimates to generate a dense map. As shown in Fig. 9, the quality of pose estimates appears to be good.

4   Related Work

Geometric BA. BA has a long and rich history in computer vision, photogrammetry and robotics [10]. BA is a large geometric minimization problem with the important property that variable interactions result in a *sparse* system of linear equations. This sparsity is key to enabling large–scale applications [52,53]. Exploiting this sparsity is also key to obtaining precise results efficiently [54,55]. The efficiency of BA has been an important research topic especially when handling large datasets [56,57] and in robotics applications [58–60]. Optimality and convergence properties of BA have been studied at length [11,61,62] and remain

Fig. 9. Dense map from Malaga dataset extract 1. The map is computed by stitching together SGM disparity with the refined camera pose.

of interest to date [63]. All the aforementioned research in geometric BA could be integrated into the proposed photometric BA framework.

Direct Multi-frame Alignment. By direct alignment we mean algorithms that estimate the parameters of interest from the image data directly and without relying on sparse features as an intermediate representation of the image [64]. The fundamental differences between direct methods (like the one proposed herein) and the commonly used feature-based pipeline is how the correspondence problem is tackled and is not related to the density of the reconstruction. In the feature-based pipeline [12], structure and motion parameters are estimated from known and fixed correspondence. In contrast, the direct pipeline to motion estimation does not use fixed correspondences. Instead, the correspondences are estimated as a byproduct of directly estimating the parameters of interest.

The use of direct algorithms for SFM applications was studied for small–scale problems [38,65–67], but feature-based alignment has proven more successful in handling wide baseline matching problems [12] as small pixel displacements is an integral assumption for direct methods. Nonetheless, with the increasing availability of high frame-rate cameras and the increasing computational power, direct methods are demonstrating great promise [5,6,9].

To date, however, the use of direct methods in VSLAM has been limited to frame–frame motion estimation. Approaches that make use of multiple frames are designed for dense depth estimation only and multi-view stereo [4,9], which assume a correct camera pose and only refine the scene structure. Other algorithms can include measurements from multiple frames, but rely on the presence of structures with strong planarity in the environment [37,68] (or equivalently restricting the motion of the camera to rotation only [69]).

In this work, in contrast to previous research in direct image-based alignment [9,38], we show that provided good initialization, it is possible to jointly refine the structure and motion parameters by minimizing the photometric error and without restricting the camera motion or the scene structure.[1]

The LSD-SLAM algorithm [5] is a recently proposed direct VSLAM algorithm. The fundamental difference in comparison to our work is that we refine

---

[1] While this work was under review, Engel *et al.* proposed a similar photometric (direct) formulation for VSLAM [70].

the parameters of motion and structure jointly in one large optimization problem. The joint optimization of motion and structure proposed herein is important in future work concerning the optimality and convergence properties of photometric structure-from-motion (SFM) and photometric, or direct, VSLAM.

Dense Multi-view Stereo (MVS). MVS algorithms aim at recovering a dense depth estimate of objects or scenes using many images with known pose [4]. To date, however, research on simultaneous refinement of motion and depth from multiple frames remains sparse. Furukawa and Ponce [15] were among the first to demonstrate that relying on minimizing the reprojection error is not always accurate enough. Recently, Delaunoy and Pollefeys [50] proposed a photometric BA approach for dense MVS. Starting from a precise initial reconstruction and a mesh model of the object, the algorithm is demonstrated to enhance MVS accuracy. The imaging conditions, however, are ideal and brightness constancy is assumed [50]. In our work, we do not require a very precise initialization and can address challenging illumination conditions. More importantly, the formulation proposed by Delaunoy and Pollefeys requires the availability of an accurate dense mesh, which is not possible to obtain in VSLAM scenarios.

5 Conclusions

In this work, we show how to improve on the accuracy of the state-of-art VSLAM methods by minimizing the photometric error across multiple views. In particular, we show that it is possible to improve results obtained by minimizing the reprojection error in a bundle adjustment (BA) framework. We also show, contrary to previous image-based minimization work [5,7,9,30,38], that the *joint* refinement of motion and structure is possible in unconstrained scenes without the need for alternation or disjoint optimization.

The accuracy of minimizing the reprojection using traditional BA is limited by the precision and accuracy of feature localization and matching. In contrast, our approach — BA without correspondences — determines the correspondences implicitly such that the photometric consistency is maximized as a function of the scene structure and camera motion parameters.

Finally, we show that accurate solutions to geometric problems in vision are not restricted to geometric primitives such as corners and edges, or even planes. We look forward to more sophisticated modeling of the geometry and photometry of the scene beyond the intensity patches used in our work.

References

1. Sun, D., Roth, S., Black, M.: Secrets of optical flow estimation and their principles. In: IEEE Conference on Computer Vision and Pattern Recognition (CVPR), pp. 2432–2439 (2010)
2. Vedula, S., Baker, S., Rander, P., Collins, R., Kanade, T.: Three-dimensional scene flow. IEEE Trans. Pattern Anal. Mach. Intell. **27**, 475–480 (2005)

3. Seitz, S., Curless, B., Diebel, J., Scharstein, D., Szeliski, R.: A comparison and evaluation of multi-view stereo reconstruction algorithms. In: IEEE Computer Society
   Conference on Computer Vision and Pattern Recognition, pp. 519–528 (2006)
4. Furukawa, Y., Hernndez, C.: Multi-view stereo: a tutorial. Found. Trends Comput.
   Graph. Vis. **9**, 1–148 (2015)
5. Engel, J., Schöps, T., Cremers, D.: LSD-SLAM: large-scale direct monocular SLAM. In: Fleet, D., Pajdla, T., Schiele, B., Tuytelaars, T. (eds.) ECCV
   2014. LNCS, vol. 8690, pp. 834–849. Springer, Heidelberg (2014). doi:10.1007/
   978-3-319-10605-2_54
6. Kerl, C., Sturm, J., Cremers, D.: Robust odometry estimation for RGB-D cameras.
   In: International Conference on Robotics and Automation (ICRA) (2013)
7. Steinbrucker, F., Sturm, J., Cremers, D.: Real-time visual odometry from dense
   RGB-D images. In: IEEE International Conference on Computer Vision, ICCV
   Workshops (2011)
8. Meilland, M., Comport, A.: On unifying key-frame and voxel-based dense visual
   SLAM at large scales. In: IEEE/RSJ International Conference on Intelligent
   Robots and Systems (IROS), pp. 3677–3683 (2013)
9. Newcombe, R., Lovegrove, S., Davison, A.: DTAM: dense tracking and mapping
   in real-time. In: IEEE International Conference on Computer Vision (ICCV), pp.
   2320–2327 (2011)
10. Triggs, B., McLauchlan, P.F., Hartley, R.I., Fitzgibbon, A.W.: Bundle adjustment
    — a modern synthesis. In: Triggs, B., Zisserman, A., Szeliski, R. (eds.) IWVA
    1999. LNCS, vol. 1883, pp. 298–372. Springer, Heidelberg (2000). doi:10.1007/
    3-540-44480-7_21
11. Hartley, R., Zisserman, A.: Multiple View Geometry in Computer Vision, 2nd edn.
    Cambridge University Press, Cambridge (2004)
12. Torr, P.H.S., Zisserman, A.: Feature based methods for structure and motion esti-
    mation. In: Triggs, B., Zisserman, A., Szeliski, R. (eds.) IWVA 1999. LNCS, vol.
    1883, pp. 278–294. Springer, Heidelberg (2000). doi:10.1007/3-540-44480-7_19
13. Kanazawa, Y., Kanatani, K.: Do we really have to consider covariance matrices for
    image features? In: Proceedings of the Eighth IEEE International Conference on
    Computer Vision, 2001. ICCV 2001, vol. 2, pp. 301–306. IEEE (2001)
14. Brooks, M.J., Chojnacki, W., Gawley, D., Van Den Hengel, A.: What value covari-
    ance information in estimating vision parameters? In: Proceedings of the Eighth
    IEEE International Conference on Computer Vision, 2001. ICCV 2001, vol. 1, pp.
    302–308. IEEE (2001)
15. Furukawa, Y., Ponce, J.: Accurate camera calibration from multi-view stereo and
    bundle adjustment. In: IEEE Conference on Computer Vision and Pattern Recog-
    nition, 2008. CVPR 2008, pp. 1–8. IEEE (2008)
16. Deriche, R., Giraudon, G.: Accurate corner detection: an analytical study. In: Pro-
    ceedings of the Third International Conference on Computer Vision, pp. 66–70.
    IEEE (1990)
17. Shimizu, M., Okutomi, M.: Precise sub-pixel estimation on area-based matching.
    In: ICCV, pp. 90–97 (2001)
18. Mur-Artal, R., Montiel, J.M.M., Tardós, J.D.: ORB-SLAM: a versatile and accu-
    rate monocular SLAM system. CoRR abs/1502.00956 (2015)
19. Milford, M., Wyeth, G.: SeqSLAM: visual route-based navigation for sunny sum-
    mer days and stormy winter nights. In: IEEE International Conference on Robotics
    and Automation (ICRA), pp. 1643–1649 (2012)
20. Reid, I.: Towards semantic visual SLAM. In: 13th International Conference on
    Control Automation Robotics Vision (ICARCV), p. 1 (2014)

21. Salas-Moreno, R., Newcombe, R., Strasdat, H., Kelly, P., Davison, A.: SLAM++: simultaneous localisation and mapping at the level of objects. In: IEEE Conference on Computer Vision and Pattern Recognition (CVPR), pp. 1352–1359 (2013)
22. Murray, R.M., Li, Z., Sastry, S.S., Sastry, S.S.: A Mathematical Introduction to Robotic Manipulation. CRC Press, Boca Raton (1994)
23. Ma, Y., Soatto, S., Kosecka, J., Sastry, S.S.: An Invitation to 3-D Vision: From Images to Geometric Models. Springer, New York (2003)
24. Hartley, R., Trumpf, J., Dai, Y., Li, H.: Rotation averaging. Int. J. Comput. Vis. **103**, 267–305 (2013)
25. Civera, J., Davison, A.J., Montiel, J.M.: Inverse depth parametrization for monocular SLAM. IEEE Trans. Robot. **24**, 932–945 (2008)
26. Zhao, L., Huang, S., Sun, Y., Yan, L., Dissanayake, G.: ParallaxBA: bundle adjustment using parallax angle feature parametrization. Int. J. Robot. Res. **34**, 493–516 (2015)
27. Lucas, B.D., Kanade, T.: An iterative image registration technique with an application to stereo vision (DARPA). In: Proceedings of the 1981 DARPA Image Understanding Workshop, pp. 121–130 (1981)
28. Horn, B.K., Schunck, B.G.: Determining optical flow. Artif. Intell. **17**, 185–203 (1981)
29. Baker, S., Matthews, I.: Lucas-Kanade 20 years on: a unifying framework. Int. J. Comput. Vis. **56**, 221–255 (2004)
30. Engel, J., Stueckler, J., Cremers, D.: Large-scale direct SLAM with stereo cameras. In: International Conference on Intelligent Robots and Systems (IROS) (2015)
31. Nocedal, J., Wright, S.J.: Numerical Optimization, 2nd edn. Springer, New York (2006)
32. Harris, C., Stephens, M.: A combined corner and edge detector. In: Alvey Vision Conference, Manchester, vol. 15, p. 50 (1988)
33. Rosten, E., Drummond, T.: Machine learning for high-speed corner detection. In: Leonardis, A., Bischof, H., Pinz, A. (eds.) ECCV 2006. LNCS, vol. 3951, pp. 430–443. Springer, Heidelberg (2006). doi:10.1007/11744023_34
34. Dellaert, F., Seitz, S.M., Thorpe, C.E., Thrun, S.: Structure from motion without correspondence. In: Proceedings of the IEEE Conference on Computer Vision and Pattern Recognition, vol. 2, pp. 557–564. IEEE (2000)
35. Meilland, M., Comport, A., Rives, P.: A spherical robot-centered representation for urban navigation. In: IROS (2010)
36. Nister, D., Naroditsky, O., Bergen, J.: Visual odometry. In: Computer Vision and Pattern Recognition (CVPR) (2004)
37. Irani, M., Anandan, P., Cohen, M.: Direct recovery of planar-parallax from multiple frames. In: Triggs, B., Zisserman, A., Szeliski, R. (eds.) IWVA 1999. LNCS, vol. 1883, pp. 85–99. Springer, Heidelberg (2000). doi:10.1007/3-540-44480-7_6
38. Stein, G., Shashua, A.: Model-based brightness constraints: on direct estimation of structure and motion. IEEE Trans. Pattern Anal. Mach. Intell. **22**, 992–1015 (2000)
39. Agouris, P., Schenk, T.: Automated aerotriangulation using multiple image multipoint matching. Photogramm. Eng. Remote Sens. **62**, 703–710 (1996)
40. Agarwal, S., Mierle, K., et al.: Ceres solver (2016). http://ceres-solver.org
41. Levenberg, K.: A method for the solution of certain non-linear problems in least squares. Q. J. Appl. Maths. **2**, 164–168 (1944)
42. Marquardt, D.W.: An algorithm for least-squares estimation of nonlinear parameters. J. Soc. Ind. Appl. Math. **11**, 431–441 (1963)

43. Snderhauf, N., Konolige, K., Lacroix, S., Protzel, P.: Visual odometry using sparse bundle adjustment on an autonomous outdoor vehicle. In: Levi, P., Schanz, M., Lafrenz, R., Avrutin, V. (eds.) Autonome Mobile Systems 2005. Informatik aktuell, pp. 157–163. Springer, Heidelberg (2006)

44. Geiger, A., Lenz, P., Urtasun, R.: Are we ready for autonomous driving? The KITTI vision benchmark suite. In: Conference on Computer Vision and Pattern Recognition (CVPR) (2012)

45. Blanco, J.L., Moreno, F.A., González-Jiménez, J.: The málaga urban dataset: high-rate stereo and lidars in a realistic urban scenario. Int. J. Robot. Res. **33**, 207–214 (2014)

46. Tardif, J.P., George, M., Laverne, M., Kelly, A., Stentz, A.: A new approach to vision-aided inertial navigation. In: IEEE/RSJ International Conference on Intelligent Robots and Systems (IROS), pp. 4161–4168. IEEE (2010)

47. Badino, H., Yamamoto, A., Kanade, T.: Visual odometry by multi-frame feature integration. In: IEEE International Conference on Computer Vision Workshops (ICCVW), pp. 222–229 (2013)

48. Lindeberg, T.: Scale-space Theory in Computer Vision. Springer, New York (1994)

49. Alismail, H., Browning, B., Lucey, S.: Direct visual odometry using bit-planes. CoRR abs/1604.00990 (2016)

50. Delaunoy, A., Pollefeys, M.: Photometric bundle adjustment for dense multi-view 3D modeling. In: IEEE Conference on Computer Vision and Pattern Recognition (CVPR), pp. 1486–1493. IEEE (2014)

51. Hirschmuller, H.: Accurate and efficient stereo processing by semi-global matching and mutual information. In: Computer Vision and Pattern Recognition (2005)

52. Agarwal, S., Snavely, N., Seitz, S.M., Szeliski, R.: Bundle adjustment in the large. In: Daniilidis, K., Maragos, P., Paragios, N. (eds.) ECCV 2010. LNCS, vol. 6312, pp. 29–42. Springer, Heidelberg (2010). doi:10.1007/978-3-642-15552-9_3

53. Konolige, K., Garage, W.: Sparse sparse bundle adjustment. In: BMVC, pp. 1–11 (2010)

54. Jeong, Y., Nister, D., Steedly, D., Szeliski, R., Kweon, I.S.: Pushing the envelope of modern methods for bundle adjustment. IEEE Trans. Pattern Anal. Mach. Intell. **34**, 1605–1617 (2012)

55. Engels, C., Stewnius, H., Nister, D.: Bundle adjustment rules. In: Photogrammetric Computer Vision (2006)

56. Wu, C., Agarwal, S., Curless, B., Seitz, S.M.: Multicore bundle adjustment. In: IEEE Conference on Computer Vision and Pattern Recognition (CVPR), pp. 3057–3064. IEEE (2011)

57. Ni, K., Steedly, D., Dellaert, F.: Out-of-core bundle adjustment for large-scale 3D reconstruction. In: IEEE 11th International Conference on Computer Vision, pp. 1–8 (2007)

58. Konolige, K., Agrawal, M.: FrameSLAM: from bundle adjustment to real-time visual mapping. IEEE Trans. Robot. **24**, 1066–1077 (2008)

59. Kaess, M., Ila, V., Roberts, R., Dellaert, F.: The Bayes tree: an algorithmic foundation for probabilistic robot mapping. In: Hsu, D., Isler, V., Latombe, J.C., Lin, M. (eds.) Algorithmic Foundations of Robotics IX. Springer Tracts in Advanced Robotics, vol. 68, pp. 157–173. Springer, Heidelberg (2011)

60. Kaess, M., Ranganathan, A., Dellaert, F.: iSAM: incremental smoothing and mapping. IEEE Trans. Robot. (TRO) **24**, 1365–1378 (2008)

61. Kahl, F., Agarwal, S., Chandraker, M.K., Kriegman, D., Belongie, S.: Practical global optimization for multiview geometry. Int. J. Comput. Vis. **79**, 271–284 (2008)

62. Hartley, R., Kahl, F., Olsson, C., Seo, Y.: Verifying global minima for L_2 minimization problems in multiple view geometry. Int. J. Comput. Vis. **101**, 288–304 (2013)
63. Aftab, K., Hartley, R.: LQ-bundle adjustment. In: IEEE International Conference on Image Processing (ICIP), pp. 1275–1279 (2015)
64. Irani, M., Anandan, P.: About direct methods. In: Triggs, B., Zisserman, A., Szeliski, R. (eds.) IWVA 1999. LNCS, vol. 1883, pp. 267–277. Springer, Heidelberg (2000). doi:10.1007/3-540-44480-7_18
65. Horn, B.K.P., Weldon, E.J.: Direct methods for recovering motion (1988)
66. Oliensis, J.: Direct multi-frame structure from motion for hand-held cameras. In: Proceedings of the 15th International Conference on Pattern Recognition, vol. 1, pp. 889–895 (2000)
67. Mandelbaum, R., Salgian, G., Sawhney, H.: Correlation-based estimation of egomotion and structure from motion and stereo. In: The Proceedings of the Seventh IEEE International Conference on Computer Vision, vol. 1, pp. 544–550 (1999)
68. Silveira, G., Malis, E., Rives, P.: An efficient direct approach to visual SLAM. IEEE Trans. Robot. **24**(5), 969–979 (2008). doi:10.1109/TRO.2008.2004829
69. Lovegrove, S., Davison, A.J.: Real-time spherical mosaicing using whole image alignment. In: Daniilidis, K., Maragos, P., Paragios, N. (eds.) ECCV 2010. LNCS, vol. 6313, pp. 73–86. Springer, Heidelberg (2010). doi:10.1007/978-3-642-15558-1_6
70. Engel, J., Koltun, V., Cremers, D.: Direct sparse odometry. ArXiv e-prints (2016)

Can Computer Vision Techniques be Applied to Automated Forensic Examinations? A Study on Sex Identification from Human Skulls Using Head CT Scans

Olasimbo Ayodeji Arigbabu[1], Iman Yi Liao[1(✉)], Nurliza Abdullah[2], and Mohamad Helmee Mohamad Noor[3]

[1] School of Computer Science, University of Nottingham Malaysia Campus, Semenyih, Malaysia
khyx5oaa@exmail.nottingham.edu.my, iman.liao@nottingham.edu.my
[2] Department of Forensic Medicine, Hospital Kuala Lumpur, Kuala Lumpur, Malaysia
azilrun@gmail.com
[3] Radiology Department, Hospital Kuala Lumpur, Kuala Lumpur, Malaysia
emeemd71@gmail.com

Abstract. Sex determination from human skeletal remains is a challenging problem in forensic anthropology. The human skull has been regarded as the second best predictor of sex because it contains several sexually dimorphic traits. Previous studies have shown that morphological assessment and morphometric analysis can be used to assess sex variation from dried skulls. With the availability of CT scanners, the field has seen increasing computer aided techniques in assisting these traditional forensic examinations. However, they largely remain at the level of providing a digital interface for landmarking for morphometric analysis. A recent research has applied shape analysis techniques for morphological analysis on a specific part of the skull. In this paper, we endeavor to explore the application of computer vision techniques that have prominently been used in the field of 3D object recognition and retrieval, for providing alternative method to achieve sex identification from human skulls automatically. We suggest a possible framework for the whole process including multi-region representation of the skull with 3D shape descriptors, and particularly examined the role of 3D descriptors on sex identification accuracy. The experimental results on 100 head post mortem CT scans indicate the potential of 3D descriptors for skull sex classification. To the best of our knowledge, this is the first work to have approached skull sex prediction from this novel perspective.

1 Introduction

Identification of human skeletal remains after occurrence of natural disaster or tragic event of some sort, is highly important in forensic science and anthropolopy [1,2]. In certain situations, approximating the possible identity of a person could be of prime importance, especially when no primary identification

© Springer International Publishing AG 2017
S.-H. Lai et al. (Eds.): ACCV 2016, Part IV, LNCS 10114, pp. 342–359, 2017.
DOI: 10.1007/978-3-319-54190-7_21

evidence such as DNA and/or fingerprint is available. Computer-aided methods such as facial reconstruction and superimposition are part of the commonly adopted solutions for assisting in the identification process [3,4]. According to Cattaneo [5], it has been strongly advised that the use of these methods should be restricted to excluding identity if there are signs of gross incompatibilities. Alternatively, forensic experts may seek to build statistical background profile (conceptually referred to as osteobiography) of humans by assessing the sustained skeletal remains [6]. In such cases, the highly regarded biological information worth documenting include sex, age, ancestry, and body stature [7]. In this paper, we focus on the problem of sex[1] determination from human skulls. From the literature, the pelvic region of human skeletal remains has often been studied by forensic experts to extract information about sex. Nevertheless, there are a number of cases where the complete skeleton can not be retrieved due to the degree of damage caused to the subjects or as a result of decomposition [8]. The human skull has gained more attention in forensic anthropology and regarded as the second best indicator of sex [9,10]. Moreover, in a high number of situations, the complete skull is usually available and it contains hard tissues whose sexual dimorphic traits can be examined with proper assessment methods. From forensic perspective, two standard methods have commonly been applied for sex prediction from human skull, which are morphological assessment and morphological analysis. In morphological method, forensic experts use visual assessment (guided by their wide experience in the differences between male and female cranial variations) to make prediction, while morphometric analysis involves linear or geometric measurement of anatomical landmarks. However, the performance of these methods can be affected by subjectivity of the observer, incompleteness of anatomical landmarks, and inability to generalize to populations other than the one in which the estimation technique is modelled.

In this paper, we offer a different perspective to skull sex classification through the use of 3D shape descriptors for skull representation. While, it is worth mentioning that 3D descriptors have previously been used for 3D face recognition [29,30], 3D object retrieval [31,32] gender recognition on the 3D facial data [33,34] they cannot be directly applied to skull data. Such notion has also been confirmed in characterization of skull deformity [25–28]. Essentially, 3D object retrieval is mainly concerned with identification of objects from broad categories, while face recognition depends on local primary features such as eyes, nose, and mouth, which are very informative and discriminative. These attributes are completely absent in a skull. Therefore, we propose a strategic approach to capture both subtle local geometric information and global shape characteristics of the skull for sex determination. The remainder of the paper is organized as follows. Section 2 provides a background review on skull sex prediction from forensic perspective. Section 3 details the experimental data. In Sects. 4 and 5, we describe the techniques used for pre-processing and 3D feature representation

---

[1] Sex is commonly regarded as gender in the biometrics research domain, but in forensic anthropology sex is the universally accepted term, because it refers to the biological information of a person. As a result, we prefer to use sex instead of gender.

of the skull. Section 6 presents the proposed multi-region feature extraction for skull representation. Section 7 provides the experimental results and we conclude our work in Sect. 8.

2   Related Works

In forensic anthropology, both morphological assessment and morphometric analysis are equally common cranial sex prediction methods [12]. The earliest attempts using morphological assessment were reported in [13–15]. In the 90's, 5 main regions of the skull were thoroughly studied and standardized on ordinal scale to aid quantification of the sexually dimorphic traits that can be visually assessed [16]. These traits are the robusticity of the nuchal crest, size of the mastoid process, sharpness of the supraorbital margin, prominence of the glabella, and the projection of the mental eminence. Generally to determine sex from a cranial, the traits are assessed and judged based on their similarity to the visual description provided in [16]. Over the last few years, this diagram has served as a guide for several forensic anthropologists in sex estimation using a variation of ordinal score rating with different estimation performance between 83% and 90% [15,17]. Some forensic experts have focused on specific dimorphic traits or regions of the cranial. For instance, the shape of supraorbital margin was assessed by Graw et al. [18] using Plasticine impression to form the contour shape of the supraorbital and assess the shape of trait on a 7-point scale. Their experiment presented a prediction rate of 70%. While morphological assessment is arguably the most conventional sex prediction technique, it can easily be affected by the inexperience of the observer about the population being studied. Recently, Pinto et al. [19] proposed a computer-aided method for assessing the supra-orbital region using silicone impression to cast the shape of the region and scanning of the imprinted surface with 3D scanner. The scanned data were further reconstructed into 3D mesh and 2D wavelet transform was adopted to study the shape variation of the supraorbital region. Though, the approach is an advancement on the conventional visual assessment, it still basically requires the knowledge of forensic expert to locate the supra-orbital margin.

Computer-aided techniques have also been useful in morphometric analysis, which involves linear or geometric measurement of anatomical landmarks. The selection of the amount of landmarks to use varies among the methods reported in literature. For instance Franklin et al. [20] performed morphormetric analysis of 8 measurements of the 3D landmarks. By using discriminant function analysis, it was found the facial width (bizygomatic breadth), and the length and height of the cranial vault possess higher sexual dimorphism. The accuracy rate of the method was between 77–80%. Other studies have found additional regions expressing greater sexually dimorphism. Bigoni et al. [21] presented a craniometric approach to sex determination to locate the regions of the cranium where sexual dimorphism are most pronounced. The study was conducted on 139 cranial set of Central European population. 82 ecto-cranial landmarks were annotated from 7 sub-regions of the cranial (the configuration of neurocranium, cranial base, midsagittal curve of vault, upper face, orbital region, nasal region, and palatal region). Through shape analysis with generalized procrustes analysis (GPA) they noticed no sex difference in the sample set when landmarks from the whole cranium were used. However, further partial shape examinations on each of the 7 sub-regions indicated that the midsagittal curve, the upper face, the orbital region, the nasal region, and the palatal region possess strong sexual dimorphism, but no sex variation in the cranium base and the neurocranium configuration. A performance comparison of two discriminant function analysis on 17 craniometric variables from 90 Iberian skull remains indicated that most metric variables were higher for males than females [22]. The problem with morphometrics includes the inability to perform accurate annotation of landmarks, where inter-observer error is almost 10% for most measurements [23]. Luo et al. [24] presented an approach which does not require annotating any specific anatomical landmarks with principal component analysis and linear discriminant analysis on Chinese samples. However, the method was examined on only the frontal part of the skull. The remaining back region of the skull include mastoid process and nuchal crest which equally exhibit sexually dimorphic characteristics. Though morphological assessment and morphometric analysis have a simple technical concept, the two methods are quite effective in providing solution to cranial sex estimation problem. Moreover, the concepts have been established on well grounded principles and universally agreed on, to be used as evidence in court cases and forensic investigation [23]. Nonetheless, it is essential to mention that these techniques possess several limitations. Firstly, observer subjectivity influences the confidence of prediction. Since the morphological assessment involves verbal description, it naturally implies that humans would exhibit differing conclusion in their visual perception. This influence has been well reported in previous studies as inter-observer variability, especially when forensic expert make prediction on cranial samples from a population they are unfamiliar with [21,35]. Likewise, inter-observer variability is inherent when observers select landmarks in the case of morphometric analysis. Furthermore, this influence of population difference affects the accuracy and precision of the traits used for identification [11]. As a result, an estimation method used in a specific population may not generalize to other populations [36]. Secondly, the accurate annotation of landmarks from anatomical regions of the cranial requires significant amount of time. Besides, it often requires expensive, specialized anthropometric equipment. In addition forensic expert usually face the challenge of performing measurements that capture subtle variations among cranial traits that are easy to see but very difficult to measure [17]. These aforementioned limitations are the core motivations for this study. It is hoped through automatic feature extraction and learning of compact and discriminative 3D shape features, we can bridge the representation gap between humans and the perceived cranial sexually dimorphic traits. In essence, we can eliminate the effect of subjectivity in visually assessed traits, incompleteness of anatomical landmarks, as well as minimize the time required for estimation.

2.1   Main Contributions

The contributions of this paper are listed as follows:

- This paper introduces a new conceptual framework, where the interest is to investigate the possibility of using 3D local feature descriptors for sex prediction from human skulls. Our method is completely automatic, it does not require manually annotation of landmarks or visual assessment of skull anatomical sites. Moreover, the method generalizes the prediction performance from different ancestral backgrounds.
- We propose aggregation of descriptors from multiple vertex points into compact and discriminative representation from different sub-regions of the skull. To segment the skull into smaller regions, we constrained the partitioning to the z-plane, where we can guarantee accurate distribution of features in each local sub-region.
- We provide enlightenment on, how and the extent to which, 3D shape descriptors can be used for sex determination from skulls. While, 3D descriptors have been demonstrated to be effective in gender recognition on 3D face data [33,34,37] their exploit on 3D skull with no textural skin properties and primary facial attributes such as eyes, eyebrow, nose, and mouth is yet to be investigated. To the best of our knowledge, this is the first study to approach the problem of skull sex estimation from this novel perspective.

3   Experimental Dataset

The experimental data is 100 sample set of post-mortem computed tomography (PMCT) scan slices obtained from Hospital Kuala Lumpur (HKL), which has been compiled between 01-05-2012 and 30-09-2012. Legal consent for the use of the dataset was obtained prior to commencing this research. The data is composed of 54 male and 46 female subjects between the ages of 5 to 85 years, from south east asia. The scanning device is Toshiba CT scanner with scanning settings of 1.0 slice thickness, and 0.8 slice interval. The original data contains the full body of the subjects, which average between 2000 to 4000 slices persubject. The resolution of the data is $512 \times 512 \times$ No of slices. For the purpose of this work, the slices belonging to the head region have been chosen for each subject. Therefore, the resulting input samples have resolutions between $512 \times 512 \times 261$ and $512 \times 512 \times 400$, depending on size of the skull of each subject. The framework of the sex classification is composed of four main stages: 3D data preprocessing, 3D feature extraction, compact feature representation, and classification.

4   3D Preprocessing

4.1   Noise Removal

Through compilation of consecutive CT slices, we can obtain a volumetric 3D data. This volumetric data is then filtered to minimize noise, local irregularities

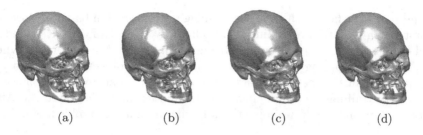

(a)   (b)   (c)   (d)

Fig. 1. Examples of reconstructed skull data. (a) Original skull (b) Downsampled to 50% (b) Downsampled to 25% (d) Downsampled to 13%.

and roughness using a discretized spline smoothing method [38]. The smoothing method uses the 3D discrete cosine transform based penalized least square regression (DCT-PLS) on equally spaced high dimension data. PLS is a fundamental concept in thin plate splines for data interpolation and smoothing. But, according to Garcia [38] we can reformulate the PLS regression problem with DCT, where the data are expressed in the form of cosine functions oscillating at different frequencies. The method can also be readily adapted to multidimensional datasets, since DCT is multidimensional.

After noise removal, the marching cubes algorithm is adopted to reconstruct the volume data into a 3D surface [39], as shown in Fig. 1. In this work, the reconstruction is achieved using an iso-value of (150) to obtain the regions containing the hard tissues. Due to the high dimension of the reconstructed skulls (for instance, the number of vertices in the original reconstructed skull is between 650000 and 850000 vertices), it is preferable to perform down-sampling in order to reduce computational complexity and time. As a result, mesh simplification [40] is performed on the original reconstructed mesh to downsample the 3D surface to 13% of the original size. The resulting downsampled skulls have <130000 vertices, but the structural details of the surface are still well preserved after downsampling, as it can be seen in Fig. 1.

4.2   Background Object Removal

Considering the fact that our data samples are real data obtained from hospital, we noticed some dynamic background objects, which have the same isovalue as the hard tissue, around some skull samples (around 27 samples) after surface reconstruction. To remove these unwanted or noisy background objects, we propose to learn an online sequential linear systems with gaussian mixture model sample initialization. Firstly, 8 samples of clean skulls and 8 samples of noisy skull data are randomly selected. Manual annotation is perform on the noisy skull data to cut out the noisy objects. The set of segmented noisy objects are used to form the negative training samples, while the clean skulls are regarded as the positive training samples. Therefore, the training set contains 8 clean skulls and 8 segmented noisy object regions resulting in ~750000 vertex points for the two classes. Secondly, due to the size of the training set, we used a sequential learning method where each time the training is performed in batches. However, the first batch of data for training is initialized by using GMM to fit a mixture model composing of K components. Given a training set $A = \{v_1, \ldots v_i\}$, where A is a matrix composing of clean and segmented noisy objects, and v_i is a vertex point, we cluster the set A into K-clusters with mixture of gaussians. To initialize the training process a cluster is selected as the initial batch A_0. After selecting the initial batch, we solve the following minimizing problem of a linear system as:

$$\text{Minimize: } ||A_0 x_0 - b_0||^2 + \lambda ||x_0||^2 \tag{1}$$

where x_0 is the learned coefficients with the smallest norm, λ is a regularization value, and $b_0 = \{+1, -1\}$ are the labels ($+1$ for clean skull, -1 for noisy object) of the initial subsets. The initial solution x_0 can be obtained analytically using:

$$x_0 = \left(A_0^\top A_0 + \lambda \mathrm{I}\right)^{-1} A_0^\top b_0 \tag{2}$$

Formally, the coefficient x_0 is regarded as the least square solution to the minimization problem in Eq. (1). The inverse matrix M_0^{-1} can be obtained as: $M_0^{-1} = \left(A_0^\top A_0 + \lambda \mathrm{I}\right)^{-1}$ and the output coefficient $x_0 = M_0^{-1} A_0^\top b_0$.

Updating x_{k+1}. Now, we would like to train the remaining subset of data in batches also, but without keeping the previously trained set in order to reduce computational time and complexity. Therefore, it is preferable to find an efficient approach to update the initially learned coefficient, while solving new linear systems minimization problem [41]. Assuming the least square solution to the minimization problem previously initialized with GMM clustering is obtained in the form of x_0, one well recognized method to subsequently update x_0 is the sherman-morrison woodbury inverse formula [42]. When new sample subsets $A_{k+1} = \{v_1^{k+1}, \ldots, v_i^{k+1}\}$ are learned in batches we can update x_0 from Eq.(2) as:

$$x_{k+1} = x_k + M_{k+1}^{-1} A_{k+1}^T \left(b_{k+1} - A_{k+1} x_k\right) \tag{3}$$

$$M_{k+1}^{-1} = M_k^{-1} - M_k^{-1} A_{k+1} \left(\mathrm{I} + A_{k+1}^T M_k^{-1} A_{k+1}\right)^{-1} A_{k+1}^T M_k^{-1} \tag{4}$$

where x_k is the output coefficient in the previous step, b_{k+1} is the newly collected labels and $M_{k+1}^{-1} = \left(A_{k+1}^T A_{k+1} + \lambda \mathrm{I}\right)^{-1}$. Once x_{k+1} has been learned, each new skull sample containing noise, which were not included in the training phase are used to test the model. The results obtained are depicted in Fig. 2.

4.3   Skull Alignment

To centralize the pose of the subjects, skull registration is performed using iterative closest point (ICP) method [43]. ICP normally requires a set of predefined interest points from a model and target surface to iteratively perform the alignment by minimizing the sum of square error between the two data. The approach in this work is to generate the interest points by using the difference of gaussians

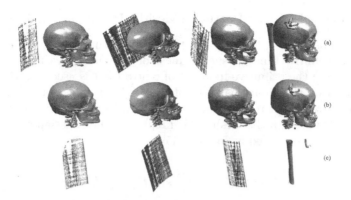

Fig. 2. Noisy objects removal (a) data with noise (b) clean and aligned skulls after object removal (c) the removed noise

(DoG) scale space technique, computed on a shape primitive. Considering a set of vertices $V = \{v_1, \ldots, v_i\}$ points from a triangulated surface with N neighborhood connectivity defined on V. We build a scale space by smoothing the surface with different gaussians at each scale (we considered 20 scales in this experiment). The surface is filtered using gaussian filter with different smoothing parameter σ_s, which increase exponentially by $\sigma_s = 2^s$. Then, gaussian curvature $G_c - k_{max} \cdot k_{min}$ at each scale is computed, where k_{max} and k_{min} are the maximum and minimum curvature of a surface point. Further, the difference between the gaussian curvature of the smoothed surface obtained at different scales is computed, and finally the extremas in a neighborhood are selected as the detected interest points. Once the interest points have been detected, a sample skull is selected as the model and every other sample skull is subsequently registered to the model using ICP to obtain the transformation parameters (rotation matrix and translation vector), which are used to transform the target data. Examples of registered skulls are depicted in Fig. 2. The registered skulls are used in the next stage for feature extraction with different 3D shape descriptors.

5 3D Shape Representation

The representative features computed from the registered mesh data are mesh local binary pattern (MeshLBP), Spin Image, Local Depth SIFT (LD-SIFT), and scale invariant heat kernel signature (SIHKS). These descriptors are chosen because they are state-of-the-art feature based descriptors, which have commonly been used in 3D face recognition and 3D shape retrieval. Out of the 4 techniques only MeshLBP and SIHKS are extracted on the entire skull because the other 2 methods are keypoint based.

5.1   Mesh Local Binary Pattern (MeshLBP)

MeshLBP is based on extracting a set of ordered ring facets (ORF) [44] and calculating a primitive function such as mean or gaussian curvature for each facet in the ring. Then the primitive function of centre facet is used as a thresholding function to obtain the binary patterns of the neighboring facets [45]. It has been used in shape retrieval and 3D face recognition [30,45] Consider a primitive function $p(f)$ defined on the mesh, which represent the gaussian curvature. The MeshLBP can be computed as follows [45]:

$$MeshLBP^r_m(f_c) = \sum_{k=0}^{m-1} s(p(f^r_k) - p(f_c)) \cdot \alpha(k) \tag{5}$$

$$s = \begin{cases} 1, & > 0 \\ 0, & \leq 0 \end{cases}$$

where r is the ring number, and m is the number of facets in the ring.

Fig. 3. Example MeshLBP feature distribution by mapping the vertex patterns to different color map.

The two parameters control the radial resolution and azimuthal quantization. The discrete function $\alpha(k)$ allows computation of other LBP variants. To extract MeshLBP in this work, 10-ring neighbors are computed for each unique facet (we chose 10 rings to offer trade-off between covering a large neighbourhood and less computational time). Then the MeshLBP is extracted by comparing the gaussian curvature of centre facet to the neighboring facets. An example of MeshLBP feature representation is illustrated in Fig. 3. The output of MeshLBP is V-by-10 matrix for each skull sample, with V denoting the number of vertex points.

5.2   Spin Image

Spin Image is a state-of-the-art descriptor proposed by Johnson and Hebert [46] for object retrieval. The descriptor is invariant to rotation and translation. The main idea of spin image is to generate a 2D histogram for mesh point composing of representation of the object geometry. In essence, it uses representation of object's coordinate system rather than viewer's coordinate system. Oriented points are computed for each point on the mesh based on the position of the vertex p and surface normal vector n. Based on this, a 2D basis (p, n) is formulated. Then to obtain the coordinate system, a tangent plane P passing through p is formed which is perpendicular to the normal n and line L through p that is parallel to n. This computation results into a cylindrical coordinate system (α, β), with α being a non-negative distance to L and β a signed distance perpendicular to P. These oriented points are used to construct a spin map S_m, which is simply a projection of 3D points x on a mesh to 2D coordinates (α, β). The 2D points are further accumulated into discrete points which are incrementally updated for each new 2D points. The size (i_{max}, j_{max}) of a spin image is determined by the size of the bin and the maximum size of the object expressed in the spin map coordinates. We used a cell size of 10 and histogram bin of 10 for extracting the features, which resulted in 10×10 spin image for each vertex.

5.3   Local Depth SIFT (LD-SIFT)

LD-SIFT is an extension of SIFT to mesh surface, presented by Darom and keller [32]. Given an interest point that is a local maxima resulting from difference of gaussians (DoG) operation on the mesh. A sphere support is constructed around it covering a neighboring region. LD-SIFT estimates a tangent plane T within the support region, and compute the distance from each point in that region to the plane to create a 2D array of depth map. Then, it sets the viewport size to match the feature scale, as detected by the DoG detector, which enables the construction to be scale invariant. Further, it computes the principal component analysis (PCA) of the points surrounding the interest point, and use their dominant direction as the local dominant angle. Similar to the standard SIFT by Lowe [47], the depth map are rotated to a canonical angle based on the dominant angle, which also makes the LD-SIFT rotation invariant. This mainly differentiate LD-SIFT from Spin Image, as the angles in Spin Image are obtained based on the normal vectors. From the resulting depth maps, the standard SIFT feature descriptors are computed to create the LD-SIFT feature descriptor. The features are mainly representation of 8-bins gradient histograms distributed in the local cells of 4-by-4 depth map. Thus the final feature is a 128-dimensional $(4 \times 4 \times 8)$ vector for each detected vertex (interest) point.

5.4   Scale Invariant Heat Kernel Signature (SIHKS)

Heat kernel signature (HKS) is a spectral shape representation that provides the point signature of a specific point in a mesh via deformable shape analysis [48]. The concept is based on modeling shapes as Riemannian manifold and use their heat conduction properties as descriptor. The original heat kernel signature for shape representation was proposed in [48], which is invariant to isometry and its has been popularly used for shape retrieval in computer vision applications. However HKS has the limitation of not being scale invariant, thus SIHKS was recently introduced to remove scale effect by sampling each point logarithimically in time $(t = \alpha^\tau)$ [49] and then calculating the derivative with respect to the scale to undo the additive function with respect to the scale. Furthermore, they used

Fourier transform to cancel out the shift variation. To extract SIHKS, we utilized a logarithmic scale-space with base $\alpha = 2$ ranging from $\tau = 1 : 20$ with increments of $1/5$. We then selected the first 10 frequencies for local description. Each of the feature point in the 10 length vector is the heat kernel signature at a particular scale.

6   Multi-region Feature Representation

6.1   Region Partitioning and Feature Aggregation

When features are extracted on the entire or substantial keypoints subset of the skull, the popular techniques that could be considered is to either concatenate the keypoints into a long-tailed vector or form compact representation with bag of words model. However, in our experiments, we have empirically discovered that these techniques generally result in a high possibility of features from a particular class having dissimilar representation and as a direct consequence, features from two different classes may exhibit the same or similar values which reduced the informative characteristics of such methods. Therefore, to minimize such inseparability, in this work we divided the skull into different sub-regions (we explore 3, 5, and 7 sub-regions). Due to the fact that, we do not have any information about the location of the anatomical sites of the skull, we partitioned the skull with equal spacing along the z-axis, as this geometric constraint could guarantee proper sub-region distributions. Then, the entire features from the vertices in each sub-region are aggregated into a single descriptor by stacking those representations. For example, assuming spin image of 10×10 size is extracted from $V = 10000$ vertices, we stack the spin image from each vertex point on one another and calculate the aggregate, which results in a final spin image of 10×10 descriptor size and feature vector of 100 dimensions, as shown in Fig. 4. This makes the representation more distinctive than the long-tailed concatenated or bag of word representation. Similar approach is used for LD-SIFT by stacking the gradient features from the keypoints and taking aggregate, which results in a final feature vector of 128-dimension. For MeshLBP, the descriptor extracted is 10 scalar values from 10 rings for each vertex point, making a $V \times 10$ descriptor matrix, thus a 32 histogram bin is aggregated for each ring along the vertices and the final feature vector is the concatenation of the histograms from the 10 rings ($32 \times 10 = 320$ feature descriptor). We used 32 bins to keep the dimension of the feature vector to a reasonable length. Also, SIHKS is compactly represented in this fashion, using 32 bins histogram bins to represent each frequency, which yields $32 \times 10 = 320$ descriptor. Examples of region partitioning of different sizes are shown in Fig. 5. The results attained using the skull as a whole and multi-region partitioning are reported in Sect. 7. To further obtain comprehensive representation from the extracted features we adopted (1) Kernel Principal Component Analysis (KPCA) [50] which is a dimensionality reduction method that generalizes the standard PCA to non-linear feature representation and (2) K-SVD [51], a dictionary learning technique.

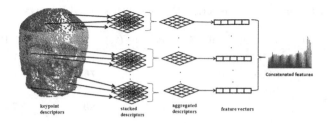

Fig. 4. An example of multi-region representation with LD-SIFT. Each color denotes a partitioned sub-region of the skull.

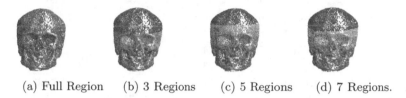

(a) Full Region   (b) 3 Regions   (c) 5 Regions   (d) 7 Regions.

Fig. 5. Illustration of multi-region feature representations of different partition size.

7 Experimental Result

This section presents the experimental results and performance comparison of the described multi-region feature extraction and compact representation techniques. We utilized support vector machine (SVM) [52], kernel extreme learning machine (KELM) [53], Sparse Representation Classifier (SRC) [54] for sex classification, which are popularly used classifiers in computer vision applications. In these experiments, the input datasets are divided into 60% and 40% as training and testing set respectively by random sampling without replacement. On the training set, a 5-fold cross validation is performed in order to determine the best model regularization value C (between 2^{-10} and 2^{10}) for SVM, KELM, SRC. After training is completed, the sex classification is achieved by testing the model with the separate unused testing set. Each experiment was repeated 10 times and the average taken as the final result.

7.1 Sex Determination

Initially, the evaluation is performed to investigate the sex prediction accuracy obtainable when the whole skull is used for feature representation (the prediction results are indicated in column "Full"). Hence, we examined each of the local feature representation described in previous section: MeshLBP, Spin Image, LD-SIFT, and SIHKS with KELM, SVM, and SRC predictive models. With KELM classifier, LD-SIFT obtained a prediction rate of 66%, Spin Image provided 74%, the result of SIHKS is 45.3%, while MeshLBP obtained a prediction rate of 63.5%, as shown in Table 1. The results of SVM and SRC are close to that of KELM with slight deviation in their performances. We further used KPCA to

Table 1. Sex prediction Results with stacked features

Feature extraction	Classifier	Full (%)	3 Regions (%)	5 Regions (%)	7 Regions (%)
LD-SIFT	KELM	66	68.5	**76.5**	72.5
	SVM	63.5	68	**76.5**	73
	SRC	68	73	73.75	**74.5**
Spin Image	KELM	74	76	**80**	75
	SVM	74	74.5	**82**	76
	SRC	72.25	71.5	**80.5**	73
SIHKS	KELM	45.3	51.1	56.6	50
	SVM	46.2	48.9	54.4	56.7
	SRC	63.5	56.5	62	60.5
MeshLBP	KELM	63.5	74	**80.25**	70
	SVM	58	74.5	**82**	64
	SRC	62.5	73	**80**	60.5
LD-SIFT + KPCA	KELM	65	67.5	68	**72.5**
	SVM	68	75	**77.75**	74
	SRC	68	67.5	72	**74.5**
Spin Image + KPCA	KELM	77	73	**82.5**	75
	SVM	77.5	81.5	**83**	76.5
	SRC	71	73.5	**81.75**	73
SIHKS + KPCA	KELM	51.1	56	52.5	55
	SVM	53.5	47	52	53.5
	SRC	52	52	54.3	51
MeshLBP + KPCA	KELM	58.5	80.5	**85.25**	66
	SVM	59.5	77	**86**	70
	SRC	60.5	70	**81.3**	70
LD-SIFT + KSVD	KELM	67.5	74.5	75	67
	SVM	60	68	68.5	67
	SRC	55	64	68	64.5
Spin Image + KSVD	KELM	75	76	**83**	75
	SVM	60	74.5	**75.5**	71
	SRC	59.5	61	**76**	65
SIHKS + KSVD	KELM	44.5	57.6	51	60
	SVM	56	58	48	51.5
	SRC	61	55	49.5	59.5
MeshLBP + KSVD	KELM	49	69.5	**83**	70.5
	SVM	51	60.5	**80**	55
	SRC	51.5	65.5	**74.5**	55

learn better subspace representation from the original features, but the results were still not too promising. It was noticed that the full skull representation could not produce distinctive representation of different geometric region of skull, hence the lower prediction results attained. By partitioning the skull into 3 sub-regions, we noticed a slight improvement over the results of full representation.

For instance, performance of MeshLBP with KELM increased to 74%. More so, with subspace representation of KPCA, we obtained an improved prediction rate of 80.5%. Likewise, spin image yielded a boost in performance to 81.5% with SVM classifier, which was the best result for 3 sub-region skull representation. By further partitioning the skull into 5 sub-regions, the prediction results even increased significantly for all descriptors except SIHKS. We observe significant improvement in the results of spin image and LD-SIFT with the three classifiers. Without KPCA or K-SVD, the sex of a skull sample could be predicted with an accuracy of 82% using MeshLBP and SVM. Through dictionary learning of spin image descriptors using K-SVD, we obtained a prediction rate of 83% with KELM classifier and MeshLBP produced similar prediction rate with KSVD and KELM. Whereas, by finding good subspace representation for MeshLBP using KPCA, we obtained the best prediction performance of 85.25% and 86% with KELM and SVM classifiers respectively. We observed that 5 sub-region partitioning of the skull could provide a reliable sex determination performance as shown in Table 1. However, further partitioning the skull into 7 sub-regions resulted in drastic reduction in prediction results.

8   Discussion and Conclusion

Prediction of sex is very crucial in forensic examination of skeletal remains. It has received vast amount of attention over decades. The conventional methods are based on the study of both morphological and geometric variations of the skull. These methods visually assess specific anatomical regions of the cranial or make use of a few set of landmarks that are carefully localized. In this paper, we diverged from the conventional approach and suggested a possible framework for sex prediction using modern feature representation and learning techniques that have gained prominence in the computer vision community. We proposed multi-region based representation and compact aggregation of features from each region with 3D shape descriptors. From our experiment, there are indications that these computer vision techniques could possibly be useful in automated sex prediction. The best result attained in this work is 86% through KPCA subspace representation of compact MeshLBP features from 5 sub-regions of the skull. Our result is within and comparable to the commonly reported sex prediction range (70%–90%) using morphometric or morphological assessment by forensic anthropologists [8,17]. This has opened the door to using alternative but more efficient methods in forensic anthropology, and potentially bridging the gap between visual assessment and perceived dimorphic traits. Moreover, it relaxes the challenge of incompleteness of anatomical landmarks, which affects the performance of morphometric analysis. From our experiments, it can be suggested that multi-region representation should be considered for feature extraction instead of full representation of the skull data. However, careful selection of 3D shape descriptors is required, as not all descriptors produced distinctive description of the skulls. For instance, SIHKS which is a standard benchmark in 3D shape retrieval is not very good in sex estimation. We loosely attribute this to the fact the difference or variation between the skull of the two sexes is very subtle (even forensic experts do have high inter-observer error when performing morphological assessment [35]). Whereas, most 3D shape retrieval applications are based on classification of objects from broad categories. In future work, we will investigate region partitioning techniques that take into consideration the anatomical locations of the skull.

Acknowledgement. This research is sponsored by the eScienceFund grant 01-02-12-SF0288, Ministry of Science, Technology, and Innovation (MOSTI), Malaysia. The project has received full ethical approval from the Medical Research & Ethics Committee (MREC), Ministry of Health, Malaysia (ref: NMRR-14-1623-18717) and from the University of Nottingham Malaysia Campus (ref: IYL170414). Iman Yi Liao would like to thank the National Institute of Forensic Medicine (NIFM), Hospital Kuala Lumpur, for providing the PMCT data, and is grateful to Dr. Ahmad Hafizam Hasmi (NIFM) and Ms. Khoo Lay See (NIFM) for their assistance in coordinating the data preparation. The authors are appreciative of the valuable comments made by the anonymous reviewers for improving the research reported in this paper.

References

1. Scheuer, L.: Application of osteology to forensic medicine. Clin. Anat. **15**, 297–312 (2002)
2. Naikmasur, V.G., Shrivastava, R., Mutalik, S.: Determination of sex in South Indians and immigrant Tibetans from cephalometric analysis and discriminant functions. Forensic Sci. Int. **197**, 122 (2010)
3. Aulsebrook, W.A., Ican, M.Y., Slabbert, J.H., Becker, P.: Superimposition and reconstruction in forensic facial identification: a survey. Forensic Sci. Int. **75**, 101–120 (1995)
4. Ibáñez, O., Cordón, O., Damas, S.: A cooperative coevolutionary approach dealing with the skullface overlay uncertainty in forensic identification by craniofacial superimposition. Soft Comput. **16**, 797–808 (2011)
5. Cattaneo, C.: Forensic anthropology: developments of a classical discipline in the new millennium. Forensic Sci. Int. **165**, 185–193 (2007)
6. Schmitt, A., Cunha, E., Pinheiro, J.: Forensic Anthropology and Medicine (2006)
7. Giurazza, F., Del Vescovo, R., Schena, E., Battisti, S., Cazzato, R.L., Grasso, F.R., Silvestri, S., Denaro, V., Zobel, B.B.: Determination of stature from skeletal and skull measurements by CT scan evaluation. Forensic Sci. Int. **222**, 398.e1–398.e9 (2012)
8. Garvin, H.M., Sholts, S.B., Mosca, L.A.: Sexual dimorphism in human cranial trait scores: effects of population, age, and body size. Am. J. Phys. Anthropol. **154**, 259–269 (2014)
9. Spradley, M.K., Jantz, R.L.: Sex estimation in forensic anthropology: Skull versus postcranial elements. J. Forensic Sci. **56**, 289–296 (2011)
10. Ramamoorthy, B., Pai, M.M., Prabhu, L.V., Muralimanju, B.V., Rai, R.: Assessment of craniometric traits in South Indian dry skulls for sex determination. J. Forensic Leg. Med. **37**, 8–14 (2015)
11. Ramsthaler, F., Kreutz, K., Verhoff, M.A.: Accuracy of metric sex analysis of skeletal remains using Fordisc® based on a recent skull collection. Int. J. Leg. Med. **121**, 447–482 (2007)

12. Franklin, D., Cardini, A., Flavel, A., Kuliukas, A.: The application of traditional and geometric morphometric analyses for forensic quantification of sexual dimorphism: Preliminary investigations in a Western Australian population. Int. J. Leg. Med. **126**, 549–558 (2012)
13. Krogman, W.M.: The Human Skeleton in Forensic Medicine. Charles C. Thomas, Springfield (1962)
14. Stewart, T.D.: Essentials of forensic anthropology (1979)
15. Konigsberg, L.W., Hens, S.M.: Use of ordinal categorical variables in skeletal assessment of sex from the cranium. Am. J. Phys. Anthropol. **107**, 97–112 (1998)
16. Buikstra, J.E., Ubelaker, D.H.: Standards for data collection from human skeletal remains. In: Proceedings of a seminar at the Field Museum of Natural History (Arkansas Archaeology Research Series 44), pp. 1704–1711 (1994)
17. Walker, P.L.: Sexing skulls using discriminant function analysis of visually assessed traits. Am. J. Phys. Anthropol. **136**, 39–50 (2008)
18. Graw, M., Czarnetzki, A., Haffner, H.T.: The form of the supraorbital margin as a criterion in identification of sex from the skull: Investigations based on modern human skulls. Am. J. Phys. Anthropol. **108**, 91–96 (1999)
19. Pinto, S.C.D., Urbanová, P., Cesar, R.M.: Two-dimensional wavelet analysis of supraorbital margins of the human skull for characterizing sexual dimorphism. IEEE Trans. Inf. Forensics Secur. **11**, 1542–1548 (2016)
20. Franklin, D., Freedman, L., Milne, N.: Sexual dimorphism and discriminant function sexing in indigenous South African crania. HOMO- J. Comp. Hum. Biol. **55**, 213–228 (2005)
21. Bigoni, L., Velemínská, J., Bržek, J.: Three-dimensional geometric morphometric analysis of cranio-facial sexual dimorphism in a Central European sample of known sex. HOMO- J. Comp. Hum. Biol. **61**, 16–32 (2010)
22. Jiménez-Arenas, J.M., Esquivel, J.A.: Comparing two methods of univariate discriminant analysis for sex discrimination in an Iberian population. Forensic Sci. Int. **228**, 175.e1–175.e4 (2013)
23. Williams, B.A., Rogers, T.L.: Evaluating the accuracy and precision of cranial morphological traits for sex determination. J. Forensic Sci. **51**, 729–735 (2006)
24. Luo, L., Wang, M., Tian, Y., Duan, F., Wu, Z., Zhou, M., Rozenholc, Y.: Automatic sex determination of skulls based on a statistical shape model. Comput. Math. Methods Med. **2013**, 1–7 (2013)
25. Shapiro, L.G., Wilamowska, K., Atmosukarto, I., Wu, J., Heike, C., Speltz, M., Cunningham, M.: Shape-based classification of 3D head data. In: Foggia, P., Sansone, C., Vento, M. (eds.) ICIAP 2009. LNCS, vol. 5716, pp. 692–700. Springer, Heidelberg (2009). doi:10.1007/978-3-642-04146-4_74
26. Atmosukarto, I., Wilamowska, K., Heike, C., Shapiro, L.G.: 3D object classification using salient point patterns with application to craniofacial research. Pattern Recogn. **43**, 1502–1517 (2010)
27. Yang, S., Shapiro, L., Cunningham, M.: Classification and feature selection for craniosynostosis. In: Proceedings of the 2nd ACM Conference on Bioinformatics, Computational Biology and Biomedicine, pp. 340–344. ACM (2011)
28. Lam, I., Cunningham, M., Speltz, M., Shapiro, L.: Use of ordinal categorical variables in skeletal assessment of sex from the cranium. In: Proceedings of IEEE Symposium on Computer-Based Medical Systems, pp. 215–220 (2014)
29. Tolga, I., Halici, U.: 3-D face recognition with local shape descriptors. IEEE Trans. Inf. Forensics Secur. **7**, 577–587 (2012)

30. Werghi, N., Tortorici, C., Berretti, S., Bimbo, A.: Boosting 3D LBP-based face recognition by fusing shape and texture descriptors on the mesh. IEEE Trans. Inf. Forensics Secur. **11**, 964–979 (2016)
31. Johnson, A.E., Hebert, M.: Using spin images for efficient object recognition in cluttered 3D scenes. IEEE Trans. Pattern Anal. Mach. Intell. **21**, 433–449 (1999)
32. Darom, T., Keller, Y.: Scale Invariant Features for 3D Mesh Models. IEEE Trans. Image Process. **21**, 1–32 (2012)
33. Han, X., Ugail, H., Palmer, I.: Gender classification based on 3D face geometry features using SVM. In: 2009 International Conference on CyberWorlds, pp. 114–118 (2009)
34. Ballihi, L., Amor, B.B., Daoudi, M., Srivastava, A., Aboutajdine, D.: Boosting 3-D-geometric features for efficient face recognition and gender classification. IEEE Trans. Inf. Forensics Secur. **7**, 1766–1779 (2012)
35. Lewis, C.J., Garvin, H.M.: Reliability of the walker cranial nonmetric method and implications for sex estimation. J. Forensic Sci. **61**, 743–751 (2016)
36. Guyomarc'h, P., Bruzek, J.: Accuracy and reliability in sex determination from skulls: a comparison of Fordisc 3.0 and the discriminant function analysis. Forensic Sci. Int. **208**, 30–35 (2011)
37. Xia, B., Ben Amor, B., Drira, H., Daoudi, M., Ballihi, L.: Combining face average-ness and symmetry for 3D-based gender classification. Pattern Recogn. **48**, 746–758 (2015)
38. Garcia, D.: Robust smoothing of gridded data in one and higher dimensions with missing values. Comput. Stat. Data Anal. **54**, 1167–1178 (2010)
39. Lorensen, W.E., Cline, H.E.: Marching cubes: a high resolution 3D surface construction algorithm. ACM SIGGRAPH Comput. Graph. **21**, 163–169 (1987)
40. Cignoni, P., Montani, C., Scopigno, R.: A comparison of mesh simplification algorithms. Comput. Graph. **22**, 37–54 (1998)
41. Do, T.N., Fekete, J.D.: Large scale classification with support vector machine algorithms. In: Proceedings of the 6th International Conference on Machine Learning and Applications (ICMLA 2007), pp. 148–153 (2007)
42. Golub, G.H., Van Loan, C.F.: Matrix Computations (1996)
43. Besl, P., McKay, N.: A method for registration of 3-D shapes (1992)
44. Werghi, N., Rahayem, M., Kjellander, J.: An ordered topological representation of 3D triangular mesh facial surface: concept and applications. EURASIP J. Adv. Sig. Process. **2012**, 1–20 (2012)
45. Werghi, N., Berretti, S.: The Mesh-LBP: a framework for extracting local binary patterns from discrete manifolds. IEEE Trans. Image Process. **24**, 220–235 (2015)
46. Johnson, A.E.: Spin-images: a representation for 3-D surface matching. Image, Rochester, N.Y., pp. 1–7 (1997)
47. Lowe, D.G.: Distinctive image features from scale-invariant keypoints. Int. J. Comput. Vis. **60**, 91–110 (2004)
48. Sun, J., Ovsjanikov, M., Guibas, L.: A concise and provably informative multi-scale signature based on heat diffusion. In: Eurographics Symposium on Geometry Processing, vol. 28, pp. 1383–1392 (2009)
49. Bronstein, M.M., Kokkinos, I.: Scale-invariant heat kernel signatures for non-rigid shape recognition. In: Proceedings of the IEEE Computer Society Conference on Computer Vision and Pattern Recognition, pp. 1704–1711 (2010)
50. Schölkopf, B., Smola, A., Müller, K.R.: Nonlinear component analysis as a kernel eigenvalue problem. Neural Comput. **10**, 1299–1319 (1998)

51. Aharon, M., Elad, M., Bruckstein, A.: K-SVD: an algorithm for designing overcomplete dictionaries for sparse representation. IEEE Trans. Sig. Process. **54**, 4311–4322 (2006)
52. Cortes, C., Vapnik, V.: Support-vector networks. Mach. Learn. **297**, 273–297 (1995)
53. Huang, G.B.: An insight into extreme learning machines: random neurons, random features and kernels. Cogn. Comput. **6**, 376–390 (2014)
54. Wright, J., Yang, A.Y., Ganesh, A., Sastry, S.S., Ma, Y.: Robust face recognition via sparse representation. IEEE Trans. Pattern Anal. Mach. Intell. **31**, 210–227 (2009)

Deep Depth Super-Resolution: Learning Depth Super-Resolution Using Deep Convolutional Neural Network

Xibin Song[1], Yuchao Dai[2(✉)], and Xueying Qin[1]

[1] School of Computer Science and Technology, Shandong University,
Jinan, China
song.sducg@gmail.com, qxy@sdu.edu.cn
[2] Research School of Engineering, Australian National University,
Canberra, Australia
yuchao.dai@anu.edu.au

Abstract. Depth image super-resolution is an extremely challenging task due to the information loss in sub-sampling. Deep convolutional neural network has been widely applied to color image super-resolution. Quite surprisingly, this success has not been matched to depth super-resolution. This is mainly due to the inherent difference between color and depth images. In this paper, we bridge up the gap and extend the success of deep convolutional neural network to depth super-resolution. The proposed deep depth super-resolution method learns the mapping from a low-resolution depth image to a high-resolution one in an end-to-end style. Furthermore, to better regularize the learned depth map, we propose to exploit the depth field statistics and the local correlation between depth image and color image. These priors are integrated in an energy minimization formulation, where the deep neural network learns the unary term, the depth field statistics works as global model constraint and the color-depth correlation is utilized to enforce the local structure in depth image. Extensive experiments on various depth super-resolution benchmark datasets show that our method outperforms the state-of-the-art depth image super-resolution methods with a margin.

1 Introduction

Recently, consumer depth cameras (*e.g.* Microsoft Kinect, ASUS Xtion Pro or the Creative Senz3D camera *etc.*) and other ToF cameras have gained significant popularity due to their affordable cost and great applicability in human computer interaction [1], computer graphics and 3D modeling [2]. However, depth image outputs from these cameras suffer from natural upper limit on spatial resolution and the precision of each depth sample, thus, making depth image

Electronic supplementary material The online version of this chapter (doi:10.1007/978-3-319-54190-7_22) contains supplementary material, which is available to authorized users.

S.-H. Lai et al. (Eds.): ACCV 2016, Part IV, LNCS 10114, pp. 360–376, 2017.
DOI: 10.1007/978-3-319-54190-7_22

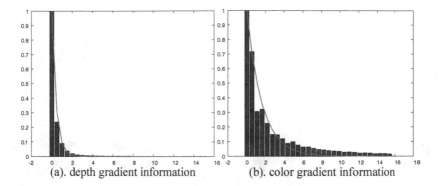

(a). depth gradient information   (b). color gradient information

Fig. 1. Histograms of gradients of depth images and color images for the Sintel dataset [3]. (a) Histogram of gradient magnitude of depth images. (b) Histogram of gradient magnitude of color images. The distribution of both depth and color gradient magnitudes can be well represented by Laplace distribution (the red line in (a) and (b)). The RMSE in model fitting for (a) is 0.0016, and for (b) is 0.0036. (Color figure online)

super-resolution (DSR) attract more and more research attentions in the community. DSR aims at obtaining a high-resolution (HR) depth image from a low-resolution (LR) depth image by inferring all the missing high frequency contents. DSR is a highly ill-posed problem as the known variables in the LR images are greatly outnumbered by the unknowns in the HR depth images.

Meanwhile, there is mounting evidences that effective features and information learned from deep convolutional neural networks (CNN) set new records for various vision applications, especially in color image super-resolution (CSR) problems. There has been considerable progress in applying deep convolutional network for CSR and new state-of-the-art has been achieved [4–6]. It has been proven that CNN has the ability to learn the nonlinear mapping from the low resolution color images to the corresponding high resolution color images. Quite surprisingly, the success of applying CNN in color image super-resolution has not been matched to depth image super-resolution. This is mainly due to the inherent differences in the acquisition of depth image and color image. In Fig. 1, we compare the statistics of gradients of depth images and color images (the color and depth images are already aligned). It is obvious to observe the difference between depth images and color images, where depth images generally contain less texture and sharp boundary, and are usually degraded by noise due to the imprecise consumer depth cameras (nevertheless time-of-flight or structured light) or difficulties in calculating the disparity (stereo vision).

Building upon the success of deep CNN in color image super-resolution, we propose to address the unique challenges with depth image super-resolution. First, to deal with different up-sampling ratios in depth super-resolution, we propose a progressive deep neural network structure, which gradually learns the high-frequency components and could be adapted to different depth

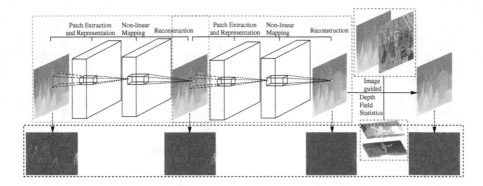

Fig. 2. Conceptual illustration of our framework, where the input is a low-resolution depth image and the output is the desired high-resolution depth image. By using convolutional neural network, our method gradually learns the high frequency components in depth image representation. The area labeled in black shows the residual between current depth map and the ground truth, which are shown in color mode. Blue to red means zeros to infinity. **Best Viewed on Screen.**

super-resolution tasks. Second, to further refine the depth image estimation, we resort to depth field statistics and color-depth correlation to constrain the global and local structures in depth image. The corresponding problem is formulated as a conditional random field (CRF) and solved via energy minimization. Our method is elegant in dealing with or without color image in super-resolving the depth image. Extensive experiments on various bench-marking datasets demonstrate the superior performance of our method compared with current state-of-the-art methods.

Our main contributions are summarized as:

(1) A progressive deep CNN framework is proposed to gradually learn high-resolution depth images from low-resolution depth images. To the best of our knowledge, this is the first deep neural network based depth super-resolution method.
(2) Depth field statistics and color-depth correlation are exploited to further refine the learned depth image as these two priors provide complement information to the deep CNN.
(3) Our method is elegant in dealing with and without high-resolution color image as guide. For depth images without high-resolution color images, the depth images themselves can be employed to refine the depth images.

2   Related Work

Image super-resolution, as one of the most active research topics in the field of computer vision and image processing, has been widely studied, and the research topic can be divided into two categories, namely, color image super-resolution (CSR) and depth image super-resolution (DSR).

2.1 Depth Image Super-Resolution

According to the information used, depth image super-resolution (DSR) methods can be classified into the following two categories, including: single DSR with additional depth map data-set, and DSR with the assistant of high resolution color image.

DSR with additional depth map datasets. Existing research has demonstrated that HR depth images can be inferred from LR depth image with prior information. Single DSR offers unique challenges compared to single CSR (*e.g.* edge preserving de-noising should also be properly tackled). Inspired by [7], Aodha et al. [8] proposed to employ a patch based MRF model in depth image super-resolution. Instead of using a prior information from an external database, Ferstl et al. [9] proposed to generate high-resolution depth edges by learning a dictionary of edges priors from an external database of high and low resolution examples. Then a variational energy model with Total Generalized Variation is employed as regularization to obtain the final high resolution depth maps. However, this method suffers blurring problems, especially areas around depth edges. What's more, Xie et al. [10] proposed to exploit a Markov random field optimization approach to generate HR depth edges from depth edges extracted from LR depth maps. Then, the HR depth edges are recognized as strong restricts to generate the final high resolution depth maps. However, the method always lose details since it is impossible to extract all depth edges from LR depth maps.

DSR with corresponding HR color image. Pre-aligned HR color images are also employed to super-resolve the depth maps as the high frequency components in color images such as edges can provide useful information to assist the process of DSR. Park et al. [11] proposed a non-local means filter (NLM) method which exploits color information to maintain the detailed structure in depth maps. Yang et al. [12] proposed an adaptive color-guided auto-regressive (AR) model for high quality depth image recovery from LR depth maps. Besides, Ferstl et al. [13] utilized an anisotropic diffusion tensor to guide the depth image super-resolution. What's more, Matsuo et al. [14] described a depth image enhancement method for consumer RGB-D cameras, where color images are used as auxiliary information to compute local tangent planes in depth images. In [15], color and depth images were exploited to generate depth areas with similar color and depth information from depth images, then a sparse representation approach was employed to super-resolve depth maps in each region. However, notwithstanding the appealing results that such approaches could generate, the lack of high resolution color images fully registered with the depth images in many cases makes the color assisted approaches less general.

2.2 Color Image Super-Resolution

Various kinds of methods have been proposed for CSR, such as MRF based methods [7], and sparse coding based methods strategies [16–19]. However, these methods tend to fail to fully express the nonlinear mapping between HR images and LR images.

Deep CNN for image super-resolution. Recently, deep CNN has extended its success in high level computer vision to low level computer vision tasks such as color image super-resolution. Dong et al. [4] proposed an end-to-end deep CNN framework to learn the nonlinear mapping between low and high resolution images. Based on [4], Kim et al. [5] proposed to use a more deeper network to represent the non-linear mapping and improved performance has been achieved. Meanwhile, in [6], a deeply-recursive convolutional network for color images super-resolution is proposed. Based on a deep recursive layer, a more accurate representation of the mapping between low and high resolution color images can be achieved, thus, better results can be generated without introducing new parameters.

In this paper, build upon the success of deep CNN for color image super-resolution, we propose a deep neural network based depth image super-resolution to effectively learn the mapping from low-resolution depth image to high-resolution depth image. Furthermore, we enforce the depth field statistics and local color-depth correlation to further refine the results.

3 Our Approach

Depth image super-resolution aims at inferring a high-resolution depth image $\mathbf{D}^H$ from a low-resolution depth image $\mathbf{D}^L$, where the up-sampling factors vary from $\times 2, \times 4$ to $\times 8$ (infer $8 \times 8 = 64$ depth values from a single depth value). Due to the information loss in sub-sampling and the nonlinear mapping between low resolution images and high-resolution images, this is a challenging task. To effectively learn the nonlinear mapping $\mathbf{F}$ from low-resolution depth image $\mathbf{D}^L$ to high-resolution depth image $\mathbf{D}^H$, we propose to use deep neural network. To deal with different up-sampling factors, our proposed deep neural network gradually learns the high frequency component in depth images and in this way progressively improves the performance. By using this progressive structure, we are able to modify the network structure adaptively for different super-resolution tasks. To further constrain the high-resolution depth image, we resort to depth field statistics and local color-depth correlation priors, which results in an energy minimization formulation.

Figure 2 provides a conceptual illustration of our proposed framework, which consists of the progressive deep neural network, depth field statistics prior and color-depth correlation prior. Given low-resolution depth image as input (interpolation to high-resolution depth images by bicubic firstly), our method outputs corresponding high-resolution depth image through deep neural network learning and energy minimization. In Fig. 3, we compare the output at different stages of our system. It could be observed that the data goes from the left side to the right side, quality of the high-resolution depth image gradually improves (as indicted by RMSE), which demonstrates that our method could gradually learn the fine details in depth image.

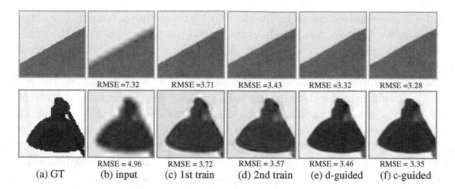

| RMSE =7.32 | RMSE =3.71 | RMSE =3.43 | RMSE =3.32 | RMSE =3.28 |

| RMSE = 4.96 | RMSE = 3.72 | RMSE = 3.57 | RMSE = 3.46 | RMSE = 3.35 |
| (a) GT | (b) input | (c) 1st train | (d) 2nd train | (e) d-guided | (f) c-guided |

Fig. 3. Typical outputs at different stages of of our method. (a) Ground truth high-resolution depth image, (b) Input low-resolution depth map (interpolated to the same size), (c) Output after the first nonlinear mapping, (d) Output after the second non-linear mapping, (e) and (f) show the depth super-resolution results with depth field statistics and color-depth correlation correspondingly. **Best Viewed on Screen.**

3.1 Learning Depth Super-Resolution by Deep CNN

Deep Convolutional Neural Networks (DCNNs) have recently shown state-of-the-art performance in high level vision tasks, such as image classification, object detection and semantic segmentation. This work brings together methods from DCNNs and probabilistic graphical models to address the task of depth image super-resolution. High-resolution depth image relates to the corresponding low-resolution one by a non-linear mapping due to the sub-sampling. The complexity of the non-linear mapping depends on the up-sampling factor as illustrated in Fig. 4. To effectively capture the high-frequency details in depth image, we propose a progressive CNN structure as demonstrated in Fig. 2. The progressive CNN structure consists of layers of depth depth super-resolution units. Each unit maps the low-resolution input $\mathbf{D}_i^L$ to the high-resolution output $\mathbf{D}_{i+1}^L$, which is the input of the next unit. In this way, our network is capable of gradually learning the high-frequency components in high-resolution depth images. As shown in the black labeled area in Fig. 2, we can see that the residual reduces, which means that high-frequency components can be learned in each step.

Low-resolution to high-resolution mapping unit: As shown in Fig. 2 (areas labeled in green and orange), for each low-resolution to high-resolution learning unit in our progressive CNN structure, we adopt the same network structure from Dong *et al.* [4]. It is worth-noting that other network structures such as [5,6] could also be used. The unit consists of three operators, namely, patch extraction and representation, non-linear mapping and reconstruction, which together learn the mapping $\mathbf{F}$ to map the input low resolution images to the high-resolution ones $\mathbf{F}(\mathbf{D}^L)$. Note that, we train the network from scratch rather than fine tune the network from [4] due to the inherent differences between depth images and color images. In principle, we could train the progressive network together from the low-resolution input to the desired high-resolution output. For the ease of training, we instead train each unit consecutively. In this way, the training procedure has been greatly simplified. A final fine tune of the whole progressive network could be implemented.

128 depth maps extracted from the Middelybury stereo dataset, the Sintel dataset and synthesis depth maps are used to construct the training data to train the deep CNN, and 14 depth maps are used as test images in the training stage. 128-depth image data-set can be decomposed into 255000 sub-images, which are extracted from original depth images with a stride of 14. For Patch Extraction and Representation, 64 convolutions are employed, and each convolution has a kernel size of 9×9. For Non-linear Mapping, 32 convolutions are used with the kernel size 1×1 in each convolution. For Reconstruction, the kernel size is 5×5.

3.2   Color Modulated Smoothness for Depth Super-Resolution

In the above section, we have demonstrated how high-resolution depth image can be learned from low-resolution depth image by using the *progressive convolutional neural network*, which already achieve comparable performance with the current state-of-the-art methods [9,10]. However, as the depth super-resolution learning is achieved in an end-to-end framework and learned across a dataset, the learned depth image could be biased by different RGB-D statistics (as illustrated by Fig. 4). As this depth map is learned from a dataset, we cannot expect the net to learn the high-frequency information or edge. Therefore we resort to two different depth super-resolution cues, which are in principle complement to the above end-to-end learning, namely, depth field statistics and color image

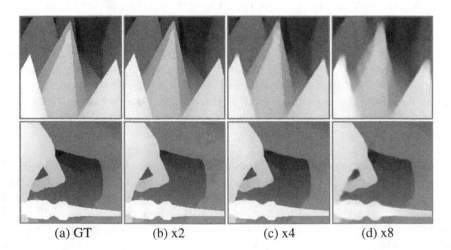

| (a) GT | (b) x2 | (c) x4 | (d) x8 |

Fig. 4. The figure shows that the learned depth images are biased by different RGB-D statistics. (a) Ground truth, (b) results of ×2, (c) results of ×4, (d) results of ×8. The first row show the part of images extracted from the Middlebury dataset, and the second row shows the part of images extracted from the Sintel data-set.

guidance. What is interesting is that we show that when the color image guidance is not available, the depth image itself could be used as guidance.

Color modulated smoothness term, aims at representing the local structure of the depth map. And depth maps for generic 3D scenes contain mainly smooth regions separated by curves with sharp boundaries. The key insight behind the color modulated smoothness term is that the depth map and the color image are locally correlated, thus the local structure of the depth map can be well represented with the guidance of the corresponding color image. The term has already employed in image colorization [20], single image depth estimation [21], depth in-painting [22] and depth image super resolution [8,23].

Denote D_u the depth value at location u in a depth map, the depth map inferred by the model can be expressed as:

$$D_u = \sum_{v \in \theta_u} \alpha_{(u,v)} D_v, \tag{1}$$

where θ_u is the neighbourhood of pixel u and $\alpha_{(u,v)}$ denotes the color modulated smoothness model coefficient for pixel v in the set of θ_u. The discrepancy between the model and the depth map (*i.e.*, the color modulated smoothness potential) can be expressed as:

$$\psi_\theta(\mathbf{D}_\theta) = \left(D_u - \sum_{r \in \theta_u} \alpha_{(u,v)} D_v \right)^2. \tag{2}$$

We need to design a local color modulated smoothness predictor α with the available high-resolution color image. To avoid incorrect depth prediction due to depth-color inconsistency, depth information is also included in α, here,

$$\alpha_{(u,v)} = \frac{1}{N_u} \alpha_{(u,v)}{}^{\overline{D}} \alpha_{(u,v)}{}^{I}, \tag{3}$$

where N_u is the normalization factor, $\overline{D}$ is the observed depth map (generated by deep CNN in our method), I is the corresponding color image, $\alpha_{(u,v)}{}^{\overline{D}} \propto \exp(-(\overline{D}_u - \overline{D}_v)^2 / 2\sigma_{\overline{D}_u}^2)$, and $\alpha_{(u,v)}{}^{I} \propto \exp(-(g_u - g_v)^2 / 2\sigma_{I_u}^2)$, where g represents the intensity value of corresponding color pixels, $\sigma_{\overline{D}_u}$ and σ_{I_u} are the variance of the depth and color intensities in the local path around u.

The window size for θ is set as 7×7 in our experiment. Note that for a depth map without corresponding color image, the depth map itself can replace the color image to construct the α^I.

3.3 Depth Super-Resolution Cue from Depth Filed Statistics

The distribution of a natural depth image $\mathbf{D}$ can often be modeled as a generalized Laplace distribution (a.k.a., generalized Gaussian distribution). As illustrated in Fig. 1, the distribution of gradient magnitude of depth images can be

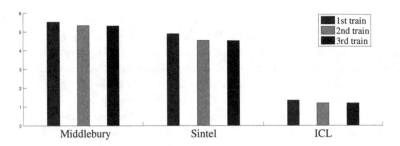

Fig. 5. The figure shows the comparison of our multi-train $RMSE$ results for upsampling factor of ×8, from which, we can see that the $RMSE$ of results generated by 2nd train is much smaller than the 1st train, while the 3rd train is similar with the 2nd train.

well approximated with Laplacian distribution. Therefore, we propose to minimize the total variation of the depth image, i.e. $\|\mathbf{D}\|_{TV} \to \min$, where the total variation could be expressed in matrix form as:

$$\|\mathbf{D}\|_{TV} = \|\mathbf{P}\text{vec}(\mathbf{D})\|_1. \tag{4}$$

3.4   An Energy Minimization Formulation

To further refine the depth super-resolution results, we integrate the depth super-resolution cue from deep progressive CNN, depth field statistics and color/depth guidance and reach the following energy minimization formulation:

$$\min_{\mathbf{D}} \frac{1}{2}\|\mathbf{D} - \overline{\mathbf{D}}\|_F^2 + \lambda_1\|\mathbf{A}\text{vec}(\mathbf{D}) - \mathbf{b}\|^2 + \lambda_2\|\mathbf{P}\text{vec}(\mathbf{D})\|_1, \tag{5}$$

where $\overline{\mathbf{D}}$ is the depth super-resolution result from deep neural network, $\mathbf{A}$ and $\mathbf{b}$ express the color/depth modulated smoothness constraint, and $\mathbf{D}$ is the final depth map we want to generate. We apply iterative reweighted least squares (IRLS) [24,25] to efficiently solve the above energy minimization problem. λ_1 and λ_2 are set as 0.7 in our experiment.

4   Experiments

In this section we evaluate the performance of different $SOTA$ methods on publicly available datasets, including the Middlebury stereo dataset [26–29], the Laser Scan dataset provided by Aodha et al. [8], the Sintel dataset [3] and the ICL dataset [30].

Error metrics: In this paper, we use three kind of error metrics to evaluate the results obtained by our method and other state-of-the-art methods, including: (1). Root Mean Squared Error ($RMSE$) of the recovered depth image with respect to the ground truth depth image; (2). Mean Absolute Error (MAE) of the recovered depth with respect to the ground truth depth image; (3). Structure Similarity of Index ($SSIM$) of the recovered depth with respect to the ground truth depth image. Note that $RMSE = \sqrt{\sum_{i=1}^{N} (X_{obs,i} - X_{gt,i})^2/N}$ and $MAE = \sum_{i=1}^{N} abs(X_{obs,i} - X_{gt,i})/N$, where X_{obs} and X_{gt} are the observed data and the ground truth respectively. As a frequently used measure, $RMSE$ represents the sample standard deviation of the differences between the predicted values and the ground truth data. Meanwhile, $SSIM$ is a n error metric evaluating the perceived quality between the obtained high resolution depth maps and the ground truth data.

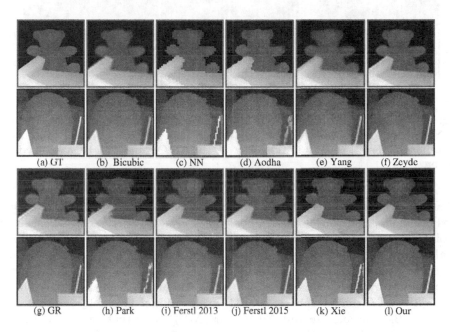

Fig. 6. Middlebury results (factor ×4). (a) Ground Truth. (b) Bicubic method. (c) Nearest neighbor method. (d) Aodha et al. [8]. (e) Yang et al. [16]. (f) Zeyde et al. [31]. (g) GR [32]. (h) Park et al. [11]. (i) Ferstl et al. [13]. (j) Ferstl et al. [9]. (k) Xie et al. [10]. (l) Our method. **Best viewed on Screen.**

Given high resolution depth maps, we sub-sample them to generate low resolution depth maps. Then we run our method to super-resolve the depth maps and evaluate the performance. Meanwhile, we set the same parameters in all of our experiments for different scales and different images.

Baseline Methods: We compare our results with the following three categories of methods. (1) State-of-the-art single depth image super resolution methods, including Aodha et al. [8], Hornacek et al. [34], Ferstl et al. [9] and Xie et al. [10];

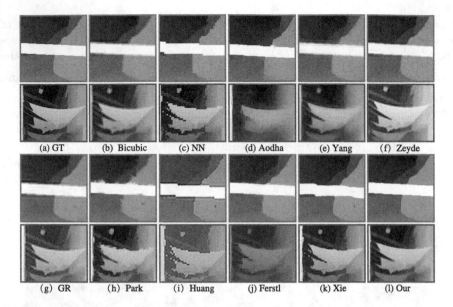

Fig. 7. ' ×4). (a) Ground Truth. (b) Bicubic. (c) Nearest neighbor. (d) Aodha et al. [8]. (e) Yang et al. [16]. (f) Zeyde et al. [31]. (g) GR [32]. (h) Park et al. [11]. (i) Huang et al. [34]. (j) Ferstl et al. [13]. (k) Xie et al. [10]. (l) Our method. **Best viewed on screen**

(2) state-of-the-art color guided depth image super resolution approaches, including Park et al. [11], Yang et al. [12], Ferstl et al. [13]; (3) Single color image super resolution approaches, including Zeyde et al. [31], Yang et al. [16] and Timofte et al. [32], which contains two kinds of methods:Global Regression (GR) and Anchored Neighborhood Regression (ANR), the neighborhood embedding methods proposed by Bevilacqua et al. [35], which contains $NE + LS$, $NE + NNLS$ and $NE + LLE$, and Huang et al. [33]; (4) Standard interpolate approaches, including Bicubic and Nearest Neighbour (NN). Note that we either use the source code provided by the authors or implement those methods by ourselves. We also select the best parameters of baseline methods by experiments.

4.1 Analysis

In this section, we present experimental analysis to the contribution of each component of our method, namely, deep regression, depth field statistics and color (or depth) guidance. Figure 3 (parts of images from the *Sintel* (first row) and the *ICL* (second row) data-set) shows the visual comparison in different stages of our method. We can see clearly that the 1st trained results are much better than the input data (obtained by bicubic), 1st trained results not only have much smaller $RMSE$, but also have better visualization, which means that high frequency depth information can be learned by the deep convolutional network (SRCNN) in this step effectively. However, the 1st trained results are not

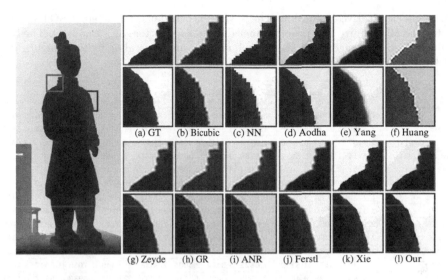

Fig. 8. Laser scan results (factor ×4). (a) Ground truth. (b) Bicubic. (c) Nearest Neighbor. (d) Aodha et al. [8]. (e) Yang et al. [16]. (f) Huang et al. [33]. (g) Zeyde et al. [31]. (h) GR [32]. (i) ANR [32]. (j) Ferstl et al. [9]. (k) Xie et al. [10]. (f) Our method. **Best viewed on screen.**

good enough and suffer from ring effect severely. And, as shown in Fig. 3(d), our 2nd trained network produces better results which generates sharper boundaries and reduces the influences of ring effect. It is obviously to find that high frequency depth information can be learned in this step. Based on Fig. 3(f), depth field statistics and the guided color images can further improve the quality of depth maps, sharper boundaries are generated and the influences of ring effect are almost eliminated. As shown in Fig. 3(e), it is interesting to see that depth maps themselves can be regarded as guidance to improve the qualities of depth maps. Generally, color image guided results are better than depth image guided results.

To evaluate the train times, experiment has been done. The average quantitative results of 1st train, 2nd train and 3rd train on different datasets in $RMSE$ for up-sampling factor of ×8 are shown in Fig. 5. We extract 4 images from Middlebury dataset, 5 images from Sintel dataset and 5 images from ICL dataset, then, the average $RMSE$ in each dataset is calculated. Obviously, the results of 2nd train are much smaller than the 1st train's, while the 3rd train's results are similar with the 2nd train's, which means much high frequency depth information is learned in the 1st and 2nd train, while less useful information is learned in the 3rd train. Hence, the train times in our method is set as 2.

4.2 Quantitative Results

In this section, we evaluate the performance of different $SOTA$ methods on publicly available benchmarks. We show the $RMSE$, MAE and $SSIM$ results on Middlebury dataset ($Cones$, $Teddy$, $Tsukuba$ and $Venus$), Sintel dataset

Table 1. Quantitative evaluation. The $RMSE$ is calculated for different SOTA methods for the Middlebury and Laserscanner data-sets for factors of ×2 and ×4. The best result of all single image methods for each data-set and up-scaling factor is highlighted and the second best is underlined.

Method	×2				×4				×4		
	Cones	Teddy	Tsukuba	Venus	Cones	Teddy	Tsukuba	Venus	Scan21	Scan30	Scan42
NN	4.4622	3.2363	9.2305	2.1298	6.0054	4.5466	12.9083	2.9333	2.6474	2.5196	5.6044
Bicubic	2.5245	1.9495	5.7828	1.3119	3.8635	2.8930	8.7103	1.9403	2.0324	1.9764	4.5813
Park et al. [11]	2.8497	2.1850	6.8869	1.2584	6.5447	4.3366	12.1231	2.2595	N/A	N/A	N/A
Yang et al. [12]	2.4214	1.8941	5.6312	1.2368	5.1390	4.0660	13.1748	2.7559	*N/A*	*N/A*	*N/A*
Ferstl et al. [13]	3.1651	2.4208	6.9988	1.4194	3.9968	2.8080	10.0352	1.6643	*N/A*	*N/A*	*N/A*
Yang et al. [16]	2.8384	2.0079	6.1157	1.3777	3.9546	3.0908	8.2713	1.9850	2.0885	2.0349	4.7474
Zeyde et al. [31]	1.9539	1.5013	4.5276	0.9305	3.2232	2.3527	7.3003	1.4751	1.6869	1.6207	3.6414
GR [32]	2.3742	1.8010	5.4059	1.2153	3.5728	2.7044	8.0645	1.8175	1.8906	1.8462	3.9809
ANR [32]	2.1237	1.6054	4.8169	1.0566	3.3156	2.4861	7.4895	1.6449	1.7334	1.6823	3.8140
NE+LS	2.0437	1.5256	4.6372	0.9697	3.2868	2.4210	7.3404	1.5225	8.6852	8.8460	8.1874
NE+NNLS	2.1158	1.5771	4.7287	1.0046	3.4362	2.4887	7.5344	1.6291	1.7313	1.6849	3.5733
NE+LLE	2.1437	1.6173	4.8719	1.0827	3.3414	2.4905	7.5528	1.6449	1.7058	1.6547	3.7975
Aodha et al. [8]	4.3185	3.2828	9.1089	2.2098	12.6938	4.1113	12.6938	2.6497	2.5983	2.6267	6.1871
Hornácek et al. [34]	3.7512	3.1395	8.8070	2.0383	5.4898	5.0212	11.1101	3.5833	2.8585	2.7243	4.5074
Huang et al. [33]	4.6273	3.4293	10.0766	2.1653	6.2723	4.8346	13.7645	3.0606	2.7097	2.6245	5.9896
Ferstl et al. [9]	2.2139	1.7205	5.3252	1.1230	3.5680	2.6474	7.5356	1.7771	1.4349	1.4298	3.1410
Xie et al. [10]	2.7338	2.4911	6.3534	1.6390	4.4087	3.2768	9.7765	2.3714	1.3993	1.4101	2.6910
Our	**1.4356**	**1.1974**	**2.9841**	**0.5592**	**2.9789**	**1.8006**	**6.1422**	**0.8796**	**1.1135**	**1.0711**	**1.6658**

(*Alley*, *Ambush*, *Bandage*, *Caves* and *Market*) and 5 images extracted from ICL dataset for up-sampling factors of ×2, ×3, ×4 and ×8. Additionally, we show the results for the real-world laser scan dataset (*Scan*21, *Scan*30, *Scan*42) provided by Aodha et al. [8] for an up-sampling factor of ×4. Tables 1, 2 and 3 show the comparisons of our proposed method in terms of different up-scaling factors with respect to the baseline methods. Numbers in bold indicate the best performance and those with underline indicate the second best performance. Note that we include the comparison results of ×3 in $RMSE$, MAE, $SSIM$ in the supplementary material, as well as the comparison results of ×2 and ×4 in MAE. What can be clearly seen is that our proposed method not only generates much smaller errors in $RMSE$, MAE, but also generates much better results in $SSIM$, which means that our method can well recover the HR depth maps well in both numerical value and structure.

4.3 Qualitative Results

We evaluate our proposed methods visually in Figs. 6, 7 and 8. Among the figures, Fig. 6 shows the results of the Middlebury data (parts of *Cones* and *Teddy*) (factor ×4), Fig. 7 shows the results of Sintel data (parts of *Alley* and *Caves*) (factor ×4), and Fig. 8 shows the result of up-scaling the depth from a laser scan with the zoomed cropped regions (factor ×4). Note that image enhancement methods are employed in these depth maps in order to show the details

Table 2. Quantitative evaluation. The $SSIM$ is calculated for different SOTA methods for the Middlebury and Laserscanner data-sets for factors of ×2 and ×4. The best result of all single image methods for each data-set and up-scaling factor is highlighted and the second best is underlined.

Method	×2				×4				×4		
	Cones	Teddy	Tsukuba	Venus	Cones	Teddy	Tsukuba	Venus	Scan21	Scan30	Scan42
NN	0.9645	0.9696	0.9423	0.9888	0.9360	0.9450	0.9003	0.9800	0.9814	0.9828	0.9679
Bicubic	0.9720	0.9771	0.9536	0.9909	0.9538	0.9619	0.9205	0.9845	0.9875	0.9879	0.9743
Park et al. [11]	0.9699	0.9767	0.9320	0.9910	0.9420	0.9553	0.8981	0.9862	N/A	N/A	N/A
Yang et al. [12]	0.9831	0.9851	0.9720	0.9944	0.9624	0.9695	0.9314	0.9879	N/A	N/A	N/A
Ferstl et al. [13]	0.9755	0.9795	0.9576	0.9938	0.9625	0.9707	0.9245	0.9901	N/A	N/A	N/A
Yang et al. [16]	0.9473	0.9564	0.9072	0.9805	0.9482	0.9566	0.9014	0.9816	0.9860	0.9866	0.9690
Zeyde et al. [31]	0.9655	0.9717	0.9438	0.9886	0.9604	0.9628	0.9147	0.9883	0.9908	0.9912	0.9830
GR [32]	0.9587	0.9656	0.9314	0.9862	0.9500	0.9592	0.9012	0.9817	0.9880	0.9885	0.9763
ANR [32]	0.9630	0.9693	0.9400	0.9879	0.9391	0.9452	0.8731	0.9806	0.9895	0.9898	0.9796
NE+LS	0.9623	0.9692	0.9391	0.9887	0.9514	0.9574	0.9042	0.9852	0.9355	0.9182	0.9302
NE+NNLS	0.9640	0.9707	0.9426	0.9883	0.9424	0.9499	0.8872	0.9820	0.9896	0.9900	0.9805
NE+LLE	0.9588	0.9658	0.9405	0.9837	0.9270	0.9331	0.8794	0.9641	0.9896	0.9896	0.9783
Aodha et al. [8]	0.9606	0.9690	0.9364	0.9874	0.9392	0.9520	0.9080	0.9822	0.9838	0.9838	0.9668
Hornácek et al. [34]	0.9696	0.9719	0.9461	0.9895	0.9501	0.9503	0.9137	0.9789	0.9814	0.9825	0.9754
Huang et al. [33]	0.9582	0.9673	0.9301	0.9875	0.9360	0.9425	0.8821	0.9784	0.9808	0.9819	0.9602
Ferstl et al. [9]	0.9866	0.9884	0.9766	0.9963	0.9645	0.9716	0.9413	0.9893	0.9918	0.9916	0.9819
Xie et al. [10]	0.9633	0.9625	0.9464	0.9852	0.9319	0.9331	0.8822	0.9730	0.9869	0.9878	0.9899
Our	**0.9989**	**0.9918**	**0.9905**	**0.9989**	**0.9783**	**0.9831**	**0.9666**	**0.9973**	**0.9948**	**0.9947**	**0.9939**

Table 3. Quantitative evaluation. The $RMSE$ and $SSIM$ are calculated for different SOTA methods for the Middlebury data-set for factors of and ×8.

Method	$RMSE$				$SSIM$			
	Cones	Teddy	Tsukuba	Venus	Cones	Teddy	Tsukuba	Venus
NN	7.5937	6.2416	18.4786	4.4645	0.8996	0.9199	0.8387	0.9634
Bicubic	5.3000	4.2423	13.3220	2.8948	0.9314	0.9442	0.8564	0.9757
Park et al. [11]	8.0078	6.3264	17.6225	3.4086	0.9231	0.9426	0.8409	0.9792
Yang et al. [12]	5.1390	4.0660	13.1748	2.7559	0.9361	0.9482	0.8624	0.9768
Huang et al. [33]	6.1629	6.6235	**10.6618**	4.1399	0.9280	0.9254	0.9027	0.9712
Our	**4.5887**	**2.8850**	11.6231	**1.7082**	**0.9510**	**0.9679**	**0.9051**	**0.9903**

more clearly. It is obvious to see that our method produces more visual appealing results than the previously reported approaches. Boundaries in the results generated by our method are sharper and smoother along the edge direction. Besides, our method can preserve the structure of the scene in regions with fine structures effectively.

5 Conclusions

In this work we propose a method for single depth image super-resolution, which is divided into two steps. First, an end to end progressive deep convolutional network is used to generate high resolution depth maps from low resolution depth maps. To our knowledge, we are the first one to use deep convolutional network to solve the problem of depth map super-resolution. As the generated high resolution depth maps may suffer from ring effect. Second, to further improve the quality of the obtained high resolution depth maps, depth statistical information and the corresponding high resolution color images are regarded as effective regularization to refine the depth maps. Besides, We find that the depth map itself can even be employed to improve the quality of depth map. By combining the two steps, we are able to generate more better depth maps than state-of-the-art approaches. In both quantitative and qualitative evaluation using widespread data-sets, we show that our method outperforms existing methods by a margin. In future, we plan to deal with noisy low-resolution depth images under our unified framework.

Acknowledgement. This work is supported by 863 program of China (No. 2015-AA016405), NSF of China (Nos. 61672326, 61420106007), ARC Grants (Nos. DE140100180, LP100100588, DP120103896) and China Scholarship Council.

References

1. Shotton, J., Sharp, T., Kipman, A., Fitzgibbon, A., Finocchio, M., Blake, A., Cook, M., Moore, R.: Real-time human pose recognition in parts from single depth images. Commun. ACM **56**, 116–124 (2013)
2. Izadi, S., Kim, D., Hilliges, O., Molyneaux, D., Newcombe, R., Kohli, P., Shotton, J., Hodges, S., Freeman, D., Davison, A., et al.: Kinectfusion: real-time 3d reconstruction and interaction using a moving depth camera. In: Proceedings of the 24th Annual ACM Symposium on User Interface Software and Technology, pp. 559–568 (2011)
3. Butler, D.J., Wulff, J., Stanley, G.B., Black, M.J.: A naturalistic open source movie for optical flow evaluation. In: Fitzgibbon, A., Lazebnik, S., Perona, P., Sato, Y., Schmid, C. (eds.) ECCV 2012. LNCS, vol. 7577, pp. 611–625. Springer, Heidelberg (2012). doi:10.1007/978-3-642-33783-3_44
4. Dong, C., Loy, C.C., He, K., Tang, X.: Learning a deep convolutional network for image super-resolution. In: Fleet, D., Pajdla, T., Schiele, B., Tuytelaars, T. (eds.) ECCV 2014. LNCS, vol. 8692, pp. 184–199. Springer, Cham (2014). doi:10.1007/978-3-319-10593-2_13
5. Kim, J., Kwon Lee, J., Mu Lee, K.: Accurate image super-resolution using very deep convolutional networks. In: Proceedings of the IEEE Conference on Computer Vision and Pattern Recognition, pp. 1646–1654 (2016)
6. Kim, J., Kwon Lee, J., Mu Lee, K.: Deeply-recursive convolutional network for image super-resolution. In: Proceedings of the IEEE Conference on Computer Vision and Pattern Recognition, pp. 1637–1645 (2016)
7. Freeman, W.T., Jones, T.R., Pasztor, E.C.: Example-based super-resolution. IEEE Comput. Graph. Appl. **22**, 56–65 (2002)
8. Mac Aodha, O., Campbell, N.D.F., Nair, A., Brostow, G.J.: Patch based synthesis for single depth image super-resolution. In: Fitzgibbon, A., Lazebnik, S., Perona, P., Sato, Y., Schmid, C. (eds.) ECCV 2012. LNCS, vol. 7574, pp. 71–84. Springer, Heidelberg (2012). doi:10.1007/978-3-642-33712-3_6

9. Ferstl, D., Ruther, M., Bischof, H.: Variational depth superresolution using example-based edge representations. In: Proceedings of the IEEE International Conference on Computer Vision, pp. 513–521 (2015)
10. Xie, J., Feris, R.S., Sun, M.T.: Edge-guided single depth image super resolution. IEEE Trans. Image Proc. **25**, 428–438 (2016)
11. Park, J., Kim, H., Tai, Y.W., Brown, M.S., Kweon, I.: High quality depth map upsampling for 3d-tof cameras. In: Proceedings of the IEEE International Conference on Computer Vision, pp. 1623–1630 (2011)
12. Yang, J., Ye, X., Li, K., Hou, C.: Depth recovery using an adaptive color-guided auto-regressive model. In: Fitzgibbon, A., Lazebnik, S., Perona, P., Sato, Y., Schmid, C. (eds.) ECCV 2012. LNCS, vol. 7576, pp. 158–171. Springer, Heidelberg (2012). doi:10.1007/978-3-642-33715-4_12
13. Ferstl, D., Reinbacher, C., Ranftl, R., Rüther, M., Bischof, H.: Image guided depth upsampling using anisotropic total generalized variation. In: Proceedings of the IEEE International Conference on Computer Vision, pp. 993–1000 (2013)
14. Matsuo, K., Aoki, Y.: Depth image enhancement using local tangent plane approximations. In: Proceedings of the IEEE Conference on Computer Vision and Pattern Recognition, pp. 3574–3583 (2015)
15. Lu, J., Forsyth, D.: Sparse depth super resolution. In: Proceedings of the IEEE Conference on Computer Vision and Pattern Recognition, pp. 2245–2253 (2015)
16. Yang, J., Wright, J., Huang, T.S., Ma, Y.: Image super-resolution via sparse representation. IEEE Trans. Image Proc. **19**, 2861–2873 (2010)
17. Wang, S., Zhang, L., Liang, Y., Pan, Q.: Semi-coupled dictionary learning with applications to image super-resolution and photo-sketch synthesis. In: Proceedings of the IEEE Conference on Computer Vision and Pattern Recognition, pp. 2216–2223 (2012)
18. Kim, K.I., Kwon, Y.: Single-image super-resolution using sparse regression and natural image prior. IEEE Trans. Pattern Anal. Mach. Intell. **32**, 1127–1133 (2010)
19. Yang, M.C., Wang, Y.C.F.: A self-learning approach to single image super-resolution. IEEE Trans. Multimedia **15**, 498–508 (2013)
20. Levin, A., Lischinski, D., Weiss, Y.: Colorization using optimization. ACM Trans. Graph. **23**, 689–694 (2004)
21. Li, B., Shen, C., Dai, Y., van den Hengel, A., He, M.: Depth and surface normal estimation from monocular images using regression on deep features and hierarchical CRFs. In: Proceedings of the IEEE Conference on Computer Vision and Pattern Recognition, pp. 1119–1127 (2015)
22. Silberman, N., Hoiem, D., Kohli, P., Fergus, R.: Indoor segmentation and support inference from RGBD Images. In: Fitzgibbon, A., Lazebnik, S., Perona, P., Sato, Y., Schmid, C. (eds.) ECCV 2012. LNCS, vol. 7576, pp. 746–760. Springer, Heidelberg (2012). doi:10.1007/978-3-642-33715-4_54
23. Diebel, J., Thrun, S.: An application of markov random fields to range sensing. In: Proceedings of the Advances in Neural Information Processing Systems, vol. 5, pp. 291–298 (2005)
24. Chartrand, R., Yin, W.: Iteratively reweighted algorithms for compressive sensing. In: IEEE International Conference on Acoustics, Speech and Signal Processing, pp. 3869–3872 (2008)
25. Ajanthan, T., Hartley, R., Salzmann, M., Li, H.: Iteratively reweighted graph cut for multi-label MRFs with non-convex priors. In: Proceedings of the IEEE Conference on Computer Vision and Pattern Recognition, pp. 5144–5152 (2015)
26. Scharstein, D., Szeliski, R.: A taxonomy and evaluation of dense two-frame stereo correspondence algorithms. Int. J. Comp. Vis. **47**, 7–42 (2002)

27. Scharstein, D., Szeliski, R.: High-accuracy stereo depth maps using structured light. In: Proceedings of the IEEE Conference on Computer Vision and Pattern Recognition, vol. 1, pp. 195–202 (2003)

28. Scharstein, D., Pal, C.: Learning conditional random fields for stereo. In: Proceedings of the IEEE Conference on Computer Vision and Pattern Recognition, pp. 1–8 (2007)

29. Hirschmüller, H., Scharstein, D.: Evaluation of cost functions for stereo matching. In: Proceedings of the IEEE Conference on Computer Vision and Pattern Recognition, pp. 1–8 (2007)

30. Handa, A., Whelan, T., McDonald, J., Davison, A.: A benchmark for RGB-D visual odometry, 3D reconstruction and SLAM. In: IEEE International Conference on Robotics and Automation, pp. 1524–1531 (2014)

31. Zeyde, R., Elad, M., Protter, M.: On single image scale-up using sparse-representations. In: Boissonnat, J.-D., Chenin, P., Cohen, A., Gout, C., Lyche, T., Mazure, M.-L., Schumaker, L. (eds.) Curves and Surfaces 2010. LNCS, vol. 6920, pp. 711–730. Springer, Heidelberg (2012). doi:10.1007/978-3-642-27413-8_47

32. Timofte, R., Smet, V., Gool, L.: Anchored neighborhood regression for fast example-based super-resolution. In: Proceedings of the IEEE Conference on Computer Vision, pp. 1920–1927 (2013)

33. Huang, J.B., Singh, A., Ahuja, N.: Single image super-resolution from transformed self-exemplars. In: Proceedings of the IEEE International Conference on Computer Vision, pp. 5197–5206 (2015)

34. Hornácek, M., Rhemann, C., Gelautz, M., Rother, C.: Depth super resolution by rigid body self-similarity in 3d. In: Proceedings of the IEEE Conference on Computer Vision and Pattern Recognition, pp. 1123–1130 (2013)

35. Bevilacqua, M., Roumy, A., Guillemot, C., Alberi-Morel, M.L.: Low-complexity single-image super-resolution based on nonnegative neighbor embedding. In: Proceedings of the British Machine Vision Conference (2012)

3D Watertight Mesh Generation
with Uncertainties from Ubiquitous Data

Laurent Caraffa[(✉)], Mathieu Brédif, and Bruno Vallet

IGN Recherche, SRIG, MATIS, Université Paris-Est, Champs-sur-Marne, France
laurent.caraffa@ign.fr

Abstract. In this paper, we propose a generic framework for watertight mesh generation with uncertainties that provides a confidence measure on each reconstructed mesh triangle. Its input is a set of vision-based or Lidar-based 3D measurements which are converted to a set of mass functions that characterize the level of confidence on the occupancy of the scene as *occupied*, *empty* or *unknown* based on Dempster-Shafer Theory. The output is a multi-label segmentation of the ambient 3D space expressing the confidence for each resulting volume element to be *occupied* or *empty*. While existing methods either sacrifice watertightness (local methods) or need to introduce a smoothness prior (global methods), we derive a per-triangle confidence measure that is able to gradually characterize when the resulting surface patches are certain due to dense and coherent measurements and when these patches are more uncertain and are mainly present to ensure smoothness and/or watertightness. The surface mesh reconstruction is formulated as a global energy minimization problem efficiently optimized with the α-expansion algorithm. We claim that the resulting confidence measure is a good estimate of the local lack of sufficiently dense and coherent input measurements, which would be a valuable input for the next-best-view scheduling of a complementary acquisition.

Beside the new formulation, the proposed approach achieves state-of-the-art results on surface reconstruction benchmark. It is robust to noise, manages high scale disparity and produces a watertight surface with a small Hausdorff distance in uncertainty area thanks to the multi-label formulation. By simply thresholding the result, the method shows a good reconstruction quality compared to local algorithms on high density data. This is demonstrated on a large scale reconstruction combining real-world datasets from airborne and terrestrial Lidar and on an indoor scene reconstructed from images.

1 Introduction

Surface reconstruction from point cloud is an important topic that has already been studied extensively. The first reason is the ill-posed characteristic of the problem. The second reason is the constantly increasing number of applications that require surface reconstruction: digital elevation model computation for flood simulation, 3D modeling, robot path planning, etc. Furthermore 3D acquisition

© Springer International Publishing AG 2017
S.-H. Lai et al. (Eds.): ACCV 2016, Part IV, LNCS 10114, pp. 377–391, 2017.
DOI: 10.1007/978-3-319-54190-7_23

Fig. 1. Result of the proposed method on scanned object and city environment. Areas with strong confidence are blue while occluded or conflict areas are red. (Color figure online)

techniques such as Lidar and multi-view stereo are now mainstream and currently harvest vast amounts of data on a daily basis. These two aspects result in multiple and various methods based on different assumptions related to the characteristics of the acquisition device and of the scene and to the user applications. This observation is pointed out by [1] that lists most of the commonly used assumptions for a given situation: smoothness prior, shape primitive reconstruction where the output is composed of a combination of primitives, visibility prior that uses the sensor information in order to detect empty areas, watertightness, etc. Because of the implicit relation between the surface definition and the smoothness prior, this assumption has been historically the most studied. Despite the important diversity of methods, the smoothness prior can be broadly categorized as local, global or piecewise. Local methods ensure that the output surface is smooth where the point cloud density is large enough. Global methods produce a watertight surface on the whole data range by filling empty holes regardless of the density. This fact is emphasized by the benchmark introduced in [2] where global methods based on indicator functions are robust with a small Hausdorff distance to the ground truth, while local methods produce a locally accurate reconstruction when the quality of the input data is good. When the density is large enough, it indicates that both approaches produce good results. Based on this consideration, we consider that rather than trying to improve the quality of the resulting mesh, we should instead characterize its quality and produce a qualified mesh where uncertain areas are localized. To this end, we use a global approach and pose the surface reconstruction problem as a volumetric segmentation of a 3D confidence space.

1.1   Background

The volumetric segmentation approach has been first used in [3] for surface reconstruction. In [4], it is used with a Delaunay triangulation and the visibility assumption for labelling tetrahedra crossed by a ray as outside and tetrahedra behind the 3D detected point as inside the surface to be reconstructed. Unlabelled tetrahedra are regularized in a global optimization framework related to a surface smoothness prior. [5] shows that this approach is mature and produces high quality results when the amount of data is large enough in the multi-view reconstruction context. In [6], the visibility prior is improved to better reconstruct surfaces having a lack of information. This approach has been greatly extended in various cases like hybrid primitive mesh reconstruction [7] or structure preserving approach where coherent crust, plane and corner are detected for an extra sampling of the point cloud [8]. In [9], a similar approach is used to produce a watertight surface from airborne Lidar with planar assumption.

When the normal orientation or the visibility prior are not available, [10] proposes a method that defines an unsigned function related to the detection of crusts around high density point, which allows to reconstruct the surface without any normal information. In [11], a method is based on the winding number that produces a piecewise mesh when a significant number of facet orientations are available. The advantage is the definition of a function in the space that tells, on each point, its probability to be *occupied* or *empty*.

Recently, from the perspective to merge multiple source data, [12] propose an approach that takes into account multiple-scale datasets. It produce high quality surface in high detailed area embedded in higher scale reconstruction.

Even if modern algorithms are robust enough to handle bad quality inputs, some conflicts will always be impossible to prevent in uncontrolled environments like moving pedestrians or cars. Outliers and conflict detection from multiple point clouds has been recently studied. [13] use the Dempster-Shafer Theory (DST) for modeling the space occupancy as *empty, occupied* or *unknown*.

1.2 Proposed Approach

One may argue that surface reconstruction in easy cases is now mature and many methods produce good results thanks to different priors in multiple cases ([5] for urban reconstruction or [14] for scanned objects). However, the amount and types of input are constantly increasing such that [1] points out acquisition ubiquity as a challenging task for the future. It is to that extent that we propose a new perspective for surface reconstruction: instead of choosing a strong surface prior and specific output, the method produces a segmentation of the space related to the confidence to be *empty* or *occupied*. The input of our algorithm is a set of mass functions that describe the occupancy of the scene with the Dempster-Shafer Theory. Instead of producing a binary segmentation, the result is a multi-label segmentation of the space where each label is a confidence to be inside or outside an object. The final surface is defined as a level set of this confidence at a given threshold of the interface between the mostly-*occupied* and mostly-*empty* volumes. Thanks to this multi-label formulation, the confidence of the output can be characterized as the gradient in occupancy confidence between the neighboring volumes of each output facet. Mesh patches composed by facets with greater confidence may be seen as patches featuring enough data support which result mainly from a local smoothing. By contrast, in uncertain areas, this confidence measure properly documents the fact that corresponding facets are

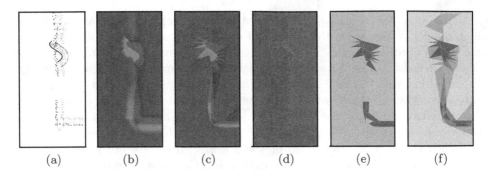

(a)   (b)   (c)   (d)   (e)   (f)

Fig. 2. Example. First, two pointsets merged with their local descriptor (a), the combined mass function of the two sets of points (b), the normalized integral of the mass function on each cell (c). The resulting segmentation with 2, 4 and 8 labels (d, e and f). (Color figure online)

only present to ensure smoothness and watertightness with little data support as it may be the case in any global surface reconstruction approaches.

In this paper, we propose two models, one based on a local descriptor introduced in [15], the second on the visibility prior for surface reconstruction when the sensor position is known for each 3D measurement. The result of the proposed method is shown in the Fig. 1 (The redder the color, the smaller the confidence). One may note that the confidence increases with the point density. However, areas that have not been scanned or with an important scale remain red (i.e. uncertain). To our knowledge, producing a multi-label result, according to the surface confidence has not been explored in the surface reconstruction literature.

Figure 2 shows an example of the proposed approach in 2D. First, a local descriptor is computed on each point cloud for extracting normal and local point cloud distribution (Fig. 2a). The space is then discretized with a 3D Delaunay triangulation (as in [4]), the mass function (Fig. 2b) is then integrated on each cell (Fig. 2c). Finally, a multi-label segmentation is performed on each tetrahedron where a label relates to the confidence of the result to be *occupied* or *empty* (Fig. 2d, e, f). This labelling relies on a global optimization that penalizes large surfaces. For the problem to be tractable, a mapping from labels {*occupied, empty, unknown*} to labels {*empty, occupied*} is performed. This allows to formalize the problem with a sub-modular energy that can be optimized globally in polynomial time. The multi-label segmentation is then conceptually cast as a binary *empty/occupied* segmentation by thresholding the label confidence at e.g. 50%, which enables the well-known extraction of the resulting mesh as the set of triangles separating an *empty*-labeled tetrahedron from a *occupied*-labeled one.

Our main contributions are:

- A method that takes an arbitrary number and type of input measurements defined by a confidence function.
- A multi-label segmentation that produce a watertight surface and enables the confidence characterization of each resulting mesh triangle.

For that, we will first define in Sect. 2 the problem statement of the multi-label segmentation of the mass function. The energy formulation of the problem is defined, then the mass function for the case of the point cloud reconstruction is introduced. In Sect. 3, technical aspect of the implementation is discussed: the space is quantification, the mass function computation and the optimization of the resulting energy. Then, the method is evaluated and compared to state of the art algorithms in Sect. 4. Section 5 shows applications and results on mesh generation on both the ability to merge heterogeneous datasets and the confidence characterization of the resulting mesh.

2   Problem Formulation

The aim of the proposed method is to find, for a set of inputs I, a segmentation of the space as *occupied* or *empty* following a suitable model. Inputs are given by a set of mass functions m_i defined by (see Sect. 2.2):

$$m_i(P) = (e, o, v) \in [0, 1]^3 \text{ s.t. } e + o + u = 1 \tag{1}$$

where e, o and u are respectively the occupancy masses for a 3D point $P \in \mathbb{R}^3$ to be *empty*, *occupied* or *unknown*. The mass functions are then merged to yield an overall mass function:

$$m_f(P) = \bigoplus_{i \in I} m_i(P) \tag{2}$$

where $m_f(P)$ is the overall mass that gives, for a point $P \in \mathbb{R}^3$, the confidence to be *empty*, *occupied* or *unknown* for the set of mass functions $(m_i)_{i \in I}$.

The problem is then defined as a multi-label segmentation of the space regarding to m_f where the labels set is $L = \{0, \frac{1}{(N-1)}, \frac{2}{(N-1)}, ..., \frac{(N-2)}{(N-1)}, 1\}$. Each label $l \in L$ represents the occupancy confidence where $l = 1$ represents a absolute confidence to be *occupied* and $l = 0$ a absolute confidence to be *empty*. The space is discretized with a 3D Delaunay triangulation, where T is its set of tetrahedra and $F \subset T^2$ the set of facets represented by pairs of adjacent tetrahedra such that the set of facets is $\{(t_1 \cap t_2) s.t. (t_1, t_2) \in F\}$. We denote as l_t the label of the tetrahedron $t \in T$ and $l_T = (l_t)_{t \in T}$ the labelling l_t of each tetrahedron in the triangulation T. Following this formulation, a per-facet confidence is defined as

$$\text{confidence}(t_1 \cap t_2) = |l_{t_1} - l_{t_2}| \tag{3}$$

Thus, what is required is to find, for each tetrahedron of the triangulation, a label l_t that encodes the confidence of the tetrahedron in being either *occupied* or *empty* for a given set of input mass functions m_i. To this end, the problem is formulated as an energy minimization framework composed of two terms: the data term $E_{data}(l_T)$, which represents how a solution is close to the overall mass function m_f and the smoothness term $E_{prior}(l_T)$, which penalizes solutions with large interfaces between labels. The final energy minimization is:

$$\min_{l_T \in L^T} E_{data}(l_T) + \lambda \, E_{prior}(l_T) \tag{4}$$

where λ is the weight balancing the prior and data terms. Now the two terms will be described in details.

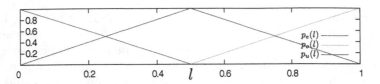

Fig. 3. Example of the related confidence of the 3 mass functions.

2.1 Energy Terms

The data term models how a label l fits to the overall mass function m_f defined in Eq. 2. To take into account the *unknown* label, the confidence of a label l to be *occupied* is decomposed into 3 functions $p_e(l), p_o(l)$ and $p_u(l)$ defined as:

$$p_j(l) = \begin{cases} p_e(l) = max(0, 1 - 2l) \\ p_o(l) = max(0, 2l - 1) \\ p_u(l) = 1 - p_o(l) - p_u(l) \end{cases} \qquad (5)$$

Figure 3 shows the mapping between labels and the 3 functions. When $l = 0.2$, the corresponding scores are $p_e(l) = 0.6$, $p_o(l) = 0$, and $p_u(l) = 0.4$. With this decomposition, the case $p_e(l) = 0.6$, $p_o(l) = 0.4$ and $p_u(l) = 0.4$ are mutually exclusive solutions for the input $l = 0.6$. It tells that the solution is a conflict or a fuzzy zone. This allows to separate the *conflict* case from the *unknown* case. However, one can imagine another cost function for conflict, change detection, accuracy measurement.

Finally, the data term is the integral of the difference between the function $p_j(l)$ and the mass function m_i for each label $i \in (e, o, u)$ over the volume of the tetrahedron, which gives:

$$E_{data}(l) = \sum_{t \in T} \sum_{j \in \{e,o,u\}} \int_t |p_j(l_t) - m_f(v)| dv \qquad (6)$$

A prior term is added that generally enforces solutions with a small surface. This idea is extended to the multi-label case by minimizing the surface multiplied by the distance between label probabilities, which leads to the following term:

$$E_{prior}(l) = \sum_{(t_1, t_2) \in F^2} \text{area}(t_1 \cap t_2) |l_{t_1} - l_{t_2}| \qquad (7)$$

where $\text{area}(t_1 \cap t_2)$ returns the area of the triangle $t_1 \cap t_2$.

2.2 Mass Function Definition

All that remains is to define the function m_e, m_o and m_u for an input set of data. When the oriented normal is known, tetrahedra in the oriented normal direction have great confidence to be *empty* and those right after the 3D point

Fig. 4. Mass function in $\mathbb{R}^2$. (Left) For a single point Q, model parameters are: a spanning parameter σ_1 and the noise parameter σ_d. σ_e and σ_o are the thickness parameter. (Right) The mass with the visibility prior in $\mathbb{R}^2$ for a point Q when the sensor center is known, model parameters are: the incertitude on the angle σ_θ and the noise parameter σ_d.

have great confidence to be *occupied*. In the case of 3D reconstruction when the center of the sensor is known, the visibility prior tells that tetrahedra crossed by the beam formed by the sensor and the 3D point have great confidence to be *empty* along the ray. This idea is encoded here in the independent definition of a mass function for each input measurement, which will be merged into a single overall function using DST.

In this section, two mass functions representing the space occupancy in $\mathbb{R}^3$ are defined: one representing the space occupancy around the 3D detected point with normal, the second representing the visibility prior when the sensor center is known.

Local mass definition: For a 3D detected point Q with a normal $\overrightarrow{n}$, an occupancy mass along the normal is defined by (o_r, o_r, u_r). Let $P \subset \mathbb{R}^3$ be a 3D point and r its signed distance to the plane $(Q, \overrightarrow{n})$. When $r > 0$, P is on the same side as the normal, it is *empty*, e_r should tends to 1 and o_r to 0. Inversely, when $r < 0$, P, it is *occupied*, then the mass e_r should tends to 0 and o_r to 1. When P is close to Q, the probability to be *occupied* or *empty* o_r and e_r should be 0.5. Finally, when P is far from Q, o_r and e_r should decrease from 1 to 0 for modeling the thickness of the scene and the empty area. One mass related to this behavior is:

$$r < 0 : e_r = 0.5 \, e^{-(\frac{|r|}{\sigma_d})^2} \; ; \; o_r = (1 - 0.5 \, e^{-(\frac{|r|}{\sigma_d})^2}) \, e^{-(\frac{|r|}{\sigma_o})^2}$$
$$r > 0 : e_r = (1 - 0.5 \, e^{-(\frac{|r|}{\sigma_d})^2}) e^{-(\frac{|r|}{\sigma_e})^2} \; ; \; o_r = 0.5 \, e^{-(\frac{|r|}{\sigma_d})^2} \tag{8}$$

where σ_d is the noise parameter, σ_o the scale of the scene thickness prior and σ_e the thickness of the empty area.

Space occupancy mass: In the 3D case, the mass of occupancy on the plane is large while decreasing as it moves away from the point, introducing a spanning parameter, the occupancy function is represented as

$$f(P, P^1, P^2) = e^{-(\frac{||P^1 - P||}{\sigma_1})^2 - (\frac{||P^2 - P||}{\sigma_2})^2} \tag{9}$$

where $(\overrightarrow{v_1}, \overrightarrow{v_2}, \overrightarrow{n})$ is a standard basis defined from the normal vector. P^1 and P^2 are the orthogonal projection on $\overrightarrow{v_1}, \overrightarrow{v_2}$. σ_1, σ_2 are the scale of the spanning parameter on the basis.

Adding the visibility prior: When the sensor center O is known, the mass of occupancy is large on the ray while decreasing as it moves away from it, for that we define

$$f(\theta) = e^{-(\frac{\theta}{\sigma_\theta})^2} \tag{10}$$

where θ is the angle formed by the ray $\overrightarrow{OQ}$ and $\overrightarrow{OP}$, σ_θ is the scale of the angle uncertainty. In this case, the scale of the *empty* area σ_e is set to ∞ and $o_r = 0$ when $r < 0$ in Eq. 8.

Mass of an input: Finally, the mass function of a 3D point is defined by:

$$m_i(P) = \begin{Bmatrix} e \\ o \\ u \end{Bmatrix} = \begin{Bmatrix} s_c \cdot f \cdot e_r \\ s_c \cdot f \cdot o_r \\ 1 - e - o \end{Bmatrix} \tag{11}$$

where e_r and o_r models the uncertainty along the normal and f the uncertainty as it moves away from the data. $s_c \in [0,1]$ is the scale coefficient. The Fig. 4 shows an example of the two mass functions in 2D.

Fig. 5. Example of fusion. First, four rays with a large uncertainty. The confidence at the intersection is much higher. Then, a fusion with both a strong confidence to be *empty* and *occupied*. The confidence at the intersection is equal to 0.5 for both *empty* and *occupied*, the *unknown* label remains small.

Overall mass: The DST is used to combine all inputs and gives the overall mass m_f. Let m_1 m_2 two mass functions, the fusion can be computed with the DST as following:

$$\begin{Bmatrix} e_1 \\ o_1 \\ u_1 \end{Bmatrix} \oplus \begin{Bmatrix} e_2 \\ o_2 \\ u_2 \end{Bmatrix} = \frac{1}{1-K} \begin{Bmatrix} e_1 \cdot e_2 + e_1 \cdot u_2 + u_1 \cdot e_2 \\ o_1 \cdot o_2 + o_1 \cdot u_2 + u_1 \cdot o_2 \\ u_1 \cdot u_2 \end{Bmatrix} \tag{12}$$

where $K = o_1 \cdot e_2 + e_1 \cdot o_2$ represents the conflict term. When K is equal to 0, there is no conflict between inputs.

The combination rule is commutative and associative, enabling an order-independent aggregation of the measurements. At a location P for a set of mass functions $m_i \in I$, the overall mass is computed following the Eq. 2. Figure 5 shows two examples of mass merging in 2D.

2.3   Setting Parameters with Local Descriptor

The mass function needs local information like normal, noise level and scale. When a parameter is not provided, it can be extracted by computing the local descriptor proposed in [15]. Based on a local Principal Component Analysis (PCA), it provides 3 eigenvectors v_1, v_2, v_3 with their eigenvalues $\sigma_1, \sigma_2, \sigma_3$. Assuming a planar surface, its normal direction $\overrightarrow{n}$ is given by v_3, σ_3 denotes its noise level. σ_1 and σ_2 are the spanning scale for each scale of standard basis component.

The scale of the point is represented by its noise level $s_c = min(\frac{\epsilon}{\sigma_3}, 1)$ where ϵ is the targeted level of details. $\sigma_3 < \epsilon$ denotes that the highest level of details is reached.

3   Algorithm

3.1   Triangulation

Thanks to the previous formulation, any 3-space partitioning could by used to discretized the aggregated mass function in order to yield a tractable combinatorial optimization problem. Ideally, this tessellation should be driven directly by the behavior of m_f, with less refined tessellation where it is homogeneous and facets orthogonal to its gradients. To limit the computing time and be robust to noise, the number of points is reduced using a K-means approach on the point cloud to approximate the tessellation. To adapt the mesh accuracy to the noise level, points are added while the distance of closest plane of the triangulation in the normal direction for each input point is above $3\sigma_d$. This approach helps to produce mesh with large facet in noisy area and small facet in detailed areas according to the noise level. We currently use the Delaunay triangulation as implemented in the Computational Geometry Algorithms Library (CGAL, http://www.cgal.org).

3.2   Score Computation

Equation 6 requires to compute the integration of the mass function on each tetrahedron. The integral $\int_t |p_j(l_t) - m_f(v)| dv$ can be estimated using Monte Carlo integration as $\frac{\text{Volume}(t)}{|S|} \sum_{P \in S} |p_j(l_t) - m_f(P)|$ where S is a set of $|S|$ uniformly sampled points in the tetrahedron t. Volume(t) is the volume of the tetrahedron t. The mass function of each point P is computed with the Eq. 2.

3.3   Optimization

According to [16], the Eq. 4 with a metric function on the prior term 7 is submodular and thus, the global minimum can be reached in polynomial time. Many methods exist for optimizing sub-modular functions. When the number of labels is small and the scene composed by large homogeneous label areas, the α-expansion algorithm first introduced in [16], converges in practice quickly on a good local minimum which is guaranteed to have an energy at most twice the globally optimal energy.

The α-expansion consists in starting from an arbitrary set of labels l_t in each tetrahedra $t \in T$ and solve iteratively Eq. 4 with a proposition $\alpha \in L$. It refers to a Boolean optimization problem studied in quadratic pseudo-Boolean optimization (QPBO). As mentioned in [17], in our case where the quadratic function on the two labels is a metric, the Eq. 4 can always be solved globally in polynomial time in the binary case with a *min-cut* reduction. We use the graph-cut code introduced in [18] to solve the binary fusion problem.

3.4 Parameter Settings

The proposed formulation requires to choose an arbitrary number of labels. For surface reconstruction, an even number of labels should be chosen in Eq. 5: an odd number of labels results in the presence of the label $l = 0.5$ in L which hinders a decision between an *empty* and an *occupied* volume. With a small number of labels, the solution is smoother according to λ. With a large number of labels, extra areas with low confidence appear thanks to the small cost between the occupied and empty volumes in ambiguous regions. In our tests, 6 labels appears to be a good compromise between the time computation, the accuracy of the global result and the number of outliers.

4 Evaluation on Benchmark

The method is evaluated on the benchmark introduced in [2], it allows to evaluate the surface according to both error metrics and topological aspects. The benchmark is dedicated to surface reconstruction from point cloud with normal. It is composed of 5 datasets generated from an implicit function with a synthetic scanner. For each dataset, 48 acquisitions are generated with different variations of the scanner's parameters such as noise or different camera positions which lead to occlusions. In order to show the advantage of computing a multi-label segmentation, a single result is computed for each file. For that, we use a confidence threshold T_c. A facet is removed when the confidence defined in Eq. 3 is lower than T_c. The resulting surface is non manifold. The larger T_c, the larger the certainty on the surface.

The result is then evaluated with two different confidence threshold $T_c = 0$ (full area) and $T_c = 0.5$ (confident area) that can be computed on the fly. In this test, the scale is set to the maximum level of detail ($s_c = 1$ for each point), the thickness to $\sigma_o = \sigma_e = 0.3$.

4.1 Evaluation on the Dataset

Figure 6 shows the result of the proposed method (wmwu) with the two thresholds ($T_c = 0$ and $T_c = 0.5$) compared to APSS [14], Fourier [19] and Scattered [20]. Two results are shown: on the first one, the resulting mesh of the proposed approach is highly confident (blue color) and results in an high quality mesh as other methods. The second case is much more uncertain because of a

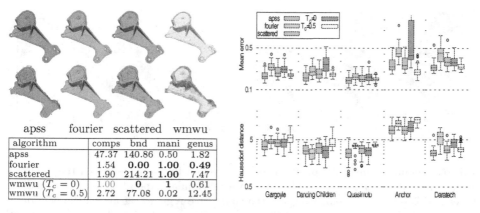

algorithm	comps	bnd	mani	genus
apss	47.37	140.86	0.50	1.82
fourier	1.54	**0.00**	**1.00**	0.49
scattered	1.90	214.21	**1.00**	7.47
wmwu ($T_c = 0$)	1.00	**0**	**1**	0.61
wmwu ($T_c = 0.5$)	2.72	77.08	0.02	12.45

Fig. 6. Result of 4 algorithms on the benchmark: APSS, Fourier, Scattered and the proposed approach: the global surface (wmwu $T_c = 0$) and with the confident threshold set to 0.5 (wmwu $T_c = 0.5$). First line shows an easy case, the second case is much more noisy with occlusions. The error mean and the Hausdorff distance are compared on the 5 datasets. Black dots show outliers when the box plot show the median and quartiles. The table shows the average shape properties over the full benchmark: *comps* refers to number of connected components, *bnd* is the length of boundary components, *mani* is whether or not a mesh is manifold, 1 being manifold and 0 otherwise, *genus* refers to the amount which deviates from the actual genus. The first section refers to the state of the art algorithm evaluated in [2], the second section shows the score of the proposed methods with both thresholds.

noisy sensor and occlusions (red areas) where other algorithms work also poorly. The global surface remains close to the ground truth. This fact is emphasized by the benchmark: when thresholding the mesh to $T_c = 0.5$, the mean error is very close to state-of-the art algorithms such as APSS or Scattered. The distribution of the means has globally a small variance and lowest maximum error on all the dataset that reflect the robustness of the proposed approach and the ability to detect uncertain facets. Like Scattered, the number of components remains low compared to APSS. Thanks to the multi-label global optimization framework that enforces the confident surface to be close, the boundary distance is 77.08 compared to 140.86 and 214.21 for APSS and Scattered.

The global mesh displays more error, this is the consequence of shapes produced by extra labels in unknown areas. The result remains good compared to other algorithms in the benchmark. The advantage of extra labels makes sense by analyzing the Hausdorff distance which is very good compare to other algorithms. The extra labels allow to better explore the unscanned area as [6] thanks to a small regularization where classic graph-cut based approaches tend to over-smooth the solution and produce holes in missing data areas. This is confirmed by the characteristic that the global surface is 100% manifold and watertight with only one component on the whole benchmark.

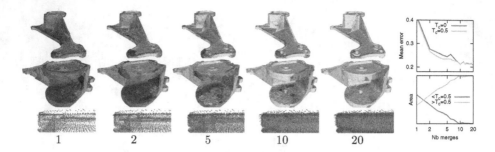

Fig. 7. Merging many data points. First and second line: the results on the union of 1, 2, 5, 10 and 20 point clouds. Third line: the accumulated point cloud. The confident threshold is set to $T_c = 0.5$. Facets in the confidence region that are on the boundary are green. Last column, the error of the global surface and the confidence surface. Then the area of regions upper and lower the threshold. (Color figure online)

4.2   Merging

The aim of the proposed approach is to merge a large amount of data. Figure 7 shows the result up to 20 point clouds processed at the same time on the Daratech dataset. For this test, the coefficient of each point cloud mass is 0.1 in order to reach the confidence threshold $T_c = 0.5$ only with a consequent number of data. We clearly show the confidence increasing while the number of files increases. Badly fitted surfaces caused by sensor drift on first iterations with high uncertainty are progressively removed and replaced by surface with high confidence. Statistics show that the confidence area is constantly increased along iterations while the mean error decreases until converging to 0.20 which is equal to the lower Quartile of best algorithms on this dataset. This test confirms the ability of the DST on managing uncertainty.

5   Results

Airborne and terrestrial Lidar reconstruction: This example shows how the DST helps for modeling different types of data. Figure 8 shows an urban scene reconstruction with both airborne and terrestrial Lidar data. Two overlapping airborne Lidar acquisition stripes cover a large area. A terrestrial point cloud is included at the center of the scene for a final 5.2 M point cloud.

Local model's parameters are fixed by the dimensionality descriptor (Sect. 2.3). Airborne and terrestrial Lidar point clouds are both represented as beams. The angle's scale of the visibility prior is set to 0.001 for the airborne and for the terrestrial data. The thickness parameter is equal to 0.07 m for terrestrial Lidar to reconstruct thin structures such as poles, cars or advertising hoarding. It is set to 30 m for airborne. This allows to reconstruct high buildings only with points recorded on the top. λ is equal to 0.005. The targeted scale is set to $\epsilon = 0.02$.

Fig. 8. Result of the proposed algorithm with terrestrial and airborne data combined.

Results show that both datasets are taken into account to produce a mesh that combines both the fine details from the ground based Lidar and a global watertight surface from the airborne data. The large thickness parameter on airborne data allows to reconstruct high building. Thanks to the small thickness and extra points on the terrestrial data, pillars are well meshed. The close up view shows that the merging between airborne and terrestrial points is well managed. We clearly see how the point cloud density affects the confidence of the resulting mesh where the confidence scale is reached. In the ground scanned place, the quality of reconstruction of the building's facade is large whereas the areas behind cars remain unknown and call for extra acquisitions.

Structure from motion data: We tested the proposed approach on indoor datasets from 3D reconstruction. A dataset of the full scene of 3.4 M point is merged to a 5.4 M points dataset representing a mechanical object. This scene has the particularity to have both high quality, noisy and unscanned areas. The targeted scale is set to $\epsilon = 0.03$. Figure 9 show the result of the proposed approach. Results show high quality reconstruction with high confidence in high detailed areas. On the contrary, uncertain areas are well smoothed with small confidence, unscanned areas are flat with the lowest confidence. The proposed approach is compared to FSSR [12], two appropriate scales are affected to each dataset. Where FSSR produces artifacts or holes, the proposed approach pro-

Fig. 9. Result of the proposed algorithm on structure from motion data. First line, 3 results of the proposed algorithm. Second line, two results with: the dataset, FSSR [12] and ours.

vides a low confidence and fills empty holes with the smallest confidence. The quality of the reconstruction is comparable in good quality areas.

6   Limitations

The benchmark shows that the resulting mesh after thresholding is not watertight and manifold anymore in many cases. More complex energies could be explored in order to manage the watertightness of the mesh at different thresholds. The algorithm is actually slow, the time computation can be highly improved by parallelizing the triangulation construction, the sampling and the score computation. Parallel heuristics can be used for the multi-label optimization.

7   Conclusion and Perspectives

A generic framework for watertight mesh generation with uncertainties is proposed for the surface reconstruction of ubiquitous data of the same scene. A confidence criterion is given on the resulting mesh. The problem is formalized as a global optimization problem that is efficiently solved with the α-expansion algorithm. The proposed approach shows results close to state of the art on both benchmark datasets and real cases like urban reconstruction from Lidar or structure from motion data. We advocate that the resulting confidence measure is a good estimate of the local lack of sufficiently dense and coherent input measurements, which would be a valuable input for the next-best view scheduling of a complementary acquisition or to detect unexplored area for terrestrial or airborne robot path planning.

Acknowledgments. This work is partly supported by the EU FP7 Project N. ICT-2011-318787(IQmulus).

References

1. Musialski, P., Wonka, P., Aliaga, D.G., Wimmer, M., van Gool, L., Purgathofer, W.: A survey of urban reconstruction. In: EUROGRAPHICS 2012 State of the Art Reports. EG STARs, pp. 1–28. Eurographics Association (2012)
2. Berger, M., Levine, J.A., Nonato, L.G., Taubin, G., Silva, C.T.: A benchmark for surface reconstruction. ACM Trans. Graph. **32**, 20:1–20:17 (2013)
3. Curless, B., Levoy, M.: A volumetric method for building complex models from range images. In: Proceedings of the 23rd Annual Conference on Computer Graphics and Interactive Techniques, SIGGRAPH 1996, pp. 303–312. ACM, New York (1996)
4. Labatut, P., Pons, J.P., Keriven, R.: Robust and efficient surface reconstruction from range data. Comput. Graph. Forum **28**, 2275–2290 (2009)
5. Vu, H.H., Labatut, P., Pons, J.P., Keriven, R.: High accuracy and visibility-consistent dense multiview stereo. IEEE Trans. Pattern Anal. Mach. Intell. **34**, 889–901 (2012)

6. Jancosek, M., Pajdla, T.: Multi-view reconstruction preserving weakly-supported surfaces. In: Proceedings of the 2011 IEEE Conference on Computer Vision and Pattern Recognition, CVPR 2011, pp. 3121–3128. IEEE Computer Society, Washington, DC (2011)

7. Lafarge, F., Keriven, R., Brédif, M., Vu, H.H.: A hybrid multiview stereo algorithm for modeling urban scenes. IEEE Trans. Pattern Anal. Mach. Intell. **35**, 5–17 (2013)

8. Lafarge, F., Alliez, P.: Surface reconstruction through point set structuring. Rapport de recherche RR-8174, INRIA (2012)

9. Kreveld, M.J., van Lankveld, T., Veltkamp, R.C.: Watertight scenes from urban lidar and planar surfaces. Comput. Graph. Forum **32**, 217–228 (2013)

10. Hornung, A., Kobbelt, L.: Robust reconstruction of watertight 3d models from non-uniformly sampled point clouds without normal information. In: Proceedings of the Fourth Eurographics Symposium on Geometry Processing, SGP 2006, pp. 41–50. Eurographics Association, Aire-la-Ville (2006)

11. Jacobson, A., Kavan, L., Sorkine-Hornung, O.: Robust inside-outside segmentation using generalized winding numbers. ACM Trans. Graph. **32**, 33:1–33:12 (2013)

12. Fuhrmann, S., Goesele, M.: Floating scale surface reconstruction. ACM Trans. Graph. **33**, 46:1–46:11 (2014)

13. Xiao, W., Vallet, B., Paparoditis, N.: Change detection in 3d point clouds acquired by a mobile mapping system. ISPRS Ann. Photogram. Remote Sens. Spat. Inf. Sci. **II**(5/W2), 331–337 (2013)

14. Guennebaud, G., Gross, M.: Algebraic point set surfaces. In: ACM SIGGRApPH 2007 Papers, SIGGRAPH 2007. ACM, New York (2007)

15. Demantké, J., Mallet, C., David, N., Vallet, B.: Dimensionality based scale selection in 3d LIDAR point clouds. ISPRS - Int. Arch. Photogrammetry Remote Sens. Spat. Inform. Sci. **3812**, 97–102 (2011)

16. Boykov, Y., Veksler, O., Zabih, R.: Fast approximate energy minimization via graph cuts. IEEE Trans. Pattern Anal. Mach. Intell. **23**, 1222–1239 (2001)

17. Boros, E., Hammer, P.L.: Pseudo-boolean optimization. Discrete Appl. Math. **123**, 155–225 (2002)

18. Boykov, Y., Kolmogorov, V.: An experimental comparison of min-cut/max-flow algorithms for energy minimization in vision. IEEE Trans. Pattern Anal. Mach. Intell. **26**, 1124–1137 (2004)

19. Kazhdan, M.: Reconstruction of solid models from oriented point sets. In: Proceedings of the Third Eurographics Symposium on Geometry Processing, SGP 2005. Eurographics Association, Aire-la-Ville (2005)

20. Ohtake, Y., Belyaev, A., Seidel, H.P.: An integrating approach to meshing scattered point data. In: Proceedings of the 2005 ACM Symposium on Solid and Physical Modeling, SPM 2005, pp. 61–69. ACM, New York (2005)

Color Correction for Image-Based Modeling in the Large

Tianwei Shen, Jinglu Wang, Tian Fang$^{(\boxtimes)}$, Siyu Zhu, and Long Quan

Department of Computer Science and Engineering,
Hong Kong University of Science and Technology, Kowloon, Hong Kong
{tshenaa,jwangae,tianft,szhu,quan}@cse.ust.hk

Abstract. Current texture creation methods for image-based modeling suffer from color discontinuity issues due to drastically varying conditions of illumination, exposure and time during the image capturing process. This paper proposes a novel system that generates consistent textures for triangular meshes. The key to our system is a color correction framework for large-scale unordered image collections. We model the problem as a graph-structured optimization over the overlapping regions of image pairs. After reconstructing the mesh of the scene, we accurately calculate matched image regions by re-projecting images onto the mesh. Then the image collection is robustly adjusted using a non-linear least square solver over color histograms in an unsupervised fashion. Finally, a connectivity-preserving edge pruning method is introduced to accelerate the color correction process. This system is evaluated with crowdsourcing image collections containing medium-sized scenes and city-scale urban datasets. To the best of our knowledge, this system is the first consistent texturing system for image-based modeling that is capable of handling thousands of input images.

1 Introduction

The past few decades have witnessed the significant achievement of 3D reconstruction. With the help of unmanned aerial vehicles and mobile devices, we can recover the 3D structures of city-scale scenes from images with ease. Moreover, one of the advantages for image-based modeling is that the images can not only be used to recover the 3D information, but also to do texture mapping for surfaces so that the models are natural and photo-realistic. In the multi-view reconstruction setting, a set of images are captured so that they depict roughly the same object from different view points. As camera parameters, such as white balance and shutter speed, are re-calculated for each image, the same object would appear differently in images with different view angles. If the original images are used in texture mapping, the mesh may contain undesired visual artifacts, which harms the visual experience of image-based modeling (Fig. 1).

Electronic supplementary material The online version of this chapter (doi:10.1007/978-3-319-54190-7_24) contains supplementary material, which is available to authorized users.

S.-H. Lai et al. (Eds.): ACCV 2016, Part IV, LNCS 10114, pp. 392–407, 2017.
DOI: 10.1007/978-3-319-54190-7_24

(a)   (b)   (c)

Fig. 1. Color correction for image-based modeling. (a) Color region correspondences generated by the reconstructed model. The top picture shows the registered cameras and the reconstructed mesh model. The colored regions are projected from a view pair at the bottom (using different colors to represent different views). The yellow region is the common region seen by the view pair, which provides the correspondence between the two views for color correction. (b) The resulting textured model using our color corrected images. (c) Textured model using the original images. It is evident that our method can generate much more harmonious and visually agreeable textured model.

Therefore, we need an efficient pre-processing step to fit the unordered image collection in a harmonious color tone. The objective is to use this set of corrected images to render a consistent and harmonious 3D model, in terms of fewer visual artifacts. Thus, we would like this color correction process to be based on the geometric relations and the 3D scenes within the images. As an existing technique, *color balancing* (or sometimes addressed as *color transfer* by the computer graphics community) deals with transferring the color palette of the source image to the target image, thus restricting the correction to an image pair. We would like to delve deeper by applying color correction on unordered image collections.

Color discontinuity also causes troubles for panoramic stitching and many works [1–3] have been proposed to address this issue. However, these techniques can not be directly applied to the 3D stitching problem due to the following reasons. First, panoramic stitches are usually conducted in a linear fashion. Once this linear order is decided, color adjustment for an image involves only the neighboring images, which is not the case for unordered image collections that form the graph structure with loops. Second, the images for panoramic stitching are often captured roughly around a single axis of rotation, thus the color discrepancy is less drastic, compared with multi-view stereo which involves more complicated camera motions, particularly for scenes containing non-Lambertian objects.

In this paper, we propose a method to address the color inconsistency problem in the image-based 3D reconstruction. Our method globally adjusts the color of original images based on histograms of color distributions in the overlapping regions. To get a precise overlapping region, we back-project the mesh to the images. Then the luminance component and the two chrominance components are separately balanced in a graph-structured optimization framework. We use the adjusted images for texturing and further processing.

Our method does not require a reference view, which differs from previous works such as [4]. There are generally two reasons that this property is preferred. First, we usually do not know the ground-truth reference view before doing the color adjustment. Second, if several images depict the same natural scene, we consider illumination variance in different images as noise and would like the color correction process to manifest the authentic color.

2 Related Work

2.1 2-View Color Transfer

Many previous works have addressed the problem of adjusting the color tone of a source image to a target image, commonly known as color transfer. These various color transfer techniques can be generally categorized using two criterions, namely *parametric* and *non-parametric* with regard to *global* and *local*. For large-scale unordered image collections, color transfer within an image pair serves as the basic building block which can be fitted in a large-scale optimization framework. To avoid computational burden for 3D reconstruction, a global model-based method is preferred as the basic unit in color correction of image collections. Global color correction ensures that the changes made to the local regions does not create visual artifacts to the textured model, while parametric methods facilitates the combination of color transfer units.

Reinhard et al. [5] first propose a popular color transfer method based on simple statistics of color distributions in images. To get rid of the undesired correlations between different channels in RBG, their method operates in the $l\alpha\beta$ color space, which has little correlation between different axes for many natural scenes [6]. Tian et al. [7] apply color correction to panoramic imaging, by matching the histograms of the overlapping region of two images. This method is still limited to an image pair thus should be categorized to 2-view methods. Recently, Hwang et al. [8] proposed a non-linear color transfer method based on matched correspondences. Each RGB color is adjusted to the target value by an affine transformation, using a process called moving least squares. Nguyen et al. [9] propose a new method which takes into consideration the scene illumination and color gamut. After running white-balancing on both the source and target images, they transform the luminance values of the target image using Xiao et al.'s gradient preserving matching technique [10]. For a detail comparison of several methods mentioned here, readers may refer to the quantitative evaluation work of Xu et al. [11].

2.2 Ordered N-View Color Correction

Nanda and Cutler's work [12] discusses several important issues for real-time panoramas generated by omnidirectional cameras, including auto brightness correction, auto white balance correction, vignetting correction, etc. Brown et al. [1]

propose the first automatic panoramic image stitching pipeline. To get a consistent color in the panoramic image, they simply tweak the gain (a linear factor) of intensity mean and apply multi-band blending [13] to further process visible seams. Xiong et al. [3] proposes a color correction method for sequences of overlapping images, which is in particular useful for panorama stitching. To deal with pixel overflow problem (pixels may be saturated during the color correction processing), they use YCbCr color space and separately apply gamma correction for the luminance component and linear correction for the chrominance components. However, this method is restricted to sequences of images and is not suitable in the large-scale unordered settings. The latter problem is much harder than tasks such as panorama stitching since there are much more constraints between pairs of images in a graphical style, and the color consistency is with respect to the whole 3D scenes, in different view points. Yamamoto and Oi's work [14] describe a color correction method for multi-view video embedded in an energy minimization scheme. This modeless non-parametric approach requires a manual decision on a reference view and a sequential order of neighboring cameras.

2.3   Other Approaches

To generate a mesh with a consistent color tone, other methods are also extensively used. Moulon et al. [4] propose a global method based on histogram quantiles of overlapping regions. This method uses VLD filter [15] to extend the corresponding region of an image pair. While in our method, we take into consideration the mesh to generate accurate corresponding regions, which particularly benefits the consistent texture mapping. Waechter et al. [16] propose a texturing system which corrects the mesh color on a per-triangle basis. Allène et al. [17] employ a multi-band texture blending technique to adjust the color of overlapping texture regions. However, texture blending has its own limitations and is not able to cope with the drastic changes in different lighting conditions, which has been come to realize by recent literature [11]. Though we do not perform multi-band texture blending explicitly, our approach is orthogonal to it and in practice we can combine these two approaches to deliver the best result.

Perhaps the most similar approach to our method is [18]. They also exploit shared color properties over an image collection. While in our work, we use a more precise way to generate correspondences through mesh re-projection. Our method uniquely combines image collection editing with consistent mesh texturing. Meanwhile, we are able to handle large-scale image collections containing thousands of images.

3   Our Method

3.1   Overview

Our method builds atop the 3D reconstruction pipeline, which is assumed readily available since our goal is to generate high-quality triangular meshes with

Fig. 2. The workflow of our method. Our method builds upon the 3D reconstruction pipeline, which consists of *Structure-from-Motion*, *dense reconstruction* and *surface reconstruction*. After *Joint optimization*, we can obtain the corrected and harmonious image collection. The color corrected image collection is further used for consistent texturing.

no visual seams. Here we give a brief overview of the state-of-the-art image-based modeling pipeline: given an unordered image collection $\mathcal{I} = \{I_i\}$ taken under different exposure and illumination conditions, camera poses and a sparse 3D point cloud is reconstructed using Structure-from-Motion(SfM) [19,20]. The scene geometry is then reconstructed using state-of-the-art multi-view stereo techniques [21,22]. The dense stereo points are further processed and triangulated to render a high-quality triangular mesh [23]. Here the image collection $\mathcal{I} = \{I_i\}$ is registered in the same coordinate as the mesh, with all the geometric relations such as camera poses $\mathcal{P} = \{P_i\}$ already known, thus enabling the interaction of original images and textures. The color correction method is based on the statistics of histograms thus it is robust to outliers in camera parameters and the mesh.

For all image pairs with enough overlapping, we compute the common corresponding regions by back-projecting the mesh $\mathcal{M}$, which contains a set of triangular faces $\mathcal{F} = \{f_i\}$, to render two masks for a pair of overlapped images. Previous methods [5,14,18] without the access to geometry information usually generate this color correspondences using full frames or SIFT features. For mesh texturing, our method enlarges the corresponding region and well satisfies the purpose of texture editing. These corresponding regions impose constraints on the camera network in a graphical representation. These constraints form a joint optimization which aims to align the histograms of the common regions in an image pair. We use a parametric model for the color adjustment of each image and apply the color correction for each channel in YCbCr color space. The workflow of the system is shown in Fig. 2. In the following sections, we describe different components of our system in details.

3.2   Color Region Correspondences

For each image I_i that is successfully registered into the scene geometry, there is a set of triangles $\mathcal{T}_i = \{f_1^{(i)}, f_2^{(i)}, \ldots, f_{n_i}^{(i)}\}$ that can be seen by this camera. Normally a face f_i is seen by several cameras, which is the origin of visual seems since the images used for texturing are captured under non-uniform conditions. Rather than tackling this problem in the texture space on the basis of per triangle, which indirectly solves the problem but is computationally intensive, we would like to pre-process the images to set them in a uniform color tone. We begin by re-projecting the mesh to cameras and obtain the visible triangle set

$\mathcal{T}_i$. To determine the visibility of each triangle, we first render a depth map by re-projecting each triangle to the view. Each pixel value of the depth map saves the id of the nearest triangle seen by view. If a pair of images I_i and I_j share enough visibility overlapping, which is measured by the cardinality of $\mathcal{T}_i \cap \mathcal{T}_j$, the color of the common regions in the two images should be roughly the same. Suppose we denote the projection from 3D space to image I_i as Π_i, we can represent the common regions by a pair of 0–1 binary masks:

$$M_{ij} = \sum_{t \in \mathcal{T}_i \cap \mathcal{T}_j} \Pi_i(t), \quad M_{ji} = \sum_{t \in \mathcal{T}_i \cap \mathcal{T}_j} \Pi_j(t) \tag{1}$$

The two masks select the accurate corresponding regions which should possess the same color distribution. To robustly match the two regions without violating the smoothness of image, we propose to measure the discrepancy of the color histograms, denoted as $D(H(I_i * M_{ij}), H(I_j * M_{ji}))$. Here $H(I_i * M_{ij})$ is the histogram of the color distribution in image I_i selected by mask M_{ij}. To facilitate the implementation and make this method scalable to large-scale datasets, we adopt statistical measures of the color histogram, instead of matching image color pixel by pixel. Namely, we define $D(H(I_i * M_{ij}), H(I_j * M_{ji}))$ to be the sum of pixel values that correspond to the same quantiles in the cumulative distribution of color histograms.

3.3 Global Optimization of Color Distribution

We use a global transformation model to parameterize the adjustment of the color histogram. Particularly, we solve the following optimization problem

$$\underset{\{s_i\}, \{o_i\}}{\text{minimize}} \quad \sum_{i,j,k} \rho\left(\frac{(s_i Q_{ij}^{(k)} + o_i) - (s_j Q_{ji}^{(k)} + o_j)}{s_i + s_j}\right)^2 \tag{2}$$

$$\text{subject to} \quad 1 - \delta_s \le s_i \le 1 + \delta_s, -\delta_o \le o_i \le \delta_o, \forall i.$$

where $Q_{ij}^{(k)}$ is the value of one color channel which corresponds to the k-th quantile of the cumulative distribution function for the overlapping region. Simply minimizing the numerator of the objective function, $(s_i Q_{ij}^{(k)} + o_i) - (s_j Q_{ji}^{(k)} + o_j)$, would result in a set of trivial solution in which the scale factors $\{s_i\}$ are all zeros and offset factors $\{o_i\}$ are unbounded. Therefore, we cancel out this shrinkage effect by normalizing the absolute error by the scale factors $s_i + s_j$ We also set lower bounds and upper bounds for scale and offset parameters, which ensures that the corrected color does not deviate too much from the original color. Equation 2 is a form of non-linear lest square problem with bounded constraints, where ρ is the loss function used to deal with outliers. Specifically, we use the Pseudo-Huber loss which writes as

$$\rho(x) = \delta^2(\sqrt{1 + (x/\delta)^2} - 1) \tag{3}$$

Equation 3 (with $\delta = 1$) is a smooth approximation of Huber loss function which is extensively used in robust estimation. The smooth approximation version ensures the derivatives are continuous for all degrees. The optimization 2 can be solved using Levenberg-Marquardt algorithm [24,25].

Image captured under the same scene usually possess stable chromatic characteristics, while differ in luminance. On the other hand, images are often stored in RGB space, which does not separate luminance component from chromatic ones. Moreover, there are correlations between different channels of RGB color space [8,26]. Arbitrary modification to RGB channels independently would lead to out-of-gamut error. Hence we apply the color correction in YCbCr color space, in which the luminance component is separated out and the primary colors are possessed into perceptually meaningful information. The transformation from analog RGB to YCbCr can be expressed by a linear mapping:

$$\begin{bmatrix} Y \\ Cb \\ Cr \end{bmatrix} = \begin{bmatrix} 0.299 & 0.587 & 0.114 \\ -0.168736 & -0.331264 & 0.5 \\ 0.5 & -0.418688 & -0.081312 \end{bmatrix} \begin{bmatrix} R \\ G \\ B \end{bmatrix} + \begin{bmatrix} 0 \\ 128 \\ 128 \end{bmatrix} \quad (4)$$

The optimization (2) is applied to three YCbCr channels independently. We have also observed that in natural scene images, the two chrominance components usually span a limited spectrum while the luminance Y channel is distributed over the whole range. Therefore, after the global color adjustment for the whole image collection, some images may be subjected to underflow or oversaturation. We remap the luminance range using a set of linear factors with respect to all the images. Suppose that $Q_\alpha'^{(i)}$ denotes the α-percentile of the luminance histogram of the i-th image, which may be above 255 or under 0, the 5-percentile and 95-percentile luminance values are mapped to the lower limit and the upper limit correspondingly.

$$\begin{aligned} 0 &= S_g \min_i \{Q_{0.05}'^{(i)}\} + O_g \\ 255 &= S_g \max_i \{Q_{0.95}'^{(i)}\} + O_g \end{aligned} \quad (5)$$

In the end the luminance value is adjusted by the combination of these two transformations

$$l' = S_g(s_i l + o_i) + O_g \quad (6)$$

The rescaling induced by S_g changes the absolute color value difference in a homogenous way, hence it does not affect the uniformness of the optimization result. The YCbCr value are then converted back to RGB color space, which completes the color correction in the image space.

3.4   Graph Compression

As the number of images increases, color balancing becomes a complicated optimization problem on a large-scale graph structure, which is hard to solve in a reasonable time. We apply two methods to tackle the efficiency issue. First, we use the input camera graph given by SfM, thus restricting the problem size to be the same as the complexity of the scene. Second, a connectivity-preserving edge pruning algorithm is applied on the scene graph to further speedup the optimization process.

Graph simplification by edge pruning. The goal of graph simplification is two-folded: First, we would like to prune unimportant and redundant edge links so as to accelerate the color correction process while preserving the general structure of the scene. Second, we would like to rearrange the edge distribution such that the color tone does not lean toward the densely-connected and over-sampled regions.

To avoid defeating the purpose of accelerating color correction process, we design a simple connectivity-preserving edge pruning (CPEP) method similar to the spirit of [27]. The input is the scene graph $G = (\mathcal{V}, \mathcal{E})$ computed after the mask generation. Each element in $v \in \mathcal{V}$ corresponds to an image I_i in the unordered image collection $\mathcal{I}$. We define the overlapping ratio between a pair of images as $OR_{ij}(I_i) = \frac{|\{\Delta|\Delta \in I_i \cap \Delta \in I_j\}|}{|\{\Delta|\Delta \in I_i\}|}$ and $OR_{ij}(I_j) = \frac{|\{\Delta|\Delta \in I_i \cap \Delta \in I_j\}|}{|\{\Delta|\Delta \in I_j\}|}$, where $|\{\Delta|\Delta \in I_i\}|$ and $|\{\Delta|\Delta \in I_j\}|$, $|\{\Delta|\Delta \in I_i \cap \Delta \in I_j\}|$ represent respectively the number of mesh triangles seen by image I_i, by image I_j and by both I_i and I_j. Two nodes are connected if the edge weight e_{ij}, defined as the overlapping ratio $OR_{ij} = \sqrt{OR_{ij}(I_i) \cdot OR_{ij}(I_j)}$ between a pair of images, is greater than a threshold δ_{or}, for which we set as 0.25 throughout the experiments. This algorithm first sorts edges by weights and then simply prunes the weakest edges iteratively while ensuring the connectivity of the whole graph (Algorithm 1). It is assumed that the input scene graph is connected, otherwise we can first extract the largest connected component of G and then apply the graph simplification on the largest connected component.

In the extreme case, the simplest graph that preserves global connectivity is its spanning tree with $|V| - 1$ edges. A simplification ratio γ_{sim} controls the fraction of edges to be pruned. CPEP algorithm prunes $N_{rm} = \gamma_{\text{sim}}(|\mathcal{E}| - (|\mathcal{V}| - 1))$ edges. We set γ_{sim} to be 0.9, thus keeping one tenth of original edges to lower the complexity of the graph by an order of magnitude. The primary computational cost lies in testing whether the removal of an edge can disconnect the graph (lines 6–9 in Algorithm 1). This operation takes $O(|\mathcal{V}|)$ time since we can compute the number of connected components using depth-first-search on the graph with the testing edge removed. Together with the cost of sorting edges ($O(|\mathcal{E}| \log |\mathcal{E}|)$), the total computational complexity of the algorithm is $O(|\mathcal{E}| \log |\mathcal{E}| + N_{rm}|\mathcal{V}|) = O(|\mathcal{E}|(\log |\mathcal{E}| + |\mathcal{V}|))$.

4   Graph Streaming on Large-Scale Datasets

The computation of mask pairs can be conducted in either streaming or parallel fashions. In both cases, I/O is the major bottleneck since the back-projection operation involve reading image pairs and marking common regions. To speedup this process it is ideal to read a batch of images that covers a possibly large

---

Algorithm 1. Connectivity-Preserving Edge Pruning (CPEP)

Require: The connected undirected scene graph $G = (\mathcal{V}, \mathcal{E})$, simplification ratio γ_{sim}
Ensure: A connected subgraph $G' = (\mathcal{V}, \mathcal{E}')$ of G
1: Sort $\mathcal{E}$ by edge weights in ascending order
2: $N_{\text{rm}} \leftarrow \gamma_{\text{sim}}(|\mathcal{E}| - (|\mathcal{V}| - 1))$
3: $\mathcal{E}' \leftarrow \mathcal{E}$
4: $i \leftarrow 0, j \leftarrow 0$
5: **while** $i < N_{\text{rm}}$ **do**
6:      {test whether e_j is a bridge}
7:      $\mathcal{E}' \leftarrow \mathcal{E}' \setminus e_j$
8:      Compute the number of connected components of G'
9:      **if** G' has more than one connected component **then**
10:         $\mathcal{E}' \leftarrow \mathcal{E}' \cup e_j$
11:     **else**
12:         $i \leftarrow i + 1$
13:     **end if**
14:     $j \leftarrow j + 1$
15: **end while**
        return $G' = (\mathcal{V}, \mathcal{E}')$

---

number of jobs that share the common image resources. Therefore, we design a batch processing paradigm to accelerate the mask pair computation.

Since each mask computation job depends on a pair of images, these jobs and down-sampled image resources constitute a bipartite graph which we called *Job-Resource-Depend-Graph (JRDG)*. For each batch iteration, maximum number of jobs and their dependent resources are loaded up to the memory limit. The order of loaded jobs can be randomly sequential or based on heuristics from JRDG. For example, we can load the images in the decreasing order of the degree of resource nodes. The pairwise mask computation is conducted after a batch loading and the memory for image resources is released for the next iteration. The histogram correspondence data for one iteration is streaming to the hard disk in order to hold as many images as possible in the memory.

When the problem scale becomes large, removing unimportant edges by checking connectivity is cumbersome and computationally expensive. To strike the balance between efficiency and perfect consistency, we take the simplest of Algorithm 1 for large-scale datasets (>10k images), in which γ_{sim} equals 1. The benefit is that we can just extract the maximum spanning tree of the scene graph without going through every weak edges, which greatly accelerates the graph simplification process. To solve large-scale color optimization in a reasonable time, we also made the approximation by using the conjugate gradients solver on the normal equations involved in the Levenberg-Marquardt algorithm. This results in an inexact step variant [28] of the Levenberg-Marquardt algorithm.

5   Experiments

5.1   Implementation Details

In this section, we describe the implementation details as well as specific model parameters. We focus on the details of the color correction engine and omit techniques involving 3D reconstruction. For the later part readers may refer to a series of literatures and open-source implementations, such as [29,30] for Structure-from-motion, [31,32] for dense reconstruction and TexRecon [16] for texturing.

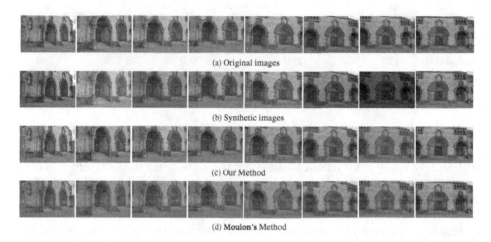

(a) Original images

(b) Synthetic images

(c) Our Method

(d) Moulon's Method

Fig. 3. Color correction result for a synthetic dataset with intense color changes. (a) the original image sequence; (b) the synthetic image sequence, modified by tuning color balance in Adobe Photoshop; (c) the result of our automatic color correction pipeline; (d) Moulon's method [4] using the unchanged image as the reference view (Color figure online).

We solve Eq. 2 using Ceres solver [33], which implements Levenberg-Marquardt algorithm [24,25] to solve the non-linear least-square problem with bounded constraints. We choose a set of different bounded constraints for the Y channel and for Cb, Cr channels, since the luminance component usually span the whole value range while the two chrominance components are concentrated on a narrow scope. We set $\delta_s = 0.4, \delta_o = 30$ for the luminance channel and $\delta_s = 0.2, \delta_o = 5$ for the chrominance channels. The number of quantiles k is fixed to be 10 across all the experiments.

5.2   Comparison Results

Small-scale benchmark dataset. We first demonstrate the color correction performance of our method on a synthetic dataset. To show that the proposed

Table 1. Computation time for different stages. Statistics shown from left to right: dataset name, number of images, number of mask pairs, number of mask pairs after graph simplification, time of computing the correspondence masks, time of optimization on the simplified graph, time of optimization on the full graph, number of iterations for loading the image collection (Sect. 4). Small-scale experiments (less than 1000 images) were running on a multi-core PC with Intel(R) Core(TM) i7-4770K processors and 32 GB main memory, while large large-scale experiments were running on a server with Intel(R) Xeon(R) E5-2630 v3 CPU and 128 GB main memory.

Dataset	#images	#mask pairs	#sim pairs	mask computation time (s)	graph sim opt time (s)	graph full opt time (s)	#batch
tunnel	370	14523	1755	95.9	48.7	372.0	1
hotel	237	11320	1345	102.4	20.8	159.5	1
monasterio	110	927	190	11.9	0.9	3.2	1
castle	200	9243	1104	173.4	27.0	196.1	1
*city**	36480	4655k	36479	6.4 hours	25	N/A	2

method can work well under drastic color tone and illumination variations, we deliberately apply color balance adjustments for the *Herz* dataset, which is consist of 8 images and widely is used in multi-view stereo benchmarks [34]. The first six images in *Herz* are each set to the extreme of one color balance axis. The color tones of them are each dominated by red, yellow, blue, green, cyan, magenta. The seventh image is modified by changing the color contrast and the last image is fixed unchanged. We show original images, modified images and the color correction results in Fig. 3. We compare our method with Moulon et al. [4], with the unchanged image as the reference view. The performance on this small-scale dataset is comparable though their method needs a pre-specified reference view. When applying on medium-sized datasets containing hundreds of images (less than 300), however, it took hours for [4] to converge because of its L-∞ formulation, while our method took less than three minutes.

Texturing performance. While the drastic color change in the above case is rarely seen in the real world, we further evaluate our method on several natural scene datasets, namely *tunnel*, *monasterio*, *castle*, and *hotel*. To the best of our knowledge, there is no previous work that demonstrates the result of color correction for an image collection on the level of mesh texturing, thus we only compare our consistent texturing results with the commonly-used texturing pipeline. Namely, the texturing pipeline can be viewed as a triangle labelling process that considers fragment quality and color discrepancy between bordering images in a MRF-based energy function [17,35]. No texture blending technique is used for all the experiments, since we would like to evaluate the consistency of the corrected image collection.

Figure 4 illustrates the work flow of our method on the *tunnel* dataset. This dataset contains 370 images taken under drastic illumination changes. The input images are first robustly registered into a common coordinate frame using SfM. Then the multi-view stereo techniques reconstruct the high-quality surface model. If we use original images to texture the model, it greatly harms the realism of the scene and yields annoying seams. Although texture blending and photo-consistency checking may partially solve this problem, these techniques are often computationally expensive and perform poorly if the input datasets are not balanced in the image space in the first place. Our method uses the corrected images for texturing and achieves a smooth and uniform color tone without visual artifacts. The color correction process finishes in less than three minutes for small-sized datasets in the aforementioned settings. The computation

(a) Input images   (b) Structure-from-Motion   (c) Dense reconstruction   (d) Surface reconstruction   (e) Color corrected texture mapping

Fig. 4. The work flow of our method shown by the *tunnel* dataset.

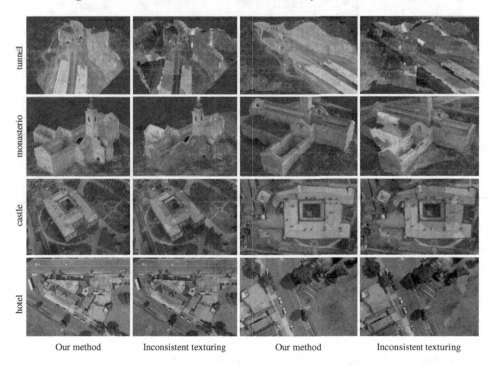

Our method   Inconsistent texturing   Our method   Inconsistent texturing

Fig. 5. Comparison of our consistent texturing and inconsistent texturing for small datasets. From top to bottom: *tunnel, monasterio, castle, hotel*.

time of different stages is showed in Table 1. The graph simplification operation greatly accelerates the optimization solver without introducing much overhead.

The second row of Fig. 5 shows the comparison result of the *monasterio* dataset which contains 110 images captured in different days and different hours. The visual seams disappear when the corrected images are used for texturing. Other comparison results for small-scale datasets are also shown in Fig. 5.

We demonstrate the performance of our method on large-scale image collections using the *city* dataset, which contains 36480 aerial images under illumination variations and occlusions. Figure 6 shows results of our method on the large-scale *city* dataset. The main computation is spent on the mask generation because of the heavy image I/O burden. Much simplification (spanning tree, inexact step) is employed on the solver side to make the computation feasible, as described in Sect. 4. Though a less accurate solver is applied on this large-scale problem, the addition of color correction improved the texture uniformness

Fig. 6. Large-scale color correction on urban dataset. From top to bottom: normal texturing without color correction; refined texturing using color corrected images; two pairs of detailed texture comparison.

significantly. For more high-resolution comparison results, readers may refer to the supplementary materials.

6   Conclusions and Discussions

In this paper, we present a global method to harmonize the color of an image collection using the scene geometry. Since this method relies on the mesh reprojection onto the original images, it is particularly useful for generating high-quality textured meshes without visual seams. In addition, our method works smoothly on large-scale datasets and imposes low computational burden for the 3D reconstruction pipeline. Moreover, our method can also be used to elevate the user experience for the display and exploration of large-scale image collections, such as the Photosynth system [29].

We have implicitly assumed that the intrinsic color of the overlapping region is view-independent, therefore for scenes with non-Lambertian objects the accuracy of our method will be affected. In fact, the modeling of non-Lambertian objects poses challenges for other aspects of multi-view reconstruction as well. Avenues for future work include incorporating other techniques such as highlight removal [36] to tackle the issues with reflective objects.

Acknowledgement. The authors would like to thank all the anonymous reviewers for their constructive feedbacks. This work is supported by Hong Kong RGC 16208614, T22-603/15N, Hong Kong ITC PSKL12EG02, and China 973 program, 2012CB316300.

References

1. Brown, M., Lowe, D.G.: Automatic panoramic image stitching using invariant features. Int. J. Comput. Vis. **74**, 59–73 (2007)
2. Eden, A., Uyttendaele, M., Szeliski, R.: Seamless image stitching of scenes with large motions and exposure differences. In: Computer Vision and Pattern Recognition (CVPR), pp. 2498–2505 (2006)
3. Xiong, Y., Pulli, K.: Color matching of image sequences with combined gamma and linear corrections. In: International Conference on ACM Multimedia, pp. 261–270 (2010)
4. Moulon, P., Duisit, B., Monasse, P.: Global multiple-view color consistency. In: Conference on Visual Media Production (CVMP) (2013)
5. Reinhard, E., Ashikhmin, M., Gooch, B., Shirley, P.: Color transfer between images. IEEE Comput. Graph. Appl. **21**, 34–41 (2001)
6. Ruderman, D.L., Cronin, T.W., Chiao, C.C.: Statistics of cone responses to natural images: implications for visual coding. JOSA A **15**, 2036–2045 (1998)
7. Tian, G.Y., Gledhill, D., Taylor, D., Clarke, D.: Colour correction for panoramic imaging. In: International Conference on Information Visualisation, pp. 483–488 (2002)
8. Hwang, Y., Lee, J.Y., Kweon, I.S., Kim, S.J.: Color transfer using probabilistic moving least squares. In: Computer Vision and Pattern Recognition (CVPR), pp. 3342–3349 (2014)

9. Nguyen, R., Kim, S., Brown, M.: Illuminant aware gamut-based color transfer. Comput. Graph. Forum **7**, 319–328 (2014)

10. Xiao, X., Ma, L.: Gradient-preserving color transfer. Comput. Graph. Forum **7**, 1879–1886 (2009)

11. Xu, W., Mulligan, J.: Performance evaluation of color correction approaches for automatic multi-view image and video stitching. In: Computer Vision and Pattern Recognition (CVPR), pp. 263–270 (2010)

12. Nanda, H., Cutler, R.: Practical calibrations for a real-time digital omnidirectional camera. CVPR Technical Sketch (2001)

13. Burt, P.J., Adelson, E.H.: A multiresolution spline with application to image mosaics. ACM Trans. Graph. (TOG) **2**, 217–236 (1983)

14. Yamamoto, K., Oi, R.: Color correction for multi-view video using energy minimization of view networks. Int. J. Autom. Comput. **5**, 234–245 (2008)

15. Liu, Z., Marlet, R.: Virtual line descriptor and semi-local matching method for reliable feature correspondence. In: British Machine Vision Conference (BMVC) (2012)

16. Waechter, M., Moehrle, N., Goesele, M.: Let there be color! large-scale texturing of 3D reconstructions. In: Fleet, D., Pajdla, T., Schiele, B., Tuytelaars, T. (eds.) ECCV 2014. LNCS, vol. 8693, pp. 836–850. Springer, Heidelberg (2014)

17. Allène, C., Pons, J.P., Keriven, R.: Seamless image-based texture atlases using multi-band blending. In: International Conference on Pattern Recognition (ICPR), pp. 1–4 (2008)

18. HaCohen, Y., Shechtman, E., Goldman, D.B., Lischinski, D.: Optimizing color consistency in photo collections. ACM Trans. Graph. (TOG) **32**, 38 (2013)

19. Agarwal, S., Snavely, N., Simon, I., Seitz, S.M., Szeliski, R.: Building Rome in a day. In: International Conference on Computer Vision (ICCV), pp. 72–79 (2009)

20. Shen, T., Zhu, S., Fang, T., Zhang, R., Quan, L.: Graph-based consistent matching for structure-from-motion. In: Leibe, B., Matas, J., Sebe, N., Welling, M. (eds.) ECCV 2016. LNCS, vol. 9907, pp. 139–155. Springer, Heidelberg (2016)

21. Goesele, M., Snavely, N., Curless, B., Hoppe, H., Seitz, S.M.: Multi-view stereo for community photo collections. In: International Conference on Computer Vision (ICCV), pp. 1–8 (2007)

22. Furukawa, Y., Curless, B., Seitz, S.M., Szeliski, R.: Towards internet-scale multi-view stereo. In: Computer Vision and Pattern Recognition (CVPR), pp. 1434–1441 (2010)

23. Kazhdan, M., Bolitho, M., Hoppe, H.: Poisson surface reconstruction. In: Eurographics Symposium on Geometry Processing (2006)

24. Levenberg, K.: A method for the solution of certain non-linear problems in least squares (1944)

25. Marquardt, D.W.: An algorithm for least-squares estimation of nonlinear parameters. J. Soc. Ind. Appl. Math. **11**, 431–441 (1963)

26. Zhang, M., Georganas, N.D.: Fast color correction using principal regions mapping in different color spaces. Real-Time Imaging **10**, 23–30 (2004)

27. Zhou, F., Mahler, S., Toivonen, H.: Simplification of networks by edge pruning. In: Berthold, M.R. (ed.) Bisociative Knowledge Discovery. LNCS (LNAI), vol. 7250, pp. 179–198. Springer, Heidelberg (2012). doi:10.1007/978-3-642-31830-6_13

28. Wright, S., Holt, J.N.: An inexact Levenberg-Marquardt method for large sparse nonlinear least squares. J. Aust. Math. Soc. Ser. B. Appl. Math. **26**, 387–403 (1985)

29. Snavely, N., Seitz, S.M., Szeliski, R.: Modeling the world from internet photo collections. Int. J. Comput. Vis. **80**, 189–210 (2008)

30. Moulon, P., Monasse, P., Marlet, R.: Global fusion of relative motions for robust, accurate and scalable structure from motion. In: International Conference on Computer Vision (ICCV), pp. 3248–3255 (2013)
31. Furukawa, Y., Ponce, J.: Accurate, dense, and robust multiview stereopsis. Pattern Anal. Mach. Intell. (PAMI) **32**, 1362–1376 (2010)
32. Lhuillier, M., Quan, L.: A quasi-dense approach to surface reconstruction from uncalibrated images. Pattern Anal. Mach. Intell. (PAMI) **27**, 418–433 (2005)
33. Agarwal, S., Mierle, K., Others: Ceres solver. (http://ceres-solver.org)
34. Strecha, C., von Hansen, W., Gool, L.V., Fua, P., Thoennessen, U.: On benchmarking camera calibration and multi-view stereo for high resolution imagery. In: Computer Vision and Pattern Recognition (CVPR), pp. 1–8 (2008)
35. Lempitsky, V., Ivanov, D.: Seamless mosaicing of image-based texture maps. In: Computer Vision and Pattern Recognition (CVPR), pp. 1–6 (2007)
36. Tan, P., Lin, S., Quan, L., Shum, H.Y.: Highlight removal by illumination-constrained inpainting. In: International Conference on Computer Vision (ICCV), pp. 164–169 (2003)

Bringing 3D Models Together: Mining Video Liaisons in Crowdsourced Reconstructions

Ke Wang[1(✉)], Enrique Dunn[2], Mikel Rodriguez[3], and Jan-Michael Frahm[1]

[1] Department of Computer Science, University of North Carolina, Chapel Hill, USA
{kewang,jmf}@cs.unc.edu
[2] Department of Computer Science, Stevens Institute of Technology, Hoboken, USA
edunn@stevens.edu
[3] Mitre Corporation, Mclean, USA
mdrodriguez@mitre.org

Abstract. The recent advances in large-scale scene modeling have enabled the automatic 3D reconstruction of landmark sites from crowdsourced photo collections. Here, we address the challenge of leveraging crowdsourced video collections to identify connecting visual observations that enable the alignment and subsequent aggregation, of disjoint 3D models. We denote these connecting image sequences as *video liaisons* and develop a data-driven framework for fully unsupervised extraction and exploitation. Towards this end, we represent video contents in terms of a histogram representation of iconic imagery contained within existing 3D models attained from a photo collection. We then use this representation to efficiently identify and prioritize the analysis of individual videos within a large-scale video collection, in an effort to determine camera motion trajectories connecting different landmarks. Results on crowdsourced data illustrate the efficiency and effectiveness of our proposed approach.

1 Introduction

Technical advances in imaging devices, digital storage, and network sharing make visual data capturing much easier and more available than ever before. For example, 300 hundred hours of videos are uploaded to YouTube every minute [17], while 1.8 billion digital images are uploaded to the Internet every single day [13]. Such crowdsourced visual data offers the potential for high redundancy in sampling at the expense of heterogeneous capture characteristics. The development of technologies for estimating 3D reconstructions from such large-scale visual media collections is an active research topic in computer vision [18,19]. Recent methods have striven to handle larger datasets [1,6], while improving model robustness and completeness [7]. However, the attained models are usually restricted to individual landmarks, as the recovery of geospatial adjacency

Electronic supplementary material The online version of this chapter (doi:10.1007/978-3-319-54190-7_25) contains supplementary material, which is available to authorized users.

S.-H. Lai et al. (Eds.): ACCV 2016, Part IV, LNCS 10114, pp. 408–423, 2017.
DOI: 10.1007/978-3-319-54190-7_25

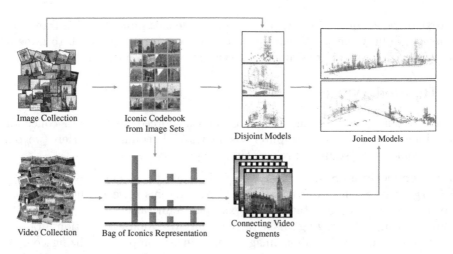

Fig. 1. Overview of our proposed algorithm.

among landmarks is still a largely unaddressed problem within crowdsourced scene modeling.

We identify two inherently interdependent challenges to the geospatial connectivity of 3D models attained from photo collections: (1) observational bias towards dominant scene elements, and (2) sampling insufficiency along intermediate structure elements between 3D models. Moreover, there are strong landmark-specific priors on both the camera spatial distribution and the viewing directions observable in a crowd-sourced photo collection. In other words, sampling tends to be highly redundant and convergent to a given landmark's most salient regions, but sampling density erodes towards the landmark's periphery. Such sampling bias causes images depicting scenes in-between modeled landmarks to appear much less frequently in crowdsourced photo collections, limiting the observational overlap required for multiview 3D reconstruction. In addition, in the absence of exhaustive pairwise connectivity analysis for an entire photo collection, under-sampled geospatial connectivities may be discarded during 3D modeling [11].

To get more complete models, auxiliary data sources are necessary to overcome the data deficiency. For example, many sight-seeing videos captured with wearable cameras and mobile phones, naturally record such missing connectivity information between landmarks. We call these geospatially connecting image sequences *video liaisons* and use them to join separate 3D models of landmarks through the alignment to the common linking camera motion trajectory. Given multiple existing landmark 3D models, our goal is to identify these video liaisons efficiently within a video collection and leverage them for inter-model 3D alignment and aggregation. An overview of the pipeline can be found in Fig. 1.

Our contributions include: (1) we introduce videos as auxiliary data sources to overcome the data deficiency problem in photo collections, thus achieving better completeness of large-scale 3D reconstructions; (2) we propose a geometric scene summarization framework based on iconic images and apply it in the context of video content analysis; (3) we leverage this representation framework to develop a fully automatic and unsupervised clustering approach to mine for observational connectivities among known 3D landmark models.

2   Related Work

Understanding and utilizing large-scale crowd-sourced visual data collections have long interested the computer vision research community. Here we only review some of the solutions relevant to our problem.

Photo collections. Starting from a few thousand images in Snavely *et al.* [18, 19], large-scale Structure-from-Motion (SfM) systems have reached milestones of processing city-scale datasets. For example, Agarwal *et al.* [1] processed 150 thousand images, leveraging image retrieval for overlapping prediction. Frahm *et al.* [6] reconstructed 3 million images on a single computer utilizing compact binary image representation for clustering. Recently, Heinly *et al.* [7] tackled a world-scale dataset by using a streaming paradigm to identify connected images in only one pass of the data.

For large-scale image collections, one of the core computational challenges and bottlenecks is efficient mining for element connectivities. We leverage techniques from state-of-the-art large-scale SfM pipelines to efficiently establish video relationships. Li *et al.* [10] introduced the concept of iconic images to model the relationship between different image clusters via iconic scene graphs. Frahm *et al.* [6] and Heinly *et al.* [7] further utilized the iconic representation for better scalability. Similarly, our method extends the concept of iconic images to represent video visual contents.

Video collections. As a dual concept to unstructured photo datasets, unordered Internet videos also exhibit sparsity and lack of structures in the dataset. In addition, the high temporal redundancy in video data makes video summarization techniques essential to achieve high scalability and throughput for real-world datasets. Video summarizations can also help humans to browse/skim large collections. Ajmal *et al.* [4] gives an anatomy of video summarization methods. With selected keyframes or shots, videos can be efficiently indexed and retrieved. For further details on video indexing and retrieval, we refer readers to Hu *et al.* [8].

To explore the structures and relationship in video collections, Tompkin *et al.* [20] proposed to identify common scenes connecting different video clips within a video collection. A connectivity graph is built with such "portals" as nodes. While Videoscapes graph is effective for interactive visualization and exploration, our work aims at creating representations for large-scale crowd-sourced video collections in a fully unsupervised manner.

Reconstructing the objects/cameras trajectories from videos or image sequences is also a challenging problem in computer vision. Zheng *et al.* [22] jointly estimates the topology of the objects motion path and reconstructs the 3D object points. Our work also needs to recover the camera trajectories of videos but is more focused on identifying such relevant video sub-sequences.

3   Methodology

We aim to further improve the completeness of 3D reconstructions from large-scale unordered Internet photo collections, by leveraging useful camera trajectories buried in massive crowd-sourced video datasets. We propose an automatic and unsupervised approach to mine such connecting video liaisons efficiently.

As shown in Fig. 1, the inputs to our algorithm include an image dataset and a video dataset. The two datasets are separately collected. Our algorithm starts by clustering and indexing the visual contents of the photo collection. Images are grouped into clusters based on visual similarity; a representative iconic image further represents each image cluster (see Sect. 3.1). Each iconic image represents some commonly captured visual structures or objects. The set of iconic images, when viewed in aggregation, forms a codebook or *"visual dictionary"* of the captured world. The contents of the video collection will be analyzed regarding their relationships to our attained codebook-based scene summarization. Our algorithm reduces the data volume of the video datasets, by selecting distinctive and representative keyframes from each video for further processing (see Sect. 3.2). Common visual elements between the image collection and the video collection can be found by matching the selected keyframes to the set of iconic images comprising the codebook. How frequently each iconic visual element occurs in a video, characterizes its visual content. To take this idea one step further, we represent videos as histograms of iconic image occurrences (see Sect. 3.3). Frequent co-occurrences of different visual elements (iconic images) in video datasets are strong indications of potential connections among the various image clusters/reconstructions. Such co-occurrence relationships are efficiently uncovered via clustering on video histogram representations (Sect. 3.4). Finally, we pick smoothly transited video sub-sequences (Sect. 3.5) to align separately reconstructed 3D models together (Sect. 3.6).

Fig. 2. Visualization of image clustering on London dataset (see Sect. 4.1). First row: iconic views for different connect components. From left to right: Big Ben, Westminster Abbey, London Eye, Buckingham Palace, Tower Bridge. Second row: selected images from one of the Big Ben image clusters. Images cropped for visualization purposes. Best view in color.

3.1   Codebook Extraction

Crowdsourced video and image datasets can have very different distributions and characteristics of visual contents. To effectively utilize video data to complete the 3D reconstruction models obtained from images, common visual elements (scenes, structures, objects, etc.) must be effectively identified between the two datasets. Iconic images, as used in Frahm *et al.* [6] and Heinly *et al.* [7] for understanding large-scale Internet photo collections, provide a compact yet informative summarization of the static image dataset. Thus we propose to use the iconic images as a common *basis* to represent the two different data modalities, and to uncover their visual connections.

Sift features [12] are extracted from each image. Each iconic image ic is represented by an augmented Bag-of-Visual-Word (BoVW) model [7], and indexed in a vocabulary tree [14] to allow fast retrieval. Given a previously unseen image I, a small set of visually similar iconic images (2 in our experiments) are retrieved using vocabulary trees. Geometric verification is performed between the new image I and every retrieved iconic image ic. Different clustering actions are taken based on registration results: (1) if the new image registers to only one iconic image, image I is assigned to that cluster; (2) if I registers to multiple iconic images, the corresponding clusters are grouped together as a connected component; (3) if the new image I fails to register to any retrieved iconic images, it will form its own new cluster with itself being the iconic image.

Such process needs great scalability and throughput to handle large-scale image datasets. Inspired by Heinly *et al.* [7], we adopted a streaming paradigm. However, streaming based image clustering approach has two minor flaws for extracting compact codebooks. Firstly, Heinly *et al.* [7] discards slowly-growing image clusters from memory to control resource consumption. Depending on the processing ordering of images, such early-discarding strategy can leave similar images in separate clusters. Different codebook elements representing the same visual content can cause confusion for mining video data. Secondly, the number of image clusters discovered from the image dataset is theoretically unbounded. Although this causes little trouble for the reconstruction problem in Heinly *et al.* [7], high dimensionality of the codebook significantly threatens the efficiency of searching large video datasets.

Thus we further regularize the codebook extraction process by running a second pass on the image data, and then thresholding on the image cluster sizes. Images are randomly shuffled into different orders before the second pass streaming process. Separated image clusters from the first pass due to ordering and discarding reasons, can be agglomerated together, thus reducing the ambiguities caused by duplicated entries in the codebook. In addition, we discard codebook iconic images with less than 200 clustered images. Examples of discovered iconic images and image clusters are shown in Fig. 2.

After the entire image collection is processed twice through the clustering pipeline with different orderings, all discovered iconic images satisfying the size constraint will together form the codebook $\mathcal{C} = \{ic_0, ic_1, \ldots, ic_m\}$.

Empirically, each cluster represents a more localized view of some certain entities, for example, buildings/landmarks/objects, while each connected component represents the ensemble of all different views of the landmarks. Like in [7], we perform densification and Structure-from-Motion on each connected components of images to get reconstructed 3D models.

3.2   Video Keyframe Selection

Compared with images, the additional temporal domain brings video with much more visual information, but also high temporal redundancy and vast data volume. Such high volume of redundant data, by itself, poses great challenges for processing, let alone mining for potential video segments linking separate reconstructions. To lessen the computation overhead, we propose to divide each video $\mathbf{v} = \{f | f \in \mathbf{v}\}$ into small non-overlapping segments $\mathbf{vs} \subseteq \mathbf{v}$ with each segment represented by one keyframe $kf \in \mathbf{vs} \subseteq \mathbf{v}$, thus achieving a balance between data redundancy and contents completeness.

We use a GPU-based KLT tracker [21] to select keyframes. Given a new video segment $\mathbf{vs}_i$ and its first frame as the corresponding keyframe kf_i, Shi-Tomasi corner points $\mathbf{x}_i$ [16] are detected within kf_i. At any given timestamp $t + 1$, the previous frame f^t, previous keypoints $\mathbf{x}^t$, and the next frame f^{t+1} are used in the KLT tracker to compute the tracked feature points $\mathbf{x}^{t+1}$. If tracking fails, or the ratio of tracked feature points $|\mathbf{x}^{t+1}|/|\mathbf{x}_i|$ falls below the pre-defined threshold 20%, the video frame f^{t+1} is selected as the new keyframe for the next video segment $\mathbf{vs}_{i+1}$. Shi-Tomasi corner points are re-detected for new keyframe kf_{i+1} and tracking is re-initialized.

We do the following processing to further increase the robustness of the keyframe selection process. Firstly, a global gain ratio β between successive frames f^t and f^{t+1} is estimated to compensate for the camera exposure changes. Given pairs of corresponding pixels $\mathbf{x}^t$ and $\mathbf{x}^{t+1}$ at successive timestamps t and $t + 1$, their pixel intensities are related by the multiplicative camera gain model: $f^{t+1}(\mathbf{x}^{t+1}) = \beta f^t(\mathbf{x}^t)$. Secondly, bogus feature tracks may be introduced for various reasons, for example, occlusions. Thus we only use feature point pairs that survive forward and backward tracking consistency check. Thirdly, watermarks on video borders can lead to constantly tracked feature points. But such feature points are not helpful to identify distinctive video keyframes. To suppress the influences of watermarks, we disable detection and tracking in the border regions of the frames. Examples of selected video frames are shown in Fig. 3.

Fig. 3. Examples of extracted keyframes. For visualization purposes, video frames are shown in grayscale and only subset of feature tracks are visualized in color.

3.3   Video Representation Extraction

With codebook $\mathcal{C}$ built from image collections and keyframes extracted from videos, we can build a global descriptor $H(\mathbf{v})$ for each video $\mathbf{v}$, by generalizing the Bag-of-Visual-Words concept. Elements from the codebook $\mathcal{C}$ (*i.e.* iconic images) are used as high-level "words" to which video keyframes kf can be assigned.

A video descriptor $H(\mathbf{v})$ is constructed by accumulating the normalized number of occurrence of each iconic view $ic \in \mathcal{C}$ within the video keyframe set into a histogram (see Eq. 1). In strict terms, occurrence means a valid geometric registration exists between an iconic image ic and a given keyframe kf.

$$H(\mathbf{v}) = [h(0), h(1), \dots, h(m)], \quad h(i) = \frac{\sum_{kf \in \mathbf{v}} GV(kf, ic_i)}{N(\mathbf{v})}, \quad ic_i \in \mathcal{C}. \quad (1)$$

where $GV(kf, ic)$ is an indicator function that returns 1 upon successful geometric verification between keyframe kf and iconic image ic, and 0 otherwise. $N(\mathbf{v})$ is the total number of registered keyframes w.r.t. iconic codebook $\mathcal{C}$.

Given the fact that codebook $\mathcal{C}$ size can be enormous for large-scale image collections, exhaustive geometric verification $GV(kf, ic)$ between every keyframe kf and iconic image ic pair is impractical. We choose to perform geometric verification for each keyframe kf only with the most visually similar iconic image retrieved using the indexed vocabulary tree (see Sect. 3.1).

With such histogram representations, the similarity between the visual content of two videos $\mathbf{v}_i$ and $\mathbf{v}_j$ can be computed as the sum of intersections between their histograms $H(\mathbf{v}_i)$ and $H(\mathbf{v}_j)$:

$$S(H(\mathbf{v}_i), H(\mathbf{v}_j)) = \sum_{k=0}^{m} \min(h_i(k), h_j(k)), \quad (2)$$

3.4   Video Representation Clustering

Intuitively, videos that capture the same set of landmarks will have similar peaks in their histogram representations, resulting in a high similarity score. If there exists such small groups of geospatially adjacent landmarks, videos depicting such landmarks would naturally form tight clusters in the feature space.

We use mean shift algorithm [5] to perform clustering on video representations to identify such landmark groups. Histogram intersection kernel (Eq. 2) is used as the weighting function. An empirical value of 0.1 is used as the clustering bandwidth d. As shown in Fig. 4, clustering videos in the descriptor space can successfully group them by geospatial proximity.

Having identified such video clusters, we need to uncover the underlying landmarks that brought these videos close to each other. By intuition, such landmarks correspond to common high peaks in the histogram representations. To suppress noise, we compute the average histogram $\tilde{H}$ of the descriptor cluster $\mathcal{H} = \{H(\mathbf{v}_1), \dots, H(\mathbf{v}_l)\}$ as:

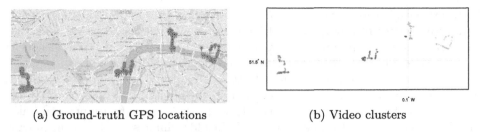

(a) Ground-truth GPS locations   (b) Video clusters

Fig. 4. Clustering results on Videoscapes video dataset (see Sect. 4.1). Codebook is extracted on the London Flickr image collection (see Sect. 4.1). Different video clusters are shown in different colors.

$$\tilde{H} = \left[\tilde{h}(0), \tilde{h}(1), \ldots, \tilde{h}(m)\right], \quad \tilde{h}(i) = \frac{\sum_{H \in \mathcal{H}} h_H(i)}{|\mathcal{H}|}. \tag{3}$$

The underlying landmarks correspond to a minimal subset of histogram bins $\{c | c \in \mathcal{C}_\mathcal{H} \subseteq \mathcal{C}\}$ that sum up to a pre-defined threshold $\sum_{c \in \mathcal{C}_\mathcal{H}} \tilde{h}(c) \geq \tau$. Without loss of generality, we sort the bins of average histogram $\tilde{H}$ into descending order H', where $h'(0) \geq h'(1) \geq \cdots \geq h'(m)$. Then we can select the minimal subset of bins $\mathcal{C}_\mathcal{H} = \{0, 1, \ldots, S\}$ such that $\sum_{i=0}^{S} h'(i) \geq \tau$, where $\tau = 0.70$.

3.5 Optimal Video Sequence Selection

To align separate landmark reconstruction models together, smooth and continuous camera trajectories are preferred. Since no temporal information is contained in the histogram based video descriptors, we need to look at the videos again to evaluate their trajectory smoothness and continuities.

Given a landmark group of interest $\mathcal{C}_\mathcal{H}$ and the corresponding video set $\{\mathbf{v}_H | H \in \mathcal{H}\}$, we first need to identify the valid subsequence for each video $\mathbf{v}_H$ that can connect the separately reconstructed 3D models. The valid subsequence of the longest consecutive sequence $Path(\mathbf{v}) = \{vs_i, vs_{i+1}, vs_{i+2}, \ldots, vs_{i+k}\}$ where the keyframes kf_i, kf_{i+k} of the ending video segments vs_i and vs_{i+k} have valid registrations with respect to the landmark set $\mathcal{C}_\mathcal{H}$. Notice that the image dataset usually has denser sampling on landmarks themselves, while the video

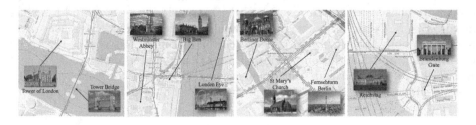

Fig. 5. Examples of identified landmark groups. Best view in color.

dataset can capture anything in between landmarks. Certain video segments may not register to any discovered landmark iconic images. Thus we relax the constraints on the in-between video segments $vs_{i+1}, \ldots, vs_{i+k-1}$ to model such registration failure due to the inherent data characteristic differences.

We sample the video sequence $Path(\mathbf{v})$ uniformly into a frame sequence $F(Path(\mathbf{v}) = [f_0, f_1, \ldots, f_M]$, to reconstruct the camera trajectory $Path(\mathbf{v})$ of the video $\mathbf{v}$. For the purpose of reconstructing the camera trajectory and linking landmarks, a good frame sequence F should exhibit smoothness in camera motion without much motion discontinuities. Thus we compute the geometric mean of the inlier ratio of the tracked features between frame pairs, as the smoothing function for each candidate frame sequences:

$$Score(F) = {}^{M+1}\sqrt{\Pi_{i=1}^{M} T(f_{i-1}, f_i)}, \tag{4}$$

where $T(f_{i_1}, f_i)$ is the ratio of tracked features between frame f_{i-1} and frame f_i, computed by the bi-directional KLT tracker as in Sect. 3.2. The KLT tracker is re-initialized for at frame f_i for each frame pair (f_i, f_{i+1}).

3.6   Model Reconstruction and Merging

One simple strategy of linking separate 3D models together is to run Structure-from-Motion together with all the images from the image components and selected video frames. Considering the potentially large number of images, such strategy can be computationally overwhelming. So we choose to reconstruct the camera trajectories by themselves and then merge multiple 3D models together.

An incremental Structure-from-motion (SfM) pipeline [15] is used to obtain 3D models on image connect components (Sect. 3.1) and video trajectories (Sect. 3.5) separately. Like in [7], we extracted camera intrinsics from the available EXIF data or assumed a horizontal 40 degrees field-of-view. As shown by [7], such heuristics work well for crowd-sourced photo collections. When building the video histogram (Eq. 1), fundamental matrices are estimated without camera intrinsics. In later SfM stages, bundle adjustment [2] is used to refine further the camera intrinsics obtained from the heuristics above.

Given a group of 3D landmark models $L_0, L_1, \ldots, L_n$ and a reference video trajectory model V, we need to estimate a similarity transformation between each L_i and V to align and merge the landmark models together. The geometric transformations needed to align the landmark models to the common camera trajectory consists of a rotation $\mathbf{R} \in \mathbb{R}^{3 \times 3}$, a translation $\mathbf{t} \in \mathbb{R}^3$, and a scaling factor $s \in \mathbb{R}$. Notice that keyframes within the selected video sequence can register to both the camera trajectory and the landmark model. Let $\mathbf{R}_i^L, \mathbf{t}_i^L$ be the rotation and translation of video frame f_i w.r.t. landmark model L, and R_i^V, t_i^V be its rotation and translation against video trajectory model V. The desired similarity transformation aligning the model L to video camera trajectory model V can be calculated as:

$$\mathbf{R} = \mathbf{R}_i^{V^T} \cdot \mathbf{R}_i^L, \quad s = \frac{\|\mathbf{c}_i^V - \mathbf{c}_j^V\|_2}{\|\mathbf{c}_i^L - \mathbf{c}_j^L\|_2}, \quad \mathbf{t} = \mathbf{c}_i^V - s\mathbf{R}\mathbf{c}_i^L. \tag{5}$$

where $\mathbf{c} \in \mathbb{R}^3$ is the camera location. Transformations obtained from multiple video frames are averaged and further optimized by bundle adjustment [2].

4 Experiments

4.1 Datasets

We evaluated the effectiveness of our proposed method with crowdsourced data. Two unordered Internet photo collections from Flickr covering different places (London and Berlin respectively) are obtained from the authors of [6] (See Table 1 for dataset statistics). Three video collections are used to align separate models. The Videoscapes dataset [20] is a manually collected video dataset, capturing major landmarks in London with ground-truth GPS trajectories. Another two sets of crowdsourced video collections for London and Berlin are obtained from YouTube by text and geo-location based queries within the "Travel&Events" video subcategory (See Table 2 for details).

4.2 Performance

We benchmark our proposed method, implemented in C++/Python, on a single computer, with three nVidia Tesla K20c GPUs, a 32-core 2 GHz Intel Xeon CPU, and 192 GB memory. Detailed timings can be found in Tables 1 and 3, respectively. To the best of our knowledge, processing such large-scale hybrid visual datasets on a single computer in a few days is unprecedented.

Compared with state-of-the-art [7], our throughput is lower for the following reasons: (1) our hardware platform has less computation capability, (2) we further process the dataset for an additional pass. In terms of registration ratio, we achieved 15% on Berlin video dataset and 13% on London video dataset, while

Table 1. Statistics on image collections (Sect. 4.1). Iconics are for clusters of size ≥ 200 (Sect. 3.1). SfM timings are reported on components with ≥ 400 images.

Dataset	Number of images			Time (hours)			
	Total	Registered	Iconics	Total	Stream	Densify	SfM
Berlin Flickr	2,661,327	865,699	37,544	25.92	18.46	1.89	5.57
London Flickr	12,036,991	3,716,916	103,290	131.67	90.75	7.09	33.83

Table 2. Statistics on two crowd-sourced video datasets. Video clusters with more than 50 videos are reported.

Dataset	Videos	Hours	Frames	Keyframes	Registered	Clusters
Berlin YouTube	17,480	2,068.41	223,388,274	4,244,377	636,689	4,135
London YouTube	19,217	2,195.96	245,586,526	5,648,490	734,303	4,937

Table 3. Processing time (in hours) of each stage of our proposed algorithm. SfM timing reported on top 30 video sub-sequences.

Dataset	Total	Keyframe	Histogram	Clustering	Ranking	SfM	Merging
Berlin YouTube	372.39	206.84	9.12	5.37	146.84	3.10	1.12
London YouTube	383.96	227.34	11.73	6.10	132.17	4.37	2.25

[7] registered 26% images on Berlin image dataset and 25% on London image dataset. The vast differences between two data modalities, and the sampling biases (Sect. 3.5) all contributed to lower the registration rate.

4.3   Discussions

Keyframe Selection. Tracking-based keyframe extraction is highly computationally efficient on GPUs. Our implementation of GPU-based KLT tracker operates at over 300 Hz per CPU thread. But as seen in Table 3, keyframe extraction takes the majority of the video processing time. We argue that high-quality video keyframe selection is critical for both controlling the dataset size and extracting meaningful representations for videos.

Firstly, notice in Table 2 the total number of raw frames exceed even the largest 100 million images dataset in [7]. Reducing videos to distinctive keyframes are necessary to make the entire pipeline practical with limited computation resources. Without video data reduction, later stages of descriptor extraction and clustering would suffer from intractable high volumes of data. Secondly, our tracking based keyframe extraction strategy achieves a good balance between computation efficiency and keyframe quality. Simple method, like uniform sampling, could have an adverse impact upon the video histogram representations. Uniform sampling in the temporal domain, though being extremely efficient can unnecessarily select redundant frames when the camera becomes stationary. It may also miss important scene contents when the camera undergoes fast movement. On the other hand, more complex methods like [3], selects better keyframes at the expenses of much more computations.

Histogram Clustering. Empirically, our method demonstrated successful model alignment for small groups of landmarks. One reason is that there are not many geo-spatially nearby landmarks. The farther away the landmarks are, the longer videos need to be to capture the entire trajectories. Such long and verbose videos are generally of less interest to tourists, thus are harder to find on the Internet. Another reason is the bandwidth parameter d used in the mean shift clustering algorithm. By using a relatively small bandwidth, we prefer more tightly distributed video clusters. Thus a larger video cluster containing multiple landmarks may be divided into separate smaller groups containing fewer landmarks. However, clustering with greater bandwidth d is more error-prone to noises in the video descriptors. Further exploration in needed on how to select the optimal bandwidth d for the purpose of grouping videos together.

In addition, we empirically notice many video clusters have a single major peak with significant magnitude. Such single mode descriptors correspond to videos that describe a single landmark. Such videos are not helpful for joining multiple 3D models, thus we discard any video groups whose largest histogram bin magnitude exceeds the threshold $h'(0) \geq \tau$. Examples of discovered landmark groups found in the London video dataset are visualized in Fig. 5.

4.4 Inter-Model Alignment Results

We present qualitative results in Figs. 6 and 7. All results in London are reported on the crowdsourced YouTube video dataset. To quantify model merging accuracy, we use registered StreetView (SV) images to our attained 3D model and leverage their associated GPS locations to obtain reference similarity transformations between separate landmark models. While many images used to reconstruct the 3D models contain geo-tags, streetview images have higher accuracy [9]. We sample the equirectangular streetview panorama from 12 uniformly distributed viewing angles to create perspective images. The resampled perspective views are then registered to the obtained 3D SfM models (from Sect. 3.6) to get ground-truth inter-model transformations.

For evaluation, we use one landmark model as the reference model. Similarity transformations (rotation $\mathbf{R}$, translation $\mathbf{t}$, and scaling s) of other landmarks models with respect to the reference model is computed similarly to Sect. 3.6. Per Table 4, our method yields adequate accuracy w.r.t. ground-truth transformations.

Table 4. Quantitative evaluations of model alignment. Euclidean distance in meters are reported for positional errors. Rotations are converted to axis-angle representation, and errors are reported as average angle differences in degrees. Relative errors in percentage are reported for scaling.

Evaluated model	London Eye	Westminster Abbey	Tower of London	Brandenburg Gate	AVG
Reference model	Big Ben	Big Ben	Tower Bridge	Reichastag	
Position error (m)	1.71	0.96	3.15	2.76	2.15
Orientation error (°)	6.94	5.46	4.38	8.34	6.28
Scaling error (%)	3.42	4.67	9.19	2.47	4.94

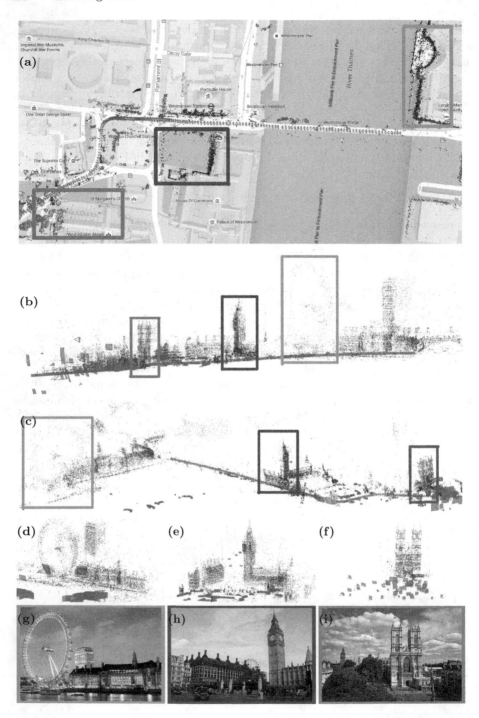

Fig. 6. Separate models, like London Eye (g), Big Ben (h), and Westminster Abbey (i) can be obtained from image collections. Our proposed method can find video segments that links these three models together, as shown in (a, b, c). Best view in color.

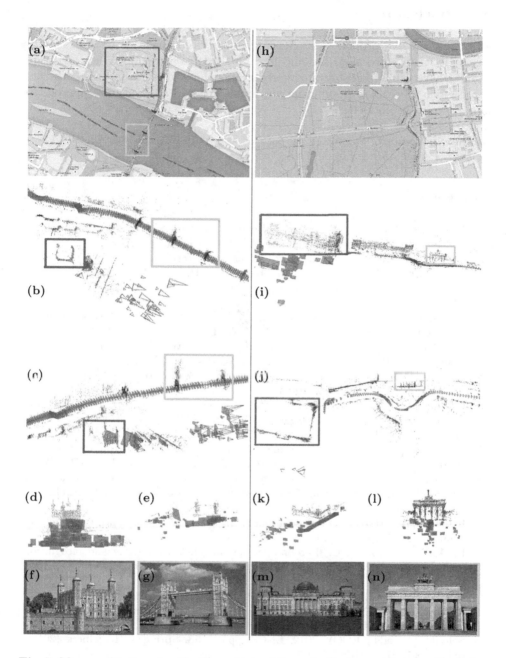

Fig. 7. More qualitative visualizations obtained from the Berlin and London YouTube dataset. Tower of London (d, f) and Tower Bridge (e, g) are aligned by the video trajectory (a) as shown in (b, c). Reichastag (k, m) and Brandenburg Gate (l, n) are aligned by video trajectory (h) as shown in (i, j). Best view in color.

5   Conclusions

To summarize, we proposed an efficient unsupervised approach to utilize auxiliary video dataset to align disjoint image 3D reconstructions. By leveraging an iconic imagery based histogram representation for videos, large-scale unstructured video collections are efficiently examined to find relevant video liaisons to join separate 3D models.

Acknowledgement. Supported in part by the NSF No. IIS-1349074, No. CNS-1405847. Partially funded by MITRE Corp.

References

1. Agarwal, S., Furukawa, Y., Snavely, N., Simon, I., Curless, B., Seitz, S.M., Szeliski, R.: Building rome in a day. Commun. ACM **54**, 105–112 (2011)
2. Agarwal, S., Mierle, K., et al.: Ceres solver. http://ceres-solver.org
3. Ahmed, M.T., Dailey, M.N., Landabaso, J.L., Herrero, N.: Robust key frame extraction for 3d reconstruction from video streams. In: VISAPP (1), pp. 231–236 (2010)
4. Ajmal, M., Ashraf, M.H., Shakir, M., Abbas, Y., Shah, F.A.: Video summarization: techniques and classification. In: Bolc, L., Tadeusiewicz, R., Chmielewski, L.J., Wojciechowski, K. (eds.) ICCVG 2012. LNCS, vol. 7594, pp. 1–13. Springer, Heidelberg (2012). doi:10.1007/978-3-642-33564-8_1
5. Comaniciu, D., Meer, P.: Mean shift: a robust approach toward feature space analysis. TPAMI **24**(5), 603–619 (2002)
6. Frahm, J.-M., et al.: Building rome on a cloudless day. In: Daniilidis, K., Maragos, P., Paragios, N. (eds.) ECCV 2010. LNCS, vol. 6314, pp. 368–381. Springer, Heidelberg (2010). doi:10.1007/978-3-642-15561-1_27
7. Heinly, J., Schonberger, J.L., Dunn, E., Frahm, J.M.: Reconstructing the world* in six days *(as captured by the yahoo 100 million image dataset). In: CVPR (2015)
8. Hu, W., Xie, N., Li, L., Zeng, X., Maybank, S.: A survey on visual content-based video indexing and retrieval. IEEE Trans. Syst. Man Cybern. Part C: Appl. Rev. **41**(6), 797–819 (2011)
9. Klingner, B., Martin, D., Roseborough, J.: Street view motion-from-structure-from-motion. In: ICCV (2013)
10. Li, X., Wu, C., Zach, C., Lazebnik, S., Frahm, J.-M.: Modeling and recognition of landmark image collections using iconic scene graphs. In: Forsyth, D., Torr, P., Zisserman, A. (eds.) ECCV 2008. LNCS, vol. 5302, pp. 427–440. Springer, Heidelberg (2008). doi:10.1007/978-3-540-88682-2_33
11. Lou, Y., Snavely, N., Gehrke, J.: MatchMiner: efficient spanning structure mining in large image collections. In: Fitzgibbon, A., Lazebnik, S., Perona, P., Sato, Y., Schmid, C. (eds.) ECCV 2012. LNCS, pp. 45–58. Springer, Heidelberg (2012). doi:10.1007/978-3-642-33709-3_4
12. Lowe, D.G.: Distinctive image features from scale-invariant keypoints. IJCV **60**, 91–110 (2004)
13. Meeker, M.: Internet trends (2016). http://www.kpcb.com/internet-trends
14. Nister, D., Stewenius, H.: Scalable recognition with a vocabulary tree. In: CVPR (2006)

15. Schonberger, J.L., Frahm, J.M.: Structure-from-motion revisited. In: CVPR (2016)
16. Shi, J., Tomasi, C.: Good features to track. In: CVPR (1994)
17. Smith, C.: By the numbers: 135 amazing youtube statistics. http://expandedramblings.com/index.php/youtube-statistics/
18. Snavely, N., Seitz, S.M., Szeliski, R.: Photo tourism: exploring photo collections in 3d. In: ACM TOG (2006)
19. Snavely, N., Seitz, S.M., Szeliski, R.: Modeling the world from internet photo collections. IJCV **80**, 189–210 (2008)
20. Tompkin, J., Kim, K.I., Kautz, J., Theobalt, C.: Videoscapes: exploring sparse, unstructured video collections. In: ACM TOG (2012)
21. Zach, C., Gallup, D., Frahm, J.M.: Fast gain-adaptive KLT tracking on the GPU. In: CVPR Workshops (2008)
22. Zheng, E., Wang, K., Dunn, E., Frahm, J.-M.: Joint object class sequencing and trajectory triangulation (JOST). In: Fleet, D., Pajdla, T., Schiele, B., Tuytelaars, T. (eds.) ECCV 2014. LNCS, vol. 8695, pp. 599–614. Springer, Cham (2014). doi:10.1007/978-3-319-10584-0_39

Planar Markerless Augmented Reality Using Online Orientation Estimation

Tatsuya Kobayashi[✉], Haruhisa Kato, and Masaru Sugano

KDDI R&D Laboratories Inc., Saitama, Japan
ts-kobayashi@kddilabs.jp

Abstract. This paper presents a fast and accurate online orientation estimation method that estimates the normal direction of an arbitrary planar target from small baseline images using efficient bundle adjustment. The estimated normal direction is used for planar metric rectification, and the rectified target images are registered as the recognition targets for markerless tracking on the fly. The conventional planar metric rectification methods estimate the normal direction in a very efficient way, and recently proposed depth map estimation methods estimate accurate normal direction. However, they suffer from either poor estimation accuracy for small baseline images or high computational cost, which degrades the usability for untrained end-users in terms of shooting procedure or waiting time for new target registration. In contrast, the proposed method achieves both high estimation accuracy and fast processing speed in order to improve the usability, by reducing the number of degrees of freedom in bundle adjustment for the conventional depth map estimation methods. We compare the proposed method with these conventional methods using both artificially generated keypoints and real camera sequences with small baseline motion. The experimental results show the high estimation accuracy of the proposed method relative to the conventional planar metric rectification methods and significantly greater speed compared to the conventional depth map estimation methods, without sacrificing estimation accuracy.

1 Introduction

The vision-based Augmented Reality (AR) technique, which recognizes predefined targets in a camera image and overlays artificial information on a display, is now widely used. Users can receive advertising and promotional information or task support information in an intuitive way through AR services [1,2]. When target frontal images or 3D models are available, camera pose relative to the target can be estimated through visual feature matching between registered targets and camera images, and direct augmentation can be achieved without using AR markers [3,4]. Specifically, when target objects are rigid planar (or approximately planar) objects or scenes, appearance change can be easily estimated by affine transformation or homography transformation. Therefore, multiple targets can be tracked simultaneously in real time on off-the-shelf mobile devices [5,6].

© Springer International Publishing AG 2017
S.-H. Lai et al. (Eds.): ACCV 2016, Part IV, LNCS 10114, pp. 424–439, 2017.
DOI: 10.1007/978-3-319-54190-7_26

On the other hand, in order to augment unknown targets, which are not registered into the database beforehand, 3D structure and camera pose have to be estimated from the camera images on site to add new targets into the database. In that case the Structure from Motion (SfM) technique, which estimates camera poses and reconstructs accurate 3D structure from a number of multiview images using bundle adjustment (BA), could be used [7]. Alternatively, the Simultaneous Localization and Mapping (SLAM) technique [8], which runs camera pose estimation and 3D reconstruction in parallel, can be applied to register targets into the database online [9]. However, since these methods require a large baseline among frames, users have to shoot targets from various viewpoints, which degrades usability for untrained end-users. In contrast, a depth map estimation approach that uses only a sequence of small baseline images to run BA and which reconstructs the 3D structure of the scene directly, on the assumption that users hold the camera still, has been proposed in recent years [10,11]. Although this approach achieves high usability in terms of camera motion, the number of Degrees of Freedom (DoF) in BA delays target registration. This is because depth values of tracked keypoints (1 DoF × number of keypoints) are estimated independently through BA even if the target is a planar object or scene. Consequently, users must hold the camera still for a considerable period.

Inspired by [10,11], we propose a novel SfM method which requires a sequence of small baseline motion. In order to achieve fast registration of planar targets, the proposed method reduces the number of DoF in BA by specializing in planar targets. It focuses on a geometric property that each depth of the keypoint on a planar target can be expressed by the orientation (normal direction) of the target surface, and estimates the normal direction (2 DoF) and camera poses (6 DoF × number of frames) directly from the trajectories of keypoints. We compare the proposed method with the conventional depth map estimation methods [10,11] and planar metric rectification methods [12,13] using both artificially generated keypoints and real camera sequences. Based on experiments, we show that the proposed method has high estimation accuracy compared to conventional planar metric rectification methods and the computational cost is significantly less than that of conventional depth map estimation methods.

In the remainder of this paper, related work on orientation estimation and markerless AR is discussed in Sect. 2. Then the proposed orientation estimation method, rectification method and markerless AR method are described in Sect. 3. We present our experimental results in Sect. 4. Section 5 presents our conclusion and future work.

2   Related Work

As with the proposed method, markerless AR approaches which register planar targets into the database on site are proposed [14–16]. On the assumption that the user shoots the targets from frontal viewpoints, orientation estimation and camera pose estimation are unnecessary and targets can be registered into the database directly [14,15]. Restricting planar targets are only on vertical or horizontal surfaces, the normal direction and camera poses can be estimated by using the accelerometers on the mobile devices, and target images can be rectified without using complex image processing [16]. Although these conventional methods are suitable for practical use in specific situations, usage is limited since the shooting procedure and targets are restricted.

There are several approaches which estimate the normal directions of planar targets in camera images by image processing, and rectify the target images (i.e. generating frontal shots of the targets by image processing) taken from arbitrary viewpoints. This eliminates the restriction in terms of the shooting procedure [12,17,18]. Utilizing geometric information such as parallel lines, orthogonal lines or exact circles in camera images [17] or machine learning [18] enables normal direction estimation from only a single image. However, the target category is restricted to artificial objects or scenes (e.g. the wall surfaces of buildings). Using multi-view images of target objects, arbitrary planar objects can be rectified. This is called planar metric rectification. Previously, planar metric rectification required computationally difficult non-linear optimization [19]. Ruiz et al. [12] propose a very efficient planar metric rectification method with non-linear optimization of 2 DoF from an image sequence by fixing inter-image homographies, and it has been demonstrated that it is possible to apply this rectification method to the planar SLAM technique [13]. However, this method requires large baselines between frames, and users have to shoot targets from various viewpoints during rectification.

In contrast, recent depth map estimation methods [10,11] can accurately estimate a 3D structure directly using BA from a sequence of small baseline images. The normal direction of a planar target can be estimated utilizing Principal Component Analysis (PCA) of the reconstructed 3D keypoints. However, a number of parameters in BA impose a high computational cost for the rectification, which degrades the usability of markerless AR applications. The proposed method requires a sequence of small baseline images as well as in [10,11], but estimates the global optimal normal directions of the planar targets efficiently using direct BA for planar metric rectification. That is, it has high usability in terms of both shooting procedure and the time required to register new targets in markerless AR applications. Lee et al. [20] also propose a fast SLAM technique with plane constraint. However, the efficiency comes from the application of local BA, which sacrifices reconstruction accuracy.

3   Proposed Method

The proposed method estimates the normal direction of the planar targets from a sequence of small baseline motion, and registers rectified target image into the markerless tracking database. After target registration, markerless tracking estimates camera poses of the registered targets in the subsequent frames and overlays 3D models onto the targets in real time. The proposed orientation estimation method is described in Sect. 3.1, and our planar metric rectification method using estimated normal direction is described in Sect. 3.2. A real-time

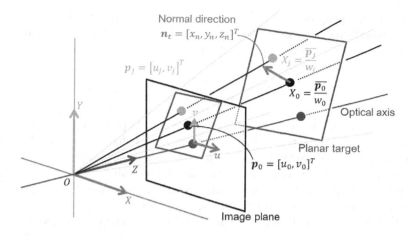

Fig. 1. Geometric relationship between a planar target and an image plane.

planar markerless tracking technique which estimates camera poses using rectified target images is described in Sect. 3.3.

3.1 Orientation Estimation

The proposed method estimates the normal direction $\mathbf{n}_t$ $[x_n, y_n, z_n]^T (|\mathbf{n}_t| = 1)$ of a planar target from N_c frames with small baseline motion. Figure 1 shows the geometric relationship between a planar target and an image plane. The 3D coordinates X_j of the keypoints $\mathbf{p}_j = [u_j, v_j]^T (j = 0, ..., N_p - 1)$ which are detected from a target object in an initial frame are expressed using the depth value $1/w_j$ of $\mathbf{p}_j$ as $X_j = \bar{\mathbf{p}}_j/w_j$ [10], where $\bar{\mathbf{p}}_j = [u_j, v_j, 1]^T$ shows homogeneous coordinates of $\mathbf{p}_j$. When the relative camera pose of i-th frame to the initial frame is $W_i = [R_i|\mathbf{t}_i](i = 1, ..., N_c - 1)$, X_j is projected onto $\mathbf{m}_{i,j} = \pi([R_i|\mathbf{t}_i]\bar{X}_j)$ in the i-th frame, where $\pi([x, y, z]^T) = [x/z, y/z]^T$ represents the projection function. On the assumption that the baseline motion among the frames is small, $R_i = E_3(=\mathrm{diag}(1,1,1))$ and $\mathbf{t}_i = [0,0,0]^T$ can be used as the initial values for the relative camera pose. As the initial value of the depth $1/w_j$ of each keypoint, random initialization [10] and constant initialization [11] are proposed. The conventional depth map estimation methods use these initial values, and estimate optimum parameters which minimize L_2-norm of the reprojection errors of X_j against keypoint trajectories $\mathbf{m}_{i,j}$ of $\mathbf{p}_j$ in a sequence utilizing BA, as shown in the following equation

$$\hat{R}_i, \hat{\mathbf{t}}_i, \hat{\mathbf{w}}_j = \underset{R_i, \mathbf{t}_i, \mathbf{w}_j}{\mathrm{argmin}} \sum_{i=1}^{N_c-1} \sum_{j=0}^{N_p-1} \|\mathbf{m}_{i,j} - \pi([R_i|\mathbf{t}_i]\bar{X}_j)\|^2. \qquad (1)$$

In contrast, the proposed method excludes the depth parameter w_j in X_j by using the equations of target plane $\mathbf{n}_t^T(X_j - X_0) = 0$ and 3D coordinate $X_j = \bar{\mathbf{p}}_j/w_j$ as shown in the following equation

$$X_j = \frac{1}{w_j}\bar{\mathbf{p}}_j = \frac{\mathbf{n}_t^T X_0}{\mathbf{n}_t^T \bar{\mathbf{p}}_j}\bar{\mathbf{p}}_j = \frac{\mathbf{n}_t^T \bar{\mathbf{p}}_0}{w_0 \mathbf{n}_t^T \bar{\mathbf{p}}_j}\bar{\mathbf{p}}_j. \tag{2}$$

where $X_0 = \bar{\mathbf{p}}_0/w_0$ is a control point on the target plane. $\mathbf{p}_0$ can be aligned with the optical axis as $\mathbf{p}_0 = [0,0]^T$, but we found that the centroid of $\mathbf{p}_j$ is a better choice when the target position in an image is distant from the image center. In that case, the effect of normalization helps stable optimization without falling into local optima. Since the absolute scale of the shooting distance cannot be estimated due to scale indeterminacy, w_0 is set to an arbitrary value (e.g. $w_0 = 1$). In the proposed method, the shooting distance is assumed to be the focal distance f of the camera, which is acquired through camera calibration beforehand, and w_0 is set to f. The initial value of the normal direction $\mathbf{n}_t$ is set to the direction parallel to the optical axis as $\mathbf{n}_z = [0,0,-1]^T$. Using the above formulation, direct BA of the proposed method can be expressed as the following equation

$$\hat{R}_i, \hat{\mathbf{t}}_i, \hat{\mathbf{n}}_t = \underset{R_i, \mathbf{t}_i, \mathbf{n}_t}{\mathrm{argmin}} \sum_{i=1}^{N_c-1} \sum_{j=0}^{N_p-1} \|\mathbf{m}_{i,j} - \pi([R_i|\mathbf{t}_i]\bar{X}_j)\|^2. \tag{3}$$

As with the conventional method [10], keypoints $\mathbf{p}_j$ in an initial frame are detected utilizing a Shi-Tomasi corner detector [21], and their trajectories $\mathbf{m}_{i,j}$ in the subsequent frames are tracked utilizing a Kanade-Lucas-Tomasi (KLT) feature tracker [22]. The number of DoF of the proposed BA process is $2+6(N_c-1)$, which is $N_p - 2$ less than that of the conventional depth map estimation ($N_p + 6(N_c - 1)$) shown in Eq. 1.

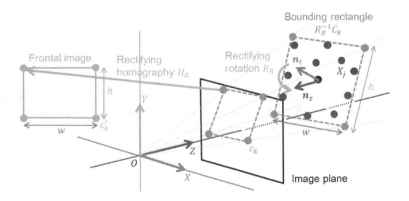

Fig. 2. Rectification of a planar target.

3.2   Planar Target Rectification

Using an estimated normal direction $\hat{\mathbf{n}}_t$, the frontal image of a target object can be generated from the initial frame where the target object is shot from an arbitrary viewpoint, removing perspective distortion. Figure 2 shows the flow of estimating homography transformation that transforms the initial frame into the frontal image of the target. First, we calculate the rectifying rotation vector $\mathbf{r}_R$, which transforms $\mathbf{n}_t$ into the Z-axis (optical axis) direction $\mathbf{n}_z$, as shown in the following equation

$$\mathbf{r}_R = \mathrm{acos}(\mathbf{n}_t^T \mathbf{n}_z) \frac{\mathbf{n}_t \times \mathbf{n}_z}{\|\mathbf{n}_t \times \mathbf{n}_z\|}. \tag{4}$$

The 3D coordinates X_j can be transformed in parallel to an image plane by rectifying the rotation matrix R_R of $\mathbf{r}_R$. The proposed method calculates a bounding rectangle of the rectified 3D coordinates $R_R X_j$, which is the target area in the rectified 3D coordinate system. The size (width w and height h) of the bounding rectangle specifies the size of the rectified image. The four vertices of the bounding rectangle $C_k (k = 0, ..., 3)$ are transformed into 3D coordinates $R_R^{-1} C_k$ on the target plane before rectification and projected onto an image plane using camera intrinsic parameter A. The projected vertices $c_k = \pi(A R_R^{-1} C_k)$ specify the target area in the initial frame.

Rectifying homography transformation matrix H_R, which transforms c_k into the four vertices of the frontal image whose size is $[w, h]$ as $(c_k' - \pi(H_R \bar{c}_k))$, is calculated using the Direct Linear Transformation (DLT) method. A rectified frontal image is generated by homography transformation of the initial frame.

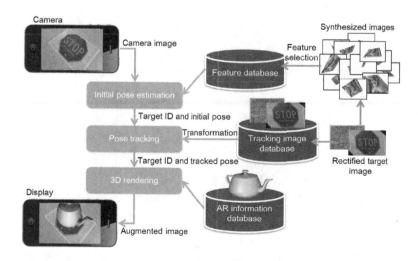

Fig. 3. Flow of markerless augmented reality processing.

3.3   Planar Markerless Augmented Reality

Visual features extracted from the rectified images of the planar targets are registered into the database of the markerless tracking technique, and real-time tracking and augmentation, which are robust to changes in viewpoint (camera motion) can be achieved. The proposed method applies a hybrid tracking framework [4] that combines an initial pose estimation (pose detection) algorithm and a consecutive pose tracking algorithm and is known as an efficient pose estimation approach.

Figure 3 shows the framework of the proposed markerless AR processing. The pose estimation method [23] is applied as the initial pose estimation method because of its robustness and registration speed. Before initial pose estimation succeeds and tracking starts, visual features (keypoints and descriptors) are detected from a camera image and matched with those in the database for each frame independently. The visual features in the database are detected from the synthesized images of the rectified target images and only robust features are selected and registered in order to improve matching performance. When sufficient feature matches are obtained, the camera pose is estimated geometrically and is used as the initial pose for the pose tracking algorithm.

The proposed method applies an efficient pose tracking method [24] that is suitable for mobile devices. The pose tracking predicts the current pose and the appearance of the registered targets from the poses estimated in previous frames, and tracks keypoints efficiently with a limited search range. The camera pose for each tracked target is updated independently. Augmentation of arbitrary planar objects can be achieved by overlaying AR information (e.g. 3D models) onto the registered targets with the estimated camera poses.

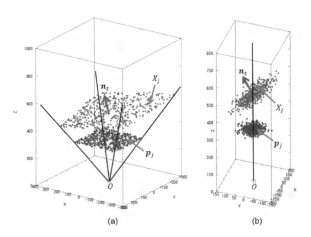

Fig. 4. Two types of the keypoint configuration $\mathbf{p}_j$ and the corresponding 3D points X_j. (a) uniform distribution. (b) Gaussian distribution.

4   Experimental Results

In our experiments, two types of datasets are used. The first dataset consists of artificial tracking points $\mathbf{m}_{i,j}$, which are generated by the projection of the randomly sampled 3D points on a virtual plane, the normal direction of which is randomly selected. The second dataset consists of real tracking points obtained from camera image sequences. A single planar target object is shown in the sequence. A Shi-Tomasi corner detector [21] and KLT feature tracker [22] are used to obtain the tracking points.

The conventional depth map estimation methods [10,11] and the planar metric rectification methods [12,13] are compared to the proposed method. We optimize the cost function of the conventional planar metric rectification and BA for both the conventional depth map estimation methods and the proposed method with Ceres Solver [25]. All the experiments are performed on a standard desktop PC with an Intel Core i7-5820K CPU.

4.1   Evaluation by Simulation

Initial keypoints $\mathbf{p}_j$ are randomly selected on an image plane. The size of the image plane is 640×480 pixels. The number of $\mathbf{p}_j$ is 500 ($N_p = 500$). 3D coordinates X_j are generated by the backprojection of $\mathbf{p}_j$ onto a target plane with normal direction of $\mathbf{n}_t$. Figure 4 shows an example of a generated set of $\mathbf{p}_j$ and X_j. Two types of keypoint configurations are evaluated: (a) uniform distribution and (b) Gaussian distribution. $\mathbf{n}_t$ is selected randomly within the range that the angular difference between $\mathbf{n}_t$ and the optical axis is less than $45°$. Relative camera poses $\mathbf{r}_i$ and $\mathbf{t}_i$ for 50 frames are generated independently. Each element of $\mathbf{r}_i$ and $\mathbf{t}_i$ is selected from the standard Gaussian distributions with zero means and small variances ($\sigma_r = 0.5°$ and $\sigma_t = 0.5\,\mathrm{cm}$, respectively). Input keypoints $\mathbf{m}_{i,j}$ are the projected points of X_j with the addition of Gaussian noise ($\sigma_m = 3$ pixels). One hundred sets of different keypoints $\mathbf{m}_{i,j}$ are generated by changing the configuration of $\mathbf{p}_j$ and $\mathbf{n}_t$. The average angle errors of the estimated normal directions and average processing times are evaluated. In the conventional depth map estimation methods [10,11], normal direction is estimated with PCA of the estimated 3D points $\hat{X}_j$. In the conventional planar metric rectification methods [12,13], normal direction is calculated directly from the estimated rectifying rotation matrix.

Figure 5(a), (b) shows the average angle errors and the processing times for the uniform distribution configuration dataset. When the number of frames N_c increases, angle errors reduce very markedly and start converging at $N_c = 10$ for all methods. In contrast, processing times keep increasing as N_c gets larger. The conventional planar metric rectification methods [12,13] are much faster than the other methods since the DoF of their cost function is constant (2 DoF) with N_c and much fewer than that of the proposed method (56 DoF at $N_c = 10$) and the conventional depth map estimation methods [10,11] (554 DoF at $N_c = 10$). However, they are less accurate than the other methods. Because they fix the inter-image homographies, they cannot optimize the relative pose

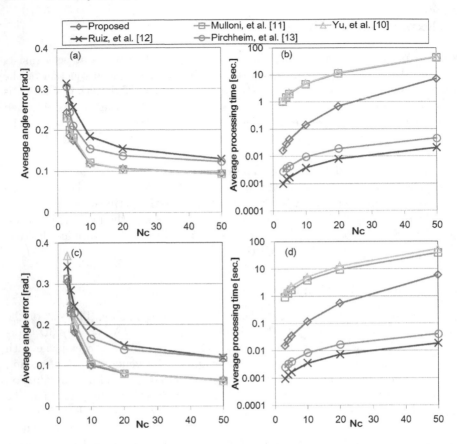

Fig. 5. (a), (b) The average angle errors and the average processing times for the dataset of a uniform distribution configuration. (c), (d) The average angle errors and the average processing times for the dataset of a Gaussian distribution configuration.

parameters. The proposed method and [10, 11] get about 30% better estimation accuracy than [12, 13]. High accuracy of the estimated orientation enables not only high rectification accuracy but also high robustness of markerless tracking against changes in viewpoint, which improves the usability of AR application. The accuracy of the proposed method and [10, 11] are close, but the proposed method is considerably faster than [10, 11] owing to the DoF reduction effect in BA. Although the improved speed becomes less with increasing N_c since N_c becomes dominant compared to N_p, the proposed method is still 25 times faster at $N_c = 10$, and 13 times faster at $N_c = 20$, respectively, compared to [10, 11].

Figure 5(c), (d) shows the result for the Gaussian distribution configuration dataset. All methods get similar results compared to the normal distribution configuration dataset. This result shows that the difference in the keypoint configuration does not have much effect on estimation accuracy.

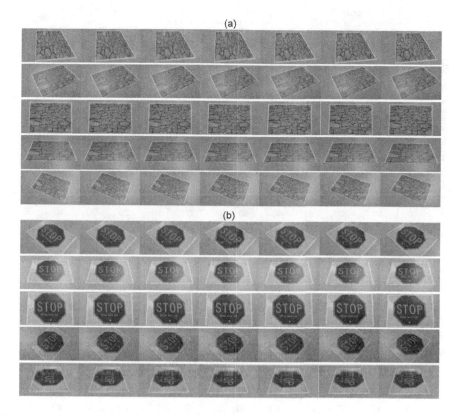

Fig. 6. Example images of real image sequences. (a) First seven frames of five sequences of image A. (b) First seven frames of image B.

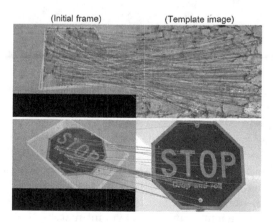

Fig. 7. Example images of ASIFT keypoints matched between the initial frames of the sequences and the template images, and pose estimation results.

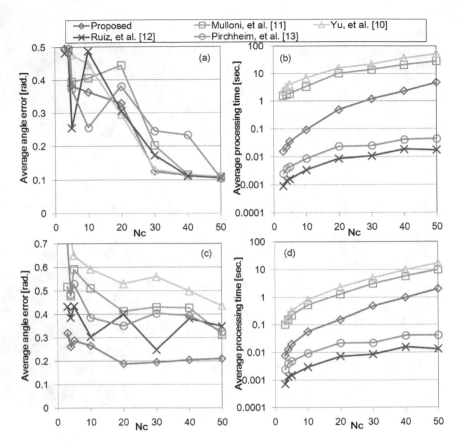

Fig. 8. (a), (b) The average angle errors and the average processing times for the dataset of image A. (c), (d) The average angle errors and the average processing times for the dataset of image B.

4.2 Evaluation on Real Sequence

For the planar target objects in the real camera sequence, two types of template images: A (Wall) and B (Stop) which were used in the evaluation in [26], are used. Each target image is captured from five different viewpoints with a smartphone camera (SONY Xperia™ Z4), and two sequences are created for each viewpoint. In total, 20 types of camera sequences are created. Figure 6(a) shows example images (first seven frames) of the image A dataset, and Fig. 6(b) shows the image B dataset. We hold the camera still in the hand while shooting, and Fig. 6 shows that only slight camera motion occurs. The frame size is 640×360 pixels, and the framerate is 30 FPS. In order to know the ground truth of the normal direction for each sequence, a sufficient number of keypoint matches between the template image and the initial frame of the sequence are obtained using ASIFT [27] as shown in Fig. 7. The relative camera pose, which minimizes the reprojection errors of the keypoints in the template image, is calculated. The outline of the template image is projected onto the initial frame with the calculated camera pose as shown in Fig. 7. The correctness of the camera pose is visually confirmed by checking that the projected outline agrees with the target object in the camera image. An accurate camera pose is estimated for the dataset of the less textured image B. The normal direction is directly calculated from the estimated camera pose, which is used as the ground truth in this experiment. We detect up to 500 keypoints from the initial frame of each sequence, and track the keypoints until the end of the sequence. Since the corner detector cannot detect maximum number of keypoints from less textured images and KLT sometimes fails to track some keypoints while tracking, N_p is equal to or less than 500 ($N_p \leq 500$). In fact, N_p is approximately 480 in the sequences of image A and is approximately 200 in the sequences of image B, respectively. The actual tracked points in some sequences are shown in Fig. 9.

Figure 8(a), (b) shows the average angle errors and the processing times for the dataset of 10 sequences of image A. Different from the simulation result, some sequences take about 30 frames to obtain stable normal direction, and the reduction in estimation error with the increase in N_c is slow. The difference is thought to be caused by the time correlation between the successive relative camera poses, which are completely independent in the simulation. In the case of the 30 FPS camera sequence containing little motion, the relative camera pose within successive frames might not be changed at all. Therefore, down-sampling of the input sequence may make the result closer to the simulation result. However, converged accuracy of the estimated normal directions and the processing times are close to those of the simulation result, which proves the effectiveness of the proposed method. The proposed method is 36 times faster at $N_c = 10$, and 20 times faster at $N_c = 20$, respectively, compared to the conventional depth map estimation methods [10,11]. The accuracy of the conventional planar metric rectification methods [12,13] is not consistent compared to the proposed method. Although they get better results at some values of N_c, the proposed method is more stable and more accurate in many cases.

Figure 8(c), (d) shows the results for the dataset of 10 sequences of image B. The result is markedly different from the simulation result and the result for image A. The converged accuracy of the proposed method is much better than the other methods [10–13]. There are no outliers in the artificial keypoints in the simulation and the tracked keypoints in the image A dataset. In contrast, since the template image of image B is not well-textured, a number of non-corner points, which are not suited for tracking, are detected as keypoints, and this results in many outliers in $\mathbf{m}_{i,j}$. For the sequence with outliers, the reconstruction errors of the outlier 3D points grow, which degrades the accuracy of the normal direction in [10,11]. The error of the estimated normal direction of the proposed method also becomes large in some sequences. However, the proposed method can estimate comparatively accurate normal directions in more sequences compared to [10–13]. On average, the proposed method gets 25–50% better estimation accuracy than [10–13] for these sequences. The average processing times decrease for the sequences of image B, compared to those of

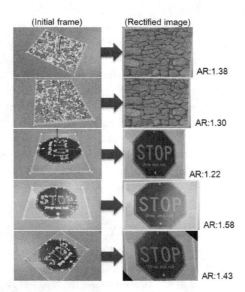

Fig. 9. The estimated normal directions for real camera sequences (red arrow: X-axis, green arrow: Y-axis, blue arrow: Z-axis) and the rectified images with the normal directions. The aspect ratio (AR) of the rectified image is shown beside the image. The correct aspect ratio of image A and image B is 1.4. (Color figure online)

image A. This is because N_p in the sequences of image B is less than that of image A due to the less-textured surface of image B. The computational costs of the proposed method and the planar metric rectification methods [12,13] are less affected by N_p than the depth map estimation methods [10,11] since their numbers of DoF do not include N_p. Although the improved speed is less compared to the sequences of image A, the proposed method is still much faster than [10,11], by 12 times where $N_c = 10$, and 8 times where $N_c = 20$.

4.3   Evaluation of Rectification Accuracy

Figure 9 shows an example rectification result of the initial frame with the estimated normal direction obtained by the proposed method. In terms of the dataset of image A, the normal direction of which is accurately estimated with an average angle error of about 0.1 rad, the rectified image is fairly similar to the original template image. In contrast, for the dataset of image B, the average angle error of which is about 0.2 rad, rectification is relatively inaccurate. Perspective distortion cannot be fully eliminated and the aspect ratio of the rectified image is not completely in accord. From the results for image B, the accuracy of the estimated normal direction against image B (average angle error of 0.2 rad) is not sufficient to generate an accurate rectified image. It is necessary for the accuracy to be improved so that it approaches the quality of image A (average angle error of 0.1 rad).

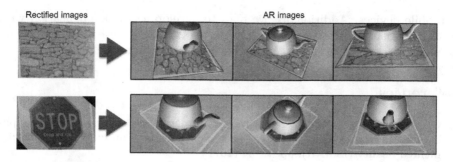

Fig. 10. Example markerless AR images with the rectified target images using the proposed method.

4.4 Evaluation of Markerless AR Accuracy

Figure 10 shows an example of markerless AR images using the proposed method. The rectified target image is registered into the markerless tracking database by the method described in Sect. 3.3, the markerless tracking method estimates the camera pose for the other sequences, the viewpoints of which are markedly different from that of the original image before rectification, and 3D teapot models are overlaid. Since registered image A is quite similar to the template image, an accurate camera pose can be estimated with the markerless tracking and the AR image is geometrically consistent. In contrast, in the case of image B, the rectification accuracy of which is lower than image A, although there are no obvious augmentation errors, the geometrical augmentation accuracy is not high. The accuracy of the camera pose estimation for image B is not so bad, but when the perspective distortion in an input image gets larger or other problems such as motion blur occur, tracking robustness degrades quickly. Improving robustness to outlier points will be the main focus of our future work.

5 Conclusion

In this paper, we proposed a fast and robust online orientation estimation method that estimates the normal direction of an arbitrary planar target shot from an arbitrary viewpoint using efficient direct BA. Experimental results show that the estimation accuracy of the proposed method is 30% better in the normal direction compared to the state-of-the-art planar metric rectification methods, and also 8–36 times faster than the conventional depth map estimation-based methods for both the artificially generated keypoints and the real camera sequences. Additionally, for the real camera sequences with many outlier tracking points, the accuracy of the proposed method is 25–50% better than all the conventional methods, which shows the robustness of the proposed method to outlier tracking points. Currently, the number of DoF of the proposed orientation estimation method increases as the number of processing frames. In the future work, we explore the efficient frame selection method to achieve further speed-up on mobile devices. Additionally, we plan to develop a mobile application with our online orientation estimation and markerless tracking methods, and evaluate the usability of the markerless AR application for untrained end-users.

References

1. Olsson, T., Salo, M.: Online user survey on current mobile augmented reality applications. In: Proceedings of IEEE International Symposium on Mixed and Augmented Reality (ISMAR), pp. 75–84 (2011)
2. Zhu, Z., Branzoi, V., Wolverton, M., Murray, G., Vitovitch, N., Yarnall, L., Acharya, G., Samarasekera, S., Kumar, R.: AR-mentor: augmented reality based mentoring system. In: Proceedings of IEEE International Symposium on Mixed and Augmented Reality (ISMAR), pp. 17–22 (2014)
3. Pauwels, K., Rubio, L., Diaz, J., Ros, E.: Real-time model-based rigid object pose estimation and tracking combining dense and sparse visual cues. In: Proceedings of IEEE Conference on Computer Vision and Pattern Recognition (CVPR), pp. 2347–2354 (2013)
4. Kyriazis, N., Argyros, A.: Scalable 3D tracking of multiple interacting objects. In: Proceedings of IEEE Conference on Computer Vision and Pattern Recognition (CVPR), pp. 3430–3437 (2014)
5. Wagner, D., Schmalstieg, D., Bischof, H.: Multiple target detection and tracking with guaranteed framerates on mobile phones. In: Proceedings of IEEE International Symposium on Mixed and Augmented Reality (ISMAR), pp. 57–64 (2009)
6. Kim, K., Lepetit, V., Woo, W.: Scalable real-time planar targets tracking for digilog books. Vis. Comput. **26**, 1145–1154 (2010)
7. Snavely, N., Seitz, S.M., Szeliski, R.: Modeling the world from internet photo collections. Int. J. Comput. Vis. **80**, 189–210 (2008)
8. Klein, G., Murray, D.: Parallel tracking and mapping for small AR workspaces. In: Proceedings of IEEE and ACM International Symposium on Mixed and Augmented Reality (ISMAR), pp. 225–234 (2007)
9. Kim, K., Lepetit, V., Woo, W.: Keyframe-based modeling and tracking of multiple 3D objects. In: Proceedings of IEEE International Symposium on Mixed and Augmented Reality (ISMAR), pp. 193–198 (2010)
10. Yu, F., Gallup, D.: 3D reconstruction from accidental motion. In: Proceedings of IEEE Conference on Computer Vision and Pattern Recognition (CVPR), pp. 3986–3993 (2014)
11. Mulloni, A., Ramachandran, M., Reitmayr, G., Wagner, D., Grasset, R., Diaz, S.: User friendly SLAM initialization. In: Proceedings of IEEE International Symposium on Mixed and Augmented Reality (ISMAR), pp. 153–162 (2013)
12. Ruiz, A., de Teruel, P.E.L., Fernandez, L.: Practical planar metric rectification. In: Chantler, M.J., Fisher, R.B., Trucco, E. (eds.) Proceedings of British Machine Vision Association (BMVC), pp. 579–588. British Machine Vision Association (2006)
13. Pirchheim, C., Reitmayr, G.: Homography-based planar mapping and tracking for mobile phones. In: Proceedings of IEEE International Symposium on Mixed and Augmented Reality (ISMAR), pp. 27–36 (2011)
14. Pilet, J., Saito, H.: Virtually augmenting hundreds of real pictures: an approach based on learning, retrieval, and tracking. In: Proceedings of IEEE International Conference on Virtual Reality (VR), pp. 71–78 (2010)

15. Kasahara, S., Heun, V., Lee, A.S., Ishii, H.: Second surface: multi-user spatial collaboration system based on augmented reality. In: Proceedings of SIGGRAPH Asia (SA 2012), pp. 20:1–20:4. ACM, New York (2012)
16. Lee, W., Park, Y., Lepetit, V., Woo, W.: Point-and-shoot for ubiquitous tagging on mobile phones. In: Proceedings of IEEE International Symposium on Mixed and Augmented Reality (ISMAR), pp. 57–64 (2010)
17. Lourakis, M.: Plane metric rectification from a single view of multiple coplanar circles. In: Proceedings of IEEE International Conference on Image Processing (ICIP), pp. 509–512 (2009)
18. Haines, O., Calway, A.: Recognising planes in a single image. IEEE Trans. Pattern Anal. Mach. Intell. **37**, 1849–1861 (2015)
19. Hartley, R.I., Zisserman, A.: Multiple View Geometry in Computer Vision, 2nd edn. Cambridge University Press, New York (2004). ISBN: 0521540518
20. Lee, G.H., Fraundorfer, F., Pollefeys, M.: MAV visual SLAM with plane constraint. In: IEEE International Conference on Robotics and Automation (ICRA), pp. 3139–3144 (2011)
21. Shi, J., Tomasi, C.: Good features to track. In: Proceedings of IEEE Computer Society Conference on Computer Vision and Pattern Recognition (CVPR), pp. 593–600 (1994)
22. Baker, S., Matthews, I.: Lucas-Kanade 20 years on: a unifying framework. Int. J. Comput. Vis. **56**, 221–255 (2004)
23. Kobayashi, T., Kato, H., Yanagihara, H.: Novel keypoint registration for fast and robust pose detection on mobile phones. In: Proceedings of IAPR Asian Conference on Pattern Recognition (ACPR), pp. 266–271 (2013)
24. Kobayashi, T., Kato, H., Yanagihara, H.: Pose tracking using motion state estimation for mobile augmented reality. In: Proceedings of IEEE International Conference on Consumer Electronics (ICCE), pp. 9 10 (2015)
25. Agarwal, S., Mierle, K., et al.: Ceres solver. http://ceres-solver.org
26. Lieberknecht, S., Benhimane, S., Meier, P., Navab, N.: A dataset and evaluation methodology for template-based tracking algorithms. In: Proceedings of IEEE International Symposium on Mixed and Augmented Reality (ISMAR), pp. 145–151 (2009)
27. Yu, G., Morel, J.M.: A fully affine invariant image comparison method. In: Proceedings of IEEE International Conference on Acoustics, Speech and Signal Processing (ICASSP), pp. 1597–1600 (2009)

Simultaneous Independent Image Display Technique on Multiple 3D Objects

Takuto Hirukawa[1], Marco Visentini-Scarzanella[1], Hiroshi Kawasaki[1], Ryo Furukawa[2(✉)], and Shinsaku Hiura[2]

[1] Computer Vision and Graphics Laboratory, Kagoshima University, Kagoshima, Japan
[2] Graduate School of Information Sciences, Hiroshima City University, Hiroshima, Japan
ryo-f@hiroshima-cu.ac.jp

Abstract. We propose a new system to visualize depth-dependent patterns and images on solid objects with complex geometry using multiple projectors. The system, despite consisting of conventional passive LCD projectors, is able to project different images and patterns depending on the spatial location of the object. The technique is based on the simple principle that multiple patterns projected from multiple projectors interfere constructively with each other when their patterns are projected on the same object. Previous techniques based on the same principle can only achieve (1) low resolution volume colorization or (2) high resolution images but only on a limited number of flat planes. In this paper, we discretize a 3D object into a number of 3D points so that high resolution images can be projected onto the complex shapes. We also propose a dynamic ranges expansion technique as well as an efficient optimization procedure based on epipolar constraints. Such technique can be used to the extend projection mapping to have spatial dependency, which is desirable for practical applications. We also demonstrate the system potential as a visual instructor for object placement and assembling. Experiments prove the effectiveness of our method.

1 Introduction

Recent improvements in projectors' resolution, brightness and cost effectiveness, extended their pervasiveness beyond flat screen projection, and to projection mapping, AR and/or MR systems [1]. In those systems, the precalculated scene geometry is used to project an appropriately warped image to be mapped on the scene as an artificial texture, with impressive visual results. However, it is difficult to layout with moving objects or dynamic scenes in general. This is because the projected pattern is the same along each ray, hence what is viewed is spatially invariant up to a projective transformation. Conversely, the potential for

Electronic supplementary material The online version of this chapter (doi:10.1007/978-3-319-54190-7_27) contains supplementary material, which is available to authorized users.

S.-H. Lai et al. (Eds.): ACCV 2016, Part IV, LNCS 10114, pp. 440–455, 2017.
DOI: 10.1007/978-3-319-54190-7_27

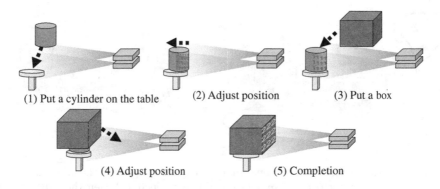

(1) Put a cylinder on the table   (2) Adjust position   (3) Put a box

(4) Adjust position   (5) Completion

Fig. 1. Basic scheme to align two objects at the right position by visual feedback using two projectors. Note that projected patterns are static.

practical applications could be significantly broadened if different patterns can be projected at different depths simultaneously; such volume displays are now intensely investigated [2–5]. However, those techniques are still under research and are still not suitable for practical systems. Recently, a simpler system consisting of just two projectors which can simultaneously project independent images on screens at different depths has been proposed [6]. The system demonstrated the ability to project simultaneous movies on multiple semi-transparent screens at different depths with no particular setup requirements, which is a promising research avenue to explore.

In this paper, we significantly extend the system [6] by (1) removing the planarity assumption: in this work, we are able to project depth-dependent patterns on complex surfaces with arbitrary geometry, (2) dynamic range expansion by adding a constraint in pattern optimization, and (3) introducing epipolar constraint to keep the problem realistic size. With our extension, the potential application is significantly broadened. For instance, we can use the system for object placement or assembling purposes as shown in Fig. 1: since the pattern is sharply visible only at predefined depths, human or robotic workers can precisely align complex parts using only qualitative visual information, without distance scanners. While the visibility of the pattern depends on its 3D position and orientation in space, translational alignment can also be achieved by including markers in the pattern. The new system also allows dynamic projection mapping on complex and semi-transparent objects. In the experiments, we show the effectiveness of the proposed technique, which is also demonstrated on multiple dynamic 3D solid objects.

2   Related Work

Applications of multiple video projectors have a long history in VR systems such as wide or surround screen projections like CAVE [7]. These types of projectors needs precise calibration between projectors, *i.e.*, the geometrical calibration to

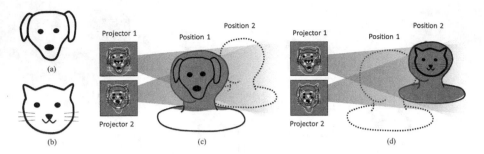

Fig. 2. Projection of virtual masks on dynamic subjects. To project the masks in images (a) and (b), specially generated images should be projected. When the subject is at position 1, only mask (a) is shown, whereas (b) is suppressed, as shown in (c). The situation is inverted in (d).

establish the correspondences between the projected images and the screens, and the photometric calibration to compensate for nonlinearities between the projected intensities and the pixel values. To this end, automated calibration techniques based on projector-camera feedback systems were developed [8,9]. Since some of the screens considered was curved, some of these works inevitably dealt with non-planar screens. In other works, multiple projectors were tiled together for improving the resolutions of the projected images [10,11]. Godin *et al.*, inspired by human vision system, proposed a dual-resolution display where the central part of the projection is projected in high-resolution, while the peripheral area is projected in low-resolution [12]. Other works on multi-projector systems focused on increasing the depth-of-field, since this is normally narrow and can cause defocusing issues on non-planar screens. Bimber and Emmerling [13] proposed to widen the depth-of-field by using multiple projectors with different focal planes. Nagase *et al.* [14] used an array of mirrors, which is equivalent to multiple projectors with different focal planes, for correcting defocus, occlusion and stretching artifacts. Levoy *et al.* also used an array of mirrors and a projector [15]. The array of mirrors was used to avoid occlusions from objects placed in front of the screen. Each of the aforementioned works is intended to project a single image onto a single display surface, which may or may not be planar. Conversely, the proposed method projects multiple, independent images onto surfaces placed at different depths, which may have a complex non-planar geometry.

Multiple projectors are also used for applications in light-field displays [3,4]. For these applications, in order to create the large number of rays needed for the light field, each ray is projected separately for a specific viewpoint and is not intended to be mixed with other rays. This is in contrast with our proposed method, where the multiple independent images are created at the intended depths and on the intended surfaces exactly by leveraging the mixing properties of rays from the projectors.

A few works exist that have explored the concept of "depth-dependent projection"; Kagami [16] projected Moire patterns to visually show the depth of a scene, similarly to active-stereo methods for range finding. Nakamura *et al.* [17] used a linear algebra formulation to highlight predetermined volume sections with specific colors. The technique assumes the volume is discretized into multiple parallel planes and is not able to produce detailed patterns or images on non planar surfaces. Moreover, similarly to the work below, the underlying mathematical formulation suffers from a limited dynamic range. Recently, Visentini-Scarzanella *et al.* proposed a method to display detailed images on distinct planes in space by actively exploiting interference patterns from multiple projectors [6]. In their experiments, two independent videos are simultaneously streamed on two screens using semi-transparent material. However, the matrix factorisation used is similar to [17], so the method suffers from a limited dynamic range. Moreover, the purpose is to project images on planar screens, and thus, applications are limited. These limitations are removed in our proposed method, were the patterns can be projected onto arbitrary shapes, and our novel optimization procedure addresses the issues with dynamic range.

3   Algorithm Overview

We provide an outline of the algorithm using the example of projecting virtual face masks on dynamic subjects as shown in Fig. 2. In Figs. 2a and b two different virtual masks are shown, which are to be projected on the subject's face when this changes position from Position 1 to Position 2 as shown in Figs. 2c and d respectively. The fundamental research question is what images should be generated for Projector 1 and Projector 2 so that the virtual masks would recombine into the patterns in Figs. 2a and b at the desired position: at position 1, only mask of Fig. 2a should appear, with no traces of mask of Fig. 2b. It should be noted that similar problem was raised for multiple LCDs and efficiently solved by [18].

Visentini-Scarzanella *et al.* realised a similar system [6], but planar screens were assumed. Moreover, significant reduction in the dynamic range was observed due to their formulation. Because of the sensitivity of [6] to the exact screen placement, it is not possible to directly apply the method to project onto objects with complex geometry by simply considering a piecewise planar approximation.

In order for the desired images to appear at the desired locations, three tasks are necessary. First, the mapping between points on the object surfaces and the pixels in the projector images are obtained (*i.e.*, geometrical calibration). Then, given the mappings, generation of the projection images can be cast as a constrained optimization problem. In [6], this was solved globally with a sparse matrix solver that distributes the error throughout the images. In this paper, we propose an efficient optimization method to solve the problem locally only for related rays, allowing to impose additional constraints on the solution resulting in improved dynamic range, as well as to allow parallelisation. Finally, the generated patterns have to be post-processed, according to the photometric characteristics obtained in the process of photometric calibration, to correct non-linearity of the projector. The main phases of the algorithm are shown in Fig. 3b.

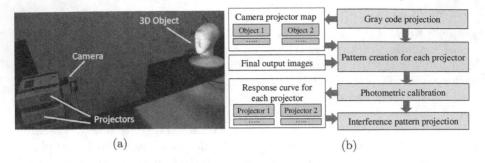

Fig. 3. (a) Configuration of our system with two projectors. (b) Overview of the algorithm.

The system consists of two standard LCD projectors stacked vertically as shown in Fig. 3a with 3D objects for the projection target that can be placed at arbitrary positions.

4   Depth-Dependent Simultaneous Image Projections

4.1   System Calibration

To achieve depth-dependent projections onto arbitrarily shaped objects, it is necessary to estimate the pixel correspondences between projector images and the object surfaces. Contrarily to the simple planar case in [6] where the mapping can be calculated as a homography using only four corresponding points, we need to obtain correspondences between the projector pixels and the points on complex-shaped 3D surfaces. To this end, we use the Gray code pattern projection [19].

The actual process is as follows. First, an additional camera is placed in the scene. Then, the Gray code patterns are projected onto the object and the projections are synchronously captured by the camera as shown in Figs. 4a, b, d and e. By decoding the pattern from the captured image sequences, the correspondences between projected patterns $\mathbf{P}_j$ and the camera image coordinates $\mathbf{C}$ are obtained. The decoded values are shown in Figs. 4c and f where pixel value represents the projector coordinate for x and y direction, respectively. The decoded results are represented by a map $f_j : \mathbf{C} \to \mathbf{P}_j$, which consists of projector coordinate values embedded at each pixel of the camera image as shown in Figs. 4c, f. An inverse map is also obtained $f_j^{-1} : \mathbf{P}_j \to \mathbf{C}$ to efficiently conduct the extraction of corresponding points along the epipolar lines as described in the following section. Note that because of mismatch between the resolutions of the image and the projection on the object, many correspondences are inevitably dropped from the maps, and thus, the final results are degraded. Such artifacts are mostly solved by preparing a high resolution map if projectors and camera can be placed close to each other. The remaining holes are in the order of a few pixels, and are removed with a simple hole filling algorithm.

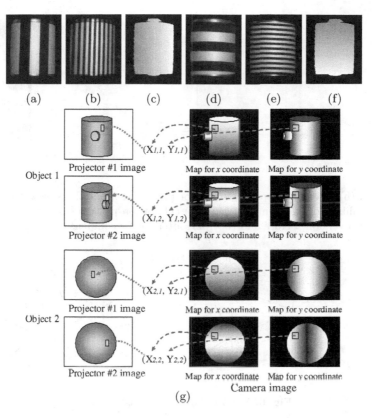

(a)   (b)   (c)   (d)   (e)   (f)

Fig. 4. (a), (b) An object of 3D screen projected by Gray code patterns, (c) projector coordinates observed in the camera coordinates for x direction, (d), (e), (f) for y direction, and (g) mapping between projector image coordinates (the left column) and the camera image obtained from Gray code (the two right columns).

In the process of estimating the projected image, we assume a linear relationship between the nominal intensity and the actual projected intensity. To achieve this, a linearly increasing gray scale pattern in the $[0, 255]$ intensity range (Fig. 5c (top)) is captured by a camera with a known linear response. The recorded values are plotted against their nominal intensity, as shown in Fig. 5a, b, which are then fitted to a $f(x) = ax^b + c$ model, where x is the intensity value. The function is then inverted and kept for compensating the generated pattern prior projection. The calibration pattern superimposed with its own mirrored version should be constant intensity. By intensity correction, this constraint is shown to be fulfilled(Fig. 5c (middle and bottom)).

4.2   A Simple Linear Algebra-Based Pattern Generation Method

We model the problem by first extending the formulation in [6] to the case of non-planar surfaces. The variables involved are shown in Fig. 6. While for clarity

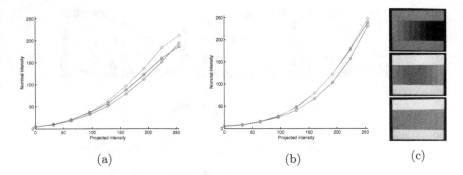

Fig. 5. (a), (b) Intensity response curves for projectors 1 and 2 respectively. (c) (top) Projected calibration pattern, and (middle) calibration pattern superimposed with its own mirrored version, before and (bottom) after colour compensation.

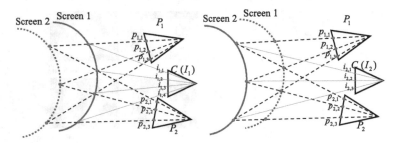

Fig. 6. Variables of linear constraints.

we illustrate the process in the case of two projectors and two different images projected onto two different objects, the system can be extended to a higher number of projectors and objects.

The projected patterns from J projectors are denoted as $\mathbf{P}_j$ where $j \in \{1, \cdots, J\}$ ($J = 2$ in case of Fig. 6), and K images projected on K 3D objects are depicted as $\mathbf{I}_k$ where $k \in \{1, \cdots, K\}$ ($K = 2$ in case of Fig. 6). Let pixels on $\mathbf{P}_j$ be expressed as $p_{j,1}, p_{j,2}, \cdots, p_{j,m}, \cdots, p_{j,M}$ and let pixels on $\mathbf{I}_k$ be $i_{k,1}, i_{k,2}, \cdots, i_{k,n}, \cdots, i_{k,N}$.

We also need to provide the mapping from the desired input images $\mathbf{I}_k$ to the camera coordinates $\mathbf{C}$, which is $g_k : \mathbf{I}_k \to \mathbf{C}$. Practically, in this paper, we use identity map for g_k, which means we use simple "projection mapping" as shown in case of Fig. 6, where the coordinates of $\mathbf{I}_1$ and $\mathbf{I}_2$ is the same as camera image of $\mathbf{C}$.

In the calibration step, the mappings f_j have been obtained. Using f_j and g_k, we can map between the pixels of the projected image $\mathbf{P}_j$ and the desired input image $\mathbf{I}_k$ with $(f_j \circ g_k) : \mathbf{I}_k \to \mathbf{P}_j$ and $(f_j \circ g_k)^{-1} : \mathbf{P}_j \to \mathbf{I}_k$.

From these assumptions, we can define an inverse projection mapping q, where, if $i_{k,n}$ is illuminated by $p_{j,m}$ (*i.e.*, pixels of $i_{k,n}$ and $p_{j,m}$ are mapped with $(f_j \circ g_k)$), $q(k, n, j)$ is defined as m, and if $i_{k,n}$ is not illuminated by any

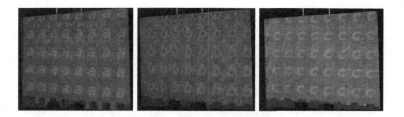

Fig. 7. Simultaneous pattern projection at three different depths.

pixels of P_j, $q(k,n,j)$ is defined as 0. In the example of Fig. 6, $q(1,2,1) = 2$ and $q(1,2,2) = 1$ since $i_{1,2}$ is illuminated by $p_{1,2}$ and $p_{2,1}$. $q(1,1,2) = 0$ since $i_{1,1}$ is not illuminated by P_2.

Let us define two imaginary pixels $p_{1,0} = p_{2,0} = 0$ to simplify formulas. Then, using these definitions, the constraints of the projections are expressed as follows:

$$i_{k,n} = \frac{(d_{k,n,1})^2}{\mathbf{L}_{k,n,1} \cdot \mathbf{N}_{k,n}} p_{1,q(k,n,1)} + \frac{(d_{k,n,2})^2}{\mathbf{L}_{k,n,2} \cdot \mathbf{N}_{k,n}} p_{2,q(k,n,2)}. \tag{1}$$

where $d_{k,n,j}$ is the distance between a pixel on the object and the projector in order to compensate for the light fall-off and $\mathbf{L}_{k,n,j} \cdot \mathbf{N}_{k,n}$ is the angle between the surface normal $\mathbf{N}$ of $i_{k,n}$ and the incoming light vector $\mathbf{L}$ at pixel n from $\mathbf{P}_j$ to compensate the Lambertian reflectance of the matte plane. If $q(k,n,j) = 0$, then we define $d_{k,n,j} = 0$ and, $\mathbf{L}_{k,n,j} = \mathbf{N}_{k,n} = (1,0,0)$.

By collecting these equations, linear equations

$$\mathbf{i}_1 = \mathbf{A}_{1,1}\mathbf{p}_1 + \mathbf{A}_{1,2}\mathbf{p}_2 \tag{2}$$

$$\mathbf{i}_2 = \mathbf{A}_{2,1}\mathbf{p}_1 + \mathbf{A}_{2,2}\mathbf{p}_2 \tag{3}$$

follow, where $\mathbf{p}_j$ is a vector $[p_{j,1}, p_{j,2}, \cdots, p_{j,M}]$, and $\mathbf{i}_k$ is a vector $[i_{k,1}, i_{k,2}, \cdots, i_{k,N}]$, and the matrix $\mathbf{A}_{k,j}$ is defined by its (m,n)-elements as

$$\mathbf{A}_{k,j}(n,m) = \begin{cases} \frac{(d_{k,n,j})^2}{\mathbf{L}_{k,n,j} \cdot \mathbf{N}_{k,n}} & (q(k,n,j) = m) \\ 0 & (otherwise) \end{cases}, \tag{4}$$

By using $\mathbf{i} \equiv \begin{bmatrix} \mathbf{i}_1 \\ \mathbf{i}_2 \end{bmatrix}$, $\mathbf{p} \equiv \begin{bmatrix} \mathbf{p}_1 \\ \mathbf{p}_2 \end{bmatrix}$ and $\mathbf{A} \equiv \begin{bmatrix} \mathbf{A}_{1,1} & \mathbf{A}_{1,2} \\ \mathbf{A}_{2,1} & \mathbf{A}_{2,2} \end{bmatrix}$, we get our complete linear system

$$\mathbf{i} = \mathbf{A}\mathbf{p}. \tag{5}$$

This system can be solved with linear algebra techniques as in [6]. Simple patterns are used to confirm that our algorithm can simultaneously project more than two planes as shown in Fig. 7.

In this paper, we assume there are no defocus blur of the projectors, which is true only if the surfaces of 3D screens are near the focal planes. In case that the

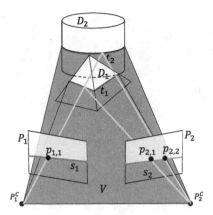

Fig. 8. Epipolar relationship between projectors for the proposed optimization.

defocus blur is not neglectable, the projected images becomes a convolution of the original image and the blur kernel. Let us assume a typical setup, where we fix a plane in 3D space, place two projectors with the same aperture size at the same distances from the fixed plane, and adjust the focuses of the two projectors onto the fixed plane. Note that these conditions are often approximately fulfilled in real setup. In this setup, the sizes of the blur kernel becomes the same for both projectors, even for the off-focus surfaces, thus, the projected images from both the projectors are convolved by the same blur kernel. Then, because of distributivity property of convolutions, the added image becomes the convolution of the non-blurred added image and the blur kernel. This means that, in this case, the off-focus blur on each projected images does not disturb the image addition process of Eq. (5), but only blur the resulting projected images. In reality, we did not find a high divergence between the simulations with non-blur assumptions and the real experiments with blur.

4.3   Problem Reduction Using Epipolar Constraints and Constraint-Aware Optimization

The problem to be solved is to obtain $\mathbf{p}$ given $\mathbf{i}$ and $\mathbf{A}$. The length of vector $\mathbf{p}$ is $M \cdot J$, while the length of vector $\mathbf{i}$ is $N \cdot K$. The matrix $\mathbf{A}$ is a very large sparse matrix. To model the real system, this simple linear model has two problems. First, it implies a global solution through pseudo-inversion of a very large matrix. Second, since $\mathbf{i}$ and $\mathbf{p}$ are images, their elements should be non-negative values with a fixed dynamic range. However, the lack of positivity constraints in the solution of the sparse system means that $\mathbf{p}$ may include negative or very large positive elements. This was solved in [6] by normalising $\mathbf{p}$ so that the elements are in the range of $[0, 255]$ using a sparse matrix linear algebra solver. However, the effect of this is a compression of the resulting dynamic range and a lowering of the contrast.

	PSNR			SSIM		
	LF	EO^{255}_{-100}	EO^{255}_0	LF	EO^{255}_{-100}	EO^{255}_0
Lena/Mandrill	10.9 / 9.2	10.8 / 11.3	**19.6 / 20.9**	-0.04 / 0.20	0.04 / 0.37	**0.66 / 0.71**
Lena/Peppers	10.4 / 7.8	11.2 / 10.7	**19.4 / 20.4**	-0.03 / -0.09	0.07 / 0.24	**0.71 / 0.73**
Peppers/House	8.0 / 8.2	9.4 / 11.1	**18.3 / 20.0**	-0.02 / 0.11	0.11 / 0.14	**0.58 / 0.67**
Peppers/Lena	8.7 / 8.0	10.3 / 10.8	**19.7 / 20.0**	0.02 / -0.16	0.20 / 0.02	**0.77 / 0.68**

Fig. 9. Top: numerical PSNR and SSIM results for validation of the system prototype according to the method tested. For each data entry, the accuracy measures are indicated for each depth plane individually in the form P/Q, where P and Q are the accuracy measures at the first and second position respectively. Best performance is highlighted in bold.

Here, we consider the case that $J = 2$, thus, we assume the two projector scenario shown in Fig. 8 with two objects D_1 and D_2 onto which the images are to be projected. Given the optical centers of the two projectors P^C_1 and P^C_2 as well as any pixel on the first pattern $p_{1,1}$ (without loss of generality), the epipolar plane V defined by the three points will intersect projected patterns $\mathbf{P}_1$ and $\mathbf{P}_2$ at lines s_1, s_2, and 3D screens D_1 and D_2 at t_1 and t_2, respectively. We can see that the pixel compensation between $\mathbf{P}_1$ and $\mathbf{P}_2$ occurs only within epipolar lines s_1 and s_2. This means that the problems of optimizing the pixels of the projected images can be solved for each epipolar line, instead of solving entire projected pixels of $\mathbf{P}_1$ and $\mathbf{P}_2$.

To obtain a finite set of the optimized pixels, we use the following steps. For any pixel $p_{1,1}$ along s_1, this will correspond to points along the intersections t_1, t_2 of the epipolar plane with the image planes. Similarly, the projected pixels of these points to $\mathbf{P}_2$ along s_2, $p_{2,1}$ and $p_{2,2}$ in Fig. 8, are added to the list of variables. By iterating the process, we obtain the list of variables involved in the calculation of the pixel compensation with respect to s_1 and s_2. Moreover, the spacing between points in the sequence on each intersection line will depend on the distance and the angle of vergence between the two projectors.

Hence, instead of formulating the problem as a large global optimization, we decompose it into a series of small local problems of the form:

$$\check{\mathbf{i}} = \check{\mathbf{A}}\check{\mathbf{p}}, \tag{6}$$

where $\check{\mathbf{i}}$ and $\check{\mathbf{p}}$ are the image and pattern pixels related by epipolar geometry and mapped to each other by the ray-tracing matrix $\check{\mathbf{A}}$. Contrarily to the large sparse system, the number of elements involved in each local optimisation is $10^1 - 10^2$ depending on the projector setup. This allows us to expand the above equation into a series of explicit sums:

$$\check{\mathbf{i}}_j = \sum_{k=1}^{N} \check{\mathbf{A}}_{j,k}\check{\mathbf{p}}_k, \tag{7}$$

where $1 \leq j \leq M$, M is the number of points on the image planes, and N is the number of points from the patterns involved in this sequence. For each one

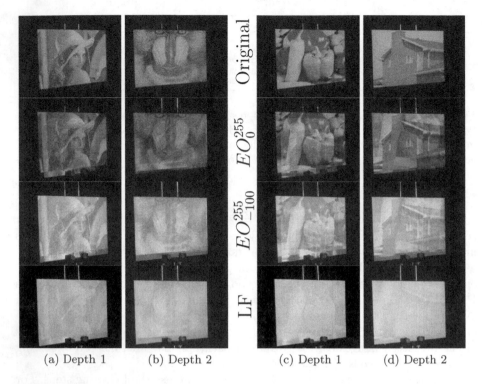

(a) Depth 1   (b) Depth 2   (c) Depth 1   (d) Depth 2

Fig. 10. Real projection results generated with proposed and linear factorisation algorithm. From top to bottom, left to right, the pairs *Lena/Mandrill* and *Peppers/House* for the methods EO_0^{255}, EO_{-100}^{255} and LF [6].

of these explicit sums, we solve for the optimal pattern pixels $\check{\mathbf{p}}_i^*$ by solving the constrained optimisation problem:

$$\check{\mathbf{p}}_i^* = \min_{\check{\mathbf{p}}_i} \sum_{k=1}^{N} \left(\check{\mathbf{i}}_j - \check{\mathbf{A}}_{j,k}\check{\mathbf{p}}_k\right)^2 \mid \check{\mathbf{p}}_k \in [a,b], \tag{8}$$

where $[a,b]$ is the allowed range of intensities during the pattern generation. With this strategy, we are able to independently solve the pattern generation problem for small sets of points at a time, which in turn allows us to impose constraints and to confidently optimise each chain with fewer chances of getting stuck in a local minimum.

5 Experiments

Our setup consists of two stacked EPSON LCD projectors (1024×768 pixels) and a single CCD camera (1600×1200 pixels) as shown in Fig. 3a. We calibrate the system for both geometrically and photometrically, and test the proposed

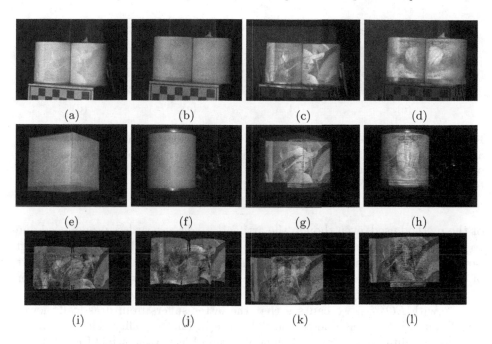

Fig. 11. *Lena/Mandrill* pairs projected on non-planar scenes placed at different positions. Projected patterns are generated with (a) (b) (e) (f) LF [6] and (c) (d) (g) (h) the proposed EO_0^{255} algorithm. (i) (j) (k) (l) are the actual generated images for each projector in the two scenarios.

system under three tasks. First, the image quality of the proposed algorithm is numerically evaluated against the Linear Factorization (LF) method in [6]. Second, the visual positioning system application scenario shown in Fig. 1 is tested. Finally, the projection results on complex surfaces are shown in the scenario of virtual mask projection of Fig. 2.

5.1 Image Quality Assessment

We first assess the quality improvement of our dynamic range expansion technique using a planar screen and choosing the combinations *Lena/Mandrill*, *Lena/Peppers*, *Peppers/House* and *Peppers/Lena* for target images. The two screens were placed at approximately 80 cm and 100 cm from the projectors. For each combination, we projected the original image on the plane, and used it as a baseline for PSNR evaluation with our proposed method denoted as EO_b^a where (a, b) is the range of allowed intensity values, as well as the LF method in [6]. Sample results are shown in Fig. 10, while exhaustive numerical results are given in Fig. 9. The results show that the result images obtained by EO_0^{255} have much wider dynamic range of colours than those of the results obtained by LF and EO_{-100}^{255}. However, in certain cases some artifacts are visible, even for results of EO_0^{255}. Interestingly, the nature of the artifacts is the same regardless

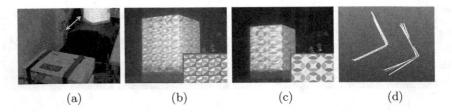

(a)   (b)   (c)   (d)

Fig. 12. (a) Test scene, projected patterns when the box is at the position (b) nearer and (c) farther away from the projectors. In (b) and (c), the target images for each object are shown inset at the bottom right corner. (d) 3D-scanned boxes during the test (white) and ground truth (red). (Color figure online)

of the method used, even though the artifacts might appear less pronounced due to the overall lower contrast of LF and EO^{255}_{-100}. This highlights the trade-off between artifacts, contrast levels and number of projectors in the scene. As part of our future work, we will investigate redundant systems with a higher number of projectors than targets to characterize better this trade-off. When comparing EO^{255}_{0} with EO^{255}_{-100}, we can see that the latter suffers from drastically lower PSNR and SSIM levels due to the difference in image quality. The techniques were also qualitatively compared on non-planar objects as shown in Fig. 11, highlighting similar improvements in dynamic range when the proposed method is used.

5.2   Visual Positioning Accuracy Evaluation

As another application scenario, we use the system to position objects at the right position and orientation based exclusively on visual feedback. Such system could be used both by human as well as robotic workers without extra sensors. For our tests, we asked several subjects to place a box in two predetermined positions using just the visual feedback from the projected pattern. The location of the box was then captured with a 3D scanner and compared with the ground truth position. The test scene, projected patterns and reconstructed shapes are shown in Fig. 12. The average RMSE was of 1.47% and 1.26% of the distance between the box and the projector for the positions closer and farther away from the projectors respectively. From the results, we can confirm that the proposed technique can be used for 3D positioning just using static passive pattern projectors.

5.3   Independent Image Projection on Multiple 3D Objects

For our third experiment, the system was tested on 3D objects with a more complex geometry, such as a mannequin head, as well as the combination of a square box and cylinders for the two scenarios mentioned in the introduction.

In the first case, we projected the virtual masks in Fig. 13a and b onto a mannequin placed at two different positions. Figure 13c and d show the results

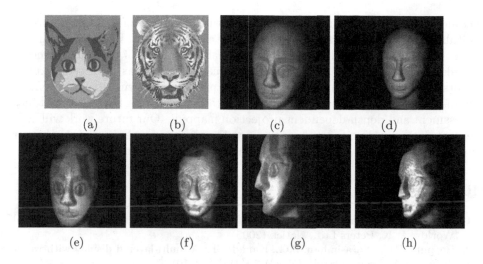

Fig. 13. Projection results on objects with complex geometry. (a) and (b) are the target images for two positions, (g), (f) are results with the LF algorithm [6], and (i) to (l) are results with our method EO_0^{255} at positions 1 and 2, respectively.

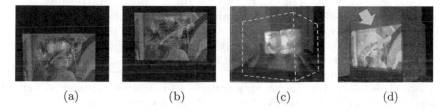

Fig. 14. Projected patterns for (a) top and (b) bottom projectors with our method EO_0^{255} and (c), (d) are the projected results.

of LF and Fig. 13e to h show the results of EO_0^{255}. The figures show that the two images projected on the mannequin are clearly visible from all angles. Moreover, our proposed optimization significantly improves the result of LF.

Finally, we show how the system can be used for the object assembly workflow shown in Fig. 1. Figure 14a and b are the calculated patterns with EO_0^{255} for the two projectors, Fig. 14c is the projected image on two cylinders and Fig. 14d is the projected image on the large box placed outside the cylinders. We can confirm *Lena/Mandrill* is clearly shown on each object, confirming that the technique has a potential to be used for correct positioning during object assembly.

6   Conclusion

In this paper, we propose a new pattern projection method which can simultaneously project independent images onto objects with complex 3D geometry at different positions. This novel system is realized by using multiple projectors with geometrical calibration using a Gray code with a simple formulation to create suitable distributed interference patterns for each projector. In addition, an efficient calculation method including additional constraints of epipolar geometry to allow parallelisation and higher color dynamic range was proposed. Experiments showed the performance of a working prototype showing its improvement against the state of the art, and two application scenarios for object distance assessment and depth-dependent projection mapping. Our future work will concentrate on extending the system to more complicated scenes involving a higher number of projectors, and studying their optimality characteristics.

References

1. Bimber, O., Raskar, R.: Spatial Augmented Reality: Merging Real and Virtual Worlds. A. K. Peters Ltd., Natick (2005)
2. Barnum, P.C., Narasimhan, S.G., Kanade, T.: A multi-layered display with water drops. ACM Trans. Graph. (TOG) **29**(4), 76 (2010)
3. Jurik, J., Jones, A., Bolas, M., Debevec, P.: Prototyping a light field display involving direct observation of a video projector array. In: IEEE Conference on Computer Vision and Pattern Recognition Workshops (CVPRW), pp. 15–20 (2011)
4. Nagano, K., Jones, A., Liu, J., Busch, J., Yu, X., Bolas, M., Debevec, P.: An autostereoscopic projector array optimized for 3d facial display. In: ACM SIGGRAPH 2013 Emerging Technologies (SIGGRAPH 2013), p. 3:1 (2013)
5. Hirsch, M., Wetzstein, G., Raskar, R.: A compressive light field projection system. ACM Trans. Graph. (TOG) **33**(4), 58 (2014)
6. Visentini-Scarzanella, M., Hirukawa, T., Kawasaki, H., Furukawa, R., Hiura, S.: A two plane volumetric display for simultaneous independent images at multiple depths. In: PSIVT Workshop Vision meets Graphics, pp. 1–8 (2015)
7. Cruz-Neira, C., Sandin, D.J., DeFanti, T.A.: Surround-screen projection-based virtual reality: the design and implementation of the cave. In: Proceedings of the 20th Annual Conference on Computer Graphics and Interactive Techniques, pp. 135–142. ACM (1993)
8. Raskar, R., Welch, G., Fuchs, H.: Seamless projection overlaps using image warping and intensity blending. In: Fourth International Conference on Virtual Systems and Multimedia, Gifu, Japan (1998)
9. Yang, R., Gotz, D., Hensley, J., Towles, H., Brown, M.S.: Pixelflex: a reconfigurable multi-projector display system. In: Proceedings of the Conference on Visualization 2001, pp. 167–174. IEEE Computer Society (2001)
10. Chen, Y., Clark, D.W., Finkelstein, A., Housel, T.C., Li, K.: Automatic alignment of high-resolution multi-projector display using an un-calibrated camera. In: Proceedings of the Conference on Visualization 2000, pp. 125–130. IEEE Computer Society Press (2000)
11. Schikore, D.R., Fischer, R.A., Frank, R., Gaunt, R., Hobson, J., Whitlock, B.: High-resolution multiprojector display walls. IEEE Comput. Graph. Appl. **20**(4), 38–44 (2000)
12. Godin, G., Massicotte, P., Borgeat, L.: High-resolution insets in projector-based display: principle and techniques. In: SPIE Proceedings: Stereoscopic Displays and Virtual Reality Systems XIII, vol. 6055 (2006)
13. Bimber, O., Emmerling, A.: Multifocal projection: a multiprojector technique for increasing focal depth. IEEE Trans. Vis. Comput. Graph. **12**(4), 658–667 (2006)

14. Nagase, M., Iwai, D., Sato, K.: Dynamic defocus and occlusion compensation of projected imagery by model-based optimal projector selection in multi-projection environment. Virtual Reality **15**(2–3), 119–132 (2011)
15. Levoy, M., Chen, B., Vaish, V., Horowitz, M., McDowall, I., Bolas, M.: Synthetic aperture confocal imaging. ACM Trans. Graph. (TOG) **23**, 825–834 (2004)
16. Kagami, S.: Range-finding projectors: visualizing range information without sensors. In: IEEE International Symposium on Mixed and Augmented Reality (ISMAR), pp. 239–240 (2010)
17. Nakamura, R., Sakaue, F., Sato, J.: Emphasizing 3D structure visually using coded projection from multiple projectors. In: Kimmel, R., Klette, R., Sugimoto, A. (eds.) ACCV 2010. LNCS, vol. 6493, pp. 109–122. Springer, Heidelberg (2011). doi:10.1007/978-3-642-19309-5_9
18. Wetzstein, G., Lanman, D., Hirsch, M., Raskar, R.: Tensor displays: compressive light field synthesis using multilayer displays with directional backlighting. ACM Trans. Graph. (Proc. SIGGRAPH) **31**(4), 1–11 (2012)
19. Sato, K., Inokuchi, S.: Three-dimensional surface measurement by space encoding range imaging. J. Robot. Syst. **2**, 27–39 (1985)

ZigzagNet: Efficient Deep Learning for Real Object Recognition Based on 3D Models

Yida Wang[1(✉)], Can Cui[2], Xiuzhuang Zhou[3], and Weihong Deng[1]

[1] School of Information and Communication Engineering,
Beijing University of Posts and Telecommunications, Beijing, China
wangyida1@bupt.edu.cn
[2] School of Electronics and Information Engineering,
Beijing Jiaotong University, Beijing, China
[3] College of Information Engineering, Capital Normal University, Beijing, China

Abstract. Effective utilization on texture-less 3D models for deep learning is significant to recognition on real photos. We eliminate the reliance on massive real training data by modifying convolutional neural network in 3 aspects: synthetic data rendering for training data generation in large quantities, multi-triplet cost function modification for multi-task learning and compact micro architecture design for producing tiny parametric model while overcoming over-fit problem in texture-less models. Network is initiated with multi-triplet cost function establishing sphere-like distribution of descriptors in each category which is helpful for recognition on regular photos according to pose, lighting condition, background and category information of rendered images. Fine-tuning with additional data further meets the aim of classification on special real photos based on initial model. We propose a 6.2 MB compact parametric model called ZigzagNet based on SqueezeNet to improve the performance for recognition by applying moving normalization inside micro architecture and adding channel wise convolutional bypass through macro architecture. Moving batch normalization is used to get a good performance on both convergence speed and recognition accuracy. Accuracy of our compact parametric model in experiment on ImageNet and PASCAL samples provided by PASCAL3D+ based on simple Nearest Neighbor classifier is close to the result of 240 MB AlexNet trained with real images. Model trained on texture-less models which consumes less time for rendering and collecting outperforms the result of training with more textured models from ShapeNet.

1 Introduction

Object recognition has been one of the most challenging tasks in pattern recognition, some attempts on CNN [1–5] give a great relief on artificial design. Great amount of training data give a help on learning discriminant features by making net deeper [6] and wider [7,8]. Back propagation makes it possible for adjusting parameters in different training stages from particular arrangement of training data. However, this data driven way has heavy reliance on abundant training

© Springer International Publishing AG 2017
S.-H. Lai et al. (Eds.): ACCV 2016, Part IV, LNCS 10114, pp. 456–471, 2017.
DOI: 10.1007/978-3-319-54190-7_28

data which might be difficult to be collected. Work on learning descriptor from CNN on synthetic data [9] shows that the performance of learning based on 3D models could also satisfy the need for recognition on real images in laboratory condition, but results of AlexNet [6], GoogleNet [1], ResNet [5] and LDO [9] directly trained with rendered samples also show that making net deeper still struggles to model objects as expected because some difference between rendered images from 3D models and real photos is hard to be eliminated through traditional parametric model. On the other hand, wire frames, plain texture [10] together with some real data also increase the accuracy on real photos, these methods are effective with additional information which is captured by special equipment apart from visible light or adding real images for training.

(a) Real photos   (b) Pure objects (c) Regular shots (d) Close shots

Fig. 1. Real images and three types of our data for fast convergence, net initiation with triplet and fine-tuning separately.

In this paper, we introduce an efficient residual network called ZigzagNet which is concatenated of repetitive modules composed of three 1×1 and one 3×3 convolutional layers in a zigzag style of expanding and squeezing. We eliminate the reliance on massive real training data by modifying convolutional neural network in 3 aspects: synthetic data rendering for training data generation in large quantities, multi-triplet cost function modification for multi-task learning and compact end-to-end micro architecture design for producing tiny parametric model while overcoming over-fit problem in texture-less models. We use textureless 3D models [11] for training data generation to get rid of dependence on large amount of models or real photos. Two loss functions are used sequentially for fine-tuning on regular images and close shots producing a sphere-like manifold distribution of descriptors which is helpful to recognition. For balance of efficiency and performance, both macro and micro CNN architectures are designed as compact as possible like SqueezeNet [4] to solve over-fitting problem triggered by discarded textures of models instead of applying quantization for parameters like BinaryNet [12,13] and Ristretto [14]. Geometric properties of visualized models could represent the core discriminant features of items in spite of diversified texture on real objects, so we use gray surface for a representation of each 3D object rather than plenty of well rendered models in [15] to gain a better performance on recognition for real photos. Training samples are generated from anchors of view points on a sphere (also representing object poses). All rendered images are put in front of randomly selected background images from Flickr [16] without particular object instances.

Our parametric model is learned in two stages using multi-triplet loss. Labels of poses are utilized to make feature distribution similar to manifold learning for real images. Poses are used as information for constructing descriptor of object itself and categories are regarded as relationship with others. Multi-triplet loss based on these information is applied on the initial training to ensure that the learned model have a better concept on 3D models using only 2D images and keep the ability to distinguish different objects from each others also. Fine-tuning lead by softmax loss with additional close shot samples further improves the ability for recognition. Our similarity in triplet is defined as square of Euclidean distance, making it easier to apply Nearest Neighbor (NN) [17] classifier on descriptors. Softmax lead fine-tuning using class labels alone with more special data which are not suitable for the arrangement of triplet does classification more efficiently and tells different objects apart more evenly.

Result on PASCAL3D+ database [11] proves that our pipeline is efficient on CNN learning based on texture-less 3D models without training with great amount of real images and gets rid of depth information [9,18]. Convergence also speeds up using optional pair wise term with triplet applied on images without background. Accuracy is 9% higher on ImageNet [19] samples than training from real images using the simplest NN classifier which achieves 47.2% accuracy on average, even on par with the result from descriptors of AlexNet trained from real images of PASCAL VOC 2012.

As shown in Fig. 2, synthetic data driven learning based on ZigzagNet aimed at doing recognition on real photos is introduced in 3 aspects: data rendering, cost function design for manifold learning and compact micro architecture design.

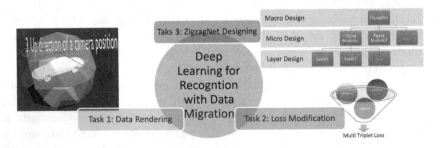

Fig. 2. Overall Pipeline of deep learning based on ZigzagNet.

2   Training Data Generation

We exploit information in geometric characteristics of 3D models using poses and lighting conditions together with class labels to simulate real photos without information of texture on object surface [15] to construct an efficient database based on few models collected from PASCAL3D+ [11]. Randomly selected background images crawled from Flickr are used to generate scene based images. The whole database is prepared for a two stage fine-tuning process on CNN for producing discriminant descriptors of real photos on object recognition. Subsets of our synthetic data are used in order to speed up convergence and improve the classification accuracy.

2.1   Object Rendering Based on Interest of View

We train CNN only using rendered images from texture-less 3D models whose poses are manifested by the camera parameters. Foreground objects are shot from positions on a sphere net combined with local triangles recursively subdivided from larger triangles [20] which are initiated from a regular icosahedron. We focus on these texture-less models rather than great amount of well painted models in [15] to make a balance on efficiency and variability. Origin of coordinate is fixed in the center of each object's bottom and z axis is set to be vertical to the regular plane where such objects usually exist in real world. Camera positions, focal points and distance from camera to the center of objects are adjusted to simulate scenes in real photos. As range of view positions is limited in real world, we cut the whole sphere of points cloud along 2 meridians which are vertical to z axis on whole sphere for most objects. View positions for images of big machines such as cruises are limited to $V_z \in (-0.1max(V_X), 0.2max(V_X))$ and ones for ordinary objects are usually satisfy range of $V_z \in (0, 0.6max(V_X))$ where (V_x, V_y, V_z) is the coordinate on sphere. Inspired by two common approaches for shooting, regular photos and close shots are used for training in two stages. Focal points on all objects are set on the center of each model for regular photos. For special objects with large ratio of length and width shown in Fig. 1 such as bus and boat, we expand the database by shifting focal points towards the head of the object from the center for a proportional distance to the offset degree from the front view axis.

2.2   Simulation on External Factors

Background images crawled from Flickr [16] among which inappropriate images are shaken out afterwards to ensure that target objects are not included. As shown in Fig. 1, samples rendered by our method could efficiently simulate photos in real world, problem caused by the absence of texture information is solved in our CNN initiation process using multi-triplet loss towards poses. Close shots are added in fine-tuning process to further improve the ability on accommodating special photos in real world as they will confuse the triplet arrangement.

3   Two-Stage Training

We increase the mutual information between features of our training data and real photos by exploiting discriminant factors to minimize over-fitting problem triggered by absence of textures. Based on additional information apart from category, ZigzagNet is pre-trained using a multi-triplet loss joint with an optional pair wise term for speeding up convergence based on labels of poses and lighting conditions. Then additional data with different focal points is used for fine-tuning with softmax loss to imitate more real scenes. Our training method is introduced for the aim of extracting descriptors for classification using NN classifier without other feature transformation such as softmax and SVM classifier which doesn't suit for data migration between synthetic data and real images. We evaluate the performance of our work on output of the penultimate layer of the net before the last inner product matrix. Our discriminant descriptor satisfies NN classifier better than models trained from real images and is even similar to the performance of AlexNet [6] descriptors which is also trained from realistic data using larger parametric model.

3.1   Multi-triplet Based Pre-training

We take advantage of our training data generated from two camera modes separately using multi-triplet cost and softmax cost. As background images are added for keeping environmental deviation related to real photos, training with category labels based on texture-less models will lead to over-fitting on distinguishing edges of objects against background so that objects without clear boundary against background in real images won't be recognized well. We introduce a special triplet cost to utilize information contained in labels of poses, lighting conditions of items.

Reference   Positive   Negative(1 pose negative + 2 class negative)

Fig. 3. Triplet set with 5 samples where reference and positive samples are fixed and 3 negative samples are used for multi task learning on camera pose related descriptor distribution and classification.

Our multi-triplet loss function which is modified from work of [9] together with an optional pair wise term is used in the initiation process to enhance the ability for recognizing physical structure characteristics of particular objects both about themselves and towards others. Figure 4 shows that pose related triplet training has a sphere-like distribution of descriptors from 3D models and 4 Nearest Neighbor Prediction indicates that trained parametric model has ability on both identification on object categories and distribution regression on object poses. Sub-figure(a) in Fig. 4 shows that learning in early stage over fits the blue surface of Ape which is bad for both classification and feature distribution. Our multi-triplet loss solves this problem on surface texture at last which is not only capable for a multi task learning on poses, but also beneficial for prediction on

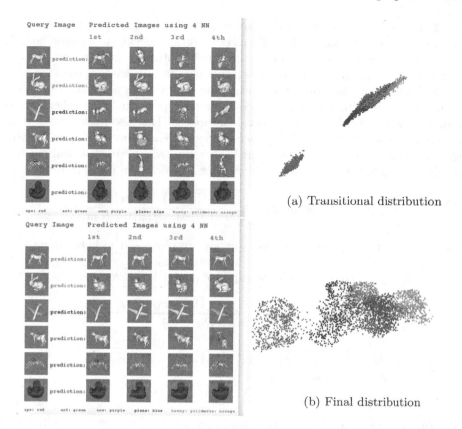

(a) Transitional distribution

(b) Final distribution

Fig. 4. First 4 nearest neighbors representation and 3D feature visualization with PCA matrix calculated from descriptors of our jointed loss. (Color figure online)

categories. Triplet makes it possible to train a model guided with substantial labels including object poses, lighting conditions and camera modes which are accurately recorded with training sample to fully exploit the information in objects themselves rather than raw information given by just huge amount of background in real images. As triplet cost doesn't use label as an supervision, we arrange samples in proper sequence for the model construction process. A particular loss set is composed of a reference sample, a positive sample and several negative samples:

$$\mathcal{L}_{tri-set} = \mathcal{L}_{pair}(s_i, s_j) + \sum_{(s_i, s_j, s_k) \in \mathcal{T}} \mathcal{L}_{tri}(s_i, s_j, s_k) \qquad (1)$$

where $\mathcal{L}_{pair}(s_i, s_j) = ||f(x_i) - f(x_j)||_2^2$. As shown in Fig. 3, our triplet set is composed of 5 samples, the positive sample is selected as the one with close pose from reference sample or only has a different lighting condition which is the same pose and 3 negative samples are selected as one differs more in pose from the same class and two from other classes. Inspired of [9], the problem of gradient vanish in traditional triplet [21] is solved. Correspondingly, large variance in background images from Flickr [16] makes the gradient easy to explode because every channel varies much, so we further apply natural logarithms on loss of a triplet. Feature distance in triplet loss is modified from Euclidean distance to its square form where distribution of learned descriptors has a manifold of sphere as shown in Fig. 4(b) which is related to camera positions on a sphere rather than a manifold of cubic in low dimensional spaces. This also makes the learning process more stable for samples which differs little, so tiny difference between reference and positive sample without background which are close will not lead to a NaN gradient which might appear in use of Euclidean distance in [9].

$$\mathcal{L}_{tri}(s_i, s_j, s_k) = \ln(\max(1, 2 - \frac{||f(x_i) - f(x_k)||_2^2}{||f(x_i) - f(x_j)||_2^2 + m})) \tag{2}$$

where $f(x)$ is the input of the loss layer for sample x and m is the margin for triplet. Denote that $D_{ij} = ||f(x_i) - f(x_j)||_2^2$ and $D_{ik} = ||f(x_i) - f(x_k)||_2^2$, so the partial differential equations for the input of triplet loss layer are:

$$\frac{\partial \mathcal{L}_{tri}}{\partial f(x_i)} = \frac{D_{ik}(f(x_i) - f(x_j)) - (D_{ij} + m)(f(x_i) - f(x_k))}{\mathcal{L}_{tri}(D_{ij} + m)^2}$$

$$\frac{\partial \mathcal{L}_{tri}}{\partial f(x_j)} = \frac{D_{ik}(f(x_j) - f(x_i))}{\mathcal{L}_{tri}(D_{ij} + m)^2}$$

$$\frac{\partial \mathcal{L}_{tri}}{\partial f(x_k)} = \frac{f(x_i) - f(x_k)}{\mathcal{L}_{tri}(D_{ij} + m)} \tag{3}$$

For convenience on training data preparation process, we set the positive and the reference sample fixed in a triplet set and use them straightly as the pair wise term if it's needed. As triplet based training is much more slower in convergence, we train with samples without background at first and attach the pair wise term in a triplet set to make it converge quicker. The pair wise term is only used to make the descriptor robust to small variance in different lighting conditions rather than much more complex backgrounds, we remove it later in the background added training stage. We directly apply NN classifier to maintain the ability of distinguish different classes apart and have a manifold distribution on features to capture the geometry of corresponding poses at the same time. Positive and negative samples in (s_i, s_j, s_k) are carefully selected regard to the reference samples. Positive sample in a set is always selected as one in the same class with reference samples with a similar pose. The key point is that positive samples should either have a slight different viewing angle from camera compared to the reference samples or have a different lighting condition.

3.2 Fine-Tuning with Additional Data

Triplet training based on poses has a restriction of fixed focal points. We add softmax loss for fine-tuning based on close shot images with shifted focal points which are moved from the center of object to the head of objects regards to the

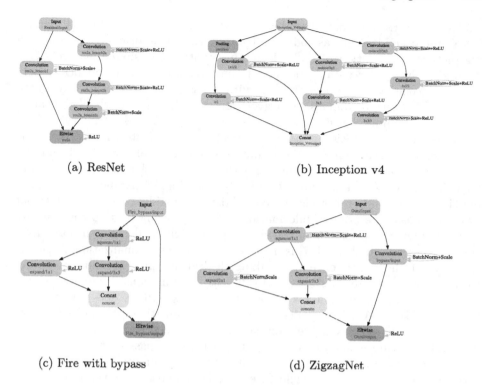

(a) ResNet

(b) Inception v4

(c) Fire with bypass

(d) ZigzagNet

Fig. 5. Core micro structures composed of several layers in ResidualNet [5], Inception v4 [7], SqueezeNet [4], SqueezeNet with bypass and our ZigzagNet.

shift angle from the axis of front view. Shift distance of focal points varies directly as the distance from the view position to horizontal axis of the object which could be represent as $(F_x', F_y', F_z') = (F_x, F_y, F_z) + 0.2((C_x, C_y, C_z) - P_{axis})$ where F represents the focal points used for the camera shots, C represents the camera position on the sphere and P_{axis} represents the intersection point of the axis of front view and the surface of the regular view sphere. Distribution of descriptors projected with 3 dimensional PCA matrix shows that special samples from real test images could be better clustered with other normal samples in the same class after fine-tuning and all samples are set apart better according to categories. Additional rendered data gives a help on reducing intra-class variability while keeping the geometry relationship of regular samples based on pre-trained model.

4   Structure of ZigzagNet

As data migration problem from synthetic training data and real testing data could not be completely eliminated, we do learnable channel-wise compression between micro architectures for deep structure to avoid over-fitting while keeping the depth and width of network to remain the ability of learning discriminant

features from training data together with loss function. SqueezeNet [4] defines
CNN micro architecture as individual layers while macro architecture as the
big-picture organization of multiple modules with some shared features. Our
ZigzagNet has tinier parametric model compared to AlexNet and better performance compared to SqueezeNet [4] with the same macro architecture where
sequence and output dimension of every micro architectures are all the same.
ZigzagNet combines SqueezeNet with residual concept [5] and getting rid of fully
connected layers with Zigzag module instead of FCN architecture [3]. It is concatenated of repetitive micro Zigzag modules composed of three 1×1 and one
3×3 convolutional layers in a zigzag style of 3 steps: channel wise squeezing
for principal representation, reception fields expanding by parallel connection of
multi-scale convolution kernels and consolidation by adding 1×1 convolutional
layer to bypass the original information before squeezing to keep the learning
process stable while remaining the size of parametric model. Batch normalization in Inception module v4 [7] is modified to a moving normalization aiming at
speeding up the convergence and making the learning process stable regarding
to different data in both stages and different loss functions. Input to a single
module is compressed by a 1×1 convolutional layer which does linear transformation like PCA [22] on 1D channel vectors within 3D input cubics to make
the following parameter model compact and representative. Output in a single module within macro architecture is joined with bypassed original information by a convolutional layer as the input of next module to keep the learning
process stable. Based on a single Fire module of bypassed SqueezeNet shown
in Fig. 5(c), ReLU operation on the expand layer before the element wise sum
layer is moved to the output of Fire module to eliminate the scale difference
between two inputs of the element wise operation layer and make information
from both branches compressed at the same time. For convenience of comparison experiments, depth of macro structure of ZigzagNet is set as the same as
AlexNet and SqueezeNet in account of micro structures with channel-wise representation: $Input(227 \times 227 \times 3) - Local\ Pool - 128 - 128 - 256 - 256 - 384 -$
$384 - 512 - 512 - Global\ Pool - Output$. All convolutional kernels in pre-training
stage are initiated as MSRA filters [23].

Moving batch normalization are used to solve over-fitting problem and accelerating the speed of convergence. Batch norm which is firstly introduced to solve
the Internal Covariate Shift [24] is used after all convolutional layers for rebal-
ancing average and scale of two data streams within a residual structure and
making fine-tuning process smoother because two cost functions lead to different distributions and scales in descriptors which will amplify layer by layer in
back propagation process of CNN learning. Such operation on inputs of batch
norm layer is done as a linear operation on channel wise normalization. We use
a moving average mean and variance to improve the speed of convergence which
vary with iterations:

$$y^{(k)} = \gamma^{(k)} \frac{x^{(k)} - E(x^{(k)})}{\sqrt{Var(x^{(k)})}} + \beta^{(k)} \tag{4}$$

$$Mean_{(t)} = \alpha BatchMean_{(t)} + (1 - \alpha)Mean_{(t-1)}$$

$$Var_{(t)} = \alpha BatchVar_{(t)} + (1 - \alpha)Var_{(t-1)} \qquad (5)$$

where $y^{(k)}$ is linear transformation of the z-score like operation result on the inputs of the batch normalization layer, k indicates channels and t indicates iterations. Triplet loss with margin of 0.01 and softmax loss for fine tuning are used in mini-batch SGD training. The momentum, base learning rate and base weight decay rate are set to be 0.9, 0.001, 0.0005 in both stages. This means that the learning rate is multiplied by 0.9 every 100 epochs.

5 Experiment

5.1 Experiments on PASCAL3D+

Advantages of our model based CNN learning against training from real images are shown in this section. PASCAL3D+ [11] database has texture-less CAD models with 3D annotations together with original real images from categories in ImageNet [19] for test. Our iteration depth of icosahedron division for camera view [9] is fixed on 3 and 4 which generates 19240 and 8020 images for regular photos and close shots respectively.

Our descriptors are evaluated on NN classifier using Euclidean distance. Making architectures deeper alone such as AlexNet [6] and GoogleNet [1] will lead to worse performance on real images caused by over-fitting problem on absence of texture and occlusion. Accuracy of AlexNet and GoogleNet trained from the same rendered data is 5.3% lower than our pipeline. As shown in Table 1, accuracy of our structure is on par with result of AlexNet [6] trained from real images which is a much larger parametric model. Result between training with real images and synthetic data also shows that our pipeline suits for NN classifier better than using real data directly. Our triplet and softmax loss have different advantages based on different prior knowledge towards various categories, so the fine-tuned result exceeds them both and is 7% higher than the one based on softmax loss alone. Performance would not be so good if the training stages are reversed because distribution related to poses is important for fine-tuning.

Some descriptors of synthetic data which are projected on PCA matrix for visualization are shown in Fig. 4. From perspective of manifolds, pre-training using triplet loss regarding to lighting, class and pose labels could reflect the distribution from the view positions of camera. Geometry of corresponding poses could give a help on telling different objects apart also, but recognition result using pose based triplet training alone in Table 1 doesn't satisfy our needs because features of different models from the same category are not set to be close to each others and images which are taken from close shots are not included in this stage due to shift of focal points. By fine-tuning with other close shots, distribution of the training samples on the final model suits for the aim of classification better than ever because images which are often taken from the front or tail of the objects are much easier to be correctly classified.

Table 1. Classification on descriptors using Nearest Neighbor classifier on ImageNet samples attached in PASCAL3D+ database. ZigzagNet is our CNN structure; t represents the triplet loss based training, s represents the softmax lead one and ts means that we use a fine-tuning process from triplet loss to softmax loss, M means that model based samples are used and R means that real images are used for training. P represent that pose information is included in triplet training. NP represent that pose information is removed in triplet training.

Method	Plane	Bike	Boat	Bottle	Bus	Car	Chair	Table	Motor	Sofa	Train	TV	Avg
ZigzagNet(ts-M)	69.5	51.3	57.7	34.9	33.8	54.2	36.5	39.9	28.5	49.5	38.2	33	47.2
ZigzagNet(s-M)	67.1	44.6	50.9	36.3	24.8	46.2	29	27.5	23.1	44.8	28.4	26.2	40.3
ZigzagNet(t-M-P)	63.3	35.4	60.1	39.3	26.7	47.6	30.4	24.1	23.8	47.6	24.8	20.8	40.3
ZigzagNet(t-M-NP)	63	37.8	56.6	32.3	21	31.8	31.7	25.5	19	43.9	23.4	17	34.9
DeepLDO(ts-M)	66.6	42	52.7	35.8	23.5	44.4	30.9	25.9	21.5	45.6	27.9	27	39.7
LDO(t-M) [9]	62.4	19.5	55.3	30.1	7.6	27.5	26.4	20	11.8	36.1	16.3	18	29.6
AlexNet(s-M) [6]	68.3	40.7	48.5	30.3	23	56.8	28.7	35.6	21.1	37.2	23.6	22.6	41.9
Alexnet(s-R) [6]	72.5	44.6	57.7	36.7	29.2	58	34	43.4	28	41.4	30.1	27.6	46.9
ZigzagNet(s-R)	62.7	31.2	57.5	29.1	21	43.9	26.2	27.2	18.7	46.8	22	25.2	37.8

Result between training using normal triplet and triplet with pose information indicates that additional information could surely give a help on distinguishing objects better with proper arrangement of triplets. Performance is improved by adding moving batch norm in our structure compared to deeper LDO [9] and exceeds the result trained from real images using the same architecture. The use of additional close shots samples for fine-tuning makes a great benefit on classification which achieves 47.2% accuracy with NN classifier similar to the result trained from real images of AlexNet.

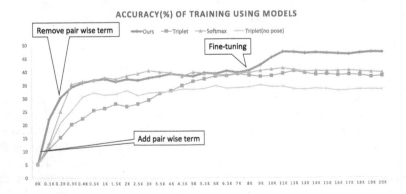

Fig. 6. Accuracy during 1.5 K iterations, our method use pure models before 0.2 K iterations and remove the pair wise term after that and using samples with background. Additional data are included after 8 K iterations using softmax loss.

As shown in Fig. 6, we make it converge faster and better than simply using any single loss function. First, pair wise term is added in triplet set applied on training images without background and removed after 200 iterations when accuracy is already as well as result of 4000 iterations using triplet alone. Then backgrounds are added back for a triplet lead learning process to mix up useful information together. Finally, the fine-tuning stage using a softmax loss is introduced after 8000 iterations using additional data to achieve the final goal of recognition on real photos.

5.2   Experiments on ShapeNet

ZigzagNet trained from texture-less models outperforms structures trained from sophisticated models in Nearest Neighbor evaluation criteria. ShapeNet [25] is a information-rich repository containing textured models while PASCAL3D+ [11] only contains texture-less models which are easier to be collected. Comparison experiments in this section show that our structure is less dependent on highly realistic 3D models and performs better in recognition on real photos. The amount of models used for training in PASCAL3D+ is only 1/6 of textured models in ShapeNet for 12 categories. We render training images in comparison experiments from 600 textured models from ShapeNet and 94 texture-less models from PASCAL3D+. Rendering strategies like [15] depend on models with textures cost nearly 5 more times producing synthetic images compared to texture-less models because surface on 3D models must be smoothed again in group of polygons for attaching texture images. On the other hand, surface color and contrast between object and background will be recognized as discriminant traits for descriptor which is not proper for real photos with rich textures on target objects if texture-less models are used directly in tasks such as pose estimation and classification on real photos. Softmax classification and nearest neighbor classification are both analyzed in this section based on descriptors trained on various methods and models. Results show that ZigzagNet pre-trained from pose related triplet loss is more effective in Nearest Neighbor classification with much fewer texture-less 3D models.

Prediction results of parametric models trained from texture-less models and textured models are shown in Tables 2 and 3. Foreground objects of PASCAL database are positioned more randomly in whole images. Experiments of training softmax classifier directly from CNN descriptors show that lack of textures makes it hard to train an applicable parametric model compared to real photos and textured 3D models. GoogleNet [1] achieves 59% in accuracy while training form textured models in ShapeNet [25] achieves 42.7% in our strategy and a texture-less based learning is much lower at 25.1%. This performance degradation is caused by severe data migration from training images to test images. Softmax classifier supervised by category labels learned from descriptors of last layer of CNN over-fits monotonous surfaces of synthetic images, so large variance between synthetic data and real photos could not be balanced well. Our strategy solves this problem by utilizing pose information for constructing a manifold distribution of descriptors for fine-tuning using Nearest Neighbor classifier.

Result using Nearest Neighbor classifier shows that performance of texture-less model based training even surpass photo based training. Accuracy of our pipeline based on few texture-less 3D models achieves 47.2% which is 1.4% higher than result based on more textured models from ShapeNet and 0.3% higher than real photos based training which means that data migration problem is almost removed. Though result of our model based strategy judged by nearest neighbor classifier has a disadvantage on accuracy compared to softmax classifier trained from other real photos, this method has great significance on data preparation that minimizes dependence on the fidelity of training data.

Table 2. Classification on ImageNet samples. t represents the triplet loss based training, s represents the softmax lead one and ts means that we use a fine-tuning process. Training data includes real photos (Photo), models from PASCAL3D+ (Pascal3D) and models from ShapeNet (ShapeNet), the testing data is always real photos. Experiments on *Nearest Neighbor* and *Softmax* Classification are introduced separately.

Classifier	Softmax			Nearest Neighbor			
Method	Data						
	Photo	Pascal3D	ShapeNet	Photo	Pascal3D	ShapeNet	Model size
ZigzagNet-*s*	55.9	24.3	41.1	37.8	40.3	40.7	6.2 MB
ZigzagNet-*ts*	–	25.1	42.7	–	47.2	45.8	6.2 MB
ResNet [5]-*s*	37.8	19.1	28.5	27.1	25.8	25.3	91 MB
Inception v4 [7]-*s*	48.4	23.7	37.2	40.9	39.2	38.1	158 MB
SqueezeNet [4]-*s*	45.5	21.4	35	33.2	37.1	35.4	2.9 MB
bypassedSqueezeNet-*s*	53.3	22.2	39.9	35.5	39.5	37.9	2.9 MB
AlexNet [6]-*s*	55.8	24	31.9	46.9	41.9	40.1	240 MB
GoogleNet [1]-*s*	59	21.9	29.9	44.6	39.5	39	40 MB
LDO [9]-*ts*	–	14.8	25.5	–	29.6	27	1.2 MB

Experiment on PASCAL database in Table 3 shows that our training strategy is also efficient on recognition for real photos where objects are not centered. Accuracy of our strategy achieves 30.4% which is 1.6% higher than result trained from real photo using AlexNet and 2.6% higher than result trained from textured models of ShapeNet. Recognition accuracy on real photos drops from ImageNet to PASCAL because foreground objects in training samples are still limited in a small range around the center of whole image, but gaps between training from real photos and synthetic images are stable at about 15% which shows that our strategy won't fail when test photos are sophisticated. Experiments of PASCAL3D and ShapeNet using softmax classifier show that additional texture information could also help improving recognition accuracy based on ZigzagNet which means that the fitting ability is still on par with InceptionNet v4 while having a compact parametric model of 6.2 MB which is 1/38 in size of AlexNet's

Table 3. Classification on photos in PASCAL. Abbreviations are the same as Table 2.

Classifier	Softmax			Nearest Neighbor			
Method	Data						
	Photo	Pascal3D	ShapeNet	Photo	Pascal3D	ShapeNet	Model size
ZigzagNet-s	39.6	14.1	25.1	25.5	26.1	24.9	6.2 MB
ZigzagNet-ts	–	16.7	27.9	–	30.4	27	6.2 MB
ResNet [5]-s	23.6	12.3	16.2	–	–	–	91 MB
Inception v4 [7]-s	35.1	16.7	20.1	26	25.1	26.3	158 MB
SqueezeNet [4]-s	32.9	14.8	19.6	18.1	22.7	21.6	2.9 MB
bypassedSqueezeNet-s	36.2	14.9	20.5	19.2	23.4	22.9	2.9 MB
AlexNet [6]-s	39.2	15.2	20.7	28.8	23.8	22.1	240 MB
GoogleNet [1]-s	42.3	14.7	17.3	28.1	24.8	22.7	40 MB
LDO [9]-t	–	11.2	18.7	–	17.2	15	1.2 MB

model and almost reaches the quantized model of SqueezeNet[1]. Our ZigzagNet designed for training with migrated data in the same micro module depth with AlexNet solvers over-fitting problem better than some other networks.

6 Conclusion

In this paper, we propose an efficient deep learning pipeline including data rendering, cost function modification and compact CNN architecture design aiming at objects recognition on real photos based on texture-less 3D models. Information in texture-less 3D models is well utilized for training according to useful internal and external factors such as object poses. Our CNN structure is both compact in size which is 6.2 MB and efficient in recognition task which surpasses some practical structures. Classification accuracy on real test photos using Nearest Neighbor classifier is 5% higher than performance of recent CNN descriptors such as Inception v4 trained on texture-less 3D models and even not lower than results trained from real photos from these methods. As for performance on softmax classifier, accuracy of ZigzagNet is even close to results of GoogleNet trained from real images which are not so large in scale and much higher than compact SqueezeNet at the cost of twice in size of trained parametric model.

Our descriptors from metric learning using multi triplet could also be used for supervised classifier learning such as SVM and LDA. These linear classifiers could be learned from descriptors of our ZigzagNet which is more accessible than large scale raw data like ImageNet as these transformation matrix have limited fitting capacity. They could be applied afterwards in replace of Nearest Neighbor classifier whose training process is less time consuming and less dependent on

---

[1] Normal 4.8 MB model is stored by 8-bits reversible quantization as 2.9 MB with float min and max ranges. Quantization isn't included in the training process.

big data compared to deep learning. CNN descriptor and simple linear or non-linear classifier could be concatenated which are trained separately according to different tasks.

Acknowledgments. This work was partially sponsored by supported by the NSFC (National Natural Science Foundation of China) under Grant No. 61375031, No. 61573068, No. 61471048, and No. 61273217, the Fundamental Research Funds for the Central Universities under Grant No. 2014ZD03-01, This work was also supported by Beijing Nova Program, CCF-Tencent Open Research Fund, and the Program for New Century Excellent Talents in University.

References

1. Szegedy, C., Liu, W., Jia, Y., Sermanet, P., Reed, S., Anguelov, D., Erhan, D., Vanhoucke, V., Rabinovich, A.: Going deeper with convolutions. CoRR abs/1409.4842 (2014)
2. Jia, Y., Shelhamer, E., Donahue, J., Karayev, S., Long, J., Girshick, R.B., Guadarrama, S., Darrell, T.: Caffe: convolutional architecture for fast feature embedding. CoRR abs/1408.5093 (2014)
3. Long, J., Shelhamer, E., Darrell, T.: Fully convolutional networks for semantic segmentation (2015)
4. Iandola, F.N., Moskewicz, M.W., Ashraf, K., Han, S., Dally, W.J., Keutzer, K.: SqueezeNet: Alexnet-level accuracy with $50\times$ fewer parameters and <1 MB model size. CoRR abs/1602.07360 (2016)
5. He, K., Zhang, X., Ren, S., Sun, J.: Deep residual learning for image recognition (2015)
6. Krizhevsky, A., Sutskever, I., Hinton, G.E.: Imagenet classification with deep convolutional neural networks. In: Advances in Neural Information Processing Systems, vol. 25, pp. 1097–1105. Curran Associates Inc. (2012)
7. Szegedy, C., Ioffe, S., Vanhoucke, V.: Inception-v4, inception-resnet and the impact of residual connections on learning. CoRR abs/1602.07261 (2016)
8. Szegedy, C., Vanhoucke, V., Ioffe, S., Shlens, J., Wojna, Z.: Rethinking the inception architecture for computer vision (2015)
9. Wohlhart, P., Lepetit, V.: Learning descriptors for object recognition and 3D pose estimation. In: Proceedings of the IEEE CVPR (2015)
10. Pepik, B., Benenson, R., Ritschel, T., Schiele, B.: What is holding back convnets for detection? CoRR abs/1508.02844 (2015)
11. Xiang, Y., Mottaghi, R., Savarese, S.: Beyond PASCAL: a benchmark for 3D object detection in the wild. In: IEEE WACV (2014)
12. Courbariaux, M., Bengio, Y.: Binarynet: training deep neural networks with weights and activations constrained to $+1$ or -1. Clinical Orthopaedics and Related Research (2016)
13. Courbariaux, M., Bengio, Y., David, J.P.: Binaryconnect: training deep neural networks with binary weights during propagations (2015)
14. Gysel, P., Motamedi, M., Ghiasi, S.: Hardware-oriented approximation of convolutional neural networks (2016)
15. Su, H., Qi, C.R., Li, Y., Guibas, L.J.: Render for CNN: viewpoint estimation in images using CNNs trained with rendered 3D model views. In: The IEEE ICCV (2015)

16. Vedantham, A.: Guides. Flickr. Overview. (2013)
17. Dean, T., Ruzon, M., Segal, M., Shlens, J., Vijayanarasimhan, S., Yagnik, J.: Fast, accurate detection of 100,000 object classes on a single machine. In: 2013 IEEE Conference on CVPR, pp. 1814–1821 (2013)
18. Hinterstoisser, S., Lepetit, V., Ilic, S., Holzer, S., Bradski, G., Konolige, K., Navab, N.: Model based training, detection and pose estimation of texture-less 3D objects in heavily cluttered scenes. In: Lee, K.M., Matsushita, Y., Rehg, J.M., Hu, Z. (eds.) ACCV 2012. LNCS, vol. 7724, pp. 548–562. Springer, Heidelberg (2013). doi:10.1007/978-3-642-37331-2_42
19. Russakovsky, O., Deng, J., Su, H., Krause, J., Satheesh, S., Ma, S., Huang, Z., Karpathy, A., Khosla, A., Bernstein, M.S., Berg, A.C., Li, F.: Imagenet large scale visual recognition challenge. CoRR abs/1409.0575 (2014)
20. Hinterstoisser, S., Benhimane, S., Lepetit, V., Fua, P., Navab, N.: Simultaneous recognition and homography extraction of local patches with a simple linear classifier. In: Proceedings of the BMVC, pp. 10.1–10.10. BMVA Press (2008)
21. Wang, J., Song, Y., Leung, T., Rosenberg, C., Wang, J., Philbin, J., Chen, B., Wu, Y.: Learning fine-grained image similarity with deep ranking. CoRR abs/1404.4661 (2014)
22. Jolliffe, I.T.: Principal component analysis. Technometrics (2014)
23. He, K., Zhang, X., Ren, S., Sun, J.: Delving deep into rectifiers: surpassing human-level performance on imagenet classification. CoRR abs/1502.01852 (2015)
24. Ioffe, S., Szegedy, C.: Batch normalization: accelerating deep network training by reducing internal covariate shift. CoRR abs/1502.03167 (2015)
25. Chang, A.X., Funkhouser, T., Guibas, L., Hanrahan, P., Huang, Q., Li, Z., Savarese, S., Savva, M., Song, S., Su, H., Xiao, J., Yi, L., Yu, F.: ShapeNet: an information-rich 3D model repository. Technical report arXiv:1512.03012 [cs.GR], Stanford University – Princeton University – Toyota Technological Institute at Chicago (2015)

Precise Measurement of Cargo Boxes for Gantry Robot Palletization in Large Scale Workspaces Using Low-Cost RGB-D Sensors

Yaadhav Raaj[1(✉)], Suraj Nair[1], and Alois Knoll[2]

[1] TUM CREATE, 1 CREATE Way, #10-02 CREATE Tower,
Singapore 138602, Singapore
{raaj.yaadhav,suraj.nair}@tum-create.edu.sg
[2] Technische Universität München (TUM), Institüt für Informatik,
Robotics and Embedded System, Munich, Germany
knoll@in.tum.de

Abstract. This paper presents a novel algorithm for extracting the pose and dimensions of cargo boxes in a large measurement space of a robotic gantry, with sub-centimetre accuracy using multiple low cost RGB-D Kinect sensors. This information is used by a bin-packing and path-planning software to build up a pallet. The robotic gantry workspaces can be up to 10 m in all dimensions, and the cameras cannot be placed top-down since the components of the gantry actuate within this space. This presents a challenge as occlusion and sensor noise is more likely.

This paper presents the system integration components on how point cloud information is extracted from multiple cameras and fused in real-time, how primitives and contours are extracted and corrected using RGB image features, and how cargo parameters from the cluttered cloud are extracted and optimized using graph based segmentation and particle filter based techniques. This is done with sub-centimetre accuracy irrespective of occlusion or noise from cameras at such camera placements and range to cargo.

1 Introduction

Motivation for this work was realized from having to build a fully automated palletizing/depalletizing gantry robot for the Air Cargo Terminal. Packing Cargo shipment into standardized containers/pallets is critical. One such pallet is the P6P (Fig. 1) measuring approx. (3 x 2.5)m, and other pallets can be found in Boeing documentation [2]. Cargo primarily consists of block-shaped shipments ranging from (0.3 to 2.0)m in 3 dimensions made of different materials from cardboard to wood, and are traditionally built up by hand onto the pallet. Our system aims to automate this process, by making use of a 4 axis Gantry Robot (Fig. 1) with approx. workspace dimensions (7.5 x 5 x 7.5)m, that can be further

Electronic supplementary material The online version of this chapter (doi:10. 1007/978-3-319-54190-7_29) contains supplementary material, which is available to authorized users.

S.-H. Lai et al. (Eds.): ACCV 2016, Part IV, LNCS 10114, pp. 472–486, 2017.
DOI: 10.1007/978-3-319-54190-7_29

extended if needed be. The end effector (EF) has 4 degrees of freedom X, Y, Z, θ and can lift cargo using a vacuum suction gripper.

1.1 Robot Workspace

Cargo is placed in a measurement space (Yellow) of the robot, where it is measured/weighed before being placed in the pallet (Grey). A bin packing and path-planning software work together to palletize and depalletize the robot. However, this component requires sub-centimetre level precision in order to optimize it's packing and prevent collision as is evident in many bin packing planners [3]. Cargo boxes are introduced at random into this space, and we do not know the dimensions of these beforehand. This is why a robust vision system with a focus in precise measurement is required. This process is not trivial as the cameras cannot be placed vertically overhead as the actuation mechanism works along that area as with most gantry robots. Hence, multiple cameras have to be placed in an inclined manner along the pillars of the robot, so as to capture maximum possible information, from which depth and color information is sent to a server for processing.

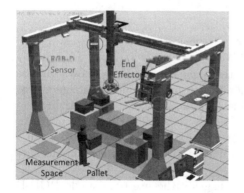
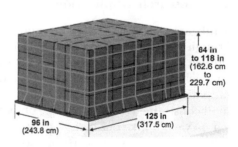

Fig. 1. Gantry Robot and P6P Pallet to be built up used in the Air Cargo Industry

Our objective in this paper, is to extract the measurements of unknown cuboid shaped cargo in the measurement space as accurately as possible, irrespective of occlusion due to the camera placement, or highly similar boxes being in close contact. Solving the problem of occlusion requires information from multiple sensors, hence having industry grade sensors may be too costly, hence we had to tailor design our method to work with commercial grade sensors that may have significant noise.

There has been not much prior work on achieving centimetre level precision of unknown models at ranges beyond 3 to 4 m using commercial grade sensors before this, with the constraint of not having top-down cameras. Hence we aim to fill this gap to achieve our objectives above. Our algorithm works with Kinect

2 RGBD cameras, by fusing point cloud data from multiple cameras, fitting models to this data, segmenting it using graph based approaches and refining this further by using edge information from the RGB cameras. This is further improved by using a 9DOF particle filter approach. Results from these are then used by the bin-packing and path-planning software to optimize and build up the pallet. The system may also be used to depalletize the pallet.

2   Related Work

2.1   Industrial Applications

There are already a significant number of industry players/systems in the automation space, such as Bosch, Universal Robotics and Z Automation. A simple search on PackExpo would give a hundred more. However, most of these systems have been designed for conveyor line scanning, where the object's model can be extracted (Fig. 2). Universal Robotic's [4] solution works with unknown models, but it's commercially available solutions advertise small workspaces, with robots having an eye-in-hand configuration (Fig. 2). There is no known solution for large gantry workspaces, where cargo can be placed anywhere in the workspace for palletization.

2.2   Research Applications

Significant work has already been done in the space of extracting known objects through feature/template matching either in the point cloud or image space, and is a well-defined problem. This is evident in many robotic depalletizing and bin packing papers, such Drost's [5] and Holz's paper [6], where he proposes comparing object surfel models with sensor data for object/pose detection, in a small constrained workspace. The objects studied there are geometrically feature rich as well, unlike our cargo boxes.

For unknown objects, Ryan Lloyd [7] suggests the use of a single RGB-D sensor, by extracting the top surface of a point cloud, after applying euclidean

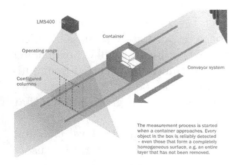

Fig. 2. SICK LMS Line Scanning and Universal Robotic's Depalletization

clustering and point cloud smoothing with Moving Least Squares (MLS) [8]. Features of the top surface are then extracted and used to train an SVM to classify different primitives, including quads. Dimensions are extracted using a Minimum Volume Bounding Box (MVBB) projected to the ground, and centimetre level precision is achieved. In this paper however, experiments are carried out with an organized point cloud due to the single camera metrology, at ranges below 1m.

Also, in Richtsfeld's seminal paper [9], he uses a combination of SVM and graph cut based segmentation to segment and get dimensions of primitive objects in cluttered scenes. The focus of this paper was not measurement, but rather precise segmentation of objects based on their color and shape information irrespective of occlusion. Much of the experiments were also conducted at ranges below 1 m with a single camera, hence spurious point cloud noise was not much of an issue.

Alternatives to SVM/Graphcuts such as SIFT matching [10] and Primitive Shape Matching [11] were also studied. Many similar papers including those above performed pose estimation using an initial SVD/PCA based seed, with an iterative refinement step. In all of these papers, a single structured light based RGB-D camera such as the Asus Xtion or Kinect 1 is used, and were generally designed for smaller workspaces such as tabletops. Many of these algorithms were also designed for single camera inputs with organized point clouds. Our workspace exceeded 5 m, and an unorganized point cloud was generated due to multiple cameras. At such ranges, the depth image also had significant Z noise, resulting in noisy point clouds, hence noisy edges.

This meant that the above methods of using MVBB, SVD/PCA to perform pose and dimension estimation or using Richtsfeld's method to perform model fitting and segmentation were challenging. To solve this issue, edge based fusion methods were also studied. Anwer's [12] paper describes using depth image refinement techniques such as depth normalization and bilateral filters. Aouada's [13] paper describes using RGB images as guidance images, where holes are filled referencing edges found in the higher resolution RGB image. These techniques are later referenced and implemented in our paper.

Lastly, large scale point cloud acquisition and measurement techniques are also studied in the form of LIDAR and Aerial Imagery fusion. It is challenging to find large scale point cloud processing methods without exploring the LIDAR domain. In Wang's [14] paper, he describes how a graph is constructed from the noisy aerial LIDAR data, how edges are extracted using iterative RANSAC techniques, and how these edges are further refined by projecting these points into the 2D Aerial camera and corrected them to edges and corners found.

3   Sensors and Systems

3.1   Camera System

A range of commercial and industrial 3D systems were explored. Industrial solutions consisted of Structured Light systems by GOM and SICK, Laser systems by Faro and Osela, and Time of Flight (TOF) systems by Basler. The GOM,

SICK and Faro systems had sub-millimetre accuracy, but had a limited range of below 1m, and were costly in large numbers. The Basler TOF system was reasonably priced, touted centimetre accuracy, and had a range of up to 5 m but it was yet to begin sales during testing time. Commercial systems tested included the Asus Xtion Pro, Intel RealSense, Microsoft Kinect 1 and 2. In the end, we decided to go with the Kinect 2, which is a TOF system with a range of up to 4.5 m, and could output 512 x 424 pixel Depth images and Full HD RGB images.

3.2   Experimental Setup

Our experimental setup consists of 4 NUC/Kinect units with a measurement space of approximately (4 x 3.5)m, with the Kinects at a height of 2.5 angled at 45 degrees. The highly reflective P6P pallet is placed in the centre of the setup and cargo boxes ranging from 30 cm to 1 m in either dimension, consisting of cardboard, styrofoam and shrink-wrapped boxes were used. Our objective is to simultaneously measure and track the boxes that are initially placed flat against the ground (though they may be in close contact), to simulate the situation where a cargo item of unknown dimensions or pose has been introduced into a robot workspace, and needs to be worked upon.

3.3   Data Extraction

RGB and Depth imagery is extracted via the libfreenect2 [15] driver through the iai-kinect2 [16] ROS (Robot Operating System) [17] interface. This driver performs several pre-processing steps. Firstly, it allows one to perform intrinsic and extrinsic calibration of the Kinect sensor by solving for intrinsics and radial distortion using Zhang's checkerboard method [18], and solving for the transformation between the IR and RGB sensors. Secondly, edge aware and bilateral filters are used to enhance the depth image. Finally, a registration step is performed to generate a depth image in the RGB frame. All of this is performed on a 6th Generation Intel NUC i5 with GPU optimization on the registration step, allowing image rates of up to 20 Hz. Each camera sends a 16 bit upscaled (540 x 960) depth image and full HD JPEG compressed RGB image to the main server via a gigabit switch.

3.4   Point Cloud Fusion

We use the approximate time scheduling policy implemented in ROS to extract a joint set of data from all cameras. Since, $^{R}T_W$ is known, points can be generated on the server, and transformed to the world frame, after which thresholding can be applied to eliminate points outside the scope of the workspace. This is done on all 2 million over points at once on the GPU. Since there are redundant points due to overlapping views, a Voxel Grid filter is applied with a distance threshold. Euclidean clustering is then applied to eliminate spacial noise, and to break

Fig. 3. ROS Integration and Experimental Setup

the point cloud into chunks that can be analysed in parallel for performance boost. These steps are done using PCL (Point Cloud Library) [20]. Finally, a Moving Least Squares (MLS) filter is applied on these chunks [8]. This is a good pre-processing step [7], as it allows for better thresholding based on normal vector values, and smoothes out noise in the point cloud spatially. Bruce Merry's GPU implementation of MLS [21] is used. This allows us to generate clean $(R, G, B, X, Y, Z, nx, ny, nz)$ point clouds of the entire 3D scene at up to 10 Hz, where (nx, ny, nz) represents the point surface normals (Fig. 3).

4   Algorithm

4.1   Model Extraction

With the point cloud made available, we are able to extract pose, dimensions and features from this. The items we are dealing with are primarily cuboid shaped cargo, that are placed against the floor of the workspace initially. These constitute 80 % of the cargo, and allows for the best packing density.

$$ax + by + cz + d = 0 \ (Plane) \quad (x, y, z) = (x_0, y_0, z_0) + t(a, b, c) \ (Line) \qquad (1)$$

$$m = \frac{1}{N} \sum_{i=1}^{N} (p_i) \qquad c = \frac{1}{N} \sum_{i=1}^{N} \left((p_i - c)(p_i - c)^T \right) \qquad (2)$$

$$C = \left(\sum_{j=1}^{N} \lambda_j p_j : \lambda_j \geq 0 \ for \ all \ j \ and \ \sum_{j=1}^{N} (\lambda_j = 1) \right) \qquad (3)$$

There are several methods to extract the pose and dimensions from cuboid objects. Applying Iterative RANSAC works for ellipsoids, cylinders, planes and lines, but not for cuboids since there is no mathematical model for it. Iteratively finding planes (Eq. 1) and their intersections through a least squares approach fails in our case as boxes being densely packed removes outward facing planes.

Applying methods such as PCA (Principle Component Analysis) by computing the mean/covariance of a set of points p_i and extracting it's min/max eigenvectors (Eq. 2), or MVBB (Minimum Volume Bounding Box) [1] by optimizing a bounding box based on various heuristics as used in [6,7] works poorly in our case (Fig. 4). This is because such methods optimize for all points, including outliers, and don't take the object model into account. Measurements derived through this method had an μ error of up to 5 cm from the ground truth, which is unacceptable for the bin packing algorithm.

Fig. 4. Contour extraction and PCA/MVBB fitting

Hence, we use a top surface contour extraction approach, to simplify measurement and pose extraction. We select only top surface points ($nz > 0.5$), fit a plane to it via RANSAC, and compute the convex hull C for points p_i (Eq. 3). A Sobel operator is applied over sections of the RGB image that C projected onto each RGB image occupies, and it's contours C_{proj} are extracted where $C_{proj} \subseteq C$. We then iteratively fit these to 3D line models (Eq. 1). Working in a multi-camera point cloud domain ensures that occlusion is not much of an issue (Fig. 5).

4.2 Segmentation Using Node Graphs

If cargo boxes are separated spatially, or have different heights, the above clustering and segmentation approaches work. However, if they are packed closely, have the same height and the same color (very likely scenario in depalletization situations), they end up getting clustered together (Fig. 6). In both cases, the contours were extracted from the same euclidean cluster. Hence splitting them requires some kind of geometric analysis.

We solve this problem graphically. First, a bi-directional connected graph $G_c = (E, V)$ is constructed. G_c contains vertices $(v_i...v_n)$, where each vertex v_i contains a corresponding point p_i, where $deg(v_i) \geq 1$. To construct this graph, we begin with our edges represented as vectors $(\overrightarrow{e_i}...\overrightarrow{e_n})$ with a point passing through them defined through the line model $(x, y, z) = (x_0, y_0, z_0) + t(a, b, c)$. An edge $\overrightarrow{e_i}$ has it's min/max points $p_i^{(min)}, p_i^{(max)}$. We construct nodes n_i by intersecting every edge $\overrightarrow{e_i} \bigcap \overrightarrow{e_j}$ ($i \neq j$). Repeat nodes are inevitable, so these

Fig. 5. Boxes here are spatially classified into one cluster due to proximity, hence requiring methods to split them up

are eliminated by averaging neighbours using a Oct-Tree search. A node v_i is created if the distance between it's parent edges $\overrightarrow{e_i}, \overrightarrow{e_j}$ is minimized, and if the angle between them are either close to 90 or 0 degrees (Eq. 4).

$$G_c = \left(\sum_{i=0}^{N} \sum_{j=i+1}^{N} \left(v_i = \overrightarrow{e_i} \bigcap \overrightarrow{e_j} \ (i \neq j) \right) \ where \ min \left(f(v_i, \overrightarrow{e_i}, \overrightarrow{e_j}) \right) \right)$$

$$f(v_i, \overrightarrow{e_i}, \overrightarrow{e_j}) = \left(\|v_i, closer(p_i^{(min)}, p_i^{(max)}))\|_2 - 0.1 \right)$$

$$+ \left(\|v_i, closer(p_j^{(min)}, p_j^{(max)}))\|_2 - 0.1 \right) + min \left((\angle(\overrightarrow{e_i}, \overrightarrow{e_j}) - 90) , (\angle(\overrightarrow{c_i}, \overrightarrow{c_j}) - 0) \right)$$

$$(4)$$

Once this graph has been constructed, we simply have to locate inner cycles $cycles_c$ within it. Such a problem can result in exponential search time, but with geometric constraints this is no longer true. This is done in a Depth-First-Search (DFS) approach. Given a set of searched connected cycles where each set N_i contains a set of nodes $N_i \epsilon(n_i^0 ... n_i^n)$ that contain their edges $E_i \epsilon(\overrightarrow{e_i^0} ... \overrightarrow{e_i^{n-1}})$, we ensure that the *angle* between them are close to 90 or 0 degrees, their *l2norm* is maximized, the vectors move in a consistent clockwise/counter-clockwise direction and that they are connected (Eq. 5).

$$cycles_c = min \left(\sum_{i=0}^{M} \sum_{j=0}^{N} \left(g(E_i \epsilon(\overrightarrow{e_i^j} ... \overrightarrow{e_i^{n-1}}) \ where \ \overrightarrow{e_i^j} = \overrightarrow{e_i^{n-1}}) \right) \right)$$

$$g(E_i) = \left(0.1 - \|\overrightarrow{e_i^j}\|_2 \right) + min \left(\left(\angle(\overrightarrow{e_i^j}, \overrightarrow{e_i^{j+1}}) - 90 \right), \left(\angle(\overrightarrow{e_i^j}, \overrightarrow{e_i^{j+1}}) - 0 \right) \right)$$

$$+ \left((\overrightarrow{e_i^j} \times \overrightarrow{e_i^{j+1}}).z - (\overrightarrow{e_i^{j+1}} \times \overrightarrow{e_i^{j+2}}).z \right)$$

$$(5)$$

This is done by using cross-product and normalized dot-product of a set of connected edges generated from the nodes as a constraint. Strongly connected component algorithms from Tarjan, Tiernan and elementary circuit discovery methods from Johnson [22] were studied and integrated for the search steps. These graph generation and analysis steps can be visualized (Fig. 6). In the left

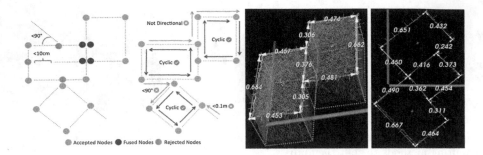

Fig. 6. Graph construction and Analysis (Color figure online)

image, one observes how a node isn't created if it's intersection with another edge isn't close enough, or it's angle doesn't satisfy the above constraints (in red). Nodes created that are too close are fused together (in green). In the right image, we can see how the search occurs, and how cycles are detected. Measurements derived through this method had an μ error of up to 2 cm from the ground truth, which is unacceptable for the bin packing algorithm.

4.3   RGB Camera Sobel Edge Fusion

One can observe the above error in measurements by projecting the fitted line into one of the camera RGB images (Fig. 7). You can see that the line doesn't fall exactly on the edges found on the box. This results in deviations from the ground truth measurement. Hence, it is observed that it might be worth using the RGB image to "correct" the contour points before fitting line models to it. This method is inspired by Wang's paper [14], where the initial model is superimposed onto a sobel edge image from the RGB camera, and then corrected. However, his solution is constrained to working from a top down view, but our scenario involves sensors that are placed inclined to the workspace, since downward facing cameras cannot be used due to obstruction with the robotic gantry components. This means that selecting a sensor to correct these contour points is not trivial, due to occlusion, and distance to the sensor. Such occlusions can be observed (Fig. 7).

Hence, the following correction algorithm is proposed. This is applied before the iterative RANSAC line search and graph generation. Given a contour $C\epsilon\{p_i..p_m\}$ with a set of points, a set of hypothesis for each p_i is generated with a 2D normal constraint in the search space (Eq. 6), where t varies to generate $p_i\epsilon(p_i^0...p_i^n)$.

$$p_i\epsilon\left(\{p_i^j..p_i^m\}\ for\ all\ j\ (p_i^0 + p_i^j.normal * t)\right) \tag{6}$$

$$min\left(\sum_{j=0}^{M}\frac{1}{N}\sum_{c=0}^{C}\left(255 - intensity(p_i^j, I_c^{RGB}) + distance(p_i^j, I_c^{Depth})\right)\right) \tag{7}$$

Fig. 7. Before and After Sobel Edge Fusion, and Correction of points along contour normals

Each hypothesis p_i^j is transformed and projected into both the RGB Sobel Edge I_c^{RGB} and Depth Images I_c^{Depth} for each camera. We then attempt to select the best p_i^j to replace p_i that minimizes the distance to the camera, the Z value on I_c^{Depth} to prevent converging on a edge/solution that is being occluded, and maximize a point falling on a edge on I_c^{RGB} (Eq. 7). We can observe the correction in (Fig. 7). Measurements derived through this method had an μ error of up to 0.5 cm from the ground truth, which is deemed suitable. You will notice that the height of the box wouldn't be optimized this way, but we found this was already made accurate during the planar RANSAC fit.

4.4 Model Constrained Pose and Sobel Edge Refinement

At this point, one can already extract the measurements from the various cargo boxes. However, we found that there were situations where several points on the contour may converge at a highly deviant position, leading to a failure in RANSAC line fitting. This is because we corrected each point individually instead of constraining it to a line model. We needed to further refine the pose in 6D and measurements and make use of the Sobel Edge information to do that in a box model constrained manner, while also making use of multiple cameras and accounting for occlusion. We also needed to track small perturbations/movements in the cargo boxes in case a worker may shift them. We found that the ideal solution to this would be a particle filter based optimization approach [23] (Fig. 8).

A particle filter with a 9 dimensional state $X_{n|n-1}^{(k)}$ is setup to represent a cuboid with position (x_w, y_w, z_w) about the centre of mass, dimensions (l_w, w_w, h_w) and 6-dof orientation $(\alpha_w, \beta_w, \gamma_w)$ in the world frame for a given particle k with 500 particles. A Brownian motion model based based on additive gaussian noise w_{n-1} is used (Eq. 8). Our weights described later are calculated using a multi-variate gaussian distribution model represented in a log-likelihood function for faster computation (Eq. 9).

$$X_{n|n-1}^{(k)} = [x_w \ \ y_w \ \ z_w \ \ l_w \ \ w_w \ \ h_w \ \ \alpha_w \ \ \beta_w \ \ \gamma_w] \qquad (8)$$

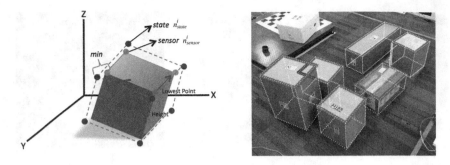

Fig. 8. Cargo Particle Filter based State optimization

$$X^{(k}_{n|n-1} = X^{(k)}_{n-1|n-1} + w_{n-1}$$

$$P(Y_n|X^{(k)}_{n|n-1}) = \frac{1}{\sqrt{2\pi\sigma^2}} \cdot exp(-\frac{d^2}{2\sigma^2})$$

$$log(P(Y_n|X^{(k)}_{n|n-1})) = -log(\sigma\sqrt{2\pi}) - \frac{0.5}{\sigma^2} \cdot d^2 \tag{9}$$

We have our nodes which were generated from the previous step represented as n^i_{sensor}. We generate a box model from a given hypothesis k, and we extract 4 nodes from it represented as n^i_{state}. We then attempt to minimize the $l2norm$ distance from each node i to a matching nearest neighbouring node, with the graph constraints in place (Eq. 10). This is done by reducing the weight through a multiplier β_{l2} of a particle whose set of matches don't belong to the same inner cycle set.

$$q^{(k)}_{l2W} = \sum_i^N \left(-log(\sigma_{l2}\sqrt{2\pi}) - \frac{0.5}{\sigma^2_{I2}} \cdot \|(n^i_{state}, n^i_{sensor})\|^2_2 \right) * \beta_{l2} \tag{10}$$

We wished to also integrate the edge information from the sobel edge images I^{RGB}_c of each camera in our particle filter step in case there were errors in the Sobel Fusion step, or if we wanted to avoid that step. This is done by sampling the top surface contour of a box $C\epsilon\{p_i..p_m\}$ generated from a given particle k, and projecting it into each camera since ^RT_W is known. We then attempt to maximize the intensity of each projected point in the sobel image space (Eq. 11). We reference the depth image I^{Depth}_c and penalize the weights of points which are occluded by other boxes through a multiplier β_I.

$$q^{(k)}_{IW} = \sum_i^N \left(-log(\sigma_I\sqrt{2\pi}) - \frac{0.5}{\sigma^2_I} \cdot (p^i_{contour} - p^i_{sobel})^2 \right) * \beta_I \tag{11}$$

To do this, once a particle k has reached a state where $\|(n^i_{state}, n^i_{sensor})\|_2$ is less than 1 cm for all n^i_{state}, we turn on additional weights that make use of the edge information. We found that this made our measurements more robust to occlusion from other cargo boxes. We then add up the weights from both $q_{l2W}^{(k)}$ and $q_{IW}^{(k)}$, apply the exponential operation on them and resample with replacement (Eq. 13).

$$q_n = \frac{P(Y_n|X_{n|n-1}^{(k)})}{\sum_k P(Y_n|X_{n|n-1}^{(k)})} \quad k \epsilon \mathbb{N} \tag{12}$$

$$X_{n+1|n}^{(k)} = h(X_{n|n}^{(k)}|q_n) \quad k \epsilon \mathbb{N} \tag{13}$$

Once the dimensions of the box have been optimized, we can simply lock (l_w, w_w, h_w) in place if we want to track pose. This can be applied for multiple boxes in the scene.

5   Results

5.1   Experimental Setup

Testing was conducted on 3 different types of cargo boxes (Fig. 9), a standard cardboard box (0.665 x 0.445 x 0.370)m, a styrofoam box with rounded edges (0.460 x 0.605 x 0.300)m and a shiny shrink wrapped box (0.360 x 0.370 x 0.440)m. They were placed in the centre of the measurement workspace [4 x 3.5]m where the kinect cameras were 2.5 m high and at a 45 degree angle. Several boxes were also placed around this to simulate occlusion (Fig. 9) where at least one camera had the box covered partially.

Fig. 9. Test boxes used and tracking of multiple boxes in a pallet

In the first experiment, we generated the corner nodes n_{sensor}^i for all 3 boxes using the MVBB method, the Node Graph Construction method with line model fitting, and the Node Graph + Sobel Edge Fusion method. We then extracted 60 samples from the most deviant dimension from the ground truth (was always in the length/breath domain) for all 3 methods. In the second experiment, we ran our particle filter algorithm on all 3 cargo boxes initialized with data extracted from the Graph Node step, one with edge fusion turned on and one without, and extracted the dimensions from the filter.

5.2   Measurement Results

For the first experiment, from (Fig. 10), we are able to see that the MVBB approaches on our system produced mean results that deviated from the ground truth by up to 5 cm, and had significant noise especially in the plastic shrink-wrapped cargo possibly due to reflectivity from the Kinect sensor. Using a RANSAC line fitting and node graph construction approach reduced that both error and noise down significantly, but still had up to a 2 cm error in some cases. Finally, the Sobel Edge Fusion pre-processing step before application of the node graph step, reduced the error down to 0.5 cm for all types of cargo, and noise to even lower levels. This level of precision is within the bounds of allowance of the gantry robot bin-packing algorithm, and can aid in further algorithms that may track the state of boxes while they are in the pallet.

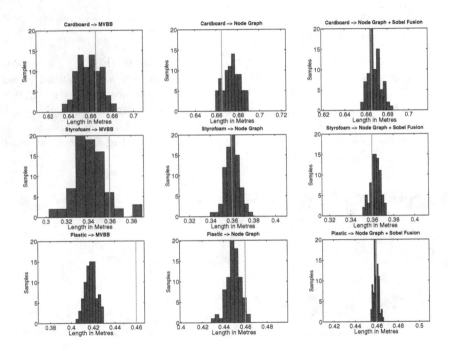

Fig. 10. Histogram of 60 samples extracted from the various methods described

For the second experiment, from (Fig. 11), you can see that for the Cardboard and Plastic cargo boxes, we see an improvement in the measurement from the ground truth in both length and breath (height was found to be always accurate). For the Styrofoam box, it was not as noticeable, possibly due to the fact that the box did not have as distinct edges in the sobel edge image. In all 3 cases however, we were able to achieve our accuracy requirement of 1 cm or below.

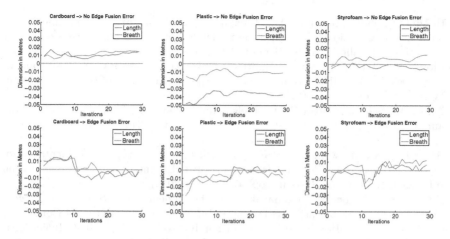

Fig. 11. Running particle filter optimization with and without edge fusion

5.3 Conclusion and Future Work

While using the Node Graph + Sobel Edge Fusion step was more accurate with enough samples taken, the Node Graph + Particle Filter with Edge fusion was more robust. Our solution allows one to place unknown cargo boxes of various materials inside a scene, and achieve sub-centimetre level precision pose and dimension extraction even with commercial grade sensors at a range of over 3.5 m, and allows a gantry robot and bin packing software to work with minimal constraints. These algorithms were run on an Intel Core i7-4790 with 4 physical cores and a NVIDIA Geforce 760 Ti. We were able to optimize up to 12 boxes in the scene at approximately 4 Hz. In the future, it may be possible to extract features from the cargo boxes, and use that to track the cargo box pose more robustly.

Acknowledgement. This work is funded by the Civil Aviation Authority of Singapore (CAAS) under the Aviation Challenge 2 grant.

References

1. Har-peled, S.: A practical approach for computing the diameter of a point set. In: SCG 2001 (2001)
2. Boeing: boeing pallets @ONLINE (2012). http://www.boeing.com/resources/boeingdotcom/company/about_bca/pdf/CargoPalletsContainers.pdf
3. Viegas, J.P.L., Vieira, S.M., Sousa, J.M.C., Henriques, E.M.P.: Metaheuristics for the 3D bin packing problem in the steel industry. In: Proceedings of the 2014 IEEE Congress on Evolutionary Computation, CEC 2014, pp. 338–343 (2014)
4. Robotics, U.: Universal robotics @ONLINE (2015). http://www.universalrobotics.com/applications/
5. Drost, B., Ulrich, M., Navab, N., Ilic, S.: Model globally, match locally: efficient and robust 3D object recognition (2012)

6. Holz, D., Behnke, S., Holz, D., Topalidou-kyniazopoulou, A.: Real-time object detection, localization and verification for fast robotic depalletizing verification for fast robotic depalletizing. In: IROS (2015)
7. Lloyd, R., McCloskey, S.: Recognition of 3D package shapes for single camera metrology. In: 2014 IEEE Winter Conference on Applications of Computer Vision, WACV 2014, pp. 99–106 (2014)
8. Alexa, M., Behr, J., Cohen-Or, D., Fleishman, S., Levin, D., Silva, C.T.: Computing and rendering point set surfaces. IEEE Trans. Vis. Comput. Graphics **9**, 3–15 (2003)
9. Richtsfeld, A., Morwald, T., Prankl, J., Zillich, M., Vincze, M.: Segmentation of unknown objects in indoor environments. In: IEEE International Conference on Intelligent Robots and Systems, pp. 4791–4796 (2012)
10. Wu, K., Ranasinghe, R., Dissanayake, G.: A fast pipeline for textured object recognition in clutter using an RGB-D sensor. In: 2014 13th International Conference on Control Automation Robotics and Vision, ICARCV 2014, pp. 1650–1655 (2014)
11. Somani, N., Cai, C., Perzylo, A., Rickert, M., Knoll, A.: Object recognition using constraints from primitive shape matching. In: 10th International Symposium on Visual Computing (ISVC 2014) (2014)
12. Anwer, A., Baig, A., Nawaz, R.: Calculating real world object dimensions from Kinect RGB-D image using dynamic resolution. In: Proceedings of 2015 12th International Bhurban Conference on Applied Sciences and Technology, IBCAST 2015, pp. 198–203 (2015)
13. Aouada, D., Ottersten, B., Mirbach, B., Garcia, F., Solignac, T.: Real-time depth enhancement by fusion for RGB-D cameras. IET Comput. Vision **7**, 335–345 (2013)
14. Wang, H., Zhang, W., Chen, Y., Chen, M., Yan, K.: Semantic decomposition and reconstruction of compound buildings with symmetric roofs from LiDAR data and aerial imagery. Remote Sens. **7**, 13945–13974 (2015)
15. libfreenect2 @ONLINE (2013). https://github.com/OpenKinect/libfreenect2
16. iai-kinect2 @ONLINE (2015). https://github.com/code-iai/iai_kinect2
17. Quigley, M., Conley, K., Gerkey, B.P., Faust, J., Foote, T., Leibs, J., Wheeler, R., Ng, A.Y.: ROS: an open-source robot operating system. In: ICRA Workshop on Open Source Software (2009)
18. Zhang, Z.: A flexible new technique for camera calibration (technical report). IEEE Trans. Pattern Anal. Mach. Intell. **22**, 1330–1334 (2002)
19. Bradski, G.: Opencv. Dr. Dobb's journal of software tools (2000)
20. Rusu, R.B., Cousins, S.: 3D is here: Point Cloud Library (PCL). In: IEEE International Conference on Robotics and Automation (ICRA), Shanghai, China (2011)
21. Merry, B., Gain, J., Marais, P.: Moving least-squares reconstruction of large models with GPUs. IEEE Trans. Vis. Comput. Graphics **20**, 249–261 (2014)
22. Johnson, D.B.: Finding all the elementary circuits of a directed graph. **4**, 77–84 (1975)
23. Fox, D., Burgard, W., Dellaert, F., Thrun, S.: Monte carlo localization: efficient position estimation for mobile robots dieter fox, wolfram burgard. In: 16th National Conference on Artificial Intelligence (AAAI99), pp. 343–349 (1999)

Visual Place Recognition Using Landmark Distribution Descriptors

Pilailuck Panphattarasap$^{(\boxtimes)}$ and Andrew Calway

Department of Computer Science, University of Bristol, Bristol, UK
pp12907@bristol.ac.uk, andrew@cs.bris.ac.uk

Abstract. Recent work by Sünderhauf et al. [1] demonstrated improved visual place recognition using proposal regions coupled with features from convolutional neural networks (CNN) to match landmarks between views. In this work we extend the approach by introducing descriptors built from landmark features which also encode the spatial distribution of the landmarks within a view. Matching descriptors then enforces consistency of the relative positions of landmarks between views. This has a significant impact on performance. For example, in experiments on 10 image-pair datasets, each consisting of 200 urban locations with significant differences in viewing positions and conditions, we recorded average precision of around 70% (at 100% recall), compared with 58% obtained using whole image CNN features and 50% for the method in [1].

1 Introduction

Visual place recognition is the task of matching a view of a place with a different view of the same place taken at a different time. If incorporated into a mapping framework, such as a topological representation of places, for example, then reliable and fast visual place recognition opens up the possibility of truly autonomous navigation, with applications in robotics and related areas. It would negate the need for a positioning infrastructure such as GPS and perhaps more interestingly, in respect of human-robot interaction, be more akin to the wayfinding techniques employed by humans.

Automated recognition of places based on visual information is however very challenging. It is highly dependent on the characteristics of places, the viewing positions and directions, and the environmental conditions in terms of light and visibility. Perspective effects, occlusions, changes in natural vegetation, differences in seasonal and day/night appearance, and the presence of transient objects such as vehicles and people, all conspire to make recognition in its most general form a very hard problem.

Research into recognising places using vision has made progress, both in the robotics and in the computer vision communities [2–5]. Broadly speaking, approaches fall into two main categories: those based on matching local features between views; and those based on comparing whole image characteristics. Of the former, techniques based around the scale-invariant feature transform (SIFT) [6] and its variants are the most common, whilst in the latter category the GIST

© Springer International Publishing AG 2017
S.-H. Lai et al. (Eds.): ACCV 2016, Part IV, LNCS 10114, pp. 487–502, 2017.
DOI: 10.1007/978-3-319-54190-7_30

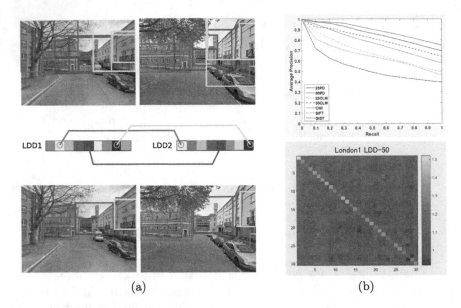

(a)   (b)

Fig. 1. Place recognition using landmark distribution descriptors (LDDs). Proposal regions from *Edge Boxes* (top left) are represented by CNN feature vectors and stacked in horizontal position order into an LDD for the view (left middle). The similarity of top matching regions within sections of the descriptor are then used as a measure of similarity between the views (bottom left). The approach outperforms comparison methods over 10 datasets each with 200 urban locations (top right) and shows excellent discrimination characteristics as illustrated by the confusion matrix in the bottom right.

descriptor [7] has found widespread use. To aid robustness, these techniques are often incorporated within some form of temporal integration, the probabilistic FAB-MAP method [8,9] being the most well-known. Other techniques aim to deal with seasonal, day/night and long term changes, see e.g. [2].

As pointed out in [2], the two categories above tend to address complementary issues: local features provide a degree of invariance to viewing position and direction, whilst global descriptors provide better invariance to changes in viewing conditions. However, neither do both. To address this, recent work, for example that described in [1,10], match local regions corresponding to salient landmarks in the scene such as buildings, trees, windows, etc. Matching these regions using global-type descriptors provides a degree of invariance to changing conditions, whilst their localised nature gives better invariance to viewing position and direction. We adopt a similar approach in this work.

1.1   Landmark Distribution Descriptors

Our main contribution is that in addition to matching landmark regions, we seek to maintain consistency of the spatial distribution of landmarks between views of a place. In doing so, we aim to reduce the impact of similar landmarks being present in different places - although individual landmarks may match, it is their relative positions across the view that characterises the place. In this work we limit ourselves to cases in which the different views of a place contain the same panorama but viewed from a different angle and distance, so that to a reasonable approximation the order of the landmarks, from left to right, say, remains the same between views. This accounts for many recognition scenarios, in which places are approached from the same general direction.

To implement this, we characterise a place using a *landmark distribution descriptor* (LDD), which consists of landmark feature vectors stacked in horizontal position order. Comparison of these descriptors then imposes the constraint of maintaining landmark order alongside matching feature vectors. We find that comparison is best achieved by identifying closest landmark pairs within vertical sections of the panorama, corresponding to subsets of adjacent feature vectors in an LDD (we used 3 sections in the experiments), and summing up distances between the respective feature vectors. We ensure view coverage by requiring sufficient numbers of proposal landmarks within each panoramic section. An example is shown in Fig. 1a.

For landmark regions and features we follow the same approach as Sünderhauf et al. [1] and use *Edge Boxes* [11] and convolutional neural network (CNN) features, specifically AlexNet [12], followed by Gaussian random projection [13] for dimensionality reduction. Our use of panoramic sections to ensure view coverage also mirrors the tiling approach adopted in [10], although it is important to note that landmark ordering was not used in that work.

To evaluate the approach, we carried out experiments using image pair datasets for places in urban environments. Each dataset consisted of 200 places, with one image pair per place taken from different viewing positions. We used Google Streetview and Bing Streetside images so that image pairs were captured at different times and in different conditions. Results demonstrate that for 10 datasets in 6 different cities, our method performs consistently and significantly better compared with that obtained using the method in [1] and whole image matching using SIFT, GIST and CNN features. For example, as shown in Fig. 1b, using 25 region proposals per view, on average over all datasets our method yielded an increase in precision (for 100% recall) of around 12% over that using whole image CNN, 20% over that using the method in [1] and 25%–30% over that using whole image SIFT and GIST.

The paper is organised as follows. In the next section we provide an overview of the system and details of the implementation of each component. Details of the datasets, the experiments and an analysis of the results is then provided in Sect. 3. We conclude with an indication of future work.

2   System Overview

In common with other place recognition systems we pose the problem as one of matching different views of the same place taken at different times. Labelled views are assumed to be held in a reference database and the task is to determine

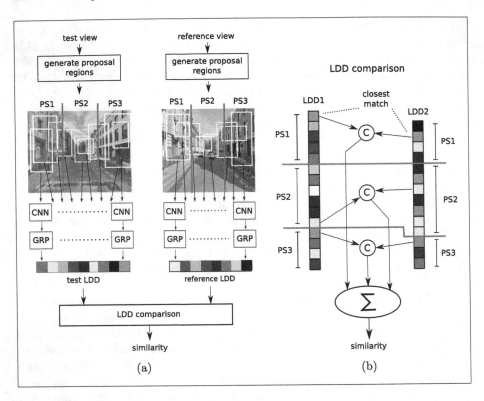

Fig. 2. Construction and comparison of landmark distribution descriptors (LDDs). (a) Landmark proposals are generated for the test and reference image using Edge Boxes [11], distributed within panoramic sections PS1-3; landmark features derived from a convolutional neural network (CNN) [12] followed by Gaussian random projection (GRP) [13] are then stacked in horizontal spatial order to form an LDD for each image; descriptors are then compared to derive a distance score between views. (b) Descriptors LDD1 and LDD2 are compared by identifying closest landmark features within each panoramic section and summing the (cosine) distances between them to derive an overall distance score.

the most likely place associated with a test view captured 'online'. In this work we opt to consider the image-pair version of this framework, in which we have one reference view per place and successful recognition corresponds to matching the test view with the one correct reference view above all others. This contrasts with the majority of other evaluations, which have been based on matching frames within videos taken along the same route and successful recognition then being defined as matching a test frame in one video with a frame from a window of frames in another reference video. We discuss this further in Sect. 3.

Given the above, we now concentrate on how we match test and reference views. There are two components to this: constructing and comparing LDDs. These are described below and Fig. 2 provides an illustration of the key elements of each.

2.1 Constructing LDDs

There are two main components to constructing an LDD for a given view as illustrated in Fig. 2a. First, proposal regions are generated, with the aim that a subset of these will correspond to salient landmarks. Second, feature vectors are computed for each of these regions, which are then combined into a single descriptor by stacking them in left-right position order.

Landmark Proposals. There are a number of algorithms available for generating proposal regions. In common with Sünderhauf et al. [1] we choose to use *Edge Boxes* as described in [11], which has found widespread use in object recognition and proved to be effective for our application. In brief, a valid edge box is identified as one in which there are a large number of contours wholly enclosed by the box. This is based on the observation that whole contours are likely to correspond to the boundary of distinct objects and hence that such boxes form good proposal regions suitable for further processing. This applies in our case as the landmarks we are interested in such as buildings, windows, trees, etc., satisfy this criterion. Also important is the fact that edge boxes can be found rapidly using fast edge detection combined with fast grouping of pixels into contours. We also make of the edge box ranking in [11] in order to limit the number of proposal landmarks and further speed up computation.

We are also interested in distributing landmark proposals across a view so that we can create a complete description. We do this by partitioning the image vertically and requiring that we select a fix number of the highest ranking landmark proposals in each section. We call these *panoramic sections* and in the experiments we used 3 sections: left, middle and right, such as that shown in Fig. 2a. In the main experiments these were positioned in a regular fashion about the image centre as shown but with overlap between sections to avoid excluding proposals which straddle a section boundary. The alternative is to align the sections according to the content of the view. We experimented with using the vanishing point (VP) as the centre and this proved effective for certain locations. We discuss this further in Sect. 3.

More formally, we denote by $L = \{l_1, l_2, \ldots, l_N\}$ the set of landmark proposals in an image discovered by the *Edge Boxes* algorithm. We then select a subset of landmarks $\hat{L}$ such that $\hat{L} \subset L$ and

$$\hat{L} = \bigcup_{s=1}^{S} \hat{L}_s \tag{1}$$

where $\hat{L}_s$ is a subset of top ranking proposals in panoramic section s and S is the number of sections, i.e. $S = 3$ in the experiments. We fix the number of top ranking proposals according to the section as described in Sect. 3 and deem a proposal to be in a section if its edge box is wholly within the section. Note that when using overlapping sections then individual landmarks can belong to two adjacent sections. This proves to be important when matching landmarks as it reduces the sensitivity to the positioning of section boundaries.

Landmark Feature Vectors. To match landmarks between views we compute feature vectors to represent the appearance of the regions associated with landmarks. As illustrated in Fig. 2a, we again take the same approach as used in [1] and make use of convolutional neural network (CNN) features [12] followed by Gaussian random projection (GRP) [13] for feature vector size reduction.

CNN features have been shown to provide high levels of invariance to different lighting conditions and viewing positions [14] and hence are ideal for place recognition. Specifically, we used the pre-trained AlexNet network [12] as provided by MatConvNet [15] and extracted the feature vector of the 3rd convolutional layer (*conv3*). Landmark regions were resized to match the required network input size of 227×227 pixels and *conv3* produces feature vectors of dimension $13 \times 13 \times 384 = 64,896$.

To reduce the computational load when comparing feature vectors, we project each vector onto a lower dimensional space using GRP [13]. This is a simple but effective method for dimensionality reduction in which feature vectors are projected onto a significantly smaller number of orthogonal random vectors in such a way that with small error the distances between vectors is maintained. This makes it ideal when matching is based on comparing those distances as in our case. For the experiments we reduced dimensionality down to 1024 for each feature vector without significant impact on performance. In the GRP we used the integer based random projection matrix suggested in [16].

For a given view, we construct feature vectors for all the landmark regions in the selected subset of proposals $\hat{L}$ and the vectors corresponding to the section subsets $\hat{L}_s$ then form the LDD for the view. In the experiments we used a total of 25 or 50 proposals per view distributed over 3 panoramic sections and thus each descriptor was of size 25×1024 or 50×1024, respectively.

2.2   Comparing LDDs

For place recognition we seek the closest LDD within the database to that of the test image. To compare LDDs we could simply use the Euclidean distance between them. However, this assumes that we have successfully detected the same landmarks in each view, which is unlikely to be the case since we are generating proposals based purely on the appearance of each view individually, and not on the likelihood that a similar landmark exists in the matching view. Hence we transfer the latter constraint into the comparison process.

As illustrated in Fig. 2b, we do this by determining the best matching feature vectors (in terms of their cosine similarity) in each of the corresponding panoramic sections. Thus, for example, given two LDDs and using 3 panoramic sections, we seek the best matching pair in each section and then compute an overall matching score corresponding to the sum of the 3 cosine similarities between the feature vectors associated with each pair. We found that using cosine similarity, again in common with [1], gave improved performance over using a straight Euclidean distance.

More formally, given two descriptors, LDD1 and LDD2, containing landmarks

$$\{\hat{L}_1^k, \hat{L}_2^k, \ldots, \hat{L}_S^k\} \tag{2}$$

for $k = 1, 2$, we seek the set of S pairs $(\hat{l}_i^1, \hat{l}_j^2)^s$, $1 \le s \le S$, such that

$$(\hat{l}_i^1, \hat{l}_j^2)^s = \underset{l_i^1 \in \hat{L}_s^1, l_j^2 \in \hat{L}_s^2}{\arg\max} \; c(\mathbf{v}_i^1, \mathbf{v}_j^2) \tag{3}$$

where $\mathbf{v}_i^1$ and $\mathbf{v}_j^2$ are the feature vectors associated with landmarks l_i^1 and l_j^2, respectively, and $c(\mathbf{u}, \mathbf{v}) = \mathbf{u}.\mathbf{v}/\|\mathbf{u}\|\|\mathbf{v}\|$ denotes the cosine similarity between two vectors $\mathbf{u}$ and $\mathbf{v}$, where '.' denotes the dot product and $\|\mathbf{u}\|$ is the length of $\mathbf{u}$. To avoid duplicating matching landmarks, we also require that no landmark can be in more than one matching pair. The overall similarity score between the two LDDs, and hence the two views, is then given by the sum of the S cosine similarities, i.e.

$$sim_{12} = \sum_{\substack{(\hat{l}_i^1, \hat{l}_j^2)^s \\ 1 \le s \le S}} c(\hat{\mathbf{v}}_i^1, \hat{\mathbf{v}}_j^2) \tag{4}$$

3   Experiments

3.1   Datasets

We evaluated our method using multiple image pair datasets taken from urban environments. Our motivation for using image-pairs in contrast to matching frames in video sequences as used by others is twofold. First, we believe that it presents a more challenging test, since recognition is based on matching with only one alternative view as opposed to matching to one of multiple frames in a video (corresponding to the vicinity of a place). Secondly, it means that we can easily create large datasets corresponding to random places using the images taken from online mapping services such as Google Streetview, Bing Streetside and Mapillary. Using images from more than one of these also enables us to evaluate performance under different viewing conditions.

For the experiments reported here we used datasets obtained from Google Streetview and Bing Streetside. Specifically, we selected 200 random locations in 6 different cities - London, Bristol, Birmingham, Liverpool, Manchester and Paris - and for each location we selected images taken in roughly the same sort of direction but displaced by between approximately 5 and 10 m. We selected one image from Streetview and one from Streetside for each location. This is ideal as the images were taken at different times and under different lighting and visibility conditions. We used datasets from different cities to enable us to evaluate the performance of the method for differing types and characteristics of architecture and urban layout. We tested the method on 10 datasets in all, using 3 from London and 3 from Bristol in order to test for any variation in performance within the same city. In total, the evaluation involved 2,000 different locations.

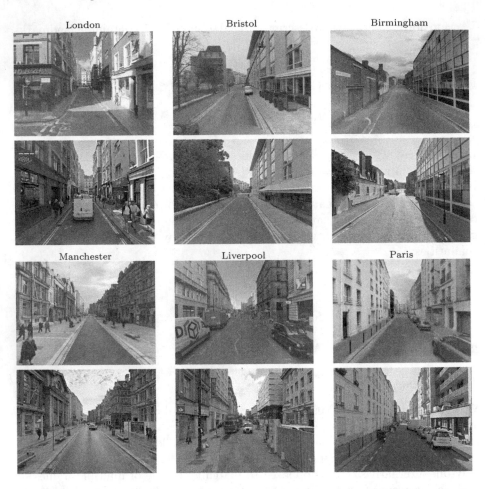

Fig. 3. Examples of view pairs from each of the 6 cities in the 10 datasets used in the experiments. The pairs are shown one above the other and there are 3 pairs per row.

Example image pairs from the different cities are shown in Fig. 3. Note that although the physical distance between the viewing positions is not great, there is a significant change in structural appearance which when coupled with the differences caused by different light and visibility conditions makes recognition far from straightforward. Notable difficulties include the presence of pedestrians and vehicles, significant changes in scale of buildings, some buildings disappearing from view whilst others come into sharper focus, and so on. However, careful observation should reveal that distribution of the key visible landmarks is maintained across the two views. It is this characteristic that we aim to exploit in this work.

3.2 Comparison Methods

We compared the performance of our method with four other methods: the CNN landmark matching method of Sünderhauf et al. [1]; whole image CNN matching [17]; whole image SIFT matching; and whole image GIST matching. Relevant details for each method are given below.

CNN Landmark matching (CLM)
    As noted earlier, the primary difference between our method and that in [1] is that matching in the latter is based on finding similar landmarks across both views, irrespective of relative position. Specifically, best matching pairs of CNN-GRP feature vectors are selected from edge box proposals based on cosine similarity and the overall similarity between two views is then the sum of the cosine similarities, weighted by a measure of similarity in box size. For comparison, we evaluated two versions of this method, one using 25 (proposals CLM-25) and one using 50 proposals (CLM-50)[1]

CNN matching (CWI)
    In this method, we used the same CNN-GRP features vectors as in [1] and in our method, but comparison between views was based on a single feature vector computed for the whole image. Cosine similarity was again used as the comparison metric. The method is similar to that used in [17].

Dense SIFT matching (SIFT)
    For this method we used a dense keypoint version of matching SIFT descriptors across both views [6]. Specifically, we used the implementation as provided in the VLFeat library [18].

GIST matching (GIST)
    Finally, we compared our method with whole image GIST matching, based on the implementation provided by Oliva and Torralba as described in [7].

3.3 Results

We compared the performance of our method against that of the comparison methods for all 10 datasets. Each dataset contained 200 view pairs from different locations, with one view taken from Streetview and the other from Steetside. In each evaluation, we used all the Streetside images as test images and the Streetview images were used as the reference images. We used precision (P) and recall (R) to measure performance, defined as $P = tp/(tp + fp)$ and $R = tp/(tp + fn)$, where tp, fp and fn denote the number of true positives, false positives and false negatives, respectively. A true positive was recorded if the

---

[1] For clarity with respect to our experiments, we should note that we found that the similarity metric provided in [1] did not give good performance and so in the interests of fairness we used a modified version which gave significantly better performance. Specifically, we modified Eqs. (2) and (3) in [1] to be $s_{ij} = 1 - (\frac{1}{2}(\frac{|w_i - w_j|}{max(w_i, w_j)} + \frac{|h_i - h_j|}{max(h_i, h_j)}))$ and $S_{ab} = \frac{1}{n_a \cdot n_b} \sum_{ij}(d_{ij} \cdot s_{ij}))$, respectively.

Table 1. Recorded precision values for 100% recall for all 10 datasets using all 7 comparison methods.

	LDD-25	CLM-25	LDD-50	CLM-50	SIFT	GIST	CWI
London1	**84.5**	55	**88**	73.5	59.5	47	66
London2	83	68	**90**	**84**	58	44	74
London3	**72**	57	**83**	69	51	58	64
Bristol1	**66.5**	51.5	**68.5**	58	51.5	33	60.5
Bristol2	**63.5**	50.5	**65.5**	59.5	40.5	26	54.5
Bristol3	59.5	47	**67**	**64.5**	48	37	61
Birmingham	**62**	44	**71.5**	60	26.5	38	44
Manchester	**69**	50.5	**71.5**	63.5	33.5	33.5	63
Liverpool	**74**	46	**75**	62	52.5	40.5	53
Paris	**61**	35	**70.5**	49	40	35	38
Avg	**69.5**	50.45	**75.05**	64.3	46.1	**39.2**	**57.8**

test image was matched with the reference image taken at the same location, a false positive was recorded if the test image was matched with a reference image taken at a different location, and a false negative was recorded if a test image was deemed not to match any of the reference images based on a threshold of the ratio between the closest and second closest matches. Variation of this threshold also enabled us to create precision-recall curves as given below. Note that our datasets do not contain any true negatives.

We evaluated two versions of our method, one using 25 landmark proposals and one using 50 landmark proposals. In each case we used 3 panoramic sections, with 50% overlap between sections. The image sizes were 640 × 480 pixels for both Streetview and Streetside and we used sections of size 320 pixels. We fixed the number of top ranked proposals selected from each section to be (5,15,5) when using 25 proposals (in left to right order) and (10,30,10) when using 50 proposals. The larger number of proposals in the central section proved to have a significant impact on performance.

Table 1 shows the precision values recorded for the different methods at 100% recall, i.e. so that all matches are accepted as positives. Note that over all datasets the LDD-50 method gives the best performance and apart from two datasets, the LDD-25 method gives the next best results. The latter is significant since the computational load is halved when using 25 proposal landmarks (the bottle neck is the computation of the CNN feature vectors) and thus it is interesting to note that good performance is still maintained from our method using the smaller number of proposals. This contrasts with the CLM method which performs significantly worse when using only 25 proposals and notably worse than using whole image CNN. We believe that this is a direct result of our method using the spatial distribution of the landmarks which provides a key characteristic to distinguish between views. Note also that the results for

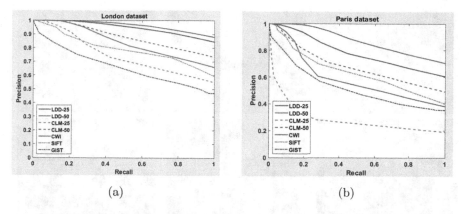

(a)   (b)

Fig. 4. Precision recall curves obtained for all comparison methods for (a) the London1 dataset and (b) the Paris dataset.

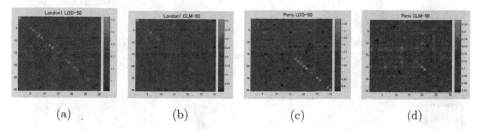

(a)   (b)   (c)   (d)

Fig. 5. Confusion matrices showing recorded similarity scores for 30 locations in the London1 and Paris datasets using (a)–(b) LDD-50 and (c)–(d) CLM.

the London datasets are noticeably better than those for the other datasets. On inspection, we found that the London places were predominantly characterised by buildings with highly distinctive appearance, in contrast, for example, to the mix of vegetation and buildings apparent in the Bristol datasets and the similarities in architecture within the Paris dataset. This can be seen from the examples in Fig. 3. An area of future work will be to investigate how we can improve performance in such cases.

Figure 4 shows the variation in precision as we reduce recall by increasing the number of false negatives via the threshold on the ratio of the closest and second closest matches for the two datasets London1 and Paris. Note that in both cases both versions of our method LDD-25 and LDD-50 outperform the other methods. Again, the difference in LDD-25 and CLM-25 is noticeable, with the former achieving almost a 30% gain in precision, corresponding to correct recognition of over 60 places compared with the latter, using the same number of proposal landmarks. This illustrates clearly the advantage of using landmark distribution to characterise views. To illustrate the distinguishing power of our method, Fig. 5 shows confusion matrices for the same two datasets using methods LDD-50 and CLM-50, where we have used 30 randomly selected location pairs

Fig. 6. Examples of correct view matches obtained using the LDD-50 method. Matches are shown one above the other and there are 3 matches per row.

rather than all 200 to aid clarity. These show the similarity scores between test and reference views. Note the high values down the main diagonal for the LDD-50 method indicating strong distinction of the correct places and contrast this with the closeness of the values obtained using CLM-50 method, especially for the Paris dataset.

To illustrate the landmarks that are being found by our method to enable correct recognition of places, Fig. 6 shows examples of views which have been correctly matched. *None of these examples were correctly matched by the other methods.* In each case, the best matching landmarks found in each panoramic section are shown in colour, where the colours indicate corresponding landmarks in each view. Pairs are shown above one another and each row shows 3 pairs. Note the difference in appearance and structure between the views, especially the changes in vegetation and building structure, but also note that with

Fig. 7. Use of the view vanishing point to center panoramic sections improves matching of landmarks (bottom) over that obtained using the image center (top).

Fig. 8. Examples of incorrectly matched views obtained using the LDD-50 method.

careful observation they can be seen to be the same places. These are challenging examples and it is encouraging that our method is able to correctly match the views.

We also experimented with adapting the positioning of the panoramic sections according to view content rather than simply dividing up the image evenly into 3 sections about the image centre. Instead, we computed the location of the vanishing point in each view, using the method described in [19], and if within the image, we used this to centre the middle section, with appropriate adaptation of the two outer sections. In many cases this had little impact since the VP was often close to the image centre. However, in a number of cases it did make a difference and resulted in correct matching of places which were previously incorrectly matched. An example is shown in Fig. 7. The top row shows a pair of views of the same place with selected landmark regions derived using panoramic sections centred about the image center. This proved not to be the best match for the left hand test image and hence resulted in an incorrect match. Clearly the detected landmarks in each view do not correspond to the same landmarks in the scene. In contrast, shifting the sections to the right in both views after detecting the VP in each, results in correspondence between the detected landmarks and this resulted in a successful match. Although encouraging, these are only provisional results and further work is needed to determine the generality of using the VP in this way.

Finally, Fig. 8 shows 3 examples in which our method fails to match the correct view. The top row shows the test images, the middle row shows the incorrectly matched view and the bottom row the correct view. Note that these are particularly challenging examples and are further complicated by landmarks being detected on vehicles which are not present in both views. How to deal with cases such as these will be the subject of further research.

4   Conclusions and Future Work

We have presented a new method for visual place recognition based on matching landmark regions represented by CNN features. The key contribution is the encoding of relative spatial position of the landmarks via the use of the landmark distribution descriptors (LDD). Although the method has aspects in common with the CLM method of Sünderhauf et al. [1], we have demonstrated that the use of LDDs has a major impact on performance, with significant gains in precision, not only over CLM but also over the other whole image techniques. It is important to point out that the gains in precision amount to significant gains in the numbers of correctly recognised places, with, for example the 19% gain in average performance of LDD-25 over CLM-25 corresponding to 38 locations.

In the future we intend to investigate the performance of the method using different datasets, including video. We will also investigate further the benefits of using VPs to better position the panoramic sections. Also of interest is the potential for extending the idea of landmark distribution matching to more general cases in which landmark positioning changes due to changes in viewpoint.

As the two are linked through geometry and motivated by the ideas and method described in [20], it may be possible to build this into a constraint for matching views which are widely disparate.

References

1. Sünderhauf, N., Shirazi, S., Jacobson, A., Dayoub, F., Pepperell, E., Upcroft, B., Milford, M.: Place recognition with convnet landmarks: viewpoint-robust, condition-robust, training-free. In: Proceedings of Robotics: Science and Systems, Rome, Italy (2015)
2. Lowry, S., Sunderhauf, N., Newman, P., Leonard, J., Cox, D., Corke, P., Milford, M.: Visual place recognition: a survey. IEEE Trans. Robot. **32**, 1–19 (2015)
3. Sattler, T., Leibe, B., Kobbelt, L.: Improving image-based localization by active correspondence search. In: Fitzgibbon, A., Lazebnik, S., Perona, P., Sato, Y., Schmid, C. (eds.) ECCV 2012. LNCS, vol. 7572, pp. 752–765. Springer, Heidelberg (2012). doi:10.1007/978-3-642-33718-5_54
4. Gronat, P., Obozinski, G., Sivic, J., Pajdla, T.: Learning and calibrating per-location classifiers for visual place recognition. In: Proceedings of the IEEE Conference on Computer Vision and Pattern Recognition, pp. 907–914 (2013)
5. Torii, A., Arandjelovic, R., Sivic, J., Okutomi, M., Pajdla, T.: 24/7 place recognition by view synthesis. In: Proceedings of the IEEE Conference on Computer Vision and Pattern Recognition, pp. 1808–1817 (2015)
6. Lowe, D.G.: Distinctive image features from scale-invariant keypoints. Int. J. Comput. Vis. **60**, 91–110 (2004)
7. Oliva, A., Torralba, A.: Modeling the shape of the scene: a holistic representation of the spatial envelope. Int. J. Comput. Vis. **42**, 145–175 (2001)
8. Cummins, M., Newman, P.: Fab-map: probabilistic localization and mapping in the space of appearance. Int. J. Rob. Res. **27**, 647–665 (2008)
9. Cummins, M., Newman, P.: Appearance-only SLAM at large scale with FAB-MAP 2.0. Int. J. Robot. Res. (2010)
10. McManus, C., Upcroft, B., Newman, P.: Scene signatures: localised and point-less features for localisation. In: Proceedings of Robotics Science and Systems (RSS), Berkeley, CA, USA (2014)
11. Zitnick, C.L., Dollár, P.: Edge boxes: locating object proposals from edges. In: Fleet, D., Pajdla, T., Schiele, B., Tuytelaars, T. (eds.) ECCV 2014. LNCS, vol. 8693, pp. 391–405. Springer, Heidelberg (2014). doi:10.1007/978-3-319-10602-1_26
12. Krizhevsky, A., Sutskever, I., Hinton, G.E.: Imagenet classification with deep convolutional neural networks. In: Pereira, F., Burges, C.J.C., Bottou, L., Weinberger, K.Q. (eds.) Advances in Neural Information Processing Systems 25, pp. 1097–1105. Curran Associates, Inc. (2012)
13. Bingham, E., Mannila, H.: Random projection in dimensionality reduction: applications to image and text data. In: Proceedings of the Seventh ACM SIGKDD International Conference on Knowledge Discovery and Data Mining, KDD 2001, pp. 245–250. ACM, New York (2001)
14. Chatfield, K., Simonyan, K., Vedaldi, A., Zisserman, A.: Return of the devil in the details: delving deep into convolutional nets. In: Proceedings of the British Machine Vision Conference (BMVC) (2014)
15. Vedaldi, A., Lenc, K.: Matconvnet - convolutional neural networks for matlab. In: Proceeding of the ACM International Conference on Multimedia (2015)

16. Achlioptas, D.: Database-friendly random projections. In: Proceedings of the Twentieth ACM SIGMOD-SIGACT-SIGART Symposium on Principles of Database Systems, PODS 2001, pp. 274–281. ACM, New York (2001)
17. Sünderhauf, N., Shirazi, S., Dayoub, F., Upcroft, B., Milford, M.: On the performance of convnet features for place recognition. In: Proceedings of the IEEE/RSJ International Conference on Intelligent Robots and Systems (IROS) (2015)
18. Vedaldi, A., Fulkerson, B.: VLFeat - an open and portable library of computer vision algorithms. In: ACM International Conference on Multimedia (2010)
19. Kong, H., Audibert, J.Y., Ponce, J.: Vanishing point detection for road detection. In: IEEE Conference on Computer Vision and Pattern Recognition, CVPR 2009, pp. 96–103. IEEE (2009)
20. Frampton, R., Calway, A.: Place recognition from disparate views. In: Proceedings of the British Machine Vision Conference (BMVC) (2013)

Real Time Direct Visual Odometry for Flexible Multi-camera Rigs

Benjamin Resch[✉], Jian Wei, and Hendrik P.A. Lensch

Computer Graphics Group, University of Tübingen, Tübingen, Germany
benjamin.resch@uni-tuebingen.de

Abstract. We present a Direct Visual Odometry (VO) algorithm for multi-camera rigs, that allows for flexible connections between cameras and runs in real-time at high frame rate on GPU for stereo setups. In contrast to feature-based VO methods, Direct VO aligns images directly to depth-enhanced previous images based on the photoconsistency of all high-contrast pixels. By using a multi-camera setup we can introduce an absolute scale into our reconstruction. Multiple views also allow us to obtain depth from multiple disparity sources: static disparity between the different cameras of the rig and temporal disparity by exploiting rig motion. We propose a joint optimization of the rig poses and the camera poses within the rig which enables working with flexible rigs. We show that sub-pixel rigidity is difficult to manufacture for 720p or higher resolution cameras which makes this feature important, particularly in current and future (semi-)autonomous cars or drones. Consequently, we evaluate our approach on own, real-world and synthetic datasets that exhibit flexibility in the rig beside sequences from established KITTI dataset.

1 Introduction

Visual Odometry (VO) and Self Location And Mapping (SLAM) systems traditionally are feature-based approaches: Distinct features are tracked over many frames. The scene reconstruction, i.e. determining for all frames the parameters of the camera poses and the positions of all scene points, is usually formulated as a least squares optimization problem that aims at minimizing the difference between the observed feature positions in the images and the reprojections of the scene points into the frames.

Direct methods instead maintain per pixel depth information for several keyframes which allows the keyframe's image to be projected into other frames, given the camera's intrinsic parameters and its relative pose to the keyframe. The camera poses of new frames can be found by minimizing the photometric error between the keyframe projected into the new frame and the new frame itself. When the relative camera transformation between the keyframe and another

Electronic supplementary material The online version of this chapter (doi:10. 1007/978-3-319-54190-7_31) contains supplementary material, which is available to authorized users.

© Springer International Publishing AG 2017
S.-H. Lai et al. (Eds.): ACCV 2016, Part IV, LNCS 10114, pp. 503–518, 2017.
DOI: 10.1007/978-3-319-54190-7_31

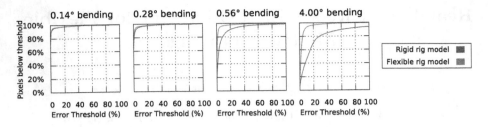

Fig. 1. ROC evaluation of the depth quality on a synthetic dataset (720p, 90° FoV camera). Note that even for small camera rotations inside the rig of 0.56°, taking the rig flexibility into account improves the results very significantly: For a 3% threshold, the recall is 49.4% (rigid model) vs. 85.9% (flexible model).

frame is known, depth information of the keyframes is improved with stereo depth estimation techniques.

With a stereo camera rig, the quality of the depth maps of the keyframes can be improved by applying not only the *dynamic* stereo between a keyframe and its subsequent frames, but also by applying *static* stereo between two images that were recorded by the rig at the same time with known rig intrinsics (poses of the cameras relative to the rig) [1].

It turns out that direct VO methods are much less forgiving to bad calibrations as feature-based VO methods. Feature-based methods optimize based on distances on the image which increase gradually when the calibration becomes bad. Direct methods optimize based on photoconsistency in a small window which can be arbitrarily bad if the calibration is just one pixel off.

While the intrinsic parameters of cameras can usually be controlled with pixel precision and the tracking of the rig's pose works just as precise, the rig intrinsics can't be assumed to be static enough for sub pixel precision over time, especially for higher resolution cameras, large baselines and/or material/cost/weight limitations of the rig. Deflection theory shows that two 0.5 kg cameras mounted on a $300 \times 25 \times 4$ mm stainless steel carrier are rotated by 0.56° against each other when an acceleration of ± 1 g is applied (see supplemental material). This already introduces significant errors if the rig flexibility is not concerned (see Fig. 1).

In this paper, we propose a method that extends existing, direct monocular or stereo VO approaches with per frame rig intrinsics tracking for multi-camera rigs. We target on high-frequency rotations of the cameras in the rig up to a few degrees which are hard to avoid but harmful to conventional direct VO methods (see Figs. 1 and 2). For each keyframe, we maintain per pixel depth information for the images of all cameras. For every new frame, we optimize for the rig's pose as well as for its intrinsics based on the photometric error between every image of the keyframe and every new image. This way, we find consistent poses for all cameras of the rig by utilizing the information that is available in all images.

In fact, we can even choose (1) the pairs of keyframe/new frame images that should be used for tracking and (2) the new frames' or keyframes' images that

Fig. 2. Comparison of the depth maps reconstructed without (center) and with (right) flexible rig optimization. The original frames are shown on the left. Without flexible tracking, tracked frames cannot be aligned consistently to the last keyframe. This leads to a bad alignment of projected depths onto image gradients and therefore to stereo errors which result in incomplete and noisy depth maps (center), while flexible tracking does not suffer from those issues (right).

should be used to improve the depth information of the keyframe images. This flexibility can be used to balance between computation speed and reconstruction quality as well as to adjust the processing to the rig configuration, e.g. avoid direct interaction of cameras that never see the same field of view (FoV).

We implemented most of our algorithm on the GPU using CUDA to address the increased computational demands compared to a monocular setup.

Please note that building a globally consistent reconstruction of a scene including loop closing and global error distribution is out of the scope of this document. However, our technique can be used as a plugin replacement for the VO part of other SLAM systems.

2   Related Work

The proposed approach reconstructs camera poses as well as the scene structure, so we consider VO/SLAM methods as well as Structure from Motion (SfM) and Multi View Stereo (MVS) techniques in this section.

2.1   Visual Odometry and SLAM

Most VO and SLAM methods are feature-based. Chiuso et al. [2] presented one of the first real time capable, monocular, feature-based SLAM systems. Nistér et al. [3] established the term "Visual Odometry" by presenting a feature tracking based system for frame-to-frame monocular or stereo camera pose estimation. Davison et al. [4] proposed MonoSLAM which is an Extended Kalman Filter (EKF) based method that creates a probabilistic but persistent map of natural landmarks and is therefore drift free in small workspaces. PTAM [5] is designed to avoid the iterative tracking and mapping per frame by separation into two concurrent threads: The mapping thread integrates all previously tracked camera poses and features into a consistent representation in the background while the tracking thread tracks every new frame against the newest available map. Paz et al. [6] combine an EKF-based framework with a stereo camera setup.

Direct methods for VO and SLAM are available for some years now and have two major benefits over the feature-based methods: First, there is no need to craft a feature detector. Second, not only the information from sparse feature points but virtually any gradient information in the images can be used for reconstruction. Direct monocular methods first appeared in the RGBD domain [7,8] where the per pixel depth information comes directly from the sensor and only the tracking has to be performed. The first direct real-time method relying solely on color images is LSD-SLAM [9] which tracks each new frame to the previous, depth-enhanced keyframe based on dense image alignment. Then, the keyframe's depth information is updated by probabilistically merging it with the depth information obtained from stereo with the previously tracked frame. LSD-SLAM was extended to stereo cameras [1] where tracking only happens on the left camera. This is in contrast to our approach which allows for tracking between multiple cameras. Consequently, in [1], a depth map is maintained only for the left camera; the right images are just used for static stereo with the corresponding left images which improves the keyframe's left depth map and introduces an absolute scale to the reconstruction. This is however not feasible if the rig intrinsics change over time. In addition, we show that we can reconstruct datasets from rigs with few overlap between the fields of view of the cameras by utilizing the information from all cameras where methods like [1] fail (see Sect. 4.4). Pillai et al. [10] accelerates the depth estimation for stereo pairs by starting from few sparse, Delaunay triangulated piecewise planar surfaces which can be refined in multiple iterations in areas where the matching cost is high, e.g. due to non-planarity. A substantially different approach is the method of Comport et al. [11] which avoids the explicit reconstruction of scene depth and uses the quadrifocal constraints from two pairs of stereo images to obtain the transformation of the camera rig instead.

2.2   Structure from Motion and Multi View Stereo

SfM (usually feature based) with subsequent MVS (dense, direct surface reconstruction) is similar to SLAM since both reconstruct consistent 3D models and camera poses from images. However, SfM+MVS methods traditionally aim more at image collections instead of videos and reconstruction quality instead of real time processing with low latency. Recently, there have been efforts to develop SfM/MVS methods that utilize the redundancy of video streams to achieve close-to-real-time performance. Resch et al. [12] have accelerated SfM by the consequent subsampling of frames, features, scene points and loop closure candidates in their Bundle Adjustment framework and employing a linear solver to keep unsampled frames consistent. The MVS methods in [13,14] compute depth maps based on the photoconsistency with up to 100 other images but compare for photoconsistency based on 1×1 pixel masks only.

Delaunoy and Pollefeys propose a Photometric Bundle Adjustment [15] which is designed to recover camera parameters and a dense surface mesh based on the photometric error between observed and model-based, generated images.

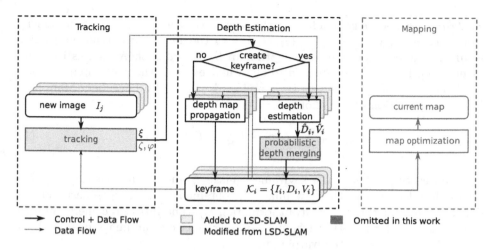

Fig. 3. Overview over the LSD-SLAM algorithm and our extensions.

3 Multi-view VO for Flexible Camera Rigs

In this section we'd like to introduce our method. Figure 3 shows an overview of the originally monocular LSD-SLAM algorithm [9] and our extensions.

LSD SLAM tracks the camera poses of new images against a previous, depth enhanced keyframe. If the current keyframe is still good for tracking (content similar to the new frame), new depth information is estimated with a stereo method between the new image and the keyframe and is merged into the keyframe. If the keyframe is not good enough anymore, it gets propagated to the current image. A mapping component finally cares about global consistency of all keyframes.

Our method allows for multiple input images from multiple cameras per frame and stores depths for all cameras of a keyframe, consequently. We extended the tracking to optimize for the rig pose and the rig intrinsics (camera poses within the rig) jointly (Sect. 3.2). For stereo cameras, we propose a reliable rig parameterization in Sect. 3.3. Tracking gives us the transformation from every keyframe image to every new frame's image, so we can generate multiple stereo observations per keyframe (Sect. 3.4). In order to utilize all that information, we extended the Bayesian depth merging approach (Sect. 3.5).

However, we'd like to start by introducing the aspects of LSD-SLAM [9,16] that are important to our method first:

3.1 LSD-SLAM

LSD-SLAM is a direct, semi-dense, monocular Self Location And Mapping approach. It consists of the following steps (see Fig. 3):

1. Tracking has the goal of finding the relative transformation ξ_{ij} of the camera from the last keyframe $\mathcal{K}_i = \{I_i, D_i, V_i\}$ to the most current image I_j. For keyframes, a semi-dense, inverse depth map D_i as well as the variances V_i of the inverse depth values are assumed to be available beside the pixel intensities I_i.

A new image is aligned to the previous keyframe by Levenberg-Marquard (LM) minimization of the photometric error

$$E(\xi_{ij}) = \sum_p (I_i(p) - I_j(\omega(p, D_i(p), \xi_{ij})))^2 \tag{1}$$

between every pixel p in the keyframe's image I_i and the intensity at the corresponding location in the new image I_j, obtained by warping p based on its depth $D_i(p)$ and the camera transformation ξ_{ij}. The intrinsic camera calibration $\pi : \mathbb{R}^3 \longrightarrow \mathbb{R}^2 \times \mathbb{R}$ transforming from view space to an image position plus depth is used to define the warping function

$$\omega(p, d, \xi) = [(\pi \circ \xi^{-1} \circ \pi^{-1})(p, d)]_p. \tag{2}$$

Note that ξ is the transformation of the camera, so ξ^{-1} has to be used to transform the points relative to the camera.

Tracking is performed on an image pyramid in a coarse-to-fine manner: A coarse representation helps to converge even with high disparities, fine details provide exact tracking at the end.

After tracking, a decision about creating a new keyframe is made based upon the distance that the camera covered since the last keyframe and the number of pixels that could be used for tracking. Depending on the decision, steps 2 and 3 or step 4 are executed.

2. Depth estimation produces a new inverse depth map $\hat{D}_i$ and a corresponding variance map $\hat{V}_i$, using a stereo method on the image intensities I_i and I_j by using the previously obtained camera transformation ξ_{ij}.

Depth $\hat{D}_i$ is obtained for all pixels that exhibit a sufficiently large gradient $\nabla I_i(p)$ along the Epipolar Line (EPL) by brute force evaluation of the Sum of Squared Differences (SSD) error in a 5×1 pixel mask (aligned with the EPL). The depth is further refined by second order Taylor approximation. The search range for each pixel p on the EPL corresponds to the depth range $(D_i(p) \pm 3V_i(p))$.

The variance values $\hat{V}_i(p)$ are estimated based on the photometric disparity error $(\approx \frac{camera\ noise}{|\nabla I_i(p)|})$ and the geometric disparity error $(\approx 1/\langle \frac{\nabla I_i(p)}{|\nabla I_i(p)|}, \frac{EPL}{|EPL|} \rangle$, i.e. it is large if the image gradient is orthogonal to the EPL direction), both determined in the image domain and transformed to the inverse depth domain.

3. Probabilistic depth merging is used to update the previous keyframe depths D_i and variances V_i with the new estimations $\hat{D}_i$ and $\hat{V}_i$. This is done by multiplying the distributions according to the update step in a Kalman filter: Given a prior distribution $\mathcal{N}_{prior} = \mathcal{N}(D_i(p), V_i(p))$ and a new observation

$\mathcal{N}_{new} = \mathcal{N}(\hat{D}_i(p), \hat{V}_i(p))$, the posterior is given by

$$\mathcal{N}_{post} = \mathcal{N}_{prior} \cdot \mathcal{N}_{new} = \mathcal{N}\left(\frac{V_i(p)\hat{D}_i(p) + \hat{V}_i(p)D_i(p)}{V_i(p) + \hat{V}_i(p)}, \frac{V_i(p)\hat{V}_i(p)}{V_i(p) + \hat{V}_i(p)} \right). \qquad (3)$$

After depth and variance have been updated, smoothing and hole filling are applied to D_i and V_i.

4. Depth map propagation establishes a new keyframe by propagating depth and variance information from the previous keyframe to the currently processed frame. A forward mapping from the keyframe to the new frame is established by using ω (see Eq. 2) and depths and variances are assigned to the closest pixel in the new frame's image.

After the depth propagation, the new depth and variance maps are subject to removal of occluded pixels, hole filling and smoothing.

5. Map Optimization every keyframe is added to the mapping component of LSD-SLAM which integrates it into a complete, globally consistent map of the reconstructed scene, caring about loop closing and tracking error distribution over the whole scene. This fifth step is out of scope of our method and just mentioned for completeness.

3.2 Tracking Flexible Multi-camera Rigs

The following subsections describe our contributions for enabling multi-camera, flexible rig VO, starting with the tracking part where we optimize for the rig pose and the rig intrinsics concurrently:

For multi-camera rigs, if the ith frame is a keyframe $\mathcal{K}_i^l = \{I_i^l, D_i^l, V_i^l\}$, it contains image intensities I, inverse depths D and inverse depth's variances V for every camera l of the rig. For tracking, we introduce the extrinsic rig transformation ζ_{ij} and the rig intrinsics $\varphi_i = (\varphi_i^{c1}, \varphi_i^{c2}, ...)$ that can be different for every frame i and expand to a rig-to-view-space transformation for every individual camera of the rig.

For using potentially all information that is available in all images of the keyframe and the frame to be tracked, the photometric error expands to

$$E(\zeta_{ij}, \varphi_j) = \sum_{c_i} \sum_{c_j} \left(f_t(c_i, c_j) \sum_p (I_i^{c_i}(p) - I_j^{c_j}(\omega(p, D_i(p), \zeta_{ij}, \varphi_i^{c_i}, \varphi_j^{c_j})))^2 \right). \tag{4}$$

c_i and c_j are iterating over the individual cameras of the keyframe's or the tracked frame's rig. The mapping $f_t : \mathbb{N}^2 \rightarrow \{0, 1\}$ is user defined and can be used to disable pairs of cameras for tracking, e.g. if their fields of view do not overlap at all or for performance reasons (see also Fig. 6).

For the extended photometric error, we need an extended warping function that projects an image position p and its depth d to the viewspace of a keyframe's camera, then to the keyframe's rig space, then to the tracked frame's rig space, to the tracked camera's view space and into the tracked camera's image:

$$\omega(p, d, \zeta, \varphi_i^{c_i}, \varphi_j^{c_j}) = [(\pi \circ \varphi_j^{c_j} \circ \zeta^{-1} \circ \varphi_i^{c_i - 1} \circ \pi^{-1})(p, d)]_p. \tag{5}$$

Note that for introducing a new rig configuration to our system, it is sufficient to specify φ_i, i.e. a rig-to-view-space transformation for each camera (for flexible rigs: based on some rig parameters which may change for each frame i).

For solving the minimization problem with the LM algorithm, we have to solve the augmented normal equations (please refer to [17] for its derivation)

$$(J^T J + \mu I)\delta_{\zeta\varphi} = J^T \epsilon \tag{6}$$

for $\delta_{\zeta\varphi} = (\delta_\zeta, \delta_\varphi)$ to improve our parameters with $\zeta_{n+1} = \zeta_n + \delta_\zeta$ and $\varphi_{n+1} = \varphi_n + \varphi_\zeta$. μ is a damping factor which controls if the updates $\delta_{\zeta\varphi}$ are Gauss Newton or small gradient descent steps. $J = (\frac{\partial E}{\partial \zeta_1}, ..., \frac{\partial E}{\partial \zeta_6}, \frac{\partial E}{\partial \varphi_1}, \frac{\partial E}{\partial \varphi_2}, ...)$ is the Jacobian matrix, containing the derivatives of the error function to all rig pose and rig intrinsics parameters. Note that each $\frac{\partial E}{\partial ...}$ is a column vector with one element for each $()^2$ term of Eq. 4. We know that the pose of one camera on the rig just has 6 degrees of freedom (DoF), so we prefer to evaluate $J_c = (\frac{\partial E}{\partial \xi_1}, ..., \frac{\partial E}{\partial \xi_6})$. To achieve this, we approximate the known transformation from rig pose ζ and rig intrinsics φ to camera pose ξ locally and linearly for every camera c with the matrix

$$J_c^{\zeta\varphi \to \xi} = (R|C_c) \qquad R = \left(\frac{\partial \xi_i}{\partial \zeta_j}\right)_{i,j=1...6} \qquad C_c = \left(\frac{\partial \xi_i}{\partial \varphi_j^c}\right)_{i=1...6; j=1...} . \tag{7}$$

By using this transformation, we can calculate $\delta_\xi = J_c^{\zeta\varphi \to \xi} \delta_{\zeta\varphi}$. From the chain rule it follows that we can replace $J = J_c(J_c^{\zeta\varphi \to \xi})^T$ in Eq. 6.

This way, we only have to evaluate the minimum number of derivatives but are able to transform them to a Jacobian for any arbitrary rig parameterization. In addition, this helps to keep the GPU code independent from the rig parameterization and allows for implementing new rigs with different camera configurations easily at one place.

3.3   A Flexible Stereo Rig Model

Generally, a stereo rig model has 6 DoF: Trivially one would use an identity camera-to-rig transformation for one of the cameras in the rig while the other one can be placed freely in the rig space. To avoid scale drifting, one might want to set the distance between the cameras fixed which leaves 5 DoF remaining. We found that this is a valid assumption even for flexible camera rigs because bending a rig by small angles leads to negligible changes in the distance between the cameras (e.g. 0.5° bending ⇒ 10^{-5}% closer).

In most stereo camera rigs, both cameras are attached to the rig in the same way and the rig is mounted or hold symmetrically, so we prefer a symmetrical

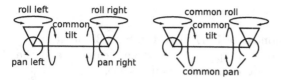

Fig. 4. Comparison of symmetric stereo rig models. The left model exhibits the theoretical minimum number of DoFs for fixed baselines. The right model is the one that we finally used due to its clearly distinct DoFs.

Fig. 5. Degeneration of the 5 DoF rig model. Cameras may shift in front of each other to reduce the projected baseline of the rig, overriding the scale fixation that it should actually introduce.

rig model as well to have a more natural mapping from real world effects to parameter space.

For our first experiments we used the rig model shown in Fig. 4 on the left. It exhibits individual parameters for the cameras' roll and pan, but a common tilt parameter. Using individual tilt parameters would introduce 6 DoF which is above the minimum and introduces an ambiguity since tilting the rig can also be accomplished by tilting both cameras.

We observed that this rig model tends to shear as shown in Fig. 5 when the scene scale starts to drift instead of fixing the scale. Therefore we switched to the 3 DoF model shown in Fig. 4 on the right which ensures that the rig's baseline stays orthogonal to the average camera's viewing direction. Although this model is not able to represent asymmetric camera rotations correctly, it provides superior reconstruction quality on exactly such input data.

Depth/Panning Ambiguity. While using the 3 DoF stereo rig model, we observed that the depth tends to get compressed in front of the cameras while the cameras start panning towards each other after a few hundred frames. Although depth and panning are not really ambiguous (i.e. the photometric error $E(\zeta_{ij}, \varphi_j)$ is always smaller with the correct panning), it seems like the constraints in the images are not strong enough to discern them clearly.

To resolve this issue, we implemented a low frequency panning correction: We assume that the camera rotations relative to the rig are high frequent and on average corresponding to the initial rig parameterization. Therefore, we keep track of the low pass filtered panning pan_{avg} and correct each optimized panning parameter for the difference between the initial and the averaged panning:

$$pan_{avg}^{n+1} = (1 - \lambda)pan_{avg}^n + \lambda pan^{n+1} \tag{8}$$

$$pan_{correction} = pan_{avg}^{n+1} - pan_{init} \tag{9}$$

$$pan_{corrected}^{n+1} = pan^{n+1} - pan_{correction} \tag{10}$$

This ensures that the panning parameter is unable to drift away for many subsequent frames. We got good results on all of our datasets with $\lambda = 0.1$.

3.4 Depth Estimation

We use the depth estimation component of LSD-SLAM in Sect. 3.1 as it is but apply it potentially once for each possible pair of a keyframe camera and the tracked frame's camera. We can simply obtain the camera-to-camera transformation

$$\xi_{ij}^{c_k c_l}(x) = \varphi_j^{c_l}(\zeta_i j(\varphi_i^{c_k-1}(x))) \qquad |x \in \mathbb{R}^3 \tag{11}$$

that is necessary for stereo depth reconstruction by concatenating the rig intrinsics and rig transformations obtained during tracking.

3.5 Probabilistic Depth Merging with Multiple Observations

In this subsection we examine how to merge multiple depth observations that come from stereo with the different images from the tracked frame into the depth and variance maps of *one* of the cameras c of keyframe. Let's consider one pixel p of the keyframe whose depth is described as a normal distribution $\mathcal{N}_{prior} = \mathcal{N}(D_i^c(p), V_i^c(p))$ and the depth distributions $\mathcal{N}_{c_j \in C} = \mathcal{N}(\hat{D}_i^{cc_j}(p), \hat{V}_i^{cc_j}(p))$ of the corresponding pixels in the stereo observations with the tracked frame's cameras c_j. To perform a Bayes filter update step, we need the joined distribution $\mathcal{D}_{new}$ of the observations which can be obtained by

$$\mathcal{D}_{new} = \sum_{c_j \in C} \mathcal{N}_{c_j}. \tag{12}$$

The update step can then be expressed by

$$\mathcal{D}_{post} = \mathcal{N}_{prior} \cdot \mathcal{D}_{new} = \mathcal{N}_{prior} \cdot \sum_{c_j \in C} \mathcal{N}_{c_j}. \tag{13}$$

Now $\mathcal{D}_{new}$ and $\mathcal{D}_{post}$ are not normal distributions but we will have to represent them as one to update the keyframe's maps. Therefore we rewrite our formulation to apply the multiplications of the distributions first (the product of two normal distributions is again a normal distribution) and approximate the sum by one single normal distribution, i.e. we do the approximation of our distribution as normal distribution as late as possible:

$$\mathcal{N}_{post} \approx \sum_{c_j \in C} \left(\mathcal{N}_{prior} \cdot \mathcal{N}_{c_j} \right) \tag{14}$$

To calculate $\mathcal{N}_{post}$, we have to extend our model of normal distributions by a weight parameter: $\mathcal{N}(d, v, w)$. It was previously omitted because we were dealing with probability distributions whose weights and integrals evaluate to one always. In the sum above, however, we need this weight since we want its terms to contribute differently to the resulting distribution, depending on how much $\mathcal{N}_{prior}$ and $\mathcal{N}_{c_j}$ overlap.

Formulas for computing the product or the merge of normal distributions can be found in [18]:

Product of normal distributions: $\mathcal{N}(d, v, w) = \mathcal{N}(d_1, v_1, w_1) \cdot \mathcal{N}(d_2, v_2, w_2)$

$$d = \frac{d_1 v_2 + d_2 v_1}{v_1 + v_2} \qquad v = \frac{v_1 v_2}{v_1 + v_2} \qquad w = \frac{\mathcal{N}(d; d_1, v_1, w_1) \cdot \mathcal{N}(d; d_2, v_2, w_2)}{\mathcal{N}(d; d, v, 1)}$$

$$(15)$$

Merge of normal distributions: $\mathcal{N}(d, v, w) \approx \sum_i \mathcal{N}(d_i, v_i, w_i)$

$$w = \sum_i w_i \qquad d = \frac{1}{w} \sum_i w_i d_i \qquad v = \sum_i \frac{w_i}{w}(v_i + (d_i - d)^2) \qquad (16)$$

Note that our implementation contains a $f_{m_d}(c_i, c_j) : \mathbb{N}^2 \to \{0, 1\}$ similar to f_t from Sect. 3.2 to allow the user to control which tracked frame cameras c_j are used to update the maps of keyframe camera c_i (see also Fig. 6).

3.6 Depth Map Propagation

We propagate depths and variances from one keyframe to the next within the same camera. When a new keyframe is created, we potentially apply one iteration of static stereo, i.e. update each camera's depth and variance maps with the stereo observations from the other cameras of the same frame. This can be user-controlled with $f_{m_s}(c_i, c_j) : \mathbb{N}^2 \to \{0, 1\}$.

4   Evaluation

We evaluated different configurations (see Fig. 6) of our algorithm on various datasets. Please note that this is a frame-to-frame VO algorithm that does not perform full SLAM including loop closing and global optimization and is therefore not directly comparable to such methods.

4.1   KITTI Dataset

Results for two scenes of the KITTI odometry benchmark [19] based on a stereo setup are shown in Fig. 7 using the configurations displayed in Fig. 6.

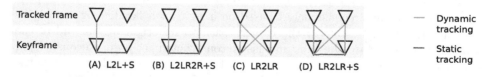

Fig. 6. Different camera tracking and mapping configurations. Dynamic tracking is either done within a single stream, within all streams or across streams. Static tracking is performed between two images captured at the same time.

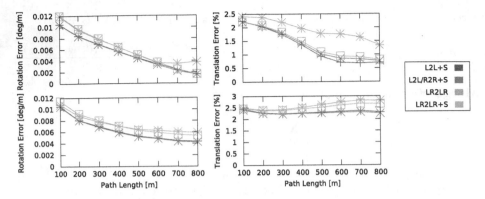

Fig. 7. Results from the KITTI odometry benchmark scenes 06 (top) and 08 (bottom). Note that static stereo seems to improve quality significantly. Tracking and depth estimation between different cameras from different frames produces worse results in the scenes because of errors that are introduced by more occlusions on longer baselines.

Dynamic tracking and mapping on the left camera only is analog to [1] and produces similar results. Worse results with other configurations of our algorithm, e.g. LR2LR+S, have also been observed by [1] and seem to be related to more outliers due to obstructions for large baselines between different cameras. However, those other configurations are crucial for flexible camera rigs (Sect. 4.2) or rigs with barely overlapping FoVs (Sect. 4.4).

Please refer to the supplemental material for four camera KITTI evaluations.

4.2 Flexible Rigs

We evaluate our estimation of intrinsic rig parameters based on synthetic and real-world data. Figure 8 compares the intrinsic rig parameters to the ground truth of a synthetic scene (highly textured, 720p, with symmetric rig intrinsics as in Sect. 3.3) and shows that our approach clearly produces correct results.

For evaluation on real world data, we recorded three scenes with a flexible stereo camera rig with a FoV of $110°$ and 2048×2048 pixels per image downsampled to 512×512 pixel images with 30 fps. The rig was bended and twisted by about $3°$ while recording. Figure 9 compares the drift of the camera pose when sequences are reconstructed based on a rigid or a flexible rig model. To evaluate the drift, we moved the camera back to its origin at the end of the sequence and compare the first and last reconstructed rig pose. Reconstructed depth maps for stereo pairs are shown in Fig. 2. Our results show that dense algorithms that don't keep track of the rig intrinsics perform significantly worse and introduce jittering to the rig pose since the tracking cannot converge on all cameras consistently. Reconstructions for this section were performed using LR2LR+S (see Fig. 6), because we observed unstable reconstructions without cross-camera tracking on some of our flexible rig scenes.

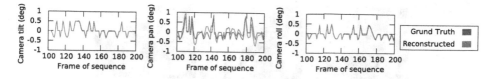

Fig. 8. Evaluation of the rig intrinsics reconstruction on synthetic data with ground truth. Note that roll and tilt are reconstructed nicely while the panning exhibits some artifacts due to our correction (Sect. 3.3) which we accept in favor of drifts or a collapsing reconstruction.

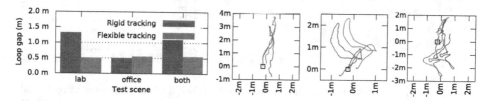

Fig. 9. Loop gap evaluation on a flexible rig dataset. Tracking with a flexible rig model clearly improves the drift in the tracking. A closer look to the plotted paths reveals that tracking a flexible rig with a rigid model leads to jittering in the reconstructed camera positions, most likely caused by convergence to one of the cameras when the cameras don't agree due to a deformed rig.

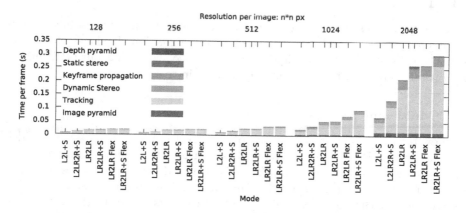

Fig. 10. Timing evaluation. The L2L+S configuration which works best for stereo rigs runs at more than 15 fps even for 4 Megapixel images. There is just small computational overhead introduced by static stereo and nonrigid tracking. Note that there is some overhead for CPU/GPU synchronization which consumes significant time for smaller resolutions. Also note that the three tasks at the top that are only performed when the keyframe is updated have a very small average per frame contribution.

Fig. 11. Test scenes for multi-camera rig evaluation: corridor (left) exhibits few gradients, repetitive ceiling texture and glossy highlights; office (center) shows many easily trackable gradients; veranda (right) is an outdoor scene with large depth variations.

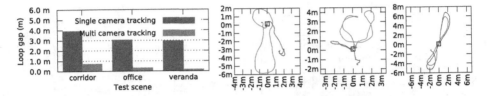

Fig. 12. Loop gap evaluation on the Point Grey Ladybug 3 which is an omnidirectional, 6 camera rig. We used static and dynamic intra-camera tracking (compare to Fig. 6, B) on a rigid rig model. Note that in contrast to stereo data with overlapping fields of view (see Fig. 7), the tracking benefits from taking all available data into account and optimizing for one consistent rig pose. The multi camera tracking reduces the error by one order of magnitude. The average processing time per frame ($6 \times 736 \times 1136\,\text{px}$) was 59 ms for single camera and 120 ms for multi camera tracking.

4.3   Timings

Figure 10 compares the per frame processing times for different resolutions, different configurations and different steps of the algorithm. It shows that there is just small overhead for nonrigid tracking. It also shows that our GPU implementation is able to process even 4 MP images in real-time.

4.4   Multi-camera Rigs

To show the multi-camera capabilities of our algorithm, we evaluated it on video streams from a Point Grey Ladybug 3 which is a 6 camera, omnidirectional camera rig with slightly overlapping FoVs. We used three test scenes shown in Fig. 11. The results in Fig. 12 indicate that our tracking approach can reconstruct valid camera paths based on the data of all cameras where methods that use just one camera for tracking like [1] fail, e.g. because of few or EPL-aligned high gradient pixels or a fixed point of expansion position in the images with leads to low disparities and therefore bad depth estimations in the surrounding.

5   Conclusion

We present a direct, semi-dense visual odometry method for flexible multi-camera rigs. Key features of our method are (1) the tracking based on the photometric error that optimizes for consistent, relative rig poses as well as the rig intrinsics at once and (2) the update of the semi-dense depth with stereo information from multiple cameras in a Bayesian framework.

Our method achieves state of the art reconstruction quality for scenes recorded with completely rigid rigs. But it even supports the reconstruction with non-rigidly connected cameras. This reduces reconstruction errors dramatically, even if only small relative camera motion is present. Additionally, we show that our method is not just suitable for stereo cameras but also for other configurations like omnidirectional camera rigs with barely overlapping views.

In the future, this work could be extended to allow also for dynamic camera intrinsics as well by making the camera intrinsics a controllable parameter in the warping function (Eq. 5).

Acknowledgements. This work was supported by Daimler AG, Germany. Real-world flexible stereo rig datasets were kindly provided by Dr. Senya Polikovsky, OSLab, Max Planck Institute for Intelligent Systems Tübingen.

References

1. Engel, J., Stueckler, J., Cremers, D.: Large-scale direct slam with stereo cameras. In: International Conference on Intelligent Robots and Systems (IROS) (2015)
2. Chiuso, A., Favaro, P., Jin, H., Soatto, S.: Structure from motion causally integrated over time. IEEE Trans. Pattern Anal. Mach. Intell. **24**, 523–535 (2002)
3. Nistér, D., Narodltsky, O., Dergen, J.: Visual odometry, pp. 652–659 (2004)
4. Davison, A.J., Reid, I.D., Molton, N.D., Stasse, O.: Monoslam: real-time single camera slam. IEEE Trans. Pattern Anal. Mach. Intell. **29**, 1052–1067 (2007)
5. Klein, G., Murray, D.: Parallel tracking and mapping for small AR workspaces. In: Proceedings of Sixth IEEE and ACM International Symposium on Mixed and Augmented Reality, ISMAR 2007, Nara, Japan (2007)
6. Paz, L.M., Piniés, P., Tardós, J.D., Neira, J.: Large scale 6-DOF slam with stereo-in-hand. IEEE Trans. Robot. **24**, 946–957 (2008)
7. Kerl, C., Sturm, J., Cremers, D.: Robust odometry estimation for RGB-D cameras. In: ICRA, pp. 3748–3754. IEEE (2013)
8. Meilland, M., Comport, A.I.: On unifying key-frame and voxel-based dense visual SLAM at large scales. In: 2013 IEEE/RSJ International Conference on Intelligent Robots and Systems, Tokyo, Japan, 3–7 November 2013, pp. 3677–3683 (2013)
9. Engel, J., Schöps, T., Cremers, D.: LSD-SLAM: Large-Scale Direct Monocular SLAM. In: Fleet, D., Pajdla, T., Schiele, B., Tuytelaars, T. (eds.) ECCV 2014. LNCS, vol. 8690, pp. 834–849. Springer, Heidelberg (2014). doi:10.1007/978-3-319-10605-2_54
10. Pillai, S., Ramalingam, S., Leonard, J.: High-performance and tunable stereo reconstruction. In: 2016 IEEE International Conference on Robotics and Automation (ICRA). IEEE (2016)
11. Comport, A.I., Malis, E., Rives, P.: Accurate quadrifocal tracking for robust 3D visual odometry. In: Proceedings 2007 IEEE International Conference on Robotics and Automation, pp. 40–45 (2007)
12. Resch, B., Lensch, H.P.A., Wang, O., Pollefeys, M., Sorkine-Hornung, A.: Scalable structure from motion for densely sampled videos. In: CVPR, pp. 3936–3944. IEEE Computer Society (2015)

13. Kim, C., Zimmer, H., Pritch, Y., Sorkine-Hornung, A., Gross, M.: Scene reconstruction from high spatio-angular resolution light fields. ACM Trans. Graph. Proc. ACM SIGGRAPH **32**, 73:1–73:12 (2013)
14. Wei, J., Resch, B., Lensch, H.P.A.: Dense and occlusion-robust multi-view stereo for unstructured videos. In: 13th Conference on Computer and Robot Vision, CRV 2016, Victoria, British Columbia, 1–3 June 2016. IEEE Computer Society (2016)
15. Delaunoy, A., Pollefeys, M.: Photometric bundle adjustment for dense multi-view 3D modeling. In: 2014 IEEE Conference on Computer Vision and Pattern Recognition (CVPR), pp. 1486–1493. IEEE (2014)
16. Engel, J., Sturm, J., Cremers, D.: Semi-dense visual odometry for a monocular camera. In: IEEE International Conference on Computer Vision (ICCV), Sydney, Australia (2013)
17. Levenberg, K.: A method for the solution of certain non-linear problems in least squares. Q. J. Appl. Math. **II**, 164–168 (1944)
18. Crouse, D.F., Willett, P., Pattipati, K., Svensson, L.: A look at Gaussian mixture reduction algorithms. In: 2011 Proceedings of the 14th International Conference on Information Fusion (FUSION), pp. 1–8 (2011)
19. Geiger, A., Lenz, P., Urtasun, R.: Are we ready for autonomous driving? the KITTI vision benchmark suite. In: Conference on Computer Vision and Pattern Recognition (CVPR) (2012)

Analysis and Practical Minimization of Registration Error in a Spherical Fish Tank Virtual Reality System

Qian Zhou[1]([✉]), Gregor Miller[1], Kai Wu[1], Ian Stavness[2], and Sidney Fels[1]

[1] Electrical and Computer Engineering, University of British Columbia,
Vancouver, BC, Canada
qzhou@ece.ubc.ca
[2] Department of Computer Science, University of Saskatchewan,
Saskatoon, SK, Canada

Abstract. We describe the design, implementation and detailed visual error analysis of a 3D perspective-corrected spherical display that uses calibrated, multiple rear projected pico-projectors. The display system is calibrated via 3D reconstruction using a single inexpensive camera, which enables both view-independent and view-dependent applications, also known as, Fish Tank Virtual Reality (FTVR). We perform error analysis of the system in terms of display calibration error and head-tracking error using a mathematical model. We found: head tracking error causes significantly more eye angular error than display calibration error; angular error becomes more sensitive to tracking error when the viewer moves closer to the sphere; and angular error is sensitive to the distance between the virtual object and its corresponding pixel on the surface. Taken together, these results provide practical guidelines for building a spherical FTVR display and can be applied to other configurations of geometric displays.

1 Introduction

As computer graphic and display technology advances rapidly, the interaction and visualization of 3D information is becoming increasingly important. Volumetric displays have shown promise [1] for interacting with and visualizing 3D data as they provide voxel-based 3D imagery. Many types of volumetric displays have been proposed in a variety of shapes [2,3]. Among those types of volumetric displays, spherical displays have seen significant interest with research prototypes and commercial products. While a true volumetric display provides pixels in actual 3D space, perspective-corrected 3D display provides this illusion by projecting the correct perspective view of the scene for the viewer on the surface of the sphere. Thus, as the viewer moves, the scene maintains the correct 3D perspective; this is known as Fish Tank Virtual Reality (FTVR).

Electronic supplementary material The online version of this chapter (doi:10. 1007/978-3-319-54190-7_32) contains supplementary material, which is available to authorized users.

S.-H. Lai et al. (Eds.): ACCV 2016, Part IV, LNCS 10114, pp. 519–534, 2017.
DOI: 10.1007/978-3-319-54190-7_32

Spherical displays embody the metaphor of a "crystal ball" and provide non-occluded views from all viewpoints around it. However, providing uniform pixels appearing on the surface of the sphere is one of the requirements for constructing spherical FTVR. One approach to achieve this effect uses an array of rear-projection projectors from within the sphere, providing a scalable system with high resolution and uniform pixel density [4]. However, geometry calibration and blending of multiple projectors on the curved screen surface to achieve seamless imagery is a challenge.

We present a practical approach for building a spherical multi-projector display to overcome this challenge. Our system works in both view-dependent and view-independent modes as we can project a seamless image over the entire surface. We introduce our system design along with the workflow for our 3D reconstruction-based, single-camera display calibration approach. We blend projected images based on geometry reconstruction, creating a seamless undistorted imagery using two-pass rendering. Using this approach, perceptual discrepancies arise when virtual objects do not remain correctly aligned based on the actual viewpoint due to error in the system pipeline. This misalignment may cause artifacts like distortions of rendered objects, making the visualization unacceptable to the viewer for a given application. To mitigate this issue, higher fidelity calibration can be performed, however, without knowing which calibration component accounts for the most perceptual error, it is difficult to know what to improve. Thus, as part of our practical guide, we provide a mathematical model of different calibration error sources. Using this model, we conduct error analysis for the spherical multi-projector FTVR system in terms of display calibration error and head-tracking error. As we discuss, we found: 1. Tracking error causes significantly more angular error than display error; 2. Angular error becomes more sensitive to tracking error when the viewer moves closer to the sphere; 3. Angular error is sensitive to the distance between the virtual object and its corresponding pixel on surface; 4. Our calibration approach is more sensitive to the error in sphere pose than projector error, making it necessary for us to improve the sphere pose estimation to reduce calibration error; 5. Calibration error is not spatially homogeneous. These results provide a guide to establish system component and calibration fidelity that can be matched to different application needs.

2   Related Work

2.1   FTVR and Spherical Displays

Fish Tank Virtual Reality (FTVR) [5] is a type of 3D head-tracked display, providing motion parallax cues to improve user's understanding of the 3D virtual scene. This technique has been widely applied to various systems and applications. The CAVE [6] is one of these well-known systems, extending the traditional FTVR by projecting on multiple screens to form a geometric shape display. Besides the CAVE, Stavness et al. [7] developed pCubee, a perspective handheld cubic display. They arranged five small LCD panels to form the sides of a cube with head-coupled perspective-corrected rendering. However, a pCubee study [8] revealed occlusion caused by the seam between screens discouraged users from changing their view from one screen to another, making the shape of the display an important form factor.

A spherical display has a promising shape for a volumetric display as it has no seam between screens. A number of spherical displays have been proposed with different implementations. One of the early work in this field is the Perspecta Spatial 3D System from Actuality Systems, Inc. [1]. It utilizes an embedded stationary projector to project imagery onto a rotating screen. Although it can generate volume-filling imagery, this type of system is expensive to build and has a limitation in resolution and scaling. As an alternative method, some systems use rear-projection directly onto a spherical screen. Sphere [9], Snowball [10], and commercial products like Pufferfish use one projector for rear-projection onto the spherical screen. While simple and fairly effective, these systems offer low resolution and lack of scalability. Spheree [4] extends this approach by utilizing multiple pico-projectors to increase resolution, making the system scalable to spherical screens with various sizes. We use the same approach to design our system, making it scalable and flexible.

2.2   Multi-projector System Calibration

The main challenge with multi-projector systems is the calibration of the system. The calibration includes geometry calibration and photometric blending to achieve seamless imagery. While there has been substantial work on calibration of planar multi-projector display using a single camera via linear homography transformations [11,12], non-planar multi-projector calibration is still under explored.

An early work on non-planar calibration was published by Raskar et al. [13], in which a stereo camera pair is used to recover projector-camera parameters and reconstruct a non-planar surface. They achieved registration using both structured light and additional surfaces. They improved their work by focusing a subset of non-planar surfaces called quadric surfaces using the stereo camera to recover a quadric transformation [14].

Harville et al. [15] proposed a calibration method for a class of shape which can be made by transforming a plane through folding, blending etc. They attached a physical checkerboard pattern to the display and used an uncalibrated camera to compute a composition of 2D-mesh-based mappings. Since the 3D geometry is not recovered, their application space is limited.

More recently, Sajadi and Majumder have published a series of papers on non-planar calibrations [16,17]. They use an uncalibrated camera to compute a rational Bezier patch for geometric correction to account for the distortion of the display surface. They extend their work for various shapes such as vertical extruded surfaces, swept surfaces, dome surfaces and CAVE-like surfaces. Their calibration methods mostly aims at large-scale displays and thus mount the camera on a pan-tilt unit to cover the entire display.

Teubl et al. [4] developed a multi-projector system library to automatically calibrate multi-projector systems for different types of display surfaces. Their approach uses fixed warping without full 3D reconstruction. For a spherical shape they use a linear assumption by using a homography transformation between camera and projector without reconstructing 3D geometry. This causes misalignment artifacts in overlapping areas.

Drawing inspirations from previous work, we also use the spherical geometry assumption as prior knowledge to calibration the system. However, our system is relatively small-scale (i.e. $< 1\,\mathrm{m}$ diameter with a hole cut at the bottom) compared to those large-scale display ($> 1\,\mathrm{m}$). Rear-projection through a small projection hole at the bottom of the sphere makes calibration difficult since the view of camera is mostly blocked by the edge of small hole. To overcome those problems, we propose a calibration pipeline similar to Raskar's work [13,14] but with modifications to make it adapted to our system.

2.3   Virtual Reality System Error Analysis

Substantial research analyzes and handles the error in virtual reality and augmented reality systems [18–21]. Holloway et al. [19] analyzed the causes of registration error for a Head Mounted Device (HMD) using a set of parameters. MacIntyre et al. [21] presented a statistical method to estimate the error and further use the estimated error to improve the AR interface. However most of them are for see-through HMD systems. Though there are many descriptions on FTVR systems and geometric displays, very little work has dealt with error analysis for these systems. Cruz et al. discussed tracking noise and delay for the CAVE system [6], but mainly aimed at comparing to HMD and normal monitor systems to show the reduced effect of these errors on the CAVE system. In this paper, we analyze errors in our spherical system in terms of display calibration error and tracking error using a mathematical model, which can be useful when choosing devices for system built-up and improvement.

3   Calibration of Multi-projector FTVR System

A basic FTVR system consists of two essential parts to make the view-dependent display work: a display system and a tracking system. The display system is made of a set of projectors (we use pico-projectors) and a spherical screen. Projectors are set under the spherical screen, rear-projecting through a projection hole onto the screen as illustrated in Fig. 1. Calibration is required to make the system work. This includes display calibration and tracker calibration.

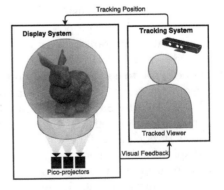

Fig. 1. System diagram of a 3D perspective-corrected spherical display.

3.1   Display Calibration

To generate seamless imagery on the spherical display, we need to reconstruct the 3D geometry of projected pixels on sphere, which is used to register projected images and then blend the intensity in overlapping areas.

As discussed in related work, cameras are commonly used to calibrate multi-projector systems. The calibration of our FTVR system is challenging and differs from other multi-projector systems in the following aspects:

1. System scale. Most multi-projector calibration methods are designed for large-scale display, where space is plenty for projectors, cameras and mirrors. In our case, we have a relatively small spherical screen of diameter 29 cm with projectors set in limited space under the screen. It is difficult to put multiple cameras in the limited space to see the screen. A single small camera is preferred in our system.
2. Visibility. The projection hole of diameter 14 cm on the spherical screen will block the view of camera, making a large portion of the projection invisible to the camera.
3. As a FTVR system, it should be able to support both view-independent and view-dependent applications to enable multi-person VR display.

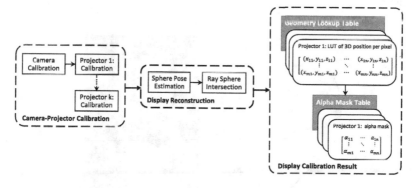

Fig. 2. Display calibration pipeline

Being able to support both types of applications forces us to reconstruct the 3D geometry of the projection on the sphere. Fixed warping with linear assumptions [4] offers low accuracy thus not applicable to our case. Although the parametric approach using quadric transformation [14] appears to be promising, it does not support view-independent applications. For view-dependent applications, we still have to update the quadric transformation for each frame since the viewpoint will be moving. The visibility problem makes most calibration methods using patch-based or mesh-based interpolations [16] not workable in our system. Finally, our system scale suggests a single camera approach.

As a result, we propose our display calibration approach as illustrated in Fig. 2. This approach begins with the calibration of camera and projectors, followed by a pose estimation of the spherical display surface, then ending with ray-sphere intersection to locate 3D position of each pixel on the surface.

Camera-Projector Calibration. Each projector is modeled as an inverse pinhole camera. We calibrate camera C and projectors P using a plane-based calibration approach [22], which is essentially an extension of Zhang's calibration technique [23] for camera-projector system. With the removal of the spherical screen, the planar pattern placed at different positions and orientations is used to avoid degenerate case for projector calibration [13]. After this step, the intrinsic and extrinsic parameters for the camera and projectors are recovered.

Sphere Pose Estimation. In this step, each projector P_i is paired with the same camera C, forming a stereo pair S_i as illustrated in the left image of Fig. 3. For each projector P_i, we rear-project an array of blobs onto the sphere, with the camera seeing part of the projected patterns. We triangulate to find the 3D position of the blobs in the camera-centered coordinate system. This procedure is repeated for each stereo pair S_i to obtain more points on the sphere. The right images of Fig. 3 show the partial projection from two projectors seen by the camera.

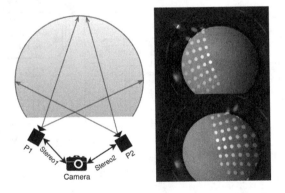

Fig. 3. Stereo pairs and projected blob patterns

We fit the sphere equation with these 3D points using Weighted Least Square [24]. The weighting matrix W comes from the re-projection error in triangulation. Points with large re-projection error will have a relative small weighting when fitting a sphere. After this step, the sphere pose with four parameters is recovered.

Ray-Sphere Intersection. Our goal is to find the 3D position for each pixel on the surface. The ray-sphere intersection is used to find the 3D position for each pixel. This begins by computing the ray equation $Ray(\lambda)$ associated to the pixel:

$$Ray(\lambda) = C_{P_i} + V_{uv} \cdot \lambda$$
$$= -R_{P_i}^{-1} T_{P_i} + \lambda (K_{P_i} R_{P_i})^{-1} \begin{pmatrix} u & v & 1 \end{pmatrix}^T , \tag{1}$$

where C_{P_i} represents the center of projector P_i in camera-centered coordinate system. V_{uv} is the vector pointing from C_{P_i} to the pixel (u, v) in projector P_i. The intrinsic matrix of projector P_i is K_{P_i} and corresponding extrinsic matrix is $(R_{P_i} \; T_{P_i})$.

$Ray(\lambda)$ is intersected with the recovered sphere for each pixel (u, v) in projector P_i to find the intersection point in the camera-centered coordinate system. This results in the 3D position of each pixel on the sphere. For the convenience of rendering, we transform those 3D points from the camera-centered coordinate system to the display coordinate system through translation. The display coordinate system has its origin set at the center of the sphere. The geometry result of each projector is stored in a look-up table as shown in Fig. 2, which is later used to compute alpha mask for blending.

Intensity Blending. For the photometric blending of the overlapping area, we compute the alpha mask for each projector based on a weighted average [13], assigning an intensity weight from 0 to 1 to each pixel in the projector. The alpha-weight $A_m(u, v)$ for the (u, v) pixel in projector m is computed:

$$A_m(u, v) = \frac{\alpha_m(m, u, v)}{\sum_i \alpha_i(m, u, v)},$$

where $\alpha_i(m, u, v) = w_i(m, u, v) \cdot d_i(m, u, v)$. $w_i(m, u, v) = 1$ if the pixel (u, v) of projector m is inside the hull of projector P_i; otherwise zero. $d_i(m, u, v)$ is the distance of the pixel (u, v) of projector m to the nearest edge of projector P_i.

$w_i(m, u, v)$ is computed based on whether the pixel is inside the view frustum of projector P_i. The distance $d_i(m, u, v)$ is computed as the length of the shortest arc that connects pixel to its nearest edge arc.

3.2 Tracker Calibration

The goal of tracker calibration is to compute a similarity transformation between tracker coordinate system and display coordinate system. We first project blob patterns onto the sphere. Then we use the tracker to detect and recover their 3D positions in tracker coordinate system. Since the display system has been calibrated, we know the 3D position for each blob in display coordinate system. We estimate the similarity transformation using SVD as an initial guess, followed by Levenberg-Marquardt method for refinement [25].

4   Error Analysis of the Spherical System

While many spherical display systems have been proposed, none has included a formal analysis of visual error to our knowledge. However, error analysis is important for two main reasons. First, when building the system, one may expect different error tolerance for each component of the system based on the purpose of application. Error analysis provides guidelines when choosing these components. Second, knowing the nature and sensitivity of the error also helps us to eliminate artifacts cause by the error. These artifacts include, but are not limited to: distortion (straight lines appear to be curved), ghosting effect (double-image) and floating effect (virtual objects that are supposed to locate at a fixed location appear to swim about as viewer moves head) [18].

4.1   Metric for FTVR System

We use eye angular error [6] as the metric for evaluating the performance of a FTVR system. Eye angular error quantifies the registration between viewer and the display surface. Qualitatively, eye angular error creates artifacts like distortions when rendering 3D objects. Quantitatively, the angular error can be defined to be the displacement between a pixel shown on the spherical screen D and its desired location $\bar{D}$ which should be perspective-corrected according to the viewer position.

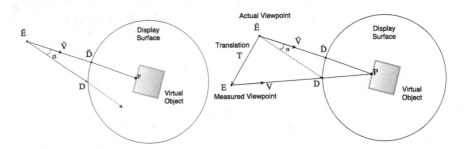

Fig. 4. (a) Eye angular error in our system (b) angular error caused by tracking error

Let $\bar{E}$ represent the viewer position, looking at a virtual point P inside the sphere as shown in Fig. 4(a). Then the angular error α can be computed using the following equation:

$$\alpha = arccos(\frac{(D - \bar{E})(P - \bar{E})}{\|(D - \bar{E})\|\|(P - \bar{E})\|}),$$ (2)

Since our FTVR system consists of display system and tracking system, it is natural to analyze the system error in terms of display error and tracking error. We now discuss how the tracking error and display error will influence α in the following sections.

4.2   Tracking Error

Tracking error represents the error between the actual viewpoint and the measured viewpoint. Sources of this error include: tracker error, tracker latency and transformation error. We use translation T to represent the tracking error between the actual viewpoint $\bar{E}$ and the measured viewpoint E in Fig. 4(b). Due to the translation T, $\bar{D}$ is the desired pixel corresponding to the actual viewpoint $\bar{E}$, while D corresponding to E is the pixel shown on the display.

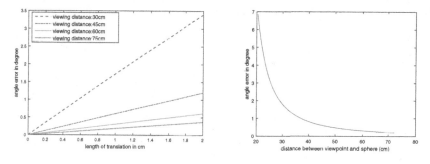

Fig. 5. (a) Tracking error caused by translation (b) by viewpoint

The ray starting at the measured viewpoint E pointing in the direction of V can be expressed as $Ray(\lambda) = E + V \cdot \lambda$. Intersecting the ray with the sphere, we can express D with respect to E and V, while the intermediate variables E and V can be expressed using $\bar{E}, P$ and T. Thus, from point D, we see the angular error α in the Eq. (2) is a function of the virtual point P, translation T and actual viewpoint $\bar{E}$.

Effect of Translation. α is maximized when the direction of T is perpendicular to the EP. In the following sections, we always assume EP is tangent to T so that we are evaluating based on the worst case. Figure 5(a) shows the simulation result of α as a function of $\|T\|$ while the viewer is looking at the virtual point P placed at the sphere center. The slope increases as the viewer gets closer to the sphere, meaning α becomes more sensitive to translation error if the viewer is closer to the display.

Effect of Viewpoint. Figure 5(b) illustrates the simulation result of how the viewpoint influences α when the viewer is looking towards a virtual point P placed at the center of the sphere. As the viewpoint $\bar{E}$ moves away from the sphere along $\bar{E}P$, α decreases dramatically. To control α, a minimum viewing distance can be established depending on the applications. For example, for interactive applications, the viewing distance is likely shorter than ones only needing visualizations. Hence, the interactive system will require a tracking device with higher accuracy.

Effect of Virtual Point. Consider that virtual point P can be placed both inside and outside the sphere. For a viewpoint $\bar{E}$, several viewing directions pointing to different virtual points P_i are illustrated in Fig. 6. As a function of the virtual point P_i, the angular error α first decreases when P_i travels towards $\bar{D}_i$, so that the virtual point away from its corresponding display pixel will cause larger α. When P_i arrives at $\bar{D}_i$, there is no angular error since D_i overlaps with $\bar{D}_i$, meaning points on the display surface towards the viewer will not cause angular error even if there is translation error in the tracker. As P_i moves out of the sphere further toward $\bar{E}$, α starts to increase rapidly, meaning points out of the sphere are more sensitive to translation error.

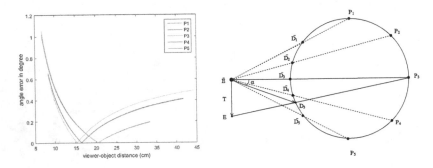

Fig. 6. Effect of the virtual point position on angular error

Since α depends on the position of virtual point P_i, there exists an "optimal" rendering region with relatively small angular error. For instance, if we want to have a maximum angular error as $0.4°$, according to Fig. 6 we can only render a virtual object inside the sphere that is less than 18 cm away from its corresponding pixel, or less than 5 cm away for an object outside the sphere, denoted as -5 cm to $+18$ cm.

Result. As a summary, the eye angular error α increases: 1. as tracking error increases, with its slope influenced by viewing distance; 2. as viewer moves closer to the sphere; 3. as the virtual point is further from the pixel on display surface (Table 1).

Based on this analysis, we have quantitative results for two tracking systems we use for our FTVR system. According to Xu et al. [26], the average joint accuracy for Kinect v2 using its joint tracking SDK is around 8 cm. If we set angular error to be no more than $2°$, the minimum viewing distance should be no less than 60 cm assuming a viewer is looking at a virtual object at the center. At this minimum viewing distance, the rendering range should be -8 cm to $+11$ cm. Naturally, improvements over the joint tracking library in Kinect v2 SDK for head tracking can improve these bounds. For the Polhemus Fastrak with a tracking accuracy of 0.2 cm within 150 cm of the transmitter, it is possible

Table 1. Quantitative result of head-tracking error

Device	Head-tracking Error	1 Angular error (object in center)	Viewing distance	Rendering range (at min view distance)
Kinect v2	8 cm [26]	< 2°	> 60 cm	−8 cm to +11 cm
Fastrak	0.2 cm [27]	< 0.5°	> 30 cm	−6 cm to +30 cm

to yield a 0.5° maximum angular error system, making it more appropriate for an *interactive* FTVR so viewers can get close enough to the display.

4.3   Display Error

Assuming the head-tracking is perfect, then the display error can be represented by the displacement between the desired 3D position of pixel $\bar{D}$ and its actual location D on the display surface in Fig. 4. This error depends on the accuracy of the calibration workflow. Ideally, with ground truth we can compute the accuracy of our calibration algorithm. Unfortunately, ground truth is hard to acquire in this system. Instead, we use covariance as a measure of the calibration accuracy [25]. Alternatively, an empirical method could be used, in which error can be measured using a camera [28].

The 3D positions of projector pixels are computed using ray-sphere intersection in the calibration pipeline, which can be expressed as:

$$X = f(x; p), \tag{3}$$

where X is the 3D position, x is the 2D projector pixel and p is a column vector that contains all parameters including projector parameters p_1 and sphere parameters p_2, with all parameters written in the form of column vector. Any error in p_1 and p_2 will disturb the 3D position of projector pixel.

According to forward propagation [25], we can compute the covariance matrix of X as: $\Sigma_X = J_{X(p)} \Sigma_p J_{X(p)}^T$, where $J_{X(p)}$ is the Jacobian matrix of the vector function $X(p)$ with respect to the parameter vector p and Σ_p is the covariance matrix of p.

Though the Jacobian matrix $J_{X(p)}$ can be computed analytically as the partial derivative of the vector function $X(p)$, the covariance matrix Σ_p is not so straight-forward. The parameter vector p contains both projector parameters and sphere parameters. The covariance matrix of projector parameters can be estimated using backward propagation [25] of the re-projection error once we calibrate the projector. The covariance matrix of sphere parameters can also be computed in a similar manner using a least square covariance computation. However, the correlation between projector parameters and sphere parameters is difficult to obtain. It is not appropriate to make the assumption that those parameters are independent since we used projector parameters to estimate sphere pose. So instead of computing a composite Σ_X in terms of all parameters, we compute Σ_X respectively to projector parameters and sphere parameters as described next.

Error in Projector Parameters. Assume p_1 is the vector that contains projector intrinsic K_{P_i} and extrinsic parameters R_{P_i}, T_{P_i}. The covariance matrix Σ_{p_1} of p_1 is computed using backward propagation from the projector Eq. (4): $\Sigma_{p_1} = (J_{m_{p_1}}{}^T \Sigma_{\mathbf{m}}^{-1} J_{m_{p_1}})^{-1}$, where Σ_m is the re-projection error during projector calibration and $J_{m_{p_1}}$ is the Jacobian matrix of the Eq. (4) with respect to the projector parameter p_1.

$$m = K_P \left(R_P \, T_P \right) M \tag{4}$$

We plug the covariance matrix Σ_{p_1} into the Eq. (3) to compute covariance matrix Σ_X. Using our calibration result, this yields to an average 3D Euclidean distance error of 0.315 mm for pixels on surface.

Error in Sphere Pose Parameters. Assume p_2 is the vector that contains sphere parameters s_1, s_2, s_3 and s_4. The covariance matrix Σ_{p_2} can be estimated [29] based on a weighted least square. Using the computed Σ_{p_2}, we get an average of 1.344 mm, indicating that error in sphere pose tends to be more influential than the error in projector calibration.

Result. From the above analysis, we see the following results. First, if we consider sphere pose and projector parameters as error sources that influence the calibration error, then our calibration is more sensitive to changes in sphere pose estimation than projector parameters. Thus, it is best to improve the sphere pose estimation to reduce calibration error.

Second, the calibration error is not spatially homogeneous. Figure 7 shows calibration errors taken at equally spaced locations on the projector using Eq. (3) when considering respectively the projector parameters and sphere pose parameters as error sources. The result is based on the implementation described in Sect. 5. The calibration error reaches peaks on the corners and falls off from the center by a factor up to ten times the error at the center. The fringe pixels have significantly more errors than pixels at the center possibly resulting in a noticeable misalignment in the overlapping projector area since overlaps happen mostly along the projection fringe. Hence, it is useful to have a pixel-by-pixel adjustment after the euclidean reconstruction to minimize those local errors.

Lastly, if we compare the display calibration error with the tracking error, the calibration error of 1 mm on the display surface only yields a $0.38°$ angular error with 30 cm viewing distance. So tracking error causes significantly more angular error than calibration error. This matches with Holloway's result [19] for an HMD system. While tracking error can be the major error source for artifacts like distortion and floating effect, the display calibration error accounts

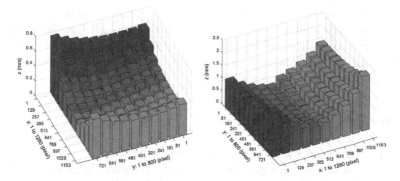

Fig. 7. Space dependency of the calibration error on pixels with (a) projector as error source (b) sphere as error source

for the ghosting effect in overlapping area. In fact, calibration error is the only error source to cause ghosting. Ghosting happens when we stitch images from adjacent projectors and is caused by the error in geometry calibration. This is independent of the viewpoint position. Thus, improving calibration error is important, especially for view-independent applications in which imagery is wall-papered on the entire sphere so that overlapping pixels are always used for rendering.

5   Implementation

Our system includes a display system and tracking system. The display system consists of two pico-projectors, an acrylic spherical display and a chassis which holds the display surface and projectors. The sphere size is 29 cm diameter and the projection hole size is 14 cm diameter. We use two ASUS P2B projectors with the resolution of 1280×800. A host with a NVIDIA Quadro K5200 graphic card directly sending rendering content to projectors. We use OpenGL to render graphics. To generate perspective-corrected images on the curved screen, we use a two-pass rendering method [11]. The two-pass rendering is chosen since the projection from 3D objects to the curved screen is non-linear. Though we evaluate the proposed approach with only two projectors, the result generalizes to more than two projectors. To add new projectors into the system requires adding another stereo pair directly registered to the world coordinate system for each projector. Doing so does not cause cascading error as the scale goes up as each projector is registered independently.

For the tracking system evaluation, we use a Kinect for Windows v2 and its SDK joint tracking APIs to track head position at 30 Hz. For comparison, we also use a Polhemus Fastrak, which is a wired magnetic tracking system with the update rate up to 120 Hz.

The calibration of the system is done once as a pre-processing step using Matlab. Figure 8(a) and (b) shows the result for a view-independent application.

This supports a wall-papered rendering for multiple users. Figure 8(c) shows the result for a view-dependent application, in which we track the viewer position and do a two-pass rendering based on the pixel geometry lookup table and viewer position. This supports a perspective-corrected view for a single viewer and scales to more than two projectors.

6   Summary and Future Work

We have presented our work for designing and implementing a spherical FTVR system. We describe several practical methods to build the system, including the display calibration, the tracker calibration and rendering procedure. We presented an error analysis for the spherical multi-projector FTVR system in terms of display calibration error and head-tracking error. Our error analysis shows the tracking error causes significantly more angular error than calibration error. Thus, based on the application needs, we can select an appropriate tracking device to match the angular error requirements. Likewise, based on the tracking error of the device used, we can establish a minimum viewing distance and rendering region to control the angular error. Although the calibration error does not incur large eye angular error, it can still cause a double-image effect in the overlap region between adjacent projectors when blending. Thus, improving calibration in terms of pose estimation and the minimization of the local misalignment error in the overlap region is important. Future work involves taking into account the compound estimation for calibration error to provide an end-to-end estimate to further improve bounds on performance for spherical FTVR display. Our results provide an important contribution to enable the construction of non-planar FTVR displays that use a matrix of projectors. We are able to demonstrate the relative error contributions from the different parts of the compound system to help designers select components to match the needs of their 3D applications.

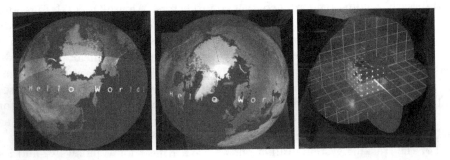

Fig. 8. View-independent application (a) after geometry calibration (b) after blending (c) View-dependent application.

Acknowledgement. We thank B-Con Engineering, NVIDIA and NSERC Canada for providing financial and in-kind support and Dr. Marcelo Zuffo and his group at University of Sao Paulo for helpful discussions.

References

1. Favalora, G.E.: Volumetric 3d displays and application infrastructure. Computer **38**, 37–44 (2005)
2. Downing, E., Hesselink, L., Ralston, J., Macfarlane, R.: A three-color, solid-state, three-dimensional display. Science **273**, 1185 (1996)
3. Blundell, B.G., Schwarz, A.J.: Volumetric Three-Dimensional Display Systems, 1st edn., p. 330. Wiley-VCH, March 2000. ISBN: 0-471-23928-3
4. Teubl, F., Kurashima, C.S., Cabral, M., Lopes, R.D., Anacleto, J.C., Zuffo, M.K., Fels, S.: Spheree: an interactive perspective-corrected spherical 3d display. In: 3DTV-Conference: The True Vision-Capture, Transmission and Display of 3D Video (3DTV-CON), pp. 1–4. IEEE (2014)
5. Arthur, K.W., Booth, K.S., Ware, C.: Evaluating 3d task performance for fish tank virtual worlds. ACM Trans. Inf. Syst. (TOIS) **11**, 239–265 (1993)
6. Cruz-Neira, C., Sandin, D.J., DeFanti, T.A.: Surround-screen projection-based virtual reality: the design and implementation of the cave. In: Proceedings of the 20th Annual Conference on Computer Graphics and Interactive Techniques, pp. 135–142. ACM (1993)
7. Stavness, I., Lam, B., Fels, S.: pCubee: a perspective-corrected handheld cubic display. In: Proceedings of the SIGCHI Conference on Human Factors in Computing Systems, pp. 1381–1390. ACM (2010)
8. Lam, B., Tang, Y., Stavness, I., Fels, S.: A 3d cubic puzzle in pcubee. In: 2011 IEEE Symposium on 3D User Interfaces (3DUI), pp. 135–136. IEEE (2011)
9. Benko, H., Wilson, A.D., Balakrishnan, R.: Sphere: multi-touch interactions on a spherical display. In: Proceedings of the 21st Annual ACM Symposium on User Interface Software and Technology, pp. 77–86. ACM (2008)
10. Bolton, J., Kim, K., Vertegaal, R.: Snowglobe: a spherical fish-tank vr display. In: CHI 2011 Extended Abstracts on Human Factors in Computing Systems, pp. 1159–1164. ACM (2011)
11. Raskar, R.: Immersive planar display using roughly aligned projectors. In: Proceedings of the Virtual Reality, pp. 109–115. IEEE (2000)
12. Raij, A., Gill, G., Majumder, A., Towles, H., Fuchs, H.: Pixelflex2: a comprehensive, automatic, casually-aligned multi-projector display. In: IEEE International Workshop on Projector-Camera Systems, Nice, France, pp. 203–211 (2003)
13. Raskar, R., Brown, M.S., Yang, R., Chen, W.C., Welch, G., Towles, H., Scales, B., Fuchs, H.: Multi-projector displays using camera-based registration. In: Proceedings of the Visualization 1999, pp. 161–522. IEEE (1999)
14. Van Baar, J., Willwacher, T., Rao, S., Raskar, R.: Seamless multi-projector display on curved screens. In: Proceedings of the Workshop on Virtual Environments, pp. 281–286. ACM (2003)
15. Harville, M., Culbertson, B., Sobel, I., Gelb, D., Fitzhugh, A., Tanguay, D.: Practical methods for geometric and photometric correction of tiled projector. In: Conference on Computer Vision and Pattern Recognition Workshop, CVPRW 2006, p. 5. IEEE (2006)

16. Sajadi, B., Majumder, A.: Automatic registration of multi-projector domes using a single uncalibrated camera. In: Computer Graphics Forum, vol. 30, pp. 1161–1170. Wiley Online Library (2011)
17. Sajadi, B., Majumder, A.: Autocalibration of multiprojector cave-like immersive environments. IEEE Trans. Vis. Comput. Graph. **18**, 381–393 (2012)
18. Azuma, R., Bishop, G.: Improving static and dynamic registration in an optical see-through hmd. In: Proceedings of the 21st Annual Conference on Computer Graphics and Interactive Techniques, pp. 197–204. ACM (1994)
19. Holloway, R.L.: Registration error analysis for augmented reality. Presence Tele-operators Virtual Environ. **6**, 413–432 (1997)
20. You, S., Neumann, U., Azuma, R.: Orientation tracking for outdoor augmented reality registration. IEEE Comput. Graph. Appl. **19**, 36–42 (1999)
21. MacIntyre, B., Coelho, E.M., Julier, S.J.: Estimating and adapting to registration errors in augmented reality systems. In: Proceedings of the Virtual Reality, pp. 73–80. IEEE (2002)
22. Falcao, G., Hurtos, N., Massich, J.: Plane-based calibration of a projector-camera system. VIBOT Master **9**, 1–12 (2008)
23. Zhang, Z.: A flexible new technique for camera calibration. IEEE Trans. Pattern Anal. Mach. Intell. **22**, 1330–1334 (2000)
24. Forbes, A.B.: Least-squares best-fit geometric elements. National Physical Laboratory Teddington (1989)
25. Hartley, R., Zisserman, A.: Multiple View Geometry in Computer Vision. Cambridge University Press, Cambridge (2003)
26. Xu, X., McGorry, R.W.: The validity of the first and second generation microsoft kinect for identifying joint center locations during static postures. Appl. Ergon. **49**, 47–54 (2015)
27. Polhemus, F.: 3space Fastrak Users Manual. F. Polhemus Inc., Colchester (1993)
28. Chen, H., Sukthankar, R., Wallace, G., Li, K.: Scalable alignment of large-format multi-projector displays using camera homography trees. In: Proceedings of the Conference on Visualization 2002, pp. 339–346. IEEE Computer Society (2002)
29. Strang, G.: Introduction to Applied Mathematics. Wellesley-Cambridge (1986)

Enhancing Direct Camera Tracking with Dense Feature Descriptors

Hatem Alismail$^{(\boxtimes)}$, Brett Browning, and Simon Lucey

The Robotics Institute, Carnegie Mellon University, Pittsburgh, USA
`halismai@cs.cmu.edu`

Abstract. Direct camera tracking is a popular tool for motion estimation. It promises more precise estimates, enhanced robustness as well as denser reconstruction efficiently. However, most direct tracking algorithms rely on the brightness constancy assumption, which is seldom satisfied in the real world. This means that direct tracking is unsuitable when dealing with sudden and arbitrary illumination changes. In this work, we propose a non-parametric approach to address illumination variations in direct tracking. Instead of modeling illumination, or relying on difficult to optimize robust similarity metrics, we propose to directly minimize the squared distance between densely evaluated local feature descriptors. Our approach is shown to perform well in terms of robustness and runtime. The algorithm is evaluated on two direct tracking problems: template tracking and direct visual odometry and using a variety of feature descriptors proposed in the literature.

1 Introduction

With the increasing availability of high frame rate cameras, direct tracking is becoming a more popular tool in myriad applications such as visual odometry [1,2], visual SLAM [3,4], augmented and virtual reality [5] and dense reconstruction [6]. Advantages of direct tracking include: (i) increased precision as much of the image could be used to estimate a few degrees of freedom [7], (ii) enhanced tracking robustness in feature-poor environments, where high frequency image content (corners and edges) are not readily available, (iii) improved ability in handling ambiguously textured scenes [8], and (iv) improved running time by exploiting the trivially parallel nature of direct tracking [6].

However, the main limitation of direct tracking is the reliance on the *brightness constancy* assumption [9,10], which is seldom satisfied in the real world. Since the seminal works of Lucas and Kanade [10] and Horn and Schunk [9], researchers have been actively seeking more robust tracking systems [11–15]. Nevertheless, the majority of research efforts have been focused on two ideas: One, is to rely on intrinsically robust objectives, such as maximizing normalized correlation [16], or the Mutual Information [13], which are inefficient to optimize and more sensitive to the initialization conditions [17]. The other, is to attempt to model the illumination parameters of the scene as part of the problem formulation [11], which is usually limited by the modeling assumptions.

© Springer International Publishing AG 2017
S.-H. Lai et al. (Eds.): ACCV 2016, Part IV, LNCS 10114, pp. 535–551, 2017.
DOI: 10.1007/978-3-319-54190-7_33

In this work, we propose the use of **densely evaluated local feature descriptor as a nonparametric means to achieving illumination invariant dense tracking**. We will show that while feature descriptors are inherently discontinuous, they are suitable for gradient-based optimization when used in a multi-channel framework. We will also show that—depending on the feature descriptor—it is possible to tackle challenging illumination conditions without resorting to any illumination modeling assumptions, which are difficult to craft correctly. Finally, we show that the change required to make use of feature descriptors in current tracking systems is minimal, and the additional computational cost is not a significant barrier.

There exists a multitude of previous work dedicated to evaluating direct tracking. For instance, Baker and Matthews [18] evaluate a range of linearization and optimization strategies along with the effects of parameterization and illumination conditions. Handa *et al.* [4] characterize direct tracking performance in terms of the frame-rate of the camera. Klose *et al.* [19] examine the effect of different linearization and optimization strategies on the precision of RGB-D direct mapping. Zia *et al.* [20] explore the parameter space of direct tracking considering power consumption and frame-rate on desktop and mobile devices. Sun *et al.* [21] evaluate different algorithms and optimization strategies for optical flow estimation. While Vogel *et al.* evaluate different data costs for optical flow [22]. Nonetheless, the fundamental question of the quantity being optimized, especially the use descriptors in direct tracking, has not yet been fully explored.

Feature descriptors, whether hand crafted [23], or learned [24], have a long and rich history in Computer Vision and have been instrumental to the success of many vision applications such as Structure-from-Motion (SFM) [25], Multi-View Stereo [26] and object recognition [27]. Notwithstanding, their use in direct tracking has been limited and is only beginning to be explored [28,29]. One could argue that this line of investigation has been hampered by the false assumption that feature descriptors, unlike pixel intensities, are non-differentiable due to their discontinuous nature. Hence, the use of feature descriptors in direct tracking has been neglected from the onset.

Among the first application of descriptors in direct tracking is the "distribution fields" work [30,31], which focused on preserving small image details that are usually lost in coarse octaves of the scale-space. Application of classical feature descriptors such as SIFT [32] and HOG [33] to Active Appearance Models have been also explored in the literature demonstrating robust alignment results [28]. The suitability of discrete descriptors for the linearization required by direct tracking has been investigated in recent work [34], where it was shown that if feature coordinates are independent, then gradient estimation of feature channels can be obtained deterministically using finite difference filters. This is advantageous as gradient-based optimization is more efficient and more precise than discrete optimization [35]. Recent work has applied descriptors to template tracking [36] in an effort to track non-Lambertian surfaces more robustly.

1.1 Contributions

In this work, we propose the use of densely evaluated feature descriptors as a means to *significantly* improve direct tracking robustness under challenging illumination conditions. We show that for dense feature descriptors to be useful for direct tracking, they must be evaluated on a small neighborhood and must have sufficient discrimination power.

We evaluate the use of dense feature descriptors using two direct tracking problems: (i) parametric motion estimation using an Affine motion model, where the warping function is linear; (ii) direct visual odometry, which is more challenging than affine template tracking due to the nonlinear warping function and its dependence on, potentially sparse, depth information.

2 Direct Camera Tracking

Let the intensity of a pixel coordinate $\mathbf{p} = (u,\ v)^\top$ in the *reference* image be given by $\mathbf{I}(\mathbf{p}) \in \mathbb{R}$. After camera motion, a new image is obtained $\mathbf{I}'(\mathbf{p}')$. The goal of direct tracking is to estimate an increment of the camera motion parameters $\varDelta\boldsymbol{\theta} \in \mathbb{R}^d$ such that the photometric error is minimized

$$\varDelta\boldsymbol{\theta}^* = \operatorname*{argmin}_{\varDelta\boldsymbol{\theta}} \sum_{\mathbf{p} \in \varOmega} \| \mathbf{I}'\left(\mathbf{w}(\mathbf{p};\boldsymbol{\theta} \boxdot \varDelta\boldsymbol{\theta})\right) - \mathbf{I}\left(\mathbf{p}\right)\|^2, \tag{1}$$

where $\varOmega$ is a subset of pixel coordinates of interest in the reference frame, $\mathbf{w}\left(\cdot\right)$ is a *warping* function that depends on the parameter vector we seek to estimate, and $\boldsymbol{\theta}$ is an initial estimate of the motion parameters. After every iteration, the current estimate of parameters is updated (*i.e* $\boldsymbol{\theta} \leftarrow \boldsymbol{\theta} \boxplus \varDelta\boldsymbol{\theta}$), where $\boxplus$ generalizes the addition operator over the optimization manifold. The process is repeated until convergence, or some termination criteria have been satisfied [10,18].

By interchanging the roles of the template and input images, Baker & Matthews devise a more efficient alignment techniques known as the Inverse Compositional (IC) algorithm [18]. Under the IC formulation we seek an update $\varDelta\boldsymbol{\theta}$ that satisfies

$$\varDelta\boldsymbol{\theta}^* = \operatorname*{argmin}_{\varDelta\boldsymbol{\theta}} \sum_{\mathbf{p} \in \varOmega} \| \mathbf{I}'\left(\mathbf{w}(\mathbf{p};\boldsymbol{\theta})\right) - \mathbf{I}\left(\mathbf{w}(\mathbf{p};\varDelta\boldsymbol{\theta})\right)\|^2. \tag{2}$$

The optimization problem in Eq. (2) is nonlinear irrespective of the form of the warping function or the parameters, as—in general—there is no linear relationship between pixel coordinates and their intensities. By equating the partial derivatives of the first-order Taylor expansion of Eq. (2) to zero, we reach at closed-form solution as the solution to the normal equations given by: $\varDelta\boldsymbol{\theta} = \left(\mathbf{J}^\top\mathbf{J}\right)^{-1}\mathbf{J}^\top\mathbf{e}$, where $\mathbf{J} = \left(\mathbf{g}(\mathbf{p}_1)^\top,\ \ldots,\ \mathbf{g}(\mathbf{p}_m)^\top\right) \in \mathbb{R}^{m \times d}$ is the matrix of first-order partial derivatives of the objective function, m is the number of pixels, and $d = |\boldsymbol{\theta}|$ is the number of parameters. Each $\mathbf{g}$ is $\in \mathbb{R}^{1 \times d}$ and is given by the chain rule as $\mathbf{g}(\mathbf{p})^\top = \nabla\mathbf{I}(\mathbf{p})\frac{\partial \mathbf{w}}{\partial \boldsymbol{\theta}}$, where $\nabla\mathbf{I} = \left(\partial\mathbf{I}/\partial u,\ \partial\mathbf{I}/\partial v\right) \in \mathbb{R}^{1 \times 2}$

is the image gradient along the u- and v-directions respectively. The quantity $\mathbf{e}(\mathbf{p}) = \mathbf{I}'(\mathtt{w}(\mathbf{p}; \boldsymbol{\theta})) - \mathbf{I}(\mathbf{p})$ is the vector of residuals. Finally, the parameters are updated via the IC rule given by $\mathtt{w}(\mathbf{p}, \boldsymbol{\theta}) \leftarrow \mathtt{w}(\mathbf{p}, \boldsymbol{\theta}) \circ \mathtt{w}(\mathbf{p}, \Delta\boldsymbol{\theta})^{-1}$.

2.1   Direct Tracking with Feature Descriptors

Direct tracking using image intensities (the brightness constraint in Eq. (1)) is known to be sensitive to illumination change. To address this limitation, we propose the use of a *descriptor constancy* assumption. Namely, we seek an update to the parameters such that

$$\Delta\boldsymbol{\theta}^* = \underset{\Delta\boldsymbol{\theta}}{\operatorname{argmin}} \|\phi(\mathbf{I}'(\mathtt{w}(\mathbf{p}; \boldsymbol{\theta} \boxplus \Delta\boldsymbol{\theta}))) - \phi(\mathbf{I}(\mathbf{p}))\|^2, \tag{3}$$

where $\phi(\cdot)$ is a multi-dimensional feature descriptor applied to both the reference and the warped input images.

The descriptor constancy objective in Eq. (3) is more complicated than its brightness counterpart in Eq. (1) as feature descriptors are high dimensional and the suitability of their linearization remains unclear. In the sequel, we will show that various descriptors linearize well and are suitable for direct tracking.

2.2   Desiderata

The goal of direct tracking is to maximize the precision of the estimated parameters. The linearization required in direct tracking implicitly assumes that we are close enough to the local minima. This fact is usually expressed by assuming small displacements between the input images. In order to maximize precision, it is important to balance the complexity of the descriptor as a function of its sampling density. Namely, descriptors with long range spatial connections such as SIFT [32] and HOG [33], while robust to a range of geometric and photometric image deformations, they contribute little to tracking precision. This is due to the increased dependencies between pixels contributing to the linear system. We will experimentally validate this hypothesis in the experimental section of this work. Hence, good descriptors for illumination invariant tracking must be: (1) locally limited with respect to their spatial extent, and (2) efficient to compute, which is desired for practical reasons. Both requirements, locality and efficiency, are closely related as most local descriptors are efficient to compute as well.

2.3   Pre-computing Descriptors for Efficiency

Descriptor constancy as stated in Eq. (3) requires re-computing the descriptors after every iteration of image warping. In addition to the extra computational cost of repeated applications of the descriptor, it is difficult to warp individual pixel locations if their motion depends on depth, such as in direct visual odometry. The difficulty arises from the lack of a 3D model that could be used to reason about occlusions and discontinuities in the image. An approximation to the descriptor constancy objective in Eq. (3) is to pre-compute the descriptors and minimize the following expression instead (using the IC formulation):

$$\min_{\Delta\theta} \sum_{\mathbf{p}\in\Omega} \sum_{i=1}^{N_c} \|\boldsymbol{\Phi}_i'(\mathbf{w}(\mathbf{p};\boldsymbol{\theta})) - \boldsymbol{\Phi}_i(\mathbf{w}(\mathbf{p};\Delta\theta))\|^2, \qquad (4)$$

where $\boldsymbol{\Phi}_i$ indicates the i-th coordinate (channel) of the pre-computed descriptor and N_c is the number of channels. This approximation incurs a loss of accuracy especially with nonlinear warps. However, we found that the loss of accuracy induced when using Eq. (4) instead of Eq. (3) to be insignificant in comparison to the computational savings and simplicity of implementation, especially when the displacements between the images are sufficiently small.

The minimization of Eq. (4) is performed similar to minimizing the sum of squared intensity residuals in Eq. (1). By taking the derivative with respect to the parameters of the 1st-order expansion of Eq. (4), we arrive at:

$$\sum_{\mathbf{p}\in\Omega} \sum_{i=1}^{N_c} \left(\frac{\partial\boldsymbol{\Phi}_i}{\partial\boldsymbol{\theta}}\right)^\top \left| \boldsymbol{\Phi}_i'(\mathbf{w}(\mathbf{p};\boldsymbol{\theta})) - \boldsymbol{\Phi}_i(\mathbf{p}) - \frac{\partial\boldsymbol{\Phi}_i}{\partial\boldsymbol{\theta}}\Delta\theta \right|, \qquad (5)$$

where the derivative of the pre-computed descriptor with respect to the parameters is given by the chain-rule per channel i as

$$\frac{\partial\boldsymbol{\Phi}_i}{\partial\boldsymbol{\theta}} = \frac{\partial\boldsymbol{\Phi}}{\partial\mathbf{u}} \frac{\partial\mathbf{w}(\mathbf{u};\boldsymbol{\theta})}{\partial\mathbf{u}}. \qquad (6)$$

Subsequently, the Gauss-Newton approximation to the Hessian is given by the summation of the partial derivatives across all pixels and across all channels as

$$\mathbf{H}(\boldsymbol{\theta}) = \sum_{\mathbf{p}\in\Omega} \sum_{i=1}^{N_c} \frac{\partial\boldsymbol{\Phi}_i(\mathbf{p};\boldsymbol{\theta})}{\partial\boldsymbol{\theta}}^\top \frac{\partial\boldsymbol{\Phi}_i(\mathbf{p};\boldsymbol{\theta})}{\partial\boldsymbol{\theta}}. \qquad (7)$$

In the next section, we review the local feature descriptor suitable for high-precision robust tracking used in this work.

3   Densely Evaluated Descriptors

We evaluate a number of descriptors suitable for dense tracking as summarized in Table 1 and visualized in Fig. 1. The descriptors are:

- **Raw intensity:** This is the trivial form of a feature descriptor, which uses the raw image intensities. We work with grayscale images, and hence it is a single channel, $\boldsymbol{\Phi}(\mathbf{I}) = \{\mathbf{I}\}$.
- **Gradient constancy:** The image gradient measures the rate of change of intensity and hence it is invariant to additive changes [37]. We found that including the raw image intensities in the optimization with the gradient constraint to work better. The descriptor is composed of three channels and is given by: $\phi = \{\mathbf{I}, \nabla_u\mathbf{I}, \nabla_v\mathbf{I}\}$.

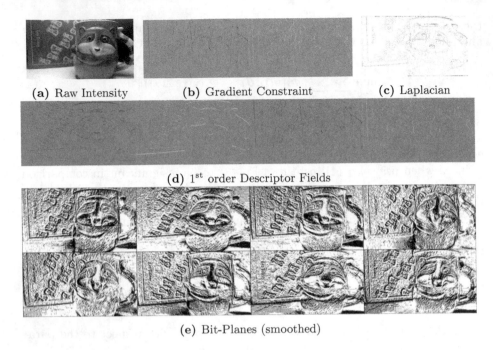

(a) Raw Intensity   (b) Gradient Constraint   (c) Laplacian

(d) 1st order Descriptor Fields

(e) Bit-Planes (smoothed)

Fig. 1. Visualization of the different descriptors. Best viewed on a screen.

- **Laplacian:** The Laplacian is based on the 2nd order derivatives of the image and, similar to the gradient constraint, it provides invariance to additive change, but using only a single channel. We found that including the raw intensities to improve results. The descriptor is given by: $\boldsymbol{\Phi} = \left\{ \mathbf{I}, |\nabla^2 \mathbf{I}| \right\}$.
- **Descriptor Fields (DF)** [36] where the idea is to separate the image gradients into different channels based on their sign. After that, a smoothing step is performed. Using first-order image gradients, denoted by DF-1, the descriptor is composed of four channels and is given by:

$$\boldsymbol{\Phi}_{\text{DF-1}}(\mathbf{I}) = \left\{ [\nabla_u \mathbf{I}]^+, \ [\nabla_u \mathbf{I}]^-, \ [\nabla_v \mathbf{I}]^+, \ [\nabla_v \mathbf{I}]^- \right\}.$$

The notation $[\cdot]^+$ indicates selecting the positive part of the gradients, or zero otherwise. The 2nd-order DF, denoted by DF-2, makes use of 2nd-order gradients as well and is composed of 10 channels. As show in Fig. 1, DF is real-valued and results in sparse channels [36].
- **Bit-Planes** [38]: is binary descriptor based on the Census Transform [39], where channels are constructed by performing local pixel comparisons. When evaluated in a 3×3 neighborhood, the descriptor results in eight channels given by: $\boldsymbol{\Phi}(\mathbf{I}) = \left\{ \mathbf{I}(\mathbf{x}) \geq \mathbf{I}(\mathbf{x} + \Delta \mathbf{x}_j) \right\}_{j=1}^{8}$, where $\mathbf{I}(\mathbf{x} + \Delta \mathbf{x}_j)$ indicates the image sampled at the j-th neighbor location in the neighborhood. We refer the reader to [38] for additional details.

Table 1. Descriptors evaluated in this work

Name	Acronym	Channels
Raw intensity	RI	1
Gradient constraint	GC	3
Laplacian	LP	2
1st order DF [36]	DF-1	4
2nd order DF [36]	DF-2	10
Bit-planes [38]	BP	8

4   Evaluation

We experiment with the different feature descriptors summarized in Table 1 on two direct tracking problems. One, is parametric motion estimation using an Affine motion model, which we use to illustrate performance on synthetically controlled illumination variations. The other, is direct visual odometry, which is more challenging as the nonlinear warping function depends on sparse depth.

4.1   Affine Template Alignment

Using the notation introduced in Sect. 2, we desire to estimate the parameters of motion between a dense descriptor designated as the template Φ_{DESC} and a dense descriptor evaluated on an input image Φ'_{DESC}, where DESC is one of the descriptors in Table 1. Under affine motion $\boldsymbol{\theta} = (\theta_1, \ldots, \theta_6)^\top \in \mathbb{R}^6$, the image coordinates of the two descriptors are related via an affine warp of the form

$$\begin{bmatrix} \mathbf{p}' \\ 1 \end{bmatrix} \equiv \mathbf{w}(\mathbf{p}; \boldsymbol{\theta}) = \mathbf{A}(\boldsymbol{\theta}) \begin{bmatrix} \mathbf{p} \\ 1 \end{bmatrix}, \tag{8}$$

where $\mathbf{A}(\boldsymbol{\theta}) \in \mathbb{R}^{2 \times 3}$ represents a 2D affine transform.

Experiments in this section are performed using a set of natural images, where the parameters of the affine transformation are randomly generated to create a synthetically input/moving image with known ground truth.

Performance under ideal conditions. While ideal imaging conditions are uncommon outside of controlled imaging applications, it is important to study the effect of any form of image deformation on the system's accuracy. Ideal conditions in this context indicate lack of appearance variations such that brightness constancy assumption is satisfied.

The question we answer in the following experiments is: *How does nonlinear deformations of the image (feature descriptors) affect estimation accuracy under ideal conditions?* Especially under the additional quantization effects caused by descriptors. The answer to question is shown in Fig. 2, where all descriptors are evaluated without additional illumination change. We conclude that are the

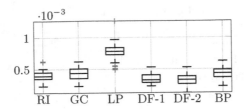

Fig. 2. Performance of dense descriptor under ideal conditions.

Fig. 3. Example illumination change according to Eq. (9). Cost surfaces for each of the evaluated descriptors for this pair of images are shown in Fig. 4. We observe similar behavior for different image content and different changes in illumination.

compared descriptors perform well under ideal conditions without incurring a significant loss of precision.

4.2 Performance Under Varying Illumination

To generate illumination variations we synthesise the input image from the template using a nonlinear intensity change model of the form

$$\mathbf{I}'(\mathbf{p}) = \texttt{floor}\left(255\left(\frac{\alpha\mathbf{I}(\mathbf{w}(\mathbf{p};\boldsymbol{\theta}))+\beta}{255}\right)^{1+\gamma}\right), \tag{9}$$

where $\boldsymbol{\theta}$ is a randomly generated vector of warp parameters, α and β are respectively multiplicative and additive terms, while $|\gamma| < 1$ is a nonlinear gamma correction term. Parameters controlling the illumination changes are also generated randomly. An example of this illumination change is shown in Fig. 3.

Results are shown in Figs. 4 and 5 using the end-point RMSE metric [18]. As expected, we observe a large RMSE when using raw intensities. No significant improvement is obtained using the gradient constraint. The Laplacian improves results only slightly. The top performing algorithms are DF, and Bit-Planes.

Extended spatial support descriptors. Another natural question to ask is whether there is any benefit from using nonlocal feature descriptors in direct tracking? By nonlocal, we mean feature descriptors that make use of nonlocal spatial information in the image, such as a making use of a large neighborhood

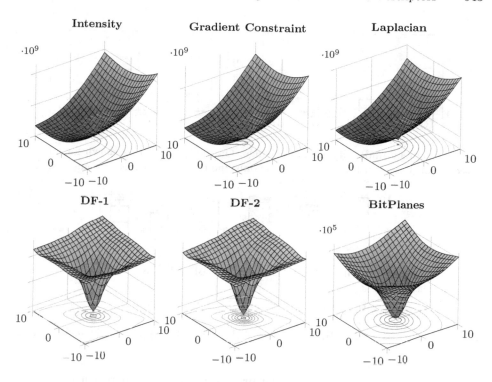

Fig. 4. Cost surfaces for each of the descriptors corresponding to the input pair shown in Fig. 3. The correct minima located at $(0,0)$. Raw intensity and the gradient constraint fail to capture the correct minima. The Laplacian correctly localizes the minima, albeit a narrow basin of convergence. The feature descriptors at the bottom row correctly identify the minima with an adequate basin of convergence.

during the descriptor's computation. Examples include SIFT [32], HOG [33], and BRIEF [40]. As shown in Fig. 6, the use of nonlocal descriptor appears to hurt performance rather than simpler local ones. In this experiment, we experiment with two possibilities of extracting channels from the BRIEF [40] descriptor. One, is extracting 128 channels similar to [38]. The other, is extracting only 16 channels, where each channel is formed of a single byte. We observed similar degradation in performance using densely evaluated SIFT, and other variations on extracting channels.

4.3 Direct Visual Odometry (VO)

Another popular application of direct tracking is estimating the 6DOF rigid-body motion of a freely moving camera. The warping function takes the form

$$\mathbf{p}' = \pi\left(\mathbf{T}(\boldsymbol{\theta})\pi^{-1}(\mathbf{p}; \mathbf{D}(\mathbf{p}))\right), \tag{10}$$

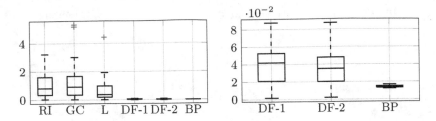

Fig. 5. Accuracy under illumination change. On the left RMSE is shown for all compared descriptors. On the right, we show only the top three for better comparison.

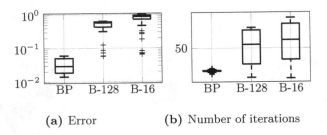

(a) Error   (b) Number of iterations

Fig. 6. Comparison with BRIEF using 128 channels (B-128) and 16 channels (B-16).

where $\pi(\cdot) : \mathbb{R}^3 \rightarrow \mathbb{R}^2$ denotes the projection onto a camera with a known intrinsic calibration, and $\pi^{-1}(\cdot, \cdot) : \mathbb{R}^2 \times \mathbb{R} \rightarrow \mathbb{R}^3$ denotes the inverse projection given the camera intrinsic parameters and the pixel's depth $\mathbf{D}(\mathbf{p})$. In our implementation, we parametrize the camera pose with the exponential map $\mathbf{T}(\boldsymbol{\theta}) = \exp(\widehat{\boldsymbol{\theta}}) \in SE(3)$ [41].

The algorithm is implemented in scale space (using 4 octaves) and the solution is obtained by Iteratively Re-Weighted Least-Squares (IRLS) using the Huber robust weighing function [42]. The maximum number of iterations per pyramid octave is set to 50, but we terminate early if the norm of the estimated parameters vector, or the relative reduction of the objective, fall below a threshold $\tau = 1 \times 10^{-6}$.

Image gradients required for linearization are implemented with central difference filters, which we found to produce more accurate results than the Sobel operator, or the Scharr filters. We also observed an improved accuracy if we selected a subset of pixels at the finest octave (the highest resolution). Pixel selection is implemented as a non maxima suppression of the absolute gradient magnitude across all descriptor channels. For the rest of the pyramid octaves, we use all pixels with non-vanishing gradient information.

The direct approach to VO has been shown to work with stereo [43], mono [3], and RGB-D data [19]. Since our focus is the study of photometric invariance rather than depth estimation, we will evaluate the approach on stereo data and simplify the step of depth estimation using standard stereo algorithms. We evaluate the performance of the different descriptors on the Tsukuba dataset [44, 45], which provides a range of illumination as shown in Fig. 7.

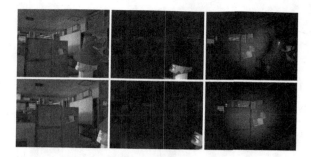

Fig. 7. Different illumination condition from the Tsukuba dataset. At the top row from left to right we have: 'fluorescent' (easy), 'lamps' (medium) and 'flashlight' (hard). The bottom row shows the form of appearance change across views.

4.4 Comparison with State-of-the-Art

We compare the top performing algorithm from the template tracking section (Bit-Planes) against two state-of-the-art VO algorithms: FOVIS [46], which is a feature-based algorithm, and DVO [1] which is a dense direct tracking method using raw pixel intensities. Qualitative results on the "lamps" dataset are shown in Figs. 8a and b. DVO using intensity only struggles to maintain tracking throughout the dataset. FOVIS's performance is slightly better as features are matched with normalized correlation, which is invariant to affine lighting change. The Bit-Planes based tracking performs the best. Quantitative evaluation for each of the descriptors per image frame is shown in Fig. 9 and summarized for the whole sequence in Table 2.

4.5 Timing Information

Experiments in the previous section were carried out using a combination of Matlab and C++. To obtain an accurate comparison of the run time of the

Table 2. Summary statistics of errors per positional degree of freedom (RMSE in mm). We use the standard right-handed coordinate convention system in vision, where the Z-axis points forward and the Y-axis points downward.

	Flashlight			Lamps		
	X	Y	Z	X	Y	Z
RI	14.34	8.14	20.94	1.45	1.01	2.16
GC	14.26	7.53	18.86	45.31	30.37	22.76
LP	13.24	6.59	18.16	0.54	0.26	0.46
DF-1	**2.03**	0.45	**0.77**	0.42	0.21	**0.40**
DF-2	2.95	0.83	1.33	2.27	1.09	4.90
BP	2.66	**0.33**	1.08	**0.37**	**0.18**	**0.40**

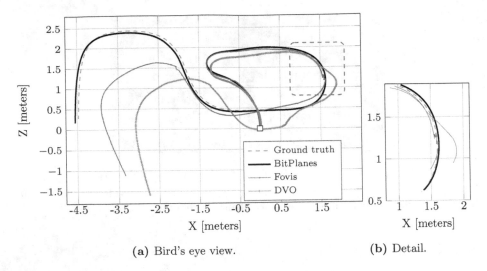

(a) Bird's eye view.   **(b)** Detail.

Fig. 8. Evaluation on the Tsukuba dataset with "lamps" illumination [44]. The figure shows a bird's eye view of the camera path. The highlighted area is shown with more details in Fig. 8b. Example images are in Fig. 7

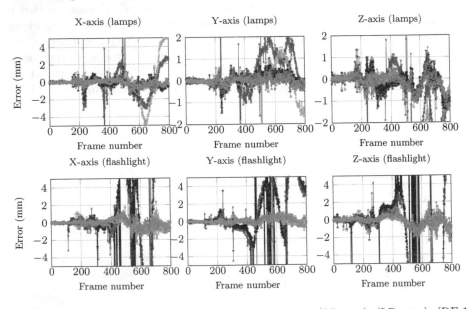

Fig. 9. Trajectory errors using the "lamps" sequence for (RI ——), (LP ——), (DF-1 ——), (DF-2 ——), and (BP ——). We truncated the plots for better visualization.

different descriptors, we implemented an optimized version in C++. In Table 3 we show the time required to compute the descriptors as a function of image resolution. All experiments were conducted with a single core i7-2640M mobile processor and an 8 GB of RAM. When using IC, computing descriptors is only

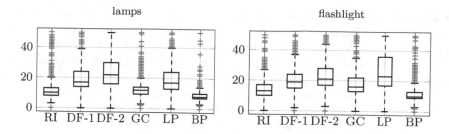

Fig. 10. Number of iterations.

Table 3. Runtime in milliseconds required to compute the descriptors as a function of image resolution

	RI	GC	LP	DF-1	DF-2	BP
80 × 60	0.00	0.01	0.03	0.09	0.20	0.08
160 × 120	0.01	0.02	0.10	0.26	0.59	0.20
320 × 240	0.04	0.11	0.37	0.93	2.09	0.60
640 × 480	0.17	0.44	1.45	4.04	9.34	3.87
1280 × 960	0.76	2.56	5.84	17.20	38.65	14.32
2560 × 1920	2.88	10.15	23.22	68.78	154.29	59.29

required when creating a new reference frame. The frequency of creating new reference frames (re-initializing the tracker) depends on the application. Based on the photometric invariance performance from the previous evaluation, we conclude that Bit-Planes is the most efficient to compute.

Image warping, however, is required at every iteration of the optimization. Image warping depends on the number of channels as shown in Fig. 11.

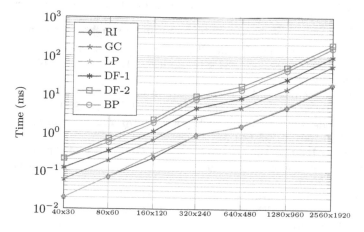

Fig. 11. Image warping running time shown in log scale.

For instance, there is virtually no difference between warping raw intensities versus the Laplacian as they differ by a single channel and since the algorithm is memory bound. Warping experiments were parallelized on two cores.

Finally, the number of iterations required for convergence for each of the feature descriptor on the two Tsukuba datasets with challenging illumination is shown in Fig. 10. Direct tracking convergence withing 20 iterations during the majority of the time. The "flashlight" illumination is more challenging, and hence all descriptors require additional iterations to converge.

5   Conclusions and Future Work

Locally evaluated dense feature descriptors are a promising avenue for non-parametric illumination invariant dense tracking. We explore various descriptors in comparison to using raw intensities and demonstrate enhanced robustness to arbitrary illumination change. More importantly, we show that local feature descriptor are suitable for the gradient-based minimization required by direct tracking. Our evaluation using two different direct tracking systems (affine template alignment, and direct visual odometry) show that the suitability of linearizing descriptors holds irrespective of the warp. Finally, the algorithmic changes required to allow existing direct tracking system to operate robustly in face of illumination variations are simple to implement without significantly comprising efficiency.

Acknowledgement. We thank the anonymous reviewers for their valuable comments.

References

1. Kerl, C., Sturm, J., Cremers, D.: Dense visual slam for RGB-D cameras. In: Proceedings of the International Conference on Intelligent Robots and Systems (2013)
2. Comport, A.I., Malis, E., Rives, P.: Real-time quadrifocal visual odometry. Int. J. Robot. Res. **29**, 245–266 (2010)
3. Engel, J., Schöps, T., Cremers, D.: LSD-SLAM: large-scale direct monocular SLAM. In: Fleet, D., Pajdla, T., Schiele, B., Tuytelaars, T. (eds.) ECCV 2014. LNCS, vol. 8690, pp. 834–849. Springer, Heidelberg (2014). doi:10.1007/978-3-319-10605-2_54
4. Handa, A., Newcombe, R.A., Angeli, A., Davison, A.J.: Real-time camera tracking: when is high frame-rate best? In: Fitzgibbon, A., Lazebnik, S., Perona, P., Sato, Y., Schmid, C. (eds.) ECCV 2012. LNCS, vol. 7578, pp. 222–235. Springer, Heidelberg (2012). doi:10.1007/978-3-642-33786-4_17
5. Salas-Moreno, R., Glocken, B., Kelly, P., Davison, A.: Dense planar SLAM. In: 2014 IEEE International Symposium on Mixed and Augmented Reality (ISMAR), pp. 157–164 (2014)
6. Newcombe, R., Lovegrove, S., Davison, A.: DTAM: Dense tracking and mapping in real-time. In: 2011 IEEE International Conference on Computer Vision (ICCV), pp. 2320–2327 (2011)

7. Irani, M., Anandan, P.: About direct methods. In: Triggs, B., Zisserman, A., Szeliski, R. (eds.) IWVA 1999. LNCS, vol. 1883, pp. 267–277. Springer, Heidelberg (2000). doi:10.1007/3-540-44480-7_18

8. Forster, C., Pizzoli, M., Scaramuzza, D.: SVO: fast semi-direct monocular visual odometry. In: Proceedings of the IEEE International Conference on Robotics and Automation (ICRA) (2014)

9. Horn, B.K., Schunck, B.G.: Determining optical flow. Artif. Intell. **17**, 185–203 (1981)

10. Lucas, B.D., Kanade, T.: An iterative image registration technique with an application to stereo vision (DARPA). In: Proceedings of the 1981 DARPA Image Understanding Workshop, pp. 121–130 (1981)

11. Bartoli, A.: Groupwise geometric and photometric direct image registration. IEEE Trans. Pattern Anal. Mach. Intell. **30**, 2098–2108 (2008)

12. Evangelidis, G.D., Psarakis, E.Z.: Parametric image alignment using enhanced correlation coefficient maximization. PAMI **30**, 1858–1865 (2008)

13. Dowson, N., Bowden, R.: Mutual information for Lucas-Kanade tracking (MILK): an inverse compositional formulation. PAMI **30**, 180–185 (2008)

14. Müller, T., Rabe, C., Rannacher, J., Franke, U., Mester, R.: Illumination-robust dense optical flow using census signatures. In: Mester, R., Felsberg, M. (eds.) DAGM 2011. LNCS, vol. 6835, pp. 236–245. Springer, Heidelberg (2011). doi:10.1007/978-3-642-23123-0_24

15. Black, M., Anandan, P.: A framework for the robust estimation of optical flow. In: 1993 Proceedings of the Fourth International Conference on Computer Vision, pp. 231–236 (1993)

16. Irani, M., Anandan, P.: Robust multi-sensor image alignment. In: 1998 Sixth International Conference on Computer Vision, pp. 959–966 (1998)

17. Nocedal, J., Wright, S.J.: Numerical Optimization, 2nd edn. Springer, New York (2006)

18. Baker, S., Matthews, I.: Lucas-kanade 20 years on: a unifying framework. Int. J. Comput. Vision **56**, 221–255 (2004)

19. Klose, S., Heise, P., Knoll, A.: Efficient compositional approaches for real-time robust direct visual odometry from RGB-D data. In: IEEE/RSJ International Conference on Intelligent Robots and Systems (2013)

20. Zia, M.Z., Nardi, L., Jack, A., Vespa, E., Bodin, B., Kelly, P.H.J., Davison, A.J.: Comparative design space exploration of dense and semi-dense SLAM. CoRR abs/1509.04648 (2015)

21. Sun, D., Roth, S., Black, M.: Secrets of optical flow estimation and their principles. In: 2010 IEEE Conference on Computer Vision and Pattern Recognition (CVPR), pp. 2432–2439 (2010)

22. Vogel, C., Roth, S., Schindler, K.: An evaluation of data costs for optical flow. In: Weickert, J., Hein, M., Schiele, B. (eds.) GCPR 2013. LNCS, vol. 8142, pp. 343–353. Springer, Heidelberg (2013). doi:10.1007/978-3-642-40602-7_37

23. Mikolajczyk, K., Schmid, C.: A performance evaluation of local descriptors. IEEE Trans. Pattern Anal. Mach. Intell. **27**, 1615–1630 (2005)

24. Krizhevsky, A., Sutskever, I., Hinton, G.E.: ImageNet classification with deep convolutional neural networks. In: Advances in Neural Information Processing Systems, pp. 1097–1105 (2012)

25. Torr, P.H.S., Zisserman, A.: Feature based methods for structure and motion estimation. In: Triggs, B., Zisserman, A., Szeliski, R. (eds.) IWVA 1999. LNCS, vol. 1883, pp. 278–294. Springer, Heidelberg (2000). doi:10.1007/3-540-44480-7_19

26. Furukawa, Y., Hernndez, C.: Multi-view stereo: a tutorial. Found. Trends Comput. Graph. Vis. **9**, 1–148 (2015)
27. Sivic, J., Zisserman, A.: Video Google: a text retrieval approach to object matching in videos. In: Proceedings of the International Conference on Computer Vision, vol. 2, pp. 1470–1477 (2003)
28. Antonakos, E., Alabort-i Medina, J., Tzimiropoulos, G., Zafeiriou, S.: Feature-based Lucas-Kanade and active appearance models. IEEE Trans. Image Process. **24**, 2617–2632 (2015)
29. Bristow, H., Lucey, S.: Regression-based image alignment for general object categories. CoRR abs/1407.1957 (2014)
30. Sevilla-Lara, L., Sun, D., Learned-Miller, E.G., Black, M.J.: Optical flow estimation with channel constancy. In: Fleet, D., Pajdla, T., Schiele, B., Tuytelaars, T. (eds.) ECCV 2014. LNCS, vol. 8689, pp. 423–438. Springer, Heidelberg (2014). doi:10.1007/978-3-319-10590-1_28
31. Sevilla-Lara, L., Learned-Miller, E.: Distribution fields for tracking. In: IEEE Conference on Computer Vision and Pattern Recognition (CVPR) (2012)
32. Lowe, D.G.: Distinctive image features from scale-invariant keypoints. Int. J. Comput. Vision **60**, 91–110 (2004)
33. Dalal, N., Triggs, B.: Histograms of oriented gradients for human detection. In: IEEE Computer Society Conference on Computer Vision and Pattern Recognition (CVPR), vol. 1, pp. 886–893 (2005)
34. Bristow, H., Lucey, S.: In defense of gradient-based alignment on densely sampled sparse features. In: Hassner, T., Liu, C. (eds.) Dense Image Correspondences for Computer Vision, pp. 135–152. Springer, Cham (2016). doi:10.1007/978-3-319-23048-1_7
35. Liu, C., Yuen, J., Torralba, A.: SIFT flow: dense correspondence across scenes and its applications. IEEE Trans. Pattern Anal. Mach. Intell. **33**, 978–994 (2011)
36. Crivellaro, A., Lepetit, V.: Robust 3D tracking with descriptor fields. In: Conference on Computer Vision and Pattern Recognition (CVPR) (2014)
37. Brox, T., Bruhn, A., Papenberg, N., Weickert, J.: High accuracy optical flow estimation based on a theory for warping. In: Pajdla, T., Matas, J. (eds.) ECCV 2004. LNCS, vol. 3024, pp. 25–36. Springer, Heidelberg (2004). doi:10.1007/978-3-540-24673-2_3
38. Alismail, H., Browning, B., Lucey, S.: Bit-Planes: Dense Subpixel Alignment of Binary Descriptors. CoRR abs/1602.00307 (2016)
39. Zabih, R., Woodfill, J.: Non-parametric local transforms for computing visual correspondence. In: Eklundh, J.-O. (ed.) ECCV 1994. LNCS, vol. 801, pp. 151–158. Springer, Heidelberg (1994). doi:10.1007/BFb0028345
40. Calonder, M., Lepetit, V., Strecha, C., Fua, P.: BRIEF: binary robust independent elementary features. In: Daniilidis, K., Maragos, P., Paragios, N. (eds.) ECCV 2010. LNCS, vol. 6314, pp. 778–792. Springer, Heidelberg (2010). doi:10.1007/978-3-642-15561-1_56
41. Murray, R.M., Li, Z., Sastry, S.S., Sastry, S.S.: A Mathematical Introduction to Robotic Manipulation. CRC Press, Boca Raton (1994)
42. Zhang, Z.: Parameter estimation techniques: a tutorial with application to conic fitting. Image Vis. Comput. **15**, 59–76 (1997)
43. Engel, J., Stueckler, J., Cremers, D.: Large-scale direct SLAM with stereo cameras. In: International Conference on Intelligent Robots and Systems (IROS) (2015)
44. Peris, M., Maki, A., Martull, S., Ohkawa, Y., Fukui, K.: Towards a simulation driven stereo vision system. In: 2012 21st International Conference on Pattern Recognition (ICPR), pp. 1038–1042 (2012)

45. Martull, S., Peris, M., Fukui, K.: Realistic CG stereo image dataset with ground truth disparity maps. In: ICPR workshop TrakMark 2012, vol. 111, pp. 117–118 (2012)
46. Huang, A.S., Bachrach, A., Henry, P., Krainin, M., Maturana, D., Fox, D., Roy, N.: Visual odometry and mapping for autonomous flight using an RGB-D camera. In: International Symposium on Robotics Research (ISRR), pp. 1–16 (2011)

Author Index

Printed in the United States
By Bookmasters